IET TELECOMMUNICATIONS SERIES 86

Advances in Communications Satellite Systems

Other volumes in this series

Volume 9	**Phase noise in signal sources** W.P. Robins
Volume 12	**Spread spectrum in communications** R. Skaug and J.F. Hjelmstad
Volume 13	**Advanced signal processing** D.J. Creasey (Editor)
Volume 19	**Telecommunications traffic, tariffs and** costs R.E. Farr
Volume 20	**An introduction to satellite communications** D.I. Dalgleish
Volume 26	**Common-channel signalling** R.J. Manterfield
Volume 28	**Very small aperture terminals (VSATs)** J.L. Everett (Editor)
Volume 29	**ATM: the broadband telecommunications solution** L.G. Cuthbert and J.C. Sapanel
Volume 31	**Data communications and networks, 3rd edition** R.L. Brewster (Editor)
Volume 32	**Analogue optical fibre communications** B. Wilson, Z. Ghassemlooy and I.Z. Darwazeh (Editors)
Volume 33	**Modern personal radio systems** R.C.V. Macario (Editor)
Volume 34	**Digital broadcasting** P. Dambacher
Volume 35	**Principles of performance engineering for telecommunication and information systems** M. Ghanbari, C.J. Hughes, M.C. Sinclair and J.P. Eade
Volume 36	**Telecommunication networks, 2nd edition** J.E. Flood (Editor)
Volume 37	**Optical communication receiver design** S.B. Alexander
Volume 38	**Satellite communication systems, 3rd edition** B.G. Evans (Editor)
Volume 40	**Spread spectrum in mobile communication** O. Berg, T. Berg, J.F. Hjelmstad, S. Haavik and R. Skaug
Volume 41	**World telecommunications economics** J.J. Wheatley
Volume 43	**Telecommunications signalling** R.J. Manterfield
Volume 44	**Digital signal filtering, analysis and restoration** J. Jan
Volume 45	**Radio spectrum management, 2nd edition** D.J. Withers
Volume 46	**Intelligent networks: principles and applications** J.R. Anderson
Volume 47	**Local access network technologies** P. France
Volume 48	**Telecommunications quality of service management** A.P. Oodan (Editor)
Volume 49	**Standard codecs: image compression to advanced video coding** M. Ghanbari
Volume 50	**Telecommunications regulation** J. Buckley
Volume 51	**Security for mobility** C. Mitchell (Editor)
Volume 52	**Understanding telecommunications networks** A. Valdar
Volume 53	**Video compression systems: from first principles to concatenated codecs** A. Bock
Volume 54	**Standard Codecs: image compression to advanced video coding, 3rd edition** M. Ghanbari
Volume 59	**Dynamic Ad Hoc Networks** H. Rashvand and H. Chao (Editors)
Volume 60	**Understanding Telecommunications Business** A Valdar and I Morfett
Volume 65	**Advances in Body-Centric Wireless Communication: Applications and State-of-the-art** Q. H. Abbasi, M. U. Rehman, K. Qaraqe and A. Alomainy (Editors)
Volume 67	**Managing the Internet of Things: Architectures, Theories and Applications** J. Huang and K. Hua (Editors)
Volume 68	**Advanced Relay Technologies in Next Generation Wireless Communications** I. Krikidis and G. Zheng
Volume 69	**5G Wireless Technologies** A. Alexiou (Editor)
Volume 70	**Cloud and Fog Computing in 5G Mobile Networks** E. Markakis, G. Mastorakis, C. X. Mavromoustakis and E. Pallis (Editors)
Volume 71	**Understanding Telecommunications Networks, 2nd edition** A. Valdar
Volume 72	**Introduction to Digital Wireless Communications** Hong-Chuan Yang
Volume 73	**Network as a service for next generation internet** Q. Duan and S. Wang (Editors)
Volume 74	**Access, Fronthaul and Backhaul Networks for 5G & Beyond** M. A. Imran, S. A. R. Zaidi and M. Z. Shakir (Editors)
Volume 76	**Trusted Communications with Physical Layer Security for 5G and Beyond** T. Q. Duong, X. Zhou and H. V. Poor (Editors)
Volume 77	**Network Design, Modelling and Performance Evaluation** Q. Vien
Volume 78	**Principles and Applications of Free Space Optical Communications** A. K. Majumdar, Z. Ghassemlooy, A. A. B. Raj (Editors)
Volume 79	**Satellite Communications in the 5G Era** S. K. Sharma, S. Chatzinotas and D. Arapoglou
Volume 80	**Transceiver and System Design for Digital Communications 5th Edition** Scott R. Bullock
Volume 81	**Applications of Machine Learning in Wireless Communications** R. He and Z. Ding (Editors)
Volume 84	**Low Electromagnetic Emission Wireless Network Technologies: 5G and beyond** M.A. Imran, F. Héliot and Y.A. Sambo (Editors)
Volume 905	**ISDN applications in education and training** R. Mason and P.D. Bacsich

Advances in Communications Satellite Systems

Proceedings of The 36th International Communications Satellite Systems Conference (ICSSC-2018)

Edited by
Ifiok Otung, Thomas Butash and Peter Garland

The Institution of Engineering and Technology

Published by The Institution of Engineering and Technology, London, United Kingdom

The Institution of Engineering and Technology is registered as a Charity in England & Wales (no. 211014) and Scotland (no. SC038698).

First published 2019

The Institution of Engineering and Technology
Michael Faraday House
Six Hills Way, Stevenage
Herts, SG1 2AY, United Kingdom

www.theiet.org

British Library Cataloguing in Publication Data
A catalogue record for this product is available from the British Library

ISBN 978-1-78561-961-8 (hardback)
ISBN 978-1-78561-962-5 (PDF)

Typeset in India by MPS Limited

Contents

Preface and Acknowledgements

The International Communications Satellite Systems Conference (ICSSC) is universally recognised as the oldest and preeminent technical conference on communications satellite systems and their innumerable applications. Its conduct has spanned more than a half century, having been first held in 1966 to commemorate the first anniversary of the launch of Early Bird (Intelsat I), the world's first commercial telecommunications satellite.

The ICSSC was held biennially in even numbered years from 1966 to 2000 and has been held annually since. The conference venue alternated between the East and West coasts of North America during the events' first three decades. Known as the "CSSC" through 1986, the adjective "International" was added in 1988 to the conference name in recognition of the global expanse of interest in, attendance and participation at the conference. Indeed, to accommodate increasing international contributions, the ICSSC was first held outside North America at Yokohama, Japan in 1998, and the conference venue has alternated between the Asia-Pacific Region and Europe/UK during all odd-numbered years since 2000.

The ICSSC was first held jointly with the Ka and Broadband Communications Conference in 2005 in Rome, Italy. Given the overwhelming success and popularity of this Joint Conference, the ICSSC and Ka-Conference were again held jointly in 2012 in Ottawa, Canada, 2013 in Florence, Italy, 2016 in Cleveland, Ohio, 2017 in Trieste, Italy and again in 2018 in Niagara Falls, Canada.

The American Institute of Aeronautics and Astronautics (AIAA) prestigious Aerospace Communications Award is presented by the AIAA Communications Systems Technical Committee at the ICSSC to the individual or individuals deemed to have made outstanding contributions to advancing the field of communications satellite systems.

The 36th International Communications Satellite Systems Conference (ICSSC 2018) was held jointly with the Ka Band conference in Canada. Having previously been held in Montreal twice – in 1976 and 2002 – and Ottawa in 2012, it was decided to hold the conference at one of Canada's iconic locations, Niagara Falls. The conference centre chosen was the Marriott on the Falls, a hotel which overlooked the Canadian Horseshoe Falls. Delegates, when they had time away from the intense conference program, were able to view the changing face of the Falls through early morning mist to late night illumination and fireworks. In choosing Niagara, the organisers realised that, although the conference was held in Canada, the United States was a bridge span away and many delegates traveled up from the

United States or traveled home via the United States to take in some of the late Fall New England scenery.

In the over fifty-year span of the ICSSC, Canada has been a regular venue for the conference. This is because, although relatively small in population, Canada has been a pioneer in many aspects of Space, particularly in the area of Satellite Communications. In September 1962 with the launch of the Alouette 1 satellite, Canada became the first country outside of the United States and Russia to build and operate a satellite. With the launch provided on a United States launch vehicle, Alouette became the first in a long series of NASA – Canadian co-operation programs, culminating in the Canada Arm for space shuttle operations and the Robot Arm for the International Space Station (ISS). In communications, Canada in 1972, became the first nation to launch and operate a Domestic Satellite ANIK A1, followed in 1976 by the communications test satellite (CTS), a satellite that pioneered much of the technology used today in the Ku Band and for Direct to Home Services.

In 1969, based on the demands of Canada's vast territory with scattered and remote populations, the Government of Canada jointly formed Telesat Canada, as the domestic Satellite operator, with the Canadian Telephone companies. Now fully privatised, Telesat Canada has grown to be one of the top four Global Satellite operators, and in 2004 together in an advanced technology program with Canadian industry, Telesat launched the first consumer internet service in Ka Band on the Anik F2 satellite. In fact, together with similar efforts in the United States, Italy and Japan, Canada's pioneering work on Ka Band was fundamental to the creation of the Ka Band conference.

Whereas Canada's evolution in Satellite Communications has followed a steady growth of the Industry over the past fifty years, recent developments have promised a paradigm shift in both technology capabilities and service development. The ICSSC 2018 conference program demonstrated the scope of these developments. Some of the key topics addressed in both the technical sessions and the Plenary Panels were:

- The evolution of the so-called high-throughput satellite systems (HTS) in the Geosynchronous orbit.
- The emergence of concepts in alternative medium or low earth orbits (MEO and LEO).
- The proposed mega-constellation ventures with their significant challenges of low production costs and state of the art manufacturing and launch capabilities.
- The evolution from digital to photonic techniques in satellite payloads.
- Advanced signal processing and phased array antenna beam forming.
- High throughput digitally transparent and regenerative repeaters.
- Low cost consumer ground terminals.
- Advanced techniques for efficient mobile satellite communications antennas.
- The integration of satellites into 5G wireless systems.

Much of the technology to support this paradigm shift was discussed in numerous technical sessions described below. Unique to the ICSSC, a full day workshop or Colloquium is held on the day prior to the opening of the conference. This year the subject chosen was "Satellites in the 5G era, towards the third decade of the 21st Century." This workshop focused on trying to define the role of satellite communications in an era that will be dominated by advanced wireless systems. The workshop reviewed potential applications that are best suited to GEO, MEO and LEO satellite systems and the correct way to integrate these satellite systems into a 5G based network. The day of the workshop was divided up into a session in the morning, from mostly users and service providers, who provided the application demands of the user and a session in the afternoon, involving both satellite and wireless manufacturers who proposed systems and technology solutions.

A tradition of the ICSSC and Ka Band conference are the Plenary sessions. An introductory panel was led by NASA and involved all the major government Space agencies. This panel reviewed the future demands for communication relay in the Lunar and Deep Space environment. Three other panels over the following days dealt with the subjects of:

- Optical Satellite Techniques, covering both free space relay communications and satellite payload optical components and sub-systems
- Mobility and Mobile Satellite Systems
- The Paradigm shift in the industry created by both advanced technology and innovative proposals for HTS and Mega Leo Constellations, addressing both the technical and business challenges

Each panel created significant interest and was populated by leading industrial authorities both from the user and the manufacturing domains. The Paradigm shift panel was arranged to be the final session of the conference and over the past two conferences this panel has built up a very strong reputation for both the quality of the opinions presented and the critical involvement of the audience. In closing the conference, this panel laid out potential developments expected in the coming year and created an expectation for a reassessment of progress in this panel at ICSSC 2019.

In terms of attendance ICSSC 2018, together with our partners in the Ka Band conference attracted a record attendance and a wide-ranging participation from the global Satellite Communications community. 230 delegates attended the conference with 125 taking up the colloquium/workshop. Significantly, at a time when the Satellite Communications industry is under some severe capital funding challenges, the delegates represented the global reach of the industry. Not surprisingly a total of 43 per cent of delegates were from Canada and the United States and 30 per cent were from Western Europe with France, Germany, Italy and the United Kingdom, representing from 5 to 9 per cent each. However, Japan represented 10 per cent of attendees and there was significant representation from Israel, South Korea and China with small delegations from other countries in Asia, Scandinavia, Eastern Europe and Russia.

An important element in the ICSSC/Ka Band program and one that contributes to a continued understanding across the global community is the social program. The conference offers a spousal attendance package, and for many, the conference offers an annual opportunity to renew long-distance friendships. Within the official social program, beyond long informal conversations late into the night, there was an exciting cocktail evening spent at the Falls pavilion with an opportunity to explore "under the Falls." A mid conference awards luncheon saw the AIAA Aerospace Communications Award presented to Dr. John Baras of the University of Maryland for lifelong contributions to Aerospace Communications. Finally, the conference banquet held at the Legends Golf Club was well attended and added to the general good feeling of the conference.

At the close of the conference, the organisers were able to announce that, as in some other years, in 2019 the two conferences will coordinate programs but hold geographically separate events. The Ka Band Conference 2019 will be held in Sorento, Italy, while the 37th ICSSC (ICSSC 2019) will take place in Okinawa, Japan. It is hoped the two conferences will come together again in 2020.

This book is based on the presentations given in the various technical sessions of ICSSC 2018. The theme of the conference, "Space based Communications, Applications and Technologies in the 5G Era: Moving forward to the Third Decade of the 21st Century", gave ample scope for papers addressing a broad range of space and terrestrial communication technologies. These papers were organised into 11 technical sessions covering the following areas:

- Multi-beam satellite systems
- Cognitive communications and propagation channel optimisation
- Flexible high-throughput satellite (HTS) systems and interference mitigation techniques
- New satellite system architectures and components
- High-speed optical communications and feeder links
- VHF data exchange system
- Mobile satellite systems and bandwidth efficient techniques
- Transmitter and modem technologies
- 5G and satellite networks integration
- Satellite networks design challenges and applications
- EM techniques advances and photonics tutorial

A selection of 40 of the presentations covering significant and hitherto unpublished contributions to advances in communications satellite systems have been written up and edited as full articles and are presented as separate Chapters of the present volume. Each Chapter is therefore a self-contained technical article with its own references. No attempt was made to integrate the contributions to eliminate any repetitive referencing or overlap in the treatment of some concepts. It is hoped that this approach preserves the individual style and perspective of the various authors and will allow readers to dip into any topic or Chapter in any order without loss of continuity. The contributing authors come from a wide range of backgrounds in academia, industry, government and regulatory bodies. The book is therefore a multi-sectoral collection of research advances which, it is hoped, will be of interest and great benefit to satellite industry practitioners, academic researchers and other technical personnel engaged in telecommunications in general.

We wish to thank all the contributing authors for the stellar effort they put into writing up their ICSSC 2018 presentations as full technical articles and for responding positively to queries and comments from reviewers and editors. We acknowledge the IET team of Olivia Wilkins, Jessica Bristow, Kruna Vukmirovic and Valerie Moliere for their patient and professional work in organising and managing this selection of the proceedings of ICSSC 2018 through to publication both online and in paper print. Organising a successful not-for-profit international conference such as ICSSC 2018 takes a large team of competent volunteers. We were fortunate to have been part of a highly experienced and dedicated conference organising committee which ensured a smooth running of all aspects of ICSSC 2018, including one colloquium, keynote speech, 4 plenary sessions, 3 social events, 11 technical sessions and general conference administration. Our special mention goes to Clotilde Canepa Fertini of FGM Events for her indefatigable work in supporting the administrative and social aspects of the conference. We also thank all the paper reviewers, technical session chairs, panel chairs, panelists, speakers and conference participants who contributed immeasurably to make ICSSC 2018 an exciting and stimulating forum for exchanging and refining the latest ideas in integrated terrestrial and satellite communications as well as space technologies. We hope that through this volume we have been able to capture and preserve some of these innovative ideas for the benefit of the global satellite communications community.

Ifiok Otung, United Kingdom
Thomas Butash, United States
Peter Garland, Canada
June 2019

Author Biographies

Prof Ifiok Otung is Professor of Satellite Communications at the University of South Wales (USW), UK. His areas of expertise include Mobile & Satellite Communication Systems and Radiowave Propagation. He is a Chartered Engineer with a broad and international experience of research and teaching at various universities in Europe and Africa. He earned his Ph.D. in Satellite Communications from the University of Surrey, UK. He has written several advanced books including *"Digital Communications Principles and Systems"* (IET, 2014). He is a member of the IET and AIAA.

Dr Thomas Butash founded *Innovative Aerospace IS* in 2011 to provide consulting services on state-of-the-art aerospace information systems (IS), with a focus on communications satellite systems design. Previously, he was a Technical Director and Engineering Fellow at BAE Systems (formerly Lockheed Martin, Loral and IBM) Space Systems & Electronics, with more than 30 years' experience in communications satellite systems development.

He has authored numerous papers, co-authored a textbook on optimal signal processing, holds four patents on digital communications signal processing and numerous professional awards. He Chairs the AIAA Communications Systems Technical Committee's ICSSC Steering Subcommittee and is an AIAA Fellow and received the BS, MS and Ph.D. EE from the University of Maryland.

Mr Peter Garland has worked on Satellite Communications in Canada for well over 35 Years. During that time he has been involved in nearly every program where the state of the art was moved forward. In the mid-nineties, he led studies that eventually resulted in the flight of an advanced Ka band digital processing payload on the Anik F2 satellite. He also led the team that first developed and promoted the DVB-RCS standard for two-way satellite communications.

His career in communications started as an apprentice in a short wave transmitting station in his native United Kingdom. This station was one of the original links in the Marconi Imperial Beam System. As a young man, he was also privileged to be part of early Satellite Communications activities at Goonhilly Radio Station in the West of England. His only time away from Communications was a spell of seven years spent at the University of Birmingham, UK, designing payloads for X-ray astronomy missions that flew on sounding rockets launched from Woomera, Australia, and on one of the first Space Shuttle flights.

He was recently presented with the 2014 American Institute of Aeronautics and Astronautics (AIAA) award for achievement in Aerospace Communications, only the second Canadian and the second UK born recipient to be presented with this international award. He is also currently Chairman of the Communications Systems Technical Committee of the AIAA.

He is currently completing a Masters in Naval History with a dissertation subject of “Wireless and Its Impact on Geopolitics and Naval Operations (1919 to 1945)”.

Section 1 – Multi-beam Satellite Systems

Symbol vs block level precoding in multi-beam satellite systems

Farbod Kayhan[1], Alireza Haqiqatnejad[1], Joel Grotz[2], and Nader Alagha[3]

[1]*SnT, University of Luxembourg, 29 Avenue JF Kennedy L-1855 Luxembourg.*
[2]*SES S.A., Betzdolf, Luxembourg.*
[3]*ESA/ESTEC, Keplerlaan 1, Postbus 299, 2200 AG Noordwijk, The Netherlands.*

May 22, 2019

Abstract

Precoding techniques for mulit-beam satellite systems have received a considerable attention in recent years as a tool to mitigate the interference among the beams, and hence increasing the throughput. Our goal is to compare two main categories of precoding schemes, namely, the conventional linear block level precoding and the symbol level precoding. Focusing on power minimization problem with signal to interference plus noise ratio (SINR) constraints, symbol level precoding (SLP) has significant gains with respect to the zero forcing (ZF). However, the lower transmit power is achieved with a price: A higher computational complexity. Therefore, several sub-optimal SLP techniques have been proposed in the literature to overcome the complexity. While ZF has the lowest complexity among the techniques chosen in this paper, it is not an optimal linear block level precoder as far as power minimization is concerned. Therefore, in order to have a more complete picture, one needs also to consider optimal block level precoders. Our results indicate that in order to have a fair comparison, one needs to consider two different scenarios, namely, low and high SINR threshold regimes. While for low SINRs the optimal linear block level precoding scheme may provide a good solution with reasonable complexity, for high SINR threshold, the SLP techniques become more attractive. Our results also indicate that the performance of SLP highly depends on the chosen constellation space, and therefore a final conclusion can be achieved only after appropriately optimizing the constellation set.

1 Introduction

The ever growing demand for data throughput has created the need for High Throughput Satellites (HTS) with a large number of beams, hence a larger frequency reuse to cope with the throughput demands for terabit/s capacity per satellite. In recent years the use of more aggressive frequency re-use schemes has attracted many research and development activities. However, the co-channel interference can hinder the potential throughput gain. A possible solution, that has been considered by several researchers, is to manage the interference using the precoding techniques [1–5]. The conventional linear block-level precoding (BLP) techniques such as zero forcing (ZF) and minimum mean square error (MMSE) have been proved to provide significant gains in aggregate throughput with respect to non-precoded systems [1]. More recently, symbol-level precoding (SLP) has also been studied in the context of SatCom systems. For example, in [5], the gains of SLP with respect to ZF and MMSE techniques has been reported, assuming a per-beam power constraint.

The main drawback of SLP is the high computational complexity (CC) at the transmitter. This has resulted in investigating sub-optimal SLP methods with possibly reduced complexity [6–8]. In [6], the SLP optimization problem minimizing the total transmit power is formulated as a non-negative least squares, which can be solved via the existing fast algorithms. For quadrature amplitude modulation (QAM) schemes, the authors in [7] have analyzed the structure of the optimal symbol-level precoder with symbol error probability constraints and proposed a heuristic low-complexity solution. More recently, the authors in [8] have proposed a sub-optimal closed form (CF) solution for SLP with distance preserving constructive interference regions (DPCIRs) which reduces considerably the complexity in a trade-off with performance. In this paper we adopt this method as it provides the state of the art.

In this paper we focus on power minimization problem under signal to interference plus noise ratio (SINR) constraints. First, we briefly review the four precoding techniques that are addressed in this paper, i.e., ZF, optimal linear BLP, optimal SLP and sub-optimal CF-SLP. The constructive interference regions for SLP methods are chosen to be DPCIRs. We provide a detailed analysis of the complexity of each technique. Then, considering a realistic multi-beam satellite scenario, we simulate the performance of each technique for various constellation sets. Our simulations indicate that optimal BLP, optimal SLP and CF-SLP all perform better than ZF as it is expected, given the fact that ZF only mitigates the interference without any optimization of transmitted power. However, the performance of optimal BLP converges to that of ZF for high SINR thresholds. On the other hand, SLP techniques provide a significant gain with respect to ZF and optimal BLP techniques for high SINR thresholds (SINR $\geq$ 8 dB). The selection between BLP and SLP in this region of SINR, is a matter of trade-off between performance and complexity and depends highly on the system parameters such as number of transmitter and receiver antennas and the constellation size and shape. For low SINR thresholds (SINR $\leq$ 7 dB), the optimal BLP outperforms the other schemes with a reasonable complexity increase compared to ZF.

While the shape of the constellation has no effect on the BLP scheme (it may only be optimized to maximize the mutual information of the channel), our results indicate that the SLP performance depends strongly on the constellation shape, hence an appropriate optimization may be needed in order to have a fair comparison between the block and symbol level precoders.

The rest of this paper is organized as follow. In Section 2 we describe the system model that we consider in this paper. Section 3 briefly reviews BLP and SLP techniques that have been considered in our simulations. We discuss in Section 4 the computational complexity of both block and symbol level precoding. The simulation results are presented in Section 5. Finally, in Section 6 we conclude the paper and propose some future studies.

Notations: To denote matrices and vectors, we use uppercase and lowercase bold-faced letters respectively. For matrices and vectors, $[\cdot]^H$ and $[\cdot]^T$ denote conjugate transpose and transpose operators, respectively. For vectors, $\|\cdot\|$ represent the l_2 norm. For any set $\mathcal{A}$, $|\mathcal{A}|$ denotes the cardinality of $\mathcal{A}$. $\mathbb{R}$ and $\mathbb{C}$ represent the sets of real and complex numbers.

2 System Model and Problem Statement

In this section we first introduce the generic system model for a downlink of multiuser unicast channel. Then, we introduce the multi-beam satellite system model which is used in Section 5 in our simulations. We also briefly describe the power minimization problem under given SINR thresholds constraints and formulate the problem.

2.1 System Model

We assume the downlink of a multiuser unicast channel where a common multiple-antenna transmitter sends independent data streams to K single-antenna users. The transmitter, which is equipped with N antennas, employs a precoder for transmission. The precoder maps independent data symbols $\{s_k\}_{k=1}^K$ onto N transmit antennas, where s_k denotes the intended symbol for the k-th user. We assume that s_k is drawn from a finite equiprobable constellation set χ with cardinality $|\chi| = M$. The signal vector to be transmitted is denoted by $\mathbf{u} = [u_1, \ldots, u_N]^T \in \mathbb{C}^{N\times 1}$, and is a function of all users' symbols $\mathbf{s} = [s_1, \ldots, s_K]^T \in \mathbb{C}^{K\times 1}$. In the presence of frequency-flat fading and additive white Gaussian noise, the received signal at the receiver of the k-th user is

$$r_k = \mathbf{h}_k\mathbf{u} + z_k, \; k = 1, ..., K, \tag{1}$$

where $\mathbf{h}_k \in \mathbb{C}^{1\times N}$ contains the channel coefficients between the transmit antennas and the single receive antenna of user k, and $z_k \sim \mathcal{CN}(0, \sigma_k^2)$ represents the complex Gaussian noise at the k-th receiver. Throughout this paper we assume that $K = N$ and the modulation set χ is the same for all users.

The channel matrix $\mathbf{H}$ is the matrix containing all channel vectors $\mathbf{h}_k$ for $k = 1, \ldots, K$. The k-th user may optimally detect s_k from r_k based on the single-user maximum-likelihood (ML) decision rule. We show the $K \times K$ precoding matrix by $\mathbf{W} = [\mathbf{w}_1, \ldots, \mathbf{w}_K]$, $\mathbf{w}_k \in \mathbb{C}^{K\times 1}$. Therefore, the precoded transmitted signal can be written as $\mathbf{u} = \mathbf{W}\mathbf{s}$ or equivalently, the symbol transmitted from antenna j is obtained as $u_j = \mathbf{w}_j\mathbf{s} = \sum_{k=1}^K w_{j,k}s_k$.

2.2 Beam Pattern Over Europe

We consider a 200-beam antenna pattern operating in the Ka band. The satellite is located on the Geostationary Earth Orbit (GEO) at longitude of $30°E$ with carrier frequency of 20 GHz and user bandwidth of 500 MHz.

The complex channel matrix is obtained as $\mathbf{H} = \mathbf{L} \circ \mathbf{B}$ where the matrix $\mathbf{B}$ contains the $K \times K$ complex value entries corresponding to the gains and phase rotations that are measured for each beam pattern. The $K \times K$ matrix $\mathbf{L}$ includes the gains and losses contributing to each pair of satellite beam user receiver (excluding the satellite antenna coefficients that are included in $\mathbf{B}$). The $\circ$ denotes the element wise multiplication of two matrices. For any n and k we denote by $\mathbf{L}_{k,n}$ the signal strength of the n-th beam's antenna received by the k-th user's antenna and is given by

$$\mathbf{L}_{k,n} = \sqrt{G_{R_k}}\left(4\pi\frac{d_{k,n}}{\lambda}\right)^{-1} \tag{2}$$

where G_{R_k} is k-th user antenna gain, $d_{k,n}$ is the distance between the two antennas, λ is the signal wavelength. The thermal noise power is computed as $\kappa T_k B_w$ where κ is the Boltzmann constant, T_k is the clear sky noise temperature of the receiving antenna and B_w is the user link bandwidth.

In Figure 1 we show the center of the beams. In our simulations we have chosen seven beams (indexed by 139, 138, 121, 120, 119, 102 and 101) in the center. The interference of the other beams are ignored in our simulations. The center of these beams and the coverage regions are depicted in Figure 2.

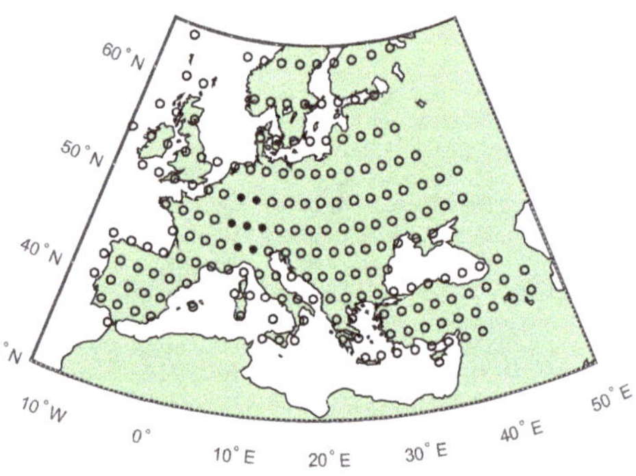

Figure 1: 200 beams covering Europe. Center of each beam is shown by a circle.

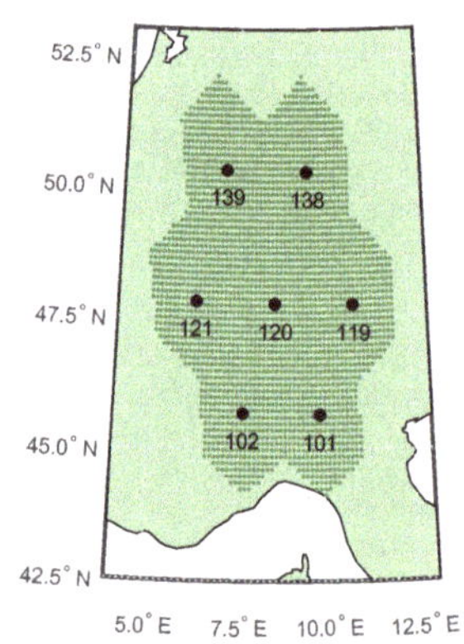

Figure 2: The seven beams selected in our simulations with their coverage regions.

2.3 Problem Statement

As mentioned before, we are interested in the power minimization problem being constrained by individual SINR requirements, i.e.,

$$\underset{\mathbf{u}}{\text{minimize}} \quad \mathbf{u}^H\mathbf{u} \tag{3}$$

$$\text{s.t.} \quad \text{SINR}_k \geq \gamma_k, \; k = 1, ..., K, \tag{4}$$

where γ_k and SINR_k are respectively the target SINR threshold and the received SINR of k-th user. The precoding problem can be formulated in several other ways depending on the desired objective function. For example, one may be interested to maximize the minimum received SINR in a system for a given average power (usually referred to as SINR balancing problem). We have chosen the power minimization problem mainly due to the fact that both optimal SLP and optimal BLP problems can be formulate easily as a convex optimization problem and hence we will be able to have a rather fair comparison between the performance and complexity of each method. For more details on SINR balancing problem for SLP precoding problem we kindly refer the reader to [9] and [10].

3 Precoding Techniques

Several precoding techniques have been proposed in the literature. As mentioned before, our goal in this paper is to have a comparison between block and symbol level precoders, both in terms of performance and complexity. In this section we briefly provide a description of the precoding schemes that are considered in this paper.

3.1 Block-Level Precoding

By BLP, we mean the precoding techniques that compute one precoding matrix $\mathbf{W}$ for any given channel matrix $\mathbf{H}$. In such cases, as long as the channel is not updated (depending on channel coherence time), one does not need to recompute the matrix $\mathbf{W}$. Notice that in BLP, only the average SINR over a given block of symbols need to be greater than or equal to the given threshold. We focus on two BLP techniques in this paper: zero forcing and optimal linear precoding.

3.1.1 Zero Forcing

ZF precoding is a method by which the multiple antenna transmitter processes the signals to eliminate the interference. By assuming that the elements of $\mathbf{H}$ are i.i.d. random variables, the channel matrix is full-rank (with high probability). In such cases, the pseudo inverse of $\mathbf{H}$ exists when $K \leq N$. In our case, where $K = N$, the matrix $\mathbf{H}$ is invertible and the precoder matrix is simply $\mathbf{W} = \mathbf{H}^{-1}$ (up to a normalization factor), eliminating the interference of all antennas. While ZF provides a perfect solution to interference mitigation, it should be noted that it is not the optimal solution of power minimization problem defined in (3). We kindly refer the readers to [11] on detailed discussion on ZF optimality in power minimization.

3.1.2 Optimal Linear Precoding

The SINR constraint power minimization problem as formulated in (3) is not convex. However, in [12] the authors propose an iterative algorithm that converges to the optimal solution. At each step of the algorithm, one only needs to compute the updated values of associated powers using a given closed form formula. As we will see later, the solution of optimal BLP converges to that of ZF for high values of SINR thresholds. This is rather intuitive, as having higher SINR thresholds implies to have lower interference. Notice that usually the power of the constellation set χ is normalized to one and therefore when one computes the average SINR over a block, the actual transmitted symbols do not have any effect on the precoding matrix.

3.2 Symbol Level Precoding

The main idea behind SLP is to exploit also the vector $\mathbf{s}$ in designing the precoder matrix. Several SLP schemes have been proposed in the literature. We adopt the definition that is based on the concept of constructive interference regions (CIRs) introduced in [13] and [14]. In particular, we focus on the so called distance preserving CIRs (DPCIRs). In [15], DPCIRs are introduced as a general family of CIRs that do not increase the symbol error rates of the users. The halfspace representation of DPCIRs is provided for generic modulation schemes based on the ML decision regions of the constellation set in [9] and [10]. It is important to notice that DPCIRs are not optimal choice in general and they can be relaxed in order to improve the performance of SLP.
The DPCIRs for an optimized 8-ary constellation (see [16]) are depicted in Figure 3.

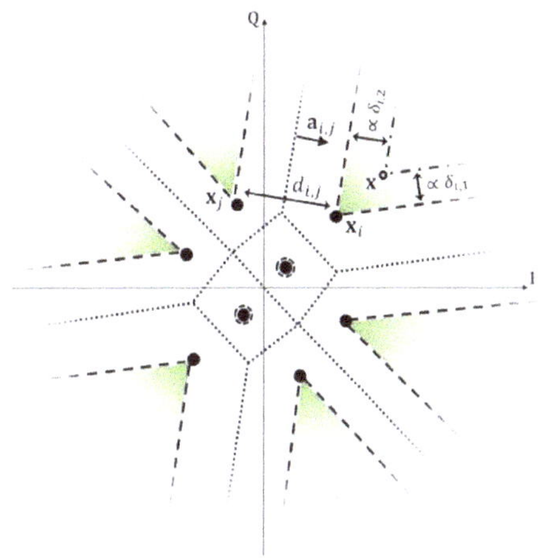

Figure 3: optimized 8-ary constellation with correspoding Voronoi regions and DPCIRs.
Given a constellation point $\mathbf{x}_i \in \chi$ with M_i neighboring points, from the representation provided in [9], it is straightforward to show that any point $\mathbf{x}$ in the DPCIR of $\mathbf{x}_i$ satisfies a linear matrix inequality (LMI) as

$$\mathbf{A}_i(\mathbf{x} - \mathbf{x}_i) \succeq \mathbf{0}, \tag{5}$$

where $\mathbf{A}_i \in \mathbb{R}^{M_i \times 2}$ is a matrix that contains the normal vectors of DPCIR boundaries (hyperplanes), given by

$$\mathbf{A}_i = \begin{bmatrix} \mathbf{a}_{i,1}^{\mathrm{T}} \\ \vdots \\ \mathbf{a}_{i,M_i}^{\mathrm{T}} \end{bmatrix} = \begin{bmatrix} (\mathbf{x}_i - \mathbf{x}_{i,1})^{\mathrm{T}} \\ \vdots \\ (\mathbf{x}_i - \mathbf{x}_{i,M_i})^{\mathrm{T}} \end{bmatrix}, \tag{6}$$

with $\mathbf{x}_{i,1}, ..., \mathbf{x}_{i,M_i}$ denoting the neighboring constellation points of $\mathbf{x}_i$. By introducing a non-negative vector $\boldsymbol{\delta} \in \mathbb{R}_{+}^{M_i \times 1}$, (5) is equivalent to

$$\mathbf{A}_i(\mathbf{x} - \mathbf{x}_i) = \boldsymbol{\delta}_i, \ \boldsymbol{\delta}_i \succeq \mathbf{0}, \tag{7}$$

which will be used as the CI constraint in our formulation of the SLP optimization problem. It is also shown that if $\mathbf{x}_i$ is a constellation point with bounded decision region, we always have $\boldsymbol{\delta}_i = \mathbf{0}$. We denote the region obtained in equation (7) by $\mathcal{D}_i$.

3.2.1 Optimal SLP

The power minimization problem for SLP can be defined as

$$\begin{aligned} &\underset{\mathbf{u}}{\text{minimize}} \quad \mathbf{u}^H\mathbf{u} \\ &\text{s.t.} \quad \mathbf{h}_k\mathbf{u} \in \sigma_k\sqrt{\gamma_k}\,\mathcal{D}_k, \ k = 1, ..., K, \end{aligned} \tag{8}$$

Notice that constellation points (transmitted signals) are present in the constrains ($\mathcal{D}_k$ is defined as a function of x_k) of the above optimization problem. In this formulation, it is assumed that all constellation points are boundary points of the constellation, for more generic formulation please see [10]. This problem can be formulated as a convex optimization problem [15]. However, the obtained precoding matrix depends on the given transmitted symbol vector $\mathbf{s}$. In our implementation, we directly optimize the vector $\mathbf{u}$ without explicitly deriving the matrix $\mathbf{W}$ as was formulated in [14]. In our simulations, we use the CVX package and MATLAB to solve the convex optimization problems.

3.2.2 Sub-optimal SLP

The optimal SLP, as formulated in the last section, provides a significant gain compared to ZF and optimal BLP for high SINR values. However, solving a convex optimization problem for each possible combination of user's symbols becomes rapidly impractical as the size of the system and/or the constellation space grow. Therefore, several sub-optimal solutions have been proposed in the literature to overcome the computational complexity of optimal SLP [6–8]. In particular, recently a closed-form (CF) sub-suboptimal solution has been proposed by the authors in [8]. The method in [8] is based on approximating the solution of the convex problem (8) by a set of linear equations. The derivation of the equations is rather technical and we kindly refer the readers to [8] for details.

4 Complexity Comparision

In this section we discuss in some details the complexity of four precoding schemes that we introduced in Section 3. In general, computing the exact CC of a precoding scheme is rather complicated (except for ZF). The difficulty mainly comes from the fact that the CC depends on several system parameters (for example coherence time, system size, constellation order, etc) and also the specific implementation of the optimization algorithms. In this paper we provide a qualitative comparison and focus only on two sources of complexity that contribute the most to the CC:

- Number of precoding matrices that are computed as a function of coherence time τ_c

- Whether solving a *convex optimization* problem is needed to obtain the precoding matrix

This simplifies the CC analysis by ignoring several other sources of complexities such as matrix computations (e.g., matrix products and matrix inverses).

As mentioned before, the main idea of SLP is to design the precoding matrix taking into the account also the transmitted signal vector $\mathbf{s}$. In other words, for each possible combination of the symbols, one needs to compute a precoding matrix. Assuming that all K users are using the same constellation χ with cardinality $|\chi| = M$, the number of possible combinations for the symbol vectors are M^K. This gives an upper bound on the number of precoding matrices that one needs to compute for any given channel $\mathbf{H}$. Assuming that N_s denotes the number of symbol vectors that are send in each coherence time period, then the number of precoding matrices to be calculated in SLP scheme is $C_{slp} = \min\{N_s, M^K\}$ (for simplicity we assume that all transmitted sequences are different). To have an idea of numerical values, for our 7×7 and QPSK constellation, there are $4^7 = 16384$ possible combinations. For 8PSK constellation this number is $\approx 2.1 \times 10^6$. As for the BLP, only one precoding matrix is needed to be calculated in each coherence time.

Now, we discuss the complexity to calculate one precoding matrix for each of the schemes. For ZF, we only need to calculate the channel pseudo inverse. This method, in general, is the least complex of all. As for optimal BLP, one needs to solve the optimization problem in 3, that can be formulated as an iterative algorithm as discussed in [12]. In each iteration one needs to compute the updated powers using a closed form formula. The number of iterations may change from one channel realization to the other and the system parameters. Usually, up to 10 iterations are sufficient in our setting. For optimal SLP we need to solve C_{slp} convex optimization problems. Therefore, optimal SLP is at least C_{slp} times more complex than optimal BLP, given the fact that solving one convex optimization problem is more complex than the proposed iterative algorithm in [12]. For CF-SLP, we need again to compute C_{slp} precoding matrices, each one needing to solve a linear system of equations. In our implementation of the CF-SLP and optimal BLP, the computation time of one optimal BLP takes around the same time as computing 100 CF-SLP matrices. In other words, if $N_s > 100$ then CF-SLP is more complex than optimal BLP. In order to have a better view of the complexity of each scheme, in Table 1 we report the complexity of each method reporting the average execution time per symbol. In this table we assume a QPSK constellation and set the SINR threshold to 5 dB. The number of simulated symbols is 50. As it can be seen, BLP and CF-SLP have almost the same order of complexity.

Table 1: Execution time of the proposed precoding schemes.

Modulation	Dimension	Execution time (ms/symbol)			
		ZFBF	OPT-BLP	OPT-SLP	CF-SLP
QPSK	7×7	0.0147	0.1375	595.64	0.1704

5 Simulation Results

In this section we present the simulation results for several constellation sets. In Table 2 we present the values of the parameters that have been used in the simulations.

Table 2: Parameters used in the simulations.

Parameter	Value
Beam Radiation Pattern	Provided by ESA (see [17])
satellite Orbit	Geostationary
Total Number of Beams	200
Number of Precoded Beams	7
Users/Beam/Time slot	1
Carrier Frequency	20 GHz
Total Bandwidth	500 MHz
Noise Variance σ_k^2	1

To generate $\mathbf{H}$, one user per beam is chosen randomly (the users are uniformly distributed inside each beam) and the elements of $\mathbf{H}$ are computed using the available measurements. In our simulations, we average the performance over 100 randomly generated transmitted symbol vectors $\mathbf{s}$ over 100 randomly generated channel matrices $\mathbf{H}$ (10^4 symbols in total).

The total transmit power in dBW as a function of SINR threshold for QPSK constellation is presented in Figure 4(up). The optimal BLP has the best performance for all SINR values from 1 to 5 dBs, which is the range of interest when QPSK constellation is employed. On the other hand, the loss of CF-SLP with respect to the optimal approach is negligible. In Figure 4(down) we plot the average SINR of each user. The SINR threshold for obtaining this plot has been fixed to 5 dB. For both ZF and optimal BLP, the threshold is satisfied closely. However, the average SINRs obtained for each user with optimal SLP and CF-SLP are higher than the requested threshold. Indeed, the formulation of SLP as (8) usually results in higher SINR than the target value.

In Figure 5 we plot the uncoded symbol error rate (SER) for all four schemes. It should be noticed that SLP based schemes have slightly lower SER with respect to block schemes. This is due to the fact that the received signals in SLP with DPCIRs usually satisfy, on average, a larger minimum distance compared to the original constellation (see for details [15]). The SLP gain in SER compared to conventional block-level precoders varies as a function of constellation shape and order, but usually it is rather small and do not provide a big advantage, hence we drop this comparison in the rest.

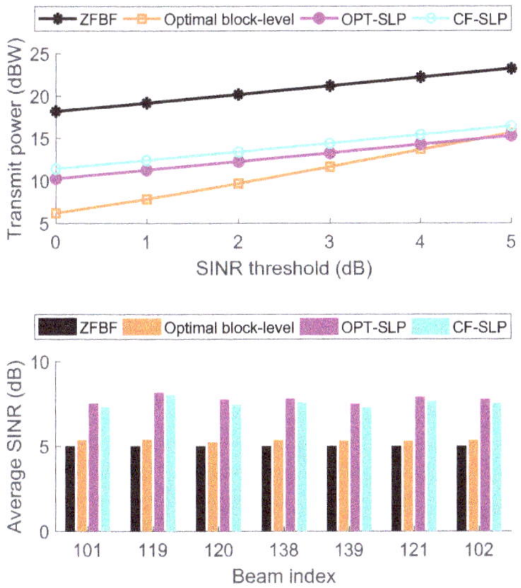

Figure 4: (Up) Average transmitted power in dBW as a function of SINR thresholds for QPSK constellation. (Down) average SINR values as a function of beam index for different schemes fixing the SINR threshold to 5 dB.

To summarize the results for QPSK constellation, both SLP schemes provide significant gains in transmit power with respect to ZF. On the other hand, the optimal BLP outperforms the SLP techniques. Notice that the optimal SLP, despite being the most computationally demanding, does not provide the best performance. However, any conclusion should be taken with utmost caution. Conceptually, SLP calculates the precoding matrix as a function of both channel and transmitted symbols, and therefore, optimal BLP is a special case of SLP. In other words, optimal BLP cannot outperform optimal symbol level precoders from the conceptual point of view. However, in this paper we are considering a specific structure for the CI regions when computing the precoding matrices, i.e., distance preserving CIRs. Moreover, due to formulation of the problem, SLP satisfies the SINR thresholds at each symbol instance instead of the average. Indeed, this is one of the reasons that average SINRs of the users are much higher than the target threshold.

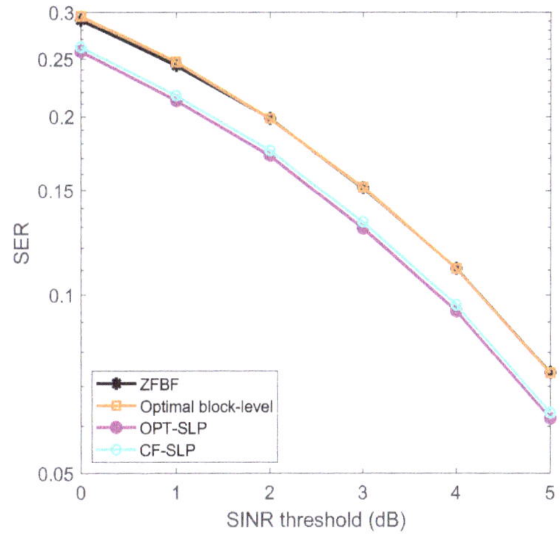

Figure 5: SER results for four different precoding schemes as a function of SINR for QPSK constellation.

Other types of CIRs may provide better results than those obtained by DPCIRs. However, the complexity of optimization problem may further increase, resulting in even more computationally demanding SLP. Moreover, these results are limited to the power minimization problem. The performance of SLP compared to block-level precoders may be different for other problem formulations, such as SINR balancing with limited power, or sum-rate maximizing problem.

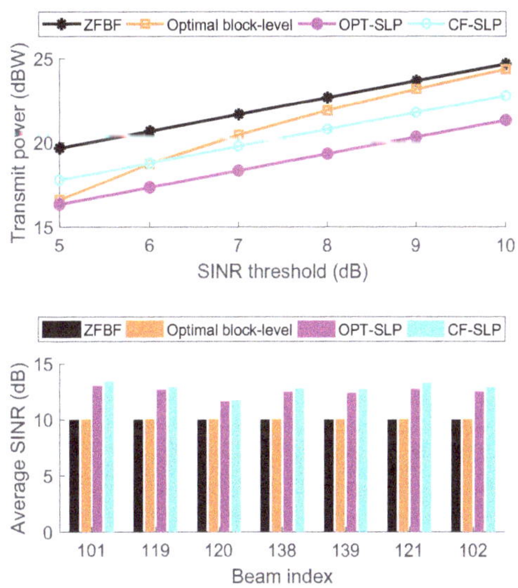

Figure 6: (Up) Average transmitted power in dBW as a function of SINR thresholds for 8-PSK constellation. (Down) average SINR values as a function of beam index for different schemes fixing the SINR threshold to 10 dB.

In Figure 6(up) and Figure 7(up) we plot, respectively, the results for 8-PSK and 8-OPT constellations for values of SINR from 5 to 10 dB. In this region, the SLP performs better than block-level precoders. Notice that the optimal BLP performance converges to that of ZF for high SINR thresholds despite a higher CC. The loss in performance of CF-SLP is higher for 8-PSK compared to that of the optimal 8 point constellation. As before, in same figures we also plot the actual obtained average SINR values, fixing the SINR threshold to 10 dB. Both SLP methods results in higher average SINR values compared to block-level schemes.

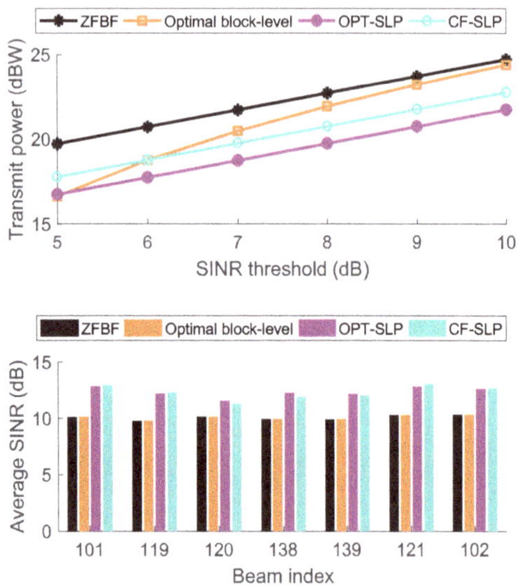

Figure 7: (Up) Average transmitted power in dBW as a function of SINR thresholds for 8-OPT constellation. (Down) average SINR values as a function of beam index for different schemes fixing the SINR threshold to 10 dB.

The CF-SLP loss does not merely depends on the order of the constellation. In order to show this, we have run similar simulations for two constellations of order 16. In Figure 8(up) we present the power minimization results for 16-QAM constellation for SINR values from 10 to 15 dB. As it can be seen, the CF-SLP performs very near to that of optimal SLP algorithm. On the other hand, in Figure 9(up) we do the same simulations for 16-QCI constellation. QCI constellations have been introduced in [18,19] and have a very good mutual information under the peak power constraint. They can be obtained as the image of QAM constellations under the radial map. For 16-QCI constellation, the CF-SLP practically has no gain compared to ZF.

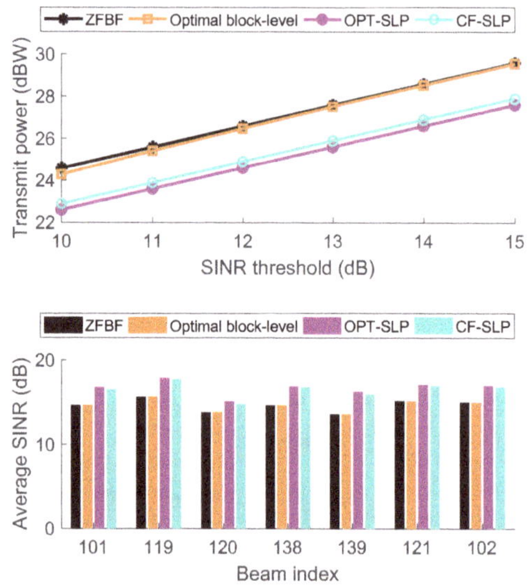

Figure 8: (Up) Average transmitted power in dBW as a function of SINR thresholds for 16-QAM constellation. (Down) average SINR values as a function of beam index for different schemes fixing the SINR threshold to 15 dB.

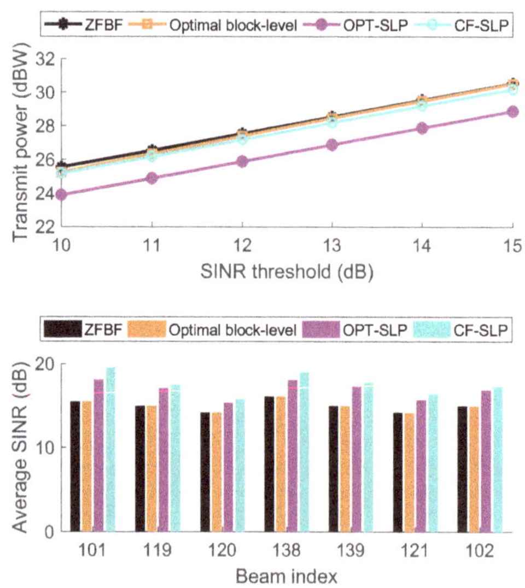

Figure 9: (Up) Average transmitted power in dBW as a funtion of SINR thresholds for 16-QCI constellation. (Down) average SINR values as a function of beam index for different schemes fixing the SINR treshold to 15 dB.

6 Conclusions and Further Research

In this paper we have compared the performance and complexity of linear ZF and optimal block-level precoder with those of optimal and sub-optimal SLP with DPCIRs in a multi-beam satellite system. We have considered a realistic channel measurements with a 200 beam pattern over Europe, minimizing the transmitted power under SINR threshold constraints. Focusing only on seven beams for precoding, we simulated the system for several constellation sets with 4, 8 and 16 signals. In general, our results indicate that further detailed studies of both schemes are needed in order to validate the gains and loss of each method. In particular, our simulations show that for small SINR values (SINR $\leq$ 7 dB), the optimal block-level precoding performs better (and have also a lower complexity) compared to SLP schemes when the constructive interference regions are chosen to be DPCIRs. On the other hand, for larger values of SINR thresholds, the SLP methods achieve lower transmitted power values, though with a higher computational complexity compared to block-level precoders. A rather detailed discussion on the complexity of each scheme is provided to ease the selection of the method. Our simulations also indicate that the performance of SLP (specially the proposed CF-SLP) strongly depends on the shape of the constellation, and therefore, the results for conventional constellation sets such as PSK and QAM may not show the full potential of the SLP scheme. Optimizing the constellation space for the symbol level precoded systems is an interesting problem and some research in this direction is ongoing.

Acknowledgments

This work has been supported by the Luxembourg National Research Fund (FNR) under CORE Junior project: C16/IS/11332341 Enhanced Signal Space opTImization for satellite comMunication Systems (ESSTIMS).

References

[1] Vazquez, M.A., Perez.Neira, A., Christopoulos, D., Chatzinotas, S., Ottersten, B., Arapoglou, P.D., et al.: 'Precoding in Multibeam Satellite Communications: Present and Future Challenges', *IEEE Wireless Communications*, 2016, **23**, (6), pp. 88–95

[2] Taricco, G.: 'Linear precoding methods for multi-beam broadband satellite systems'. In: European Wireless 2014; 20th European Wireless Conference, 2014, pp. 1–6

[3] Guidotti, A., Vanelli.Coralli, A., Taricco, G., Montorsi, G.: 'User clustering for multicast precoding in multi-beam satellite systems', *arXiv:abs/170609482*, 2017

[4] Christopoulos, D., Chatzinotas, S., Zheng, G., Grotz, J., Ottersten, B.: 'Linear and nonlinear techniques for multibeam joint processing in satellite communications', *EURASIP Journal on Wireless Communications and Networking*, 2012, p. 162

[5] Spano, D., Chatzinotas, S., Andrenacci, S., Krause, J., Ottersten, B.: 'Per-antenna power minimization in symbol-level precoding for the multibeam satellite downlink', *International Journal of Satellite Communications and Networking*, doi:10.1002/sat.1244

[6] J..Krivochiza, S.C. A. Kalantari, Ottersten, B.: 'Low complexity symbol-level design for linear precoding systems'. In: ymposium on Information Theory and Signal Processing in the Benelux, 2016, pp. 1–6

[7] Masouros, C., Sellathurai, M., Ratnarajah, T.: 'Vector perturbation based on symbol scaling for limited feedback MISO downlinks', *IEEE Trans Signal Process*, 2014, **62**, (3), pp. 562–571

[8] Haqiqatnejad, A., Kayhan, F., Ottersten, B.: 'Power minimizer symbol-level precoding: A closed-form sub-optimal solution', *arXiv:180710619*, 2018

[9] Haqiqatnejad, A., Kayhan, F., Ottersten, B.: 'Symbol-level precoding design for max-min sinr in multiuser miso broadcast channels'. In: 2018 IEEE 19th International Workshop on Signal Processing Advances in Wireless Communications (SPAWC), 2018, pp. 1–5

[10] Haqiqatnejad, A., Kayhan, F., Ottersten, B.: 'Symbol-level precoding design based on distance preserving constructive interference regions', *available on: arXiv:180400930*, 2018

[11] Peel, C.B., Hochwald, B.M., Swindlehurst, A.L.: 'A vector-perturbation technique for near-capacity multiantenna multiuser communication-part i: channel inversion and regularization', *IEEE Transactions on Communications*, 2005, **53**, (1), pp. 195–202

[12] Bengtsson, M., Ottersten, B.: 'Handbook of Antennas in Wireless Communications', chapter 18: "Optimal and suboptimal transmit beamforming", CRC Press, 2002.

[13] Masouros, C., Zheng, G.: 'Exploiting known interference as green signal power for downlink beamforming optimization', *IEEE Transactions on Signal Processing*, 2015, **63**, (14), pp. 3628–3640

[14] Alodeh, M., Chatzinotas, S., Ottersten, B.: 'Constructive multiuser interference in symbol level precoding for the miso downlink channel', *IEEE Transactions on Signal Processing*, 2015, **63**, (9), pp. 2239–2252

[15] Haqiqatnejad, A., Kayhan, F., Ottersten, B.: 'Constructive interference for generic constellations', *IEEE Signal Processing Letters*, 2018, **25**, (4), pp. 586–590

[16] Kayhan, F., Montorsi, G.: 'Constellation design for memoryless phase noise channels', *IEEE Transactions on Wireless Communications*, 2014, **13**, (5), pp. 2874–2883

[17] De Gaudenzi, R., Alagha, N., Angelone, M., Gallinaro, G.: 'Exploiting code division 7 multiplexing with decentralized multiuser detection in the satellite multibeam forward link', *International Journal of Satellite Communications and Networking*, 36(3), pp. 239–276

[18] Kayhan, F.: 'Qam to circular isomorphic constellations'. In: 2016 8th Advanced Satellite Multimedia Systems Conference and the 14th Signal Processing for Space Communications Workshop (ASMS/SPSC), 2016, pp. 1–5

[19] Kayhan, F.: 'On low complexity detection for qam isomorphic constellations'. In: 2017 IEEE Global Communications Conference (GLOBECOM), 2017, pp. 1–5

Hardware Demonstration of Precoded Communications in Multi-Beam UHTS Systems

Juan Duncan, Jevgenij Krivochiza, Stefano Andrenacci, Symeon Chatzinotas, Björn Ottersten[1]

[1]*SnT, University of Luxembourg, 29 Avenue JF Kennedy L-1855 Luxembourg*

May 19, 2019

Abstract

In this paper, we present a hardware test-bed to demonstrate closed-loop precoded communications for interference mitigation in the forward link of the multi-beam ultra-high throughput satellite systems. The hardware demonstrator is a full-chain closed-loop communication system with a multi-beam DVB-S2X compliant gateway, a satellite payload, and MIMO channel emulator and a set of DVB-S2X user terminals with real-time CSI estimation and feedback. We experimentally show the feasibility of Precoding implementation in satellite communications based on the superframe structure DVB-S2X standard. Using the test-bed we have a possibility to run real-time precoded DVB-S2X communication and benchmark its performance under realistic environment. The hardware demonstrator is suitable to perform realistic benchmarks of Block- and Symbol-level Precoding techniques for multicast and unicast user scheduling scenarios.

1 Introduction

The 5th generation of mobile radio communications systems should provide a high level of integration and flexibility between different types of telecommunication networks. Terrestrial and satellite systems historically evolved independently of each other, which results in technological diversity between the networks. The launched 5GPPP research program co-funded by the European Commission is set to work towards a definition of new common standards for 5G networks [1]. The objective of the project METIS 2020 as a part of 5GPPP is to build the foundation for a future mobile and wireless communications system for 2020 and beyond [2]. These standards will allow seamless joint operation of mobile cellular communications and satellite systems as a single service. The use cases of modern satellite communications (SATCOM) systems in 5G networks include increasing coverage of conventional terrestrial cells, facilitating caching through multicast/broadcast data transmission and providing off-load backhauling for unicast user traffic [3,4].

Although many activities focus on the integration of SATCOM into 5G using higher layers e.g. Software-defined networking (SDN), and network functions virtualisation (NFV), the business potential will be limited unless the raw channel capacity of the satellite systems can be increased proportionally to the terrestrial counterparts. Multi-beam satellites can increase their throughput capacity by utilising channel aided precoding [5–8]. Similar techniques are used in industrial standards in VDSL2 [9] and Long Term Evolution (LTE) [10] communications. In SATCOM the DVB-S2X [11] was developed as a precoding enabling standard.

MIMO Precoding techniques are based on the closed-loop approach by employing the retrieved Channel state information (CSI) from the User Terminals (UTs), requiring a feedback channel from UT back to Gateway (GW). Accuracy of the CSI estimations, which are affected by imperfections of the transmitter and all the receivers, and relevance due to the time-varying nature of a wireless channel are not perfect in real communications systems. The inability of acquiring instantaneous CSI at the GW for mobile satellite systems can be very challenging and affect the Precoding performance. However, there is potential for specific types of applications such as aeronautical/maritime systems, where the channel is predictable and there is no direct blockage of the line of sight component.

MIMO precoding techniques, which are defined as convex optimisation problems, have to be solved by time-consuming iterative convex optimisation (CVX) or Non-negative least squares (NNLS) solving methods that must fit into a relevant time frame. Further research is done on the reduction of the processing times to meet channel requirements [12,13].

On the other hand, academic research shows that Precoding techniques in SATCOM potentially allow more efficient spectral utilisation and substantially higher service availability [14–16]. To enable the efficient utilisation of satellite transponders, multiple carriers have to be relayed through a single HPA. However, the non-linear nature of the HPA results in adjacent channel interference and increased Peak-to-average power-ratio (PARP), which limits the expected performance gains [17,18]. In this context, studies on energy efficient onboard predistortion techniques, to maximise the performance of HPA by uniformly distributing the power load are conducted [19–21].

In this work, we focus on the implementation of the hardware demonstrator for the closed-loop precoded SATCOM. We describe the design and functionality of the multi-beam DVB-S2X compliant GW, the satellite MIMO Channel Emulator and the set of UTs. We validate the design requirements using reasonable software and hardware resources.

We implemented the physical layer of the DVB-S2X standard by means of software defined radio (SDR) techniques using commercial SDR platforms. Developing on SDR allows to rapidly prototype and deploy the precoded transmission in a more realistic environment rather than using only numerical simulations. Authors used the same approach in [22,23] to benchmark a novel Precoding technique using a small-scale hardware test-bed.

Notation: Upper-case and lower-case bold-faced letters are used to denote matrices and column vectors. The superscripts $(\cdot)^H$ and $(\cdot)^{-1}$ represents Hermitian matrix and inverse operations.

2 Hardware Demonstrator

In this paper, we present a hardware test-bed to demonstrate closed-loop precoded communications for interference mitigation in multi-beam ultra-high throughput satellite (UHTS) systems. We build the test-bed to demonstrate real-time precoded communications under realistic environments. For this matter, we designed a scalable architecture of the gateway and UTs compatible with the DVB-S2X superframe structure. Fig. 1 shows the block diagram of the demonstrator.

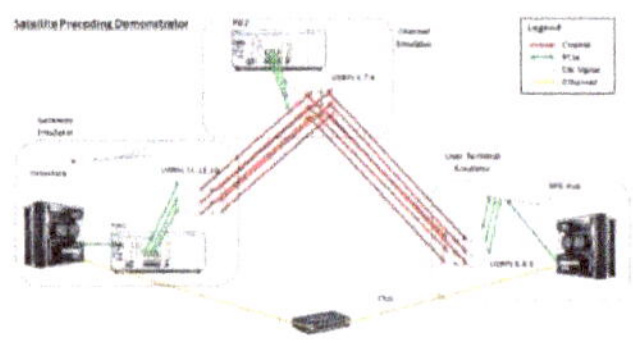

Figure 1: Block diagram of the Hardware Demonstrator

We use the commercially available SDR platform developed by National Instruments (NI). The platform consists of two NI PXI (PCI EXtension for Instruments) 1085 chassis, which allow centralised connection of the set of the NI USRP (Universal Software Radio Peripheral) 2954R and FlexRIO (Reconfigurable IO) 7976R. The NI USRP and FlexRIO have integrated FPGA (Field-Programmable Gate Array) module Kintex-7 from Xilinx.

The gateway simultaneously transmits 6 precoded signals towards 6 user terminals through a 6×6 multi-beam satellite channel emulator. The channel emulator acquires the gateway signals, applies the impairments of the satellite payload, Gaussian noise, and the multi-beam interference and transmits the signals to the UTs. The UTs estimate the CSI based on the DVB-S2X standard pilots and report the estimated values to the gateway through a dedicated feedback channel over an Ethernet link. The gateway uses this CSI data to compute a Precoding matrix. The table 1 summaries the current capabilities and the final targets of the demonstrator.

Table 1: Parameters of the hardware demonstrator

Parameter	Current	Target
Gateway IQ channels	6	16
Sampling frequency	1 MHz	80 MHz
Oversampling factor	4	4
Gateway TX freq.	1.21 GHz	-
Channel Emulator RX freq.	1.21 GHz	-
Channel Emulator TX freq.	960 MHz	-
User Terminal RX freq.	960 MHz	-
Filter roll-off factor	0.2, 0.15, 0.1, 0.05	-
Forward Error Correction	no	yes
LDPC code rate	no	1/2, 2/3, 3/4, 4/5

2.1 System Model

We consider a system model, which focuses on the forward link of a multi-beam satellite system. We assume a full frequency reuse scenario, in which all the beams transmit in the same frequency and time. The multi-user interference is mitigated by using the signal Precoding technique. The defined the number of transmitting antenna is equal to the total number of users in the coverage area. In this scenario, we consider a 6×6 MIMO channel. In the specified MIMO channel model, the received signal at the i-th terminal is given by $y_i = \mathbf{h}_i^\dagger f(\mathbf{x}) + n_i$, where $\mathbf{h}_i^\dagger$ is a 1×6 vector representing the complex channel coefficients between the i-th terminal and the 6 antennas of the transmitter, $\mathbf{x}$ is defined as the 6×1 vector of the transmitted symbols of DVB-S2X superframe at a certain symbol period and n_i is the independent complex circular symmetric (c.c.s.) independent identically distributed (i.i.d) zero mean Additive White Gaussian Noise (AWGN) inserted to the i-th terminal's receive signal. The function $f(\mathbf{x})$ represents the non-linear behaviour of the satellite channel.

Looking at the general formulation of the received signal, which includes the whole set of terminals, the signal model is

$$\mathbf{y} = \mathbf{H}f(\mathbf{x}) + \mathbf{n} = \mathbf{H}f(\mathbf{W}\mathbf{s}) + \mathbf{n}, \tag{1}$$

where $\mathbf{y} \in \mathbb{C}^{6\times 1}$, $\mathbf{n} \in \mathbb{C}^{6\times 1}$, $\mathbf{x} \in \mathbb{C}^{6\times 1}$, and $\mathbf{s} \in \mathbb{C}^{6\times 1}$ and $\mathbf{H} \in \mathbb{C}^{6\times 6}$. In this scenario, we define the Block-level Precoding matrix $\mathbf{W} \in \mathbb{C}^{6\times 6}$ as

$$\mathbf{W} = \hat{\mathbf{H}}^H \cdot (\hat{\mathbf{H}} \cdot \hat{\mathbf{H}}^H)^{-1}, \tag{2}$$

where $\widehat{\mathbf{H}} \in \mathbb{C}^{6\times 6}$. We consider the data symbols $\mathbf{s}$ to be unit variance complex vectors $|s_k| = 1$ for every $k = 1 \ldots 6$.

2.2 Gateway

The gateway operates with a central NI FlexRIO FPGA and 3 NI USRP nodes. The 3 NI USRP nodes connected to the same oscillator reference clock source. A single NI USRP has only 2 RF outputs. In order to transmit 6 independent signals on 6 RF channels simultaneously, we need to utilise 3 NI USRP nodes. It is required to have the synchronised frequency and time clocks in all the nodes while performing joint beamforming. The Fig. 2 shows the logical connections between the NI FlexRio, the NI USRP nodes and the controller (NI PXI HOST). Here the upper blue section represents the processes implemented in the host computer and the lower yellow section represents the blocks implemented in the FPGA for fast processing.

The FPGA IP block of the DVB-S2X deployed in the NI FlexRio generates 6 parallel streams of symbols of the DVB-S2X superframe. Each stream carriers terminal specific data. The streams can be independently configured through the dedicated graphical interface as shown in Fig. 3.

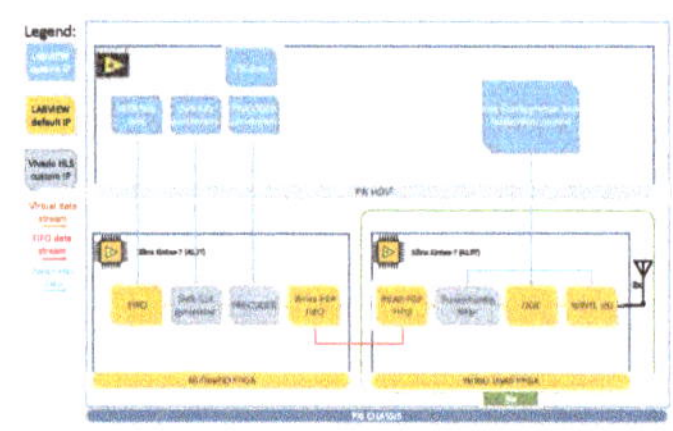

Figure 2: Block diagram of the DVB-S2X Gateway

Figure 3: DVB-S2x Gateway configuration graphical interface

We implemented the configurations, which are covered by the DVB Standards [24,25], namely: MODulation and CODing Mode (MODCOD), Super-Frame Format Indicator (SFFI), Index Stream, Index of the Walsh-Hadamard (WH) matrix, scramble flag for Pilots and Start Of Super-Frame (SOSF). We include an extra parameter indicating the Precoding type (). The Precoding type describes the type of the Precoding technique allowed to precode each stream. Rank 0 indicates no Precoding is applied, 1 - Channel based Zero Forcing (ZF) or the Minimal Means Square Error (MMSE) Precoding are used [26], 2 - reserved for the future use, 3 - Symbol-Level Precoding (SLP) techniques [27] are used if possible, otherwise - ZF and MMSE.

The streams are jointly precoded by the PRECODE IP block. The PRECODE IP block multiplies 6 symbols from a single time slot with the Precoding matrix $\mathbf{W}$. Additionally, the Precoding Mask controls Precoding behaviour over the segments of the DVB-S2X superframe as illustrated in Fig. 4.

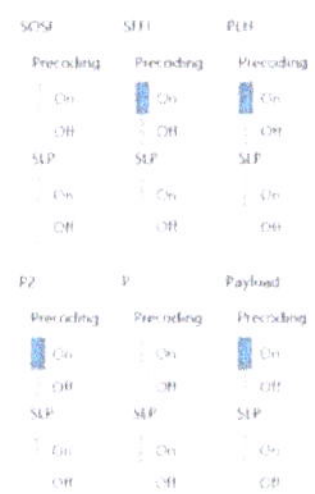

Figure 4: Precoding configaration for DVB-S2X superframe structure

Through the configuration panel we can enable and disabled Zero-Forcing and SLP Precoding techniques for each segment of the DVB-S2X superframe. We disable Precoding for SOSF and P pilots under normal operation. The SOSF is a known sequence, which can be reliably detected at a user terminal even in a high interference environment. The P pilots are not precoded because they are used by UTs to estimate the CSI ($\widehat{\mathbf{H}}$).

The streams with 6 superframes are transferred to the 3 NI USRP nodes, where the signals are oversampled using the Pulshaping filter IP block. The filter's impulse response is given by the Raised-cosine function with different roll-off factors. The roll-off factor of the filter response is configurable according to the DVB-S2X standard. In this iteration of the hardware demonstrator we implemented the roll-off factors of 0.2, 0.15, 0.1 and 0.05. We measured the actual filter's frequency response for these roll-off values as shown in Fig. 5. The oversampled signals are processed by the digital upconverter (DUC) and transmitted to the RF domain at the desired carrier frequency.

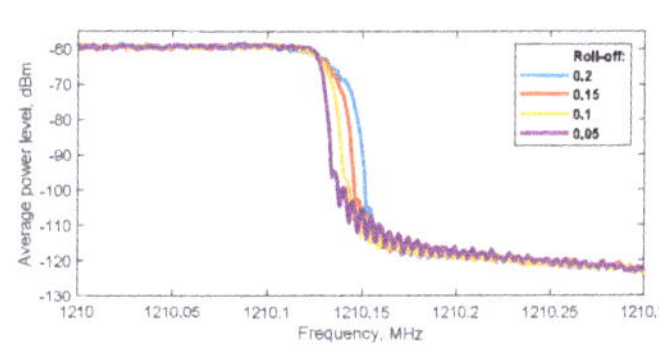

Figure 5: Frequency characteristics of the transmitted signal

2.3 Channel Emulator

The channel emulator is running on a PXI HOST controller, a NI FlexRIO module for central signal processing and 3 NI USRP nodes for the RF front-end. The nodes are interconnected as shown in Fig. 6.

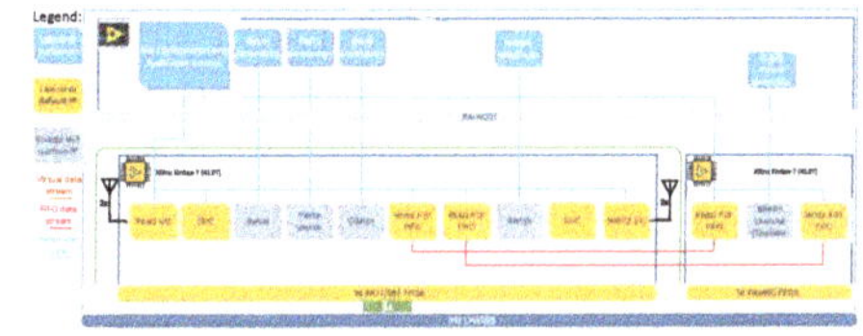

Figure 6: Block diagram of the channel emulator

Each NI USRP node acquires two streams of the sampled baseband, which are generated by the digital down converter (DDC). Therefore, by utilising 3 synchronised RIO USRP nodes the channel emulator simultaneously samples 6 independent RF baseband streams. The RF inputs and outputs of the NI USRP nodes operate in RF bands of the Low Noise Block (LNB) in the GW and the LNB on the UTs. All the effects that occur in the actual Ka-band are emulated in the channel emulator.

We designed and implemented custom IP blocks into USRP FPGA nodes to emulate the channel impairments in real time. We designed and build an additive white Gaussian noise (AWGN) generator [28] with configurable amplitude. We implemented Input Multiplexing (IMUX) and Output Multiplexing (OMUX) filtering FPGA IP blocks. The input and output characteristics of the filters are shown in Fig. 7.

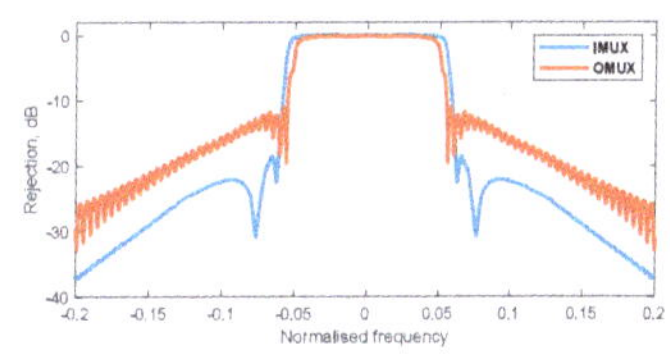

Figure 7: IMUX and OMUX filter characteristics of the Channel Emulator

We developed a custom IP block to emulate Traveling-Wave Tube Amplifier (TWTA) non-linearities. The input and output characteristics of the TWTA IP block correspond to the DVB-S2 standard specifications and are depicted in Fig. 8.

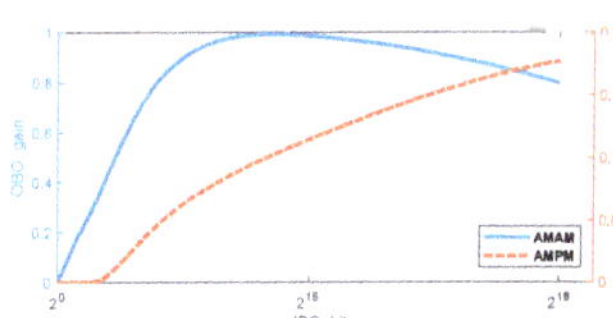

Figure 8: TWTA AM/AM and AM/PM characteristics of the Channel Emulator

The streams from the 3 USRP nodes are transferred to the FlexRIO FPGA. The channel matrix ($\mathbf{H}$) is jointly applied towards all the streams by the MIMO Channel Emulator IP block. The desired channel matrix is controlled by the PXI HOST. The 6×6 matrix of complex coefficients is based on the realistic satellite beam pattern illustrated in Fig. 9. It is called ESA71 (after the origin and the number of beams). ESA71 makes use of the Ka-band exclusive band 19.7 to 20.2 GHz. We consider a scenario of full frequency reuse, where the same frequency band is applied in every beam. We can simultaneously select up to 6 user terminals in the coverage area and generate realistic channel coefficients.

The resulting signals with added interference are transferred back to the corresponding NI USRP nodes. In every NI USRP node, the signal is upsampled in DUC and sent through the RF analogue outputs. The RF inputs and outputs of the channel emulator operate at different carrier frequencies as shown in the Table. 1. Using this configuration we decrease the mutual coupling between the transmission and reception links through the RF part of the Channel Emulator to facilitate the accuracy of the designed channel matrix.

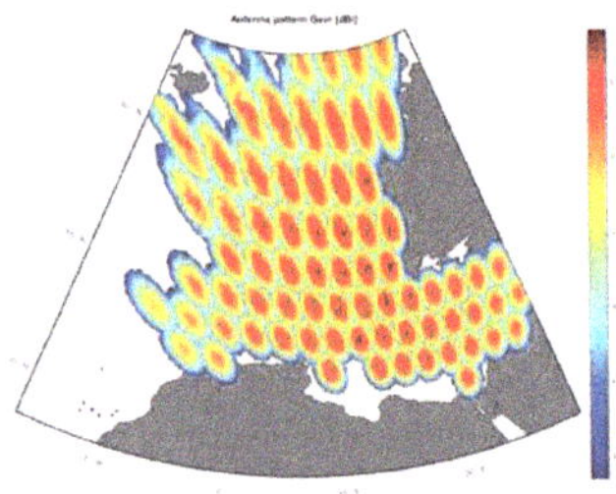

Figure 9: ESA71 beam pattern, antenna performance

2.4 User Terminal

The User Terminals (UT) are the are ground-based users which collect the transmitted waveforms and recover the transmitted data bits. The UTs are implemented using the same SDR platform used in for the Gateway and the Channel Emulator, the USRP RIO NI2944. We use each of these USRP to implement two independent UTs using the two RF inputs of the device. The FPGA inside the SDR platform performs signal processing for the two UT chains and communicates with the host computer.

Fig. 10 shows an architectural block diagram of the UT implementation. The FPGA interface the RF daughter boards, which down-convert the incoming RF signal using an analog local oscillator. The FPGA reads the ADCs sampled data coming in four sampled streams for the two IQ down-converted pairs. The sampling rate of this particular SDR platform is 200MSPS. After sampling, two DDC blocks take the two IQ sampled streams. Each DDC shifts in frequency the IQ streams using a Numerically Controlled Oscillator, and applies a decimation filter in order to produce a decimated output stream at a selectable sampling rate. The frequency shifting performed in the DDC has the advantage of avoiding the problematic zero-frequency part of the spectrum after analog down-conversion, which usually is corrupted by some leakage of the local oscillator signal.

After the DDC, the DVB-S2X block performs the recovery of the information from the digitally down-converted stream. Fig. 11 shows a simplified functional block diagram of the DVB-S2X Receiver IP block used in the UTs. Due to the efficient FPGA accelerated processing, a single USRP RIO FPGA unit is capable to simultaneously receive two DVB-S2X compliant signals. The DVB-S2X processing chain is capable of recovering the format 2 and 3 superframes from the DVB-S2X standard. These processes include frequency acquisition, matched filtering, time synchronization, frame (including Super-Frame) synchronization, fine phase tracking, and CSI estimation. The block labelled Phase Recovery on Fig.11 performs symbol recovery from the input stream with an over-sampling factor of four. The recovered symbols contain the payload data plus the pilot structure of the DVB-S2X framing. All the frame fields and the CSI information are passed to the host computer for their further utilisation. The host computer reports the CSI information to the central Gateway using a custom feedback channel.

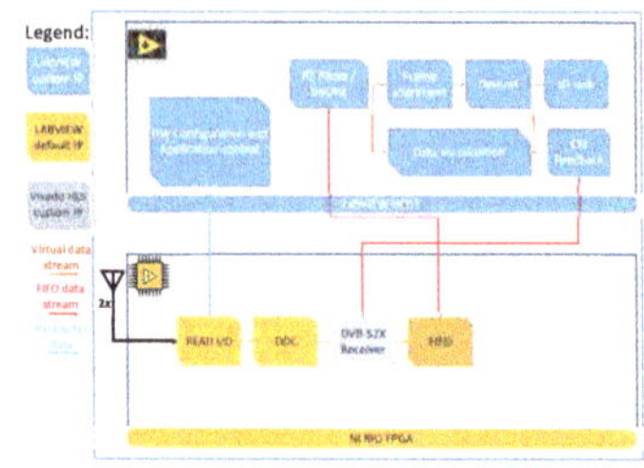

Figure 10: Architectural block diagram of the User Terminal. Note that two input RF chains are present in a single USRP.

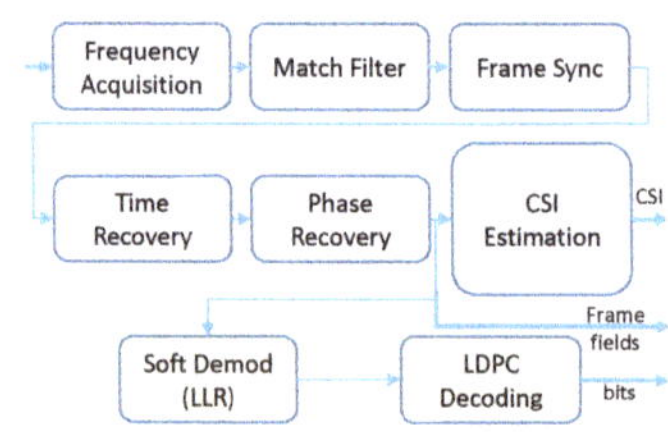

Figure 11: Functional diagram of the DVB-S2X receiver block in the UT

2.5 Resource Occupation in FPGAs

The described software equipment is heavily based on FPGA code acceleration and parallel computing. The FPGA resources in the used equipment (NI USRP 2954-R and NI FlexRIO 7976R) are limited, thus the design of the functional FPGA blocks is a compromise between the functionality, the resource occupation, and the data throughput. Table 2 is the summary of the actual FPGA resource occupation of each part of the hardware demonstrator. The occupation percentage of the Digital Signal Processors (DSP), Block RAMs (BRAMS), lookup tables (LUT) is manageable for the planned functionality. Hence, we can notice that the Slices (Each slice contains four LUTs and eight flip-flops) occupation is very hight at the user terminal. The complexity of the terminal is much higher than the one with a single receiver. It is evident that if we want to expand the functionality of the terminal we need to deploy only one DVB-S2X receiver per USRP node.

Table 2: FPGA Xilinx Kintex-7 (410TFFG-2) Resource Occupation

	DSP48E	BRAM	LUT	Slices
Gateway node	19%	18%	13%	27%
Gateway FlexRIO	34%	23%	26%	41%
Channel Emu. node	34%	56%	32%	49%
Channel Emu. FlexRio	9%	27%	31%	43%
User Terminal 1 RX	31%	22%	36%	63%
User Terminal 2 RX	40%	23%	57%	83%

3 Conclusion

In this paper, we present, to the best author knowledge, the first hardware demonstrator for Precoding in DVB-S2X systems. The hardware demonstrator is a full-chain closed-loop communication system with multi-beam gateway transmitter, MIMO channel emulator and receiver terminals with real-time CSI estimation. With the hardware demonstrator we are able to show experimentally the application of Precoding in satellite communications based on the DVB-S2X standard.

Acknowledgments

This research was supported by Luxembourg National Research Fund grant No. FNR11295545 for "SERENADE" project and AFR-PPP grant No. FNR11481283 "End-to-end Signal Processing Algorithms for Precoded Satellite Communications" project.

References

[1] 5G-Infrastructure-Association. EC H2020 5G Infrastructure PPP - Prestructuring Model RTD & INNO Strands. 2014.

[2] 5G-PPP. Mobile and Wireless Communications Enablers for Twenty-twenty (2020) Information Society. 2014.

[3] 3GPP TR 38.811 V15.0.0. 3rd Generation Partnership Project; Technical Specification Group Radio Access Network; Study on New Radio (NR) to support non terrestrial networks (Release 15). June 2018.

[4] B. Evans, O. Onireti, T. Spathopoulos, and M. A. Imran. The Role of Satellites in 5G. In *2015 23rd European Signal Processing Conference (EUSIPCO)*, pages 2756–2760, Aug 2015.

[5] G. Zheng, S. Chatzinotas, and B. Ottersten. Generic Optimization of Linear Precoding in Multibeam Satellite Systems. *IEEE Transactions on Wireless Communications*, 11(6):2308–2320, June 2012.

[6] Dimitrios Christopoulos, Symeon Chatzinotas, Gan Zheng, Joël Grotz, and Björn Ottersten. Linear and Nonlinear Techniques for Multibeam Joint Processing in Satellite Communications. *EURASIP Journal on Wireless Communications and Networking*, 2012(1):162, May 2012.

[7] M. A. Vazquez, A. Perez-Neira, D. Christopoulos, S. Chatzinotas, B. Ottersten, P. D. Arapoglou, A. Ginesi, and G. Taroccο. Precoding in Multibeam Satellite Communications: Present and Future Challenges. *IEEE Wireless Communications*, 23(6):88–95, December 2016.

[8] S. Andrenacci, D. Spano, D. Christopoulos, S. Chatzinotas, J. Krause, and B. Ottersten. Optimized Link Adaptation for DVB-S2X Precoded Waveforms Based on SNIR Estimation. In *2016 50th Asilomar Conference on Signals, Systems and Computers*, pages 502–506, Nov 2016.

[9] ITU-T. ITU-T Recommendation G.993.5 (2010), Self-FEXT Cancellation (Vectoring) for use with VDSL2 transceivers (Vectored DSL).

[10] B. Schulz. LTE Transmission Modes and Beamforming, Rhode and Schwarz white paper. 2011.

[11] Morello Alberto and Mignone Vittoria. DVB-S2X: The New Extensions to The Second Generation DVB Satellite Standard DVB-S2. *International Journal of Satellite Communications and Networking*, 34(3):323–325.

[12] Alireza Haqiqatnejad, Farbod Kayhan, and Björn E. Ottersten. Power Minimizer Symbol-Level Precoding: A Closed-Form Sub-Optimal Solution. *CoRR*, abs/1807.10619, 2018.

[13] J. Krivochiza, A. Kalantari, S. Chatzinotas, and B. Ottersten. Low Complexity Symbol-Level Design for Linear Precoding Systems. In *2017 Symposium on Information Theory and Signal Processing in the Benelux*, page 117. Delft University of Technology, 2017.

[14] N. Mazzali, S. Boumard, J. Kinnunen, B. Shankar M. R., M. Kiviranta, and N. Alagha. Enhancing Mobile Services with DVB-S2X Superframing. *International Journal of Satellite Communications and Networking*, 0(0).

[15] S. Jiang, F. Gong, and X. Chen. A Low-complexity Soft Demapper for 128APSK of DVB-S2X. In *2016 8th International Conference on Wireless Communications Signal Processing (WCSP)*, pages 1–4, Oct 2016.

[16] J. Lucciardi, P. Potier, G. Buscarlet, F. Barrami, and G. Mesnager. Non-Linearized Amplifier and Advanced Mitigation Techniques: DVB-S2X Spectral Efficiency Improvement. In *GLOBECOM 2017 - 2017 IEEE Global Communications Conference*, pages 1–7, Dec 2017.

[17] Shree Krishna Sharma, Symeon Chatzinotas, and Pantelis-Daniel Arapoglou. *Satellite Communications in the 5G Era (Telecommunications)*. The Institution of Engineering and Technology, 2018.

[18] Symeon Chatzinotas, Bjorn Ottersten, and Riccardo De Gaudenzi. *Cooperative and Cognitive Satellite Systems*. Academic Press, Inc., Orlando, FL, USA, 1st edition, 2015.

[19] Danilo Spano, Symeon Chatzinotas, Stefano Andrenacci, Jens Krause, and Björn Ottersten. Per-antenna Power Minimization in Symbol-level Precoding for the Multibeam Satellite Downlink. *International Journal of Satellite Communications and Networking*, may 2018.

[20] A. Mengali, R. B. S. Mysore, and B. Ottersten. Joint Predistortion and PAPR Reduction in Multibeam Satellite Systems. In *2016 IEEE International Conference on Communications (ICC)*, pages 1–7, May 2016.

[21] M. Baek, J. Yun, H. Lim, Y. Kim, and N. Hur. Joint Masking and PAPR Reduction for Digital Broadcasting System with Faster-than-Nyquist Signaling. In *2017 IEEE International Symposium on Broadband Multimedia Systems and Broadcasting (BMSB)*, pages 1–2, June 2017.

[22] J. C. Merlano-Duncan, J. Krivochiza, S. Andrenacci, S. Chatzinotas, and B. Ottersten. Computationally Efficient Symbol-Level Precoding Communications Demonstrator. In *2017 IEEE 28th Annual International Symposium on Personal, Indoor, and Mobile Radio Communications (PIMRC)*, pages 1–5, Oct 2017.

[23] Jevgenij Krivochiza, Juan Carlos Merlano-Duncan, Stefano Andrenacci, Symeon Chatzinotas, and Bjrn Ottersten. Computationally and Energy Efficient Symbol-Level Precoding Communications Demonstrator. *Physical Communication*, 28:108 – 115, 2018.

[24] ETSI. Digital Video Broadcasting (DVB); Implementation guidelines for the second generation system for Broadcasting, Interactive Services, News Gathering and other broadband satellite applications; Part 1: DVB-S2. 2015.

[25] ETSI. Digital Video Broadcasting (DVB); Second generation framing structure, channel coding and modulation systems for Broadcasting, Interactive Services, News Gathering and other broadband satellite applications; Part 2: DVB-S2 Extensions (DVB-S2X). 2015.

[26] C. B. Peel, B. M. Hochwald, and A. L. Swindlehurst. A Vector-Perturbation Technique for Near-Capacity Multiantenna Multiuser Communication-Part I: Channel Inversion and Regularization. *IEEE Transactions on Communications*, 53(1):195–202, Jan 2005.

[27] M. Alodeh, D. Spano, A. Kalantari, C. G. Tsinos, D. Christopoulos, S. Chatzinotas, and B. Ottersten. Symbol-Level and Multicast Precoding for Multiuser Multiantenna Downlink: A State-of-the-Art, Classification, and Challenges. *IEEE Communications Surveys Tutorials*, 20(3):1733–1757, thirdquarter 2018.

[28] C. Politis, S. Maleki, J. M. Duncan, J. Krivochiza, S. Chatzinotas, and B. Ottesten. SDR Implementation of a Testbed for Real-Time Interference Detection With Signal Cancellation. *IEEE Access*, 6:20807–20821, 2018.

ON THE CAPACITY OF ASYNCHRONOUS COOPERATIVE NOMA IN MULTIBEAM SATELLITE SYSTEMS

Nazli Ahmad Khan Beigi, Mohammad Reza Soleymani

Concordia University, Montreal, Canada
{naz_ahm,msoleyma}@encs.concordia.ca

Keywords: Cooperative NOMA, Asynchronous Channel, Multibeam Satellite Systems, Channel Capacity, Successive Interference Cancellation (SIC)

Abstract

In this paper, we investigate the effect of asynchronous reception in multibeam satellite system forward link using our previously proposed cooperative non-orthogonal multiple access (NOMA) technique. Our proposed system is in a dense frequency reuse scheme and based on the cooperation of transmitting beams by targeting one user terminal at a time. The jointly reception of data streams features multiple access channel (MAC), where both data will be recovered by devising successive interference cancellation (SIC). Herein paper we show that in case of asynchronous reception, the propagation of errors through SIC causes huge loss of data frames and throughput degradation. Hence, we propose the channel model of the asynchronous NOMA, and we investigate the system's capacity region, where the information theoretic results show that the asynchronous reception of the data streams can indeed improve the sum-rate upper bounds.

1. Introduction

The recent challenge in increasing throughput in satellites has led to research and technical efforts in extending the number of satellite beams and reusing spectrum to boost the throughput [1]. Non-orthogonal multiple access (NOMA) has recently been proposed as a solution in sharing the resources [2]. In this paper, we refer NOMA as a means of employing the same frequency band throughout all beams in a multibeam satellite system. This aggressive frequency reuse results in huge co-channel interference (CCI) on the adjacent beams.

In our proposed cooperative NOMA in multibeam satellite systems, we have shown that by managing the interference, it is possible to remove its effect, and even better, use it as an extra resource [3]. We restrict the study in this paper to the forward link in two-beams cooperating NOMA. As both incumbent and the adjacent interfering beams have access to their data sources, they can cooperate to jointly target one user at a time. Hence, the time-frequency resources can be shared by the two beams employing NOMA in power domain. The target user will receive the aggregate of signals which can both be recovered using successive interference cancellation (SIC). In SIC, the signal received with higher power is decoded and recovered first. Then the recovered signal is reconstructed to be subtracted from the aggregate received signals. This leaves us with second data stream as if there was no interference at all. Likewise, the user in the adjacent beam will be served in the remainder of the time.

Our analysis on cooperative NOMA as well as other researches till now have been based on synchronous reception of the data streams at the target user. Yet, due to many practical issues, perfect synchronization assumption is not justifiable. Previous works on asynchronous NOMA are scarce but do exist in the related literature. In [4], an interference cancellation (IC) method is proposed which a triangular pattern is followed in the IC of all interfering users for the sake of the desired user. Also, in [5][6], the effort is to use symbol asynchronism to reduce the mutual interference between NOMA users. Moreover, in [7] the authors use cyclic prefix in addition to a SIC-phase compensation technique to reduce the effect of asynchronism. These researches are all based on proposing compensation techniques to improve the SIC performance.

As we analyse in this paper, the asynchronous reception of data streams changes the channel's attributes to a channel with memory which degrades SIC's functionality. This has been investigated in detail in multi-access channels with inter-symbol interference (ISI) in [12], where the results could be applied to asynchronous multi-access channels. Thus, in order to overcome the SIC's failure in such channels, we go a step further than SIC. Indeed, by relying on the cooperative NOMA, we show that the correlation caused by channel's memory leads to higher throughput achievements.

The contribution of the present paper is to approach and extend the cooperative NOMA with asynchronous channel in the downlink of multibeam satellite systems. To the best of authors' knowledge, the proposed approach has not been previously investigated.

This paper is organized as follows. In Section 2, channel model and our proposed cooperative NOMA system model are provided. In Section 3, the effect of ISI on the capacity region of Gaussian multi-access channels (MAC) with SIC at the receiver is investigated. Section 4 pursues the discussed concepts by going further than SIC method and showing the effect of cooperation on the capacity region. Analytical results are confirmed by information theoretical evaluations, concluded in Section 5.

2. System Model

2.1. Channel Model

The system model includes multibeam satellite downlink system with full frequency reuse. The channel is AWGN. Due to the frequency reuse, each user receives the signals from all other beams. Uniform distribution of users in all beams in addition to equal number of users per beam (M) is assumed. In beam b, i.e. $b = 1, \dots, B$, the set of users is provided as $S = \{u(b,1), u(b,2), \dots, u(b,M)\}$. The term $u(b,k)$ stands for the k-th, i.e. $k = 1, \dots, M$, user located at beam b. At the gateway, each $u(b,k)$-th user data stream is channel coded and modulated independently. Considering $S_{u(b,k),b'}$ as the coded-modulated signal of user $u(b,k)$ from beam b', with average power equal to one, i.e. $E[|S_{u(b,k),b'}|^2 = 1$, and the transmitted power as $P_{u(b,k),b'}$, also assuming strict synchronization, in a multibeam satellite system with B beams, the output $\boldsymbol{Y}$ will be given by

$$\boldsymbol{Y} = \begin{bmatrix} 1 & \rho_{1,2} & \cdots & \rho_{1,B} \\ \rho_{2,1} & 1 & \cdots & \rho_{2,B} \\ & \vdots & & \\ \rho_{B,1} & \rho_{B,2} & \cdots & 1 \end{bmatrix} \boldsymbol{X} + \boldsymbol{N} \quad (1)$$

Where $\boldsymbol{X}$ is the vector of the transmitted signals, i.e. $X_1, X_2, \dots, X_B$, and $\boldsymbol{N}$ is the vector of Gaussian variables with zero mean, $E[N_k^2] = \sigma^2$. The cross correlation $\rho_{i,j}$ between signals received in beam i from beam j can be shown as

$$\rho_{i,j} = \int_0^T s_{u(b,k),i}(t) s_{u(b,k),j}(t)\, dt \quad (2)$$

2.2. Proposed Cooperative NOMA

Improving channel capacity by employing power-domain NOMA has been investigated in classic cooperative NOMA [8][9]. In classic cooperative NOMA, the user terminal with better channel condition will recover both its and the other user's data and re-transmits this data through a relay channel to the other user. Due to huge distances between user terminals in satellite beams, this scenario is not applicable, and in this paper, we investigate our previously proposed cooperative NOMA in multibeam satellite systems [3].

We have shown that data throughput will be maximized if for each user terminal, its main beam cooperates with one or two adjacent beams with the dominant interference. In other words, by employing the cooperative NOMA, not all the signals received from interfering beams degrade the system performance; whereas, the signal received from the most dominant interfering beams will be used as extra source of information.

Based on the proposed cooperative NOMA with two beams cooperation, we assume b_m to be the beam number for which the user $u(b,k)$ receives the dominant interference from, denoted by:

$$\rho_{b,b_m} = max(\boldsymbol{\rho}_b), b_m \neq b \quad (3)$$

where $\boldsymbol{\rho}_b$ is the b-th row of the cross-correlation matrix. Hence, based on cooperative NOMA, the received SNIR at user $u(b,k)$is:

$$\gamma_{u(b,k)} = \frac{\rho_{b,b}{}^2 P_{u(b,k),b} + \rho_{b,b_m}{}^2 P_{u(b,k),b_m}}{\sum_{b'=1,\neq b,b_m}^{B} \rho_{b,b'}{}^2 P_{u(b,k),b'} + n_{u(b,k)}}. \quad (4)$$

Where, the numerator is the sum of received powers from the incumbent and the most dominant interfering beams. The denominator is the addition of channel noise and summation of all interferences, except for the cooperating most dominant interference. The channel states of the incumbent, most dominant interfering, and other interfering beams could be entirely estimated as explained in [3]. The sum rate at user $u(b,k)$ will then be as:

$$R_{sum-rate} = W log_2(1 + \gamma_{u(b,k)}). \quad (5)$$

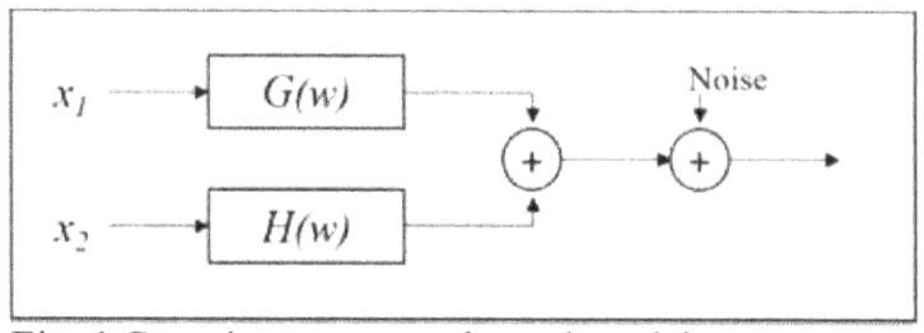

Fig. 1 Gaussian two-user channel model

3. Gaussian Multi-Access Channels with ISI

In this section, we evaluate the effect of ISI in a multi-access channel, as the results could be applied to asynchronous multi-access channels as well.

In general, the capacity region of the two source memoryless MAC can be written as [10]

$$R_1 \leq I(x_1; y | x_2) \quad (6)$$

$$R_2 \leq I(x_2; y | x_1) \quad (7)$$

$$R_1 + R_2 \leq I(x_1, x_2; y) \quad (8)$$

However, in case of asynchronous reception of the two sources, the timing offset results in each symbol transmitted by one user getting overlapped by two symbols of the other user, resulting in a channel with memory [11]. This channel with memory degrades the performance of the successive decoder in the MAC. Unlike the memoryless MAC where the channel capacity region could be found by analysing the optimal power spectral density (PSD) for each user, in this case no optimal PSD could be found for users independently as their signals interfere with each other.

The Gaussian MAC with ISI is studied in detail in [12] where asynchronous multi-access channel is one of its special cases. The capacity region is achieved through a multiuser water-filling scheme that we briefly present its results here.

Fig. 1 shows the channel model for the two Gaussian MAC with ISI, where $G(w)$ and $H(w)$ are the channel transfer functions of user 1 and 2, respectively. As the capacity of memoryless Gaussian channels are well known, the tactic to evaluate the capacity region of MAC with memory would be to decompose the channel into parallel memoryless channels.

This is not as straight forward as single user channels as the two channels' coefficients affect each other; yet, a water-filling scheme could be used to show graphically the optimal total PSD over frequency domain. Also, the successive cancellation could be modelled graphically on the water-filling scheme to find the optimal PSD of each user over frequency domain. As follows, the two-user multi-access channel with similar and dissimilar transfer functions are investigated. The case with similar transfer functions is provided for clarification, where the one with dissimilarity is our case of interest in this paper.

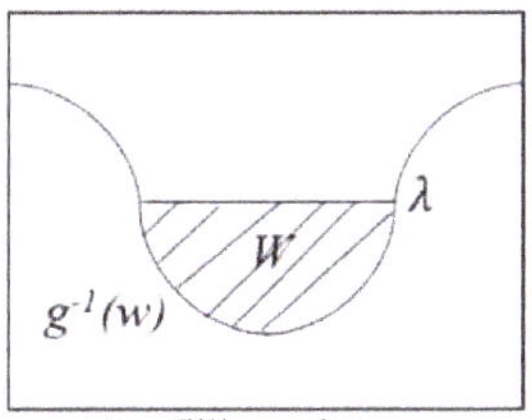

Fig. 2, Single user water-filling scheme

3.1. *Two-User Channel with $H(w) = G(w)$*

For a single user channel, the water-filling scheme is shown in Fig. 2, where $g(w) = |G(w)|^2/N(w)$, $N(w)$ is the noise PSD, and W is the total power. The $g^{-1}(w)$ is the bottom of the water-filling container and the fixed amount of water (power), W, is poured into the container. Based on water-filling scheme [10], the optimal PSD as it is shaded in Fig. 2 is the solution to below equations:

$$S(w) = [\lambda - g^{-1}(w)]^{+}, \quad (9)$$

$$W = \frac{1}{\pi}\int_0^{\pi} S(w)dw \quad (10)$$

Similar aspect is followed for finding optimal PSD of two user channel with similar transfer functions, where $T^{-1}(w) = g^{-1}(w) = h^{-1}(w)$ will be used as the bottom of the containers for $S_1(w)$, $S_2(w)$, and $S_1(w) + S_2(w)$ water-filling diagrams, separately.

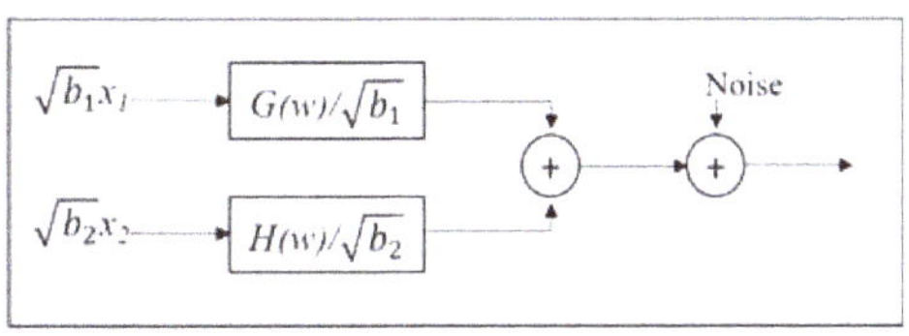

Fig. 3 Equivalent Gaussian two-user channel model

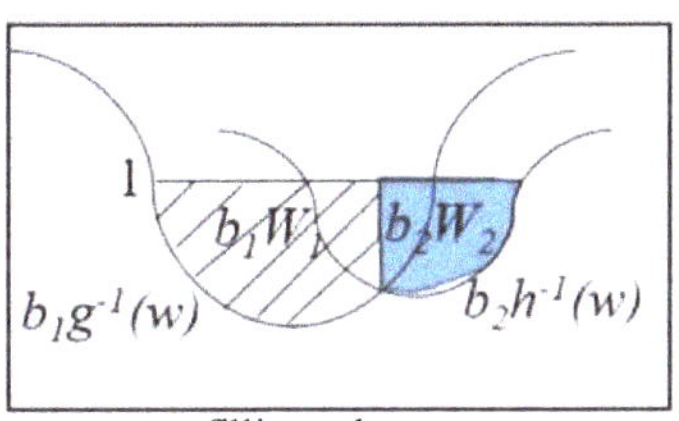

Fig. 4 Two-user water-filling scheme

3.2. *Two-User Channel with $H(w) \neq G(w)$*

In this case, the target is to find the optimal PSD for the two users that maximizes a weighted sum rate $\alpha R_1 + (1 - \alpha)R_2$, $\alpha \in [0,1]$. The procedure is based on the concept of the SIC. The user with worse channel condition is considered lower priority and is decoded first, considering the second user's signal as noise. Then the first user's signal is reconstructed and subtracted from the aggregate signal to decode the second user's signal with no effect from the first user.

The optimal PSD of two users cannot be found in two separate water-filling diagrams as they cause interference on each other. In order to solve this problem, an equivalent channel is required to be able to combine both $g^{-1}(w)$ and $h^{-1}(w)$. Using this equivalent channel allows us to find the optimal sum of the PSDs as well as individual PSDs based on the SIC scheme.

The equivalent channel is shown in Fig. 3. The idea behind this channel is to use a scaled version of the original channels to be able to combine both in one water-filling diagram. The corresponding water-filling diagram is shown in Fig. 4.

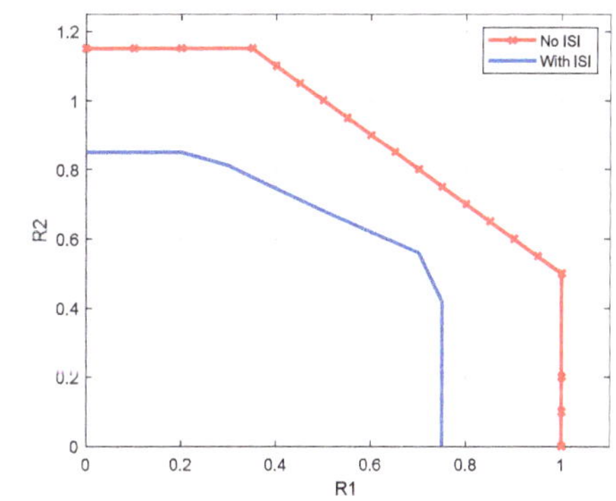

Fig. 5 Capacity region of the Two-user channel

3.3. *Simulation Results*

In order to find the optimal scales that maximize the sum rate and consequently maximizes the sum rate, we have adopted the procedure in [13]. We have simulated two-user channel with transfer functions as follows for $\alpha \in [0,1]$.

$$H(w) = (1 + 0.1e^{-j\omega})/1.005 \quad (11)$$

$$G(w) = (1 + 0.3e^{-j\omega})/1.044 \quad (12)$$

We have assumed $W_1 = 3$, $W_2 = 4$, and noise variance of one. The transfer functions are normalized to have unit energy. The capacity region is shown in Fig. 5 with the solid line. In addition, the capacity region of mentioned channels without ISI, i.e. $h(t) = g(t) = \delta(t)$ is provided. As expected, it is clear that ISI between the two channels has degraded the single-user as well as the sum rate capacities. Also, the capacity region is no more pentagonal, and it is curved at the boundary corner points. This is again due to the fact that no fixed single user PSD can maximize the sum rate due to the interfering effect of ISI.

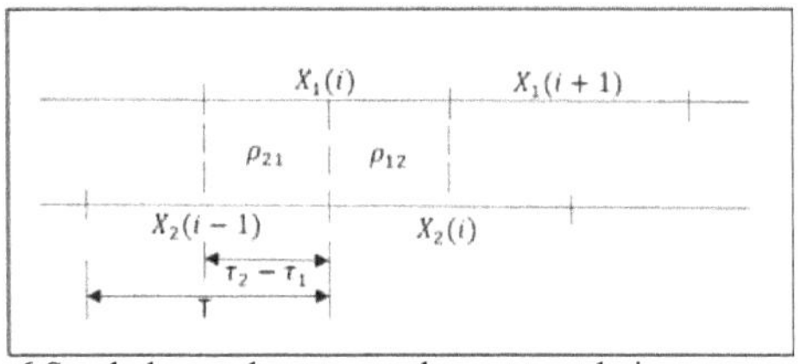

Fig. 6 Symbol asynchronous and cross correlations

4. Asynchronous Multi-Access Channel: Capacity Region

Based on previous section, in a NOMA channel based on SIC receiver, the asynchronism degrades the channel capacity by adding memory and unwanted correlation. In this section we show that this degradation is due to the nature of the SIC at the receiver. As SIC recovers and subtracts signals in different steps, it cannot benefit from the introduced memory. In this section we show that due to the correlation, the capacity region could be improved indeed. So, by using cooperative NOMA, not only channel having memory is not causing degradation, but we can benefit from the correlation that it causes.

4.1. Asynchronous MAC

Considering a two-user multi-access channel, due to the symbol asynchronism, each transmitted symbol will overlap with preceding and succeeding symbols of the other transmitter, see Fig. 6. This leads to below channel model for the two-user multi-access channel.

$$\boldsymbol{Y}[i] = \begin{bmatrix} 0 & \rho_{21} \\ 0 & 0 \end{bmatrix} \boldsymbol{X}[i-1] + \begin{bmatrix} 0 & \rho_{12} \\ \rho_{12} & 0 \end{bmatrix} \boldsymbol{X}[i] + \begin{bmatrix} 0 & 0 \\ \rho_{21} & 0 \end{bmatrix} \boldsymbol{X}[i+1] + \boldsymbol{N}[i] \qquad (13)$$

Where, $\boldsymbol{Y}[i] = [Y_1(i)\ Y_2(i)]^T$, $\boldsymbol{X}[i] = [X_1(i)\ X_2(i)]^T$, and $\boldsymbol{N}[i] = [N_1(i)\ N_2(i)]^T$. This channel is different from (1), as the asynchronism in the channel has caused a channel with memory. It is proved in [11] that the sum rate of the asynchronous channel where the transmitters are aware of their time offset can be written as

$$C = R_1 + R_2 \le \frac{1}{4\pi}\int_{-\pi}^{\pi} \log[1 + \frac{S_1(w)}{\sigma^2} + \frac{S_2(w)}{\sigma^2} + \frac{S_1(w)S_2(w)}{\sigma^4}(1-\rho^2(w))]dw \qquad (14)$$

Where $1-\rho^2(w) = 1-\rho_{12}^2-\rho_{21}^2-2\rho_{12}\rho_{21}\cos(w)$, and $S_i(w)$ is the signal i's PSD. In this equation, there is an extra term $\frac{S_1(w)S_2(w)}{\sigma^4}(1-\rho^2(w))$ which is due to the channel having memory and dependability of $cov(Y)$ and $E[N_1N_2]$ on $\rho(w)$. Yet, for $|\rho| = 1$, we go back to the channels being the same with no time delay and no memory. This extra term makes the sum rate different from (8), resulting in a bigger capacity region than a synchronous MAC. This is in contrary with results provided in Section 3, and Fig. 5, once again emphasizing on the fact that SIC is not the appropriate tool in a NOMA channel with symbol asynchronous reception of the data streams.

4.2. Asynchronous NOMA- Single Content Transmission

In cooperative NOMA, due to the transmitters cooperation, they are aware of the time differences and can arrange their timing so that the least misalignment occurs. The sum rate capacity in (14) achieves its highest value for $\tau = 0.5$. This in not efficient in our proposed system with just slight misalignments, i.e. $\tau \ll 0.5$. Hence, we prove if the transmitters cooperatively share their data and send the same content, due to the singularity of the input matrix, the sum-rate capacity is

$$C = R_1 + R_2 \le \frac{1}{4\pi}\int_{-\pi}^{\pi} \log(1 + \frac{S_1(w)}{\sigma^2} + \frac{S_2(w)}{\sigma^2} + 2\rho(w)\sqrt{\frac{S_1(w)S_2(w)}{\sigma^4}})\, dw \qquad (15)$$

Proof:

The upper bound of the mutual information can be written as

$$I(X_1, X_2; Y) \le \frac{1}{2}\log det[cov(Y)] - \frac{1}{2}\log det\left[\sigma^2 \begin{bmatrix} 1 & \rho \\ \rho & 1 \end{bmatrix}\right] \qquad (16)$$

This follows because of the differential entropy of Gaussian random variables. As we consider similar content from both sources, (16) can be shown as

$$I(X_1, X_2; Y) \le \frac{1}{2}\log \det\left[I_2 + \frac{1}{\sigma^2}\begin{bmatrix} var(X_1) & \sqrt{var(X_1)}\sqrt{var(X_2)} \\ \sqrt{var(X_1)}\sqrt{var(X_2)} & var(X_2) \end{bmatrix} \begin{bmatrix} 1 & \rho \\ \rho & 1 \end{bmatrix}\right]$$

So, considering a zero-mean n-vector Gaussian process, the mutual information between the output and the inputs is equal to

$$\lim_{n\to\infty} \frac{1}{n} I(X_1^n, X_2^n; Y^n) = \frac{1}{4\pi}\int_{-\pi}^{\pi} \log \det\left[I_2 + \frac{1}{\sigma^2}\begin{bmatrix} S_1(w) & \sqrt{S_1(w)}\sqrt{S_2(w)} \\ \sqrt{S_1(w)}\sqrt{S_2(w)} & S_2(w) \end{bmatrix} \begin{bmatrix} 1 & \rho(w) \\ \rho(w) & 1 \end{bmatrix}\right] dw$$

Which, with some calculations results in (15). ■

As discussed earlier and showed in (14), the asynchronous MAC with no cooperation achieves its highest sum rate for $\tau = \tau_2 - \tau_1 = 0.5T_s$, where T_s is the symbol time. It can be seen in (15) that as ρ increases and gets closer to one, which models the small time delays ($\tau \ll T_s$), the sum rate capacity increases. This is very desirable for systems with transmitters that are just slightly off like the satellite transmitters. Though it's not applicable in practice, for $\rho = 1$, the sum rate presents the coherence reception of the two data streams which equals to

$$C = R_1 + R_2 \le \frac{1}{4\pi}\int_{-\pi}^{\pi} \log\left(1 + \frac{\left(\sqrt{S_1(w)}+\sqrt{S_2(w)}\right)^2}{\sigma^2}\right) \quad (17)$$

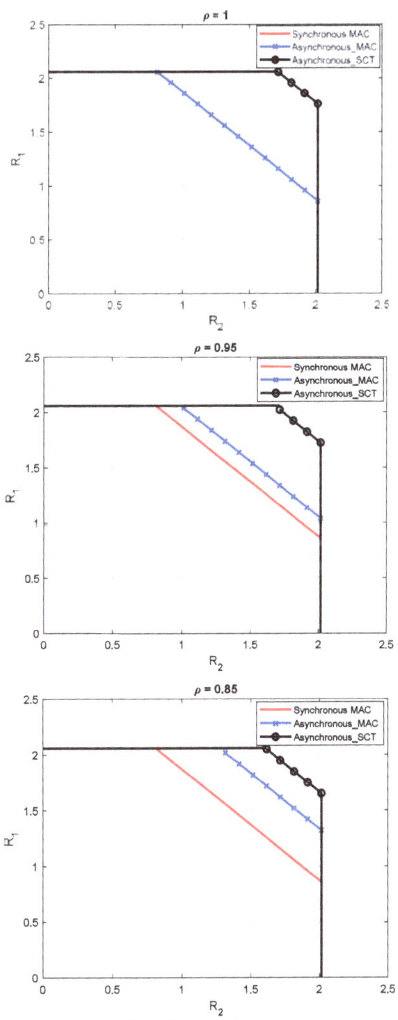

Fig. 7 Asynchronous MAC capacity regions

4.3. Simulation Results

The capacity region for synchronous MAC, asynchronous MAC as proposed in [11], and asynchronous with single content transmission (SCT) as proposed in this paper, are provided in Fig. 7. The SNR for both receivers is assumed to be 5dB, and the results for different cross correlation values are evaluated.

As expected, the single-user data rate is independent of the cross correlation between the two channels and the proposed scheme. However, in asynchronous SCT, the increase of ρ is followed by the raise in the sum rate. This is so reasonable as with the symbols aligning more, the signal detection and decoding will be accomplished with lower rates of error resulting in higher sum rates. This is in contrary with asynchronous MAC which achieves its highest sum rate for time delay equal to half the symbol time.

For $\rho = 1$, the asynchronous MAC shows its original nature and achieves the sum rate equal to synchronous MAC; whereas, due to the cooperation in transmitters, the asynchronous SCT achieves a much higher data rate than MAC due to the coherent reception of the two similar data streams.

5. Conclusion

In this paper we investigated the effect of symbol asynchronism in cooperative NOMA scheme in multibeam satellite forward link. The results show that being asynchronous not only does not degrade the performance but also outperforms the synchronous systems. This is in contrast with conventional beliefs, which originate from the implementation restrictions of SIC. The achieved gain in this paper is due to the cooperation at the satellite transmitters and taking advantage of the added correlation at the receivers. We suppose using iterative joint-detection and decoding can be the solution to benefit from the existing correlations to achieve the promised capacities. Further analysis of this issue is left as part of authors' future work.

Acknowledgement

The authors would like to express their gratitude to Dr. Nader Alagha, ESA-ESTEC, whose comments helped to considerably improve the style and presentation of the paper.

References

[1] R. Gaudenzi, N. Alagha, M. Angelone, et. Al, 'Exploiting code division multiplexing with decentralized multiuser detection in the satellite multibeam forward link', Int J sat. commun & network, 2018, vol. 36, Issue 3, pp 239-276

[2] Z. Ding, et al., 'On the performance of NOMA in 5G systems with randomly deployed users,' in IEEE Sig. Proc. Let., 2014, Vol 21, No.2

[3] N. Khan Beigi, M.R. Soleymani, 'Interference management using cooperative NOMA in multibeam satellite systems,' IEEE Int. Conf. on Commun. (ICC), May 2018

[4] H. Haci, H. Zhu, J. Wang, 'Performance of Non-Orthogonal Multiple Access with a Novel Asynchronous

Interference Cancellation Technique', IEEE Trans on Commun., 2017, Vol. 65, Issue 3, pp.1319-1335

[5] M. Yemini, A.S. Baruch, A. Leshem, 'On the multiple access channel with asynchronous cognition,' in IEEE Trans. on information theory, 2016, Vol. 62, No. 10, pp. 5643-5663.

[6] J. Cui, et al., 'Asynchronous NOMA for downlink transmission,' in IEEE Communication letters, Vol. 21, No.2, Feb.2017, pp. 402-405

[7] X. Wang, F. Labeau, L. Mei, 'Asynchronous uplink non-orthogonal multiple access (NOMA) with cyclic prefix', IEEE wireless commun. & networking conf., 2018

[8] Z. Ding, M. Peng, H. Vincent Poor, "Cooperative non-orthogonal multiple access in 5G systems," in IEEE Commun. Letters, Vol.19, No.8, Aug. 2017, pp. 1462-1465

[9] J. Kim, I. Lee, "Capacity analysis of cooperative relaying systems using non-orthogonal multiple access," in IEEE Commun. Letters, Vol.19, No.11, Nov. 2015, pp. 1949-1952.

[10] T. M. Cover and J. A. Thomas, Elements of Information Theory, 2nd ed. New York, NY, USA: Wiley-Interscience, 1991

[11] S. Verdu, 'The capacity region of symbol-asynchronous Gaussian multiple-access channel', IEEE Trans. On Information Theory, VOL. 35, No. 4, July 1989

[12] R.S. Cheng, S. Verdu, 'Gaussian multi-access channels with ISI: capacity region and multiuser water-filling', IEEE Trans. On Information Theory, VOL. 39, No.3, May 1993

[13] C. Zeng, L. Hoo, J. Cioffi, 'Optimal water-filling algorithms for a Gaussian multi-access channel with inter-symbol interference', in IEEE Intern. Conf. on Commun., June 2001

OVERLAPPING CLUSTERING FOR BEAM-HOPPING SYSTEMS

***Shigenori Tani*[1], *Shigeru Uchida*[2], *and Atsushi Okamura*[3]**

[1,2,3] *Information Technology R&D centre, Mitsubishi Electric Corporation, Kamakura, Japan*
[1]*Tani.Shigenori@eb.MitsubishiElectric.co.jp*
[2]*Uchida.Shigeru@dh.MitsubishiElectric.co.jp*
[3]*Okamura.Atsushi@bc.MitsubishiElectric.co.jp*

Keywords: BEAM-HOPPING, HTS, OPTIMIZATION PROBLEM.

Abstract

Beam hopping system sharing transmitting resources in time enables satellites and gateways to reduce their costs for increased HTS capacity. However, the conventional beam hopping system decreases frequency efficiency if the traffic demands vary among clusters because it shares their resources only in a cluster. Therefore, this paper proposes the novel beam hopping technique which overlaps each of clusters to share transmitting resources among them. The effectiveness of the proposal is verified through simulation results.

1. Introduction

Satellite communications have advantages in wide range, broadcasting, and disaster-resistant over the terrestrial communications such as cellular networks and optical networks. Therefore, they have been widely used in rural and isolated area, especially for maritime and aeronautical consumers as the only alternative. On the other hand, current High Throughput Satellites (HTS) which deliver high capacity utilizing frequency reuse in spot beams becomes a competitive and/or cooperative relationship to reduce the cost per bit for satellite operators [1]. In order to further reduce cost per bit, usage of higher frequencies including Q/V band and free space optical communication is effective to increase available bandwidth [2]. In addition, flexible satellite which is re-programmable of frequency, power, and connectivity depending on the communication demand makes it possible to improve efficiency. For example, digital channelizer improves frequency efficiency [3] and beam hopping improves space-time efficiency [4]. Therefore, these techniques can achieve required throughput with fewer satellite resources and gateways relative to conventional bent-pipe fashion.

When the system uses higher frequency, the rain attenuation also increases. Site diversity is one of the popular techniques in order to overcome rain attenuation, however, it needs a greater number of gateways to keep system availability. Therefore, frequency efficiency of feeder-link needs to be improved to save the gateway cost. Regarding the conventional HTS systems always illuminate all beams, the beam hopping system divides the coverage area into a plurality of clusters to switch their transmitting beams synchronizing super frames depending on the traffic demand. In other words, the satellites and gateways are sufficient to have transceiver not for all beams but for all clusters. However, the conventional beam hopping system controls beam hopping time plan independently for every cluster, it may cause surplus or unmet resource allocation among the clusters in case the traffic demands are unevenly distributed. In order to solve aforementioned, this paper proposes overlapping clustering for beam hopping system. The proposed method aggregates neighbouring clusters as a clustering group to share each transmitting resources in the group. Thus, two or more beams can be illuminated to desired area in a group flexibly even if the traffic demands are unevenly distributed. To increase a number of clusters configuring a group is equivalent to increasing a number of samples to pick out from a population. Therefore, slot assignment becomes efficient to enlarge clustering groups because of decreasing the standard error. Furthermore, this paper proposes optimal beam assignment algorithm using mini-batch stochastic gradient descent (SGD) to maximize system throughput considering satellite transmitting antenna pattern and traffic generated area. Hence, our proposal contributes next generation HTS systems to offer required system throughput with fewer gateways than conventional beam hopping technique when the traffic demand is unevenly distributed. Furthermore, the proposed method improves efficiency by selecting optimal beam direction considering Signal to Interference and Noise Ratio (SINR).

The remainder of this paper is organized as follows. The relevant research works about multi-beam satellite communication are surveyed and system analysis of beam hopping system is introduced in Section 2. Section 3 describes problem statement of conventional beam hopping system. In addition, our proposed overlapping clustering beam hopping system is introduced for the problem. Section 4 presents simulation results of throughput performance when the traffic demands are unevenly distributed. Finally, this paper is concluded in Section 5.

2. Related research works

2.1. *Frequency efficiency improvements for HTS*

The HTS systems utilize space, frequency and time division multiplexing techniques to improve system capacity. Space division multiplexing techniques enable to use the same frequencies and polarizations repeatedly across a system by using narrow beams pointed to different geographic areas. As a typical system architecture of HTS is shown in Fig. 1, a user-

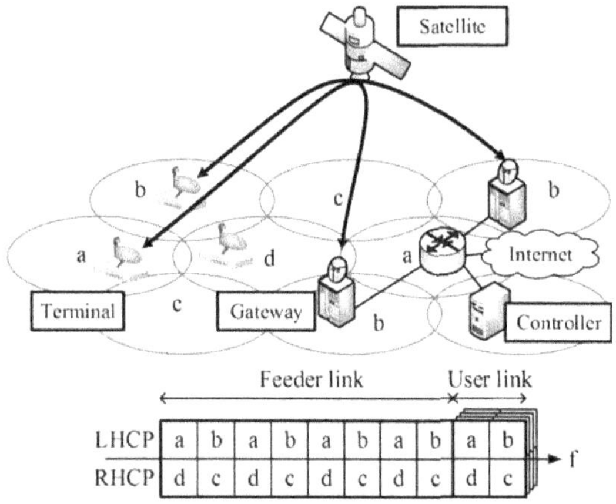

Fig. 1 System architecture of HTS

link consists of four colours, indicating two sub-bands and two polarizations (Left Hand Circular Polarization, Right Hand Circular Polarization), and each beam uses one of the four colours. The remaining bandwidths are used at a feeder-link [5].

Frequency division multiplexing assigns a separate portion of the frequency to each of the individual beams and users. Since static frequency allocation strategies does not utilize the shared channel efficiently, dynamic frequency allocation by using digital channelizer is effective to improve frequency efficiency [3]. Also, beam forming network (BFN) and multi-port amplifier (MPA) achieve flexible power allocation to coordinate gain and phase of inputs of amplifiers and combining their outputs [6], [7].

Time division multiplexing assigns a fraction of the time to each of the individual users as used in Digital Video Broadcasting Satellite – Second Generation Extensions (DVB-S2X) [8]. In Ka-band HTS system, rain attenuation is severe problem to keep the required Quality of Service (QoS), the modulation scheme and coding rate are controlled according to the link condition for each time slot. Furthermore as shown in Fig. 2, while the current HTS system assigns resources statically or semi-statically, the beam hopping system divides the coverage area into a plurality of clusters to share their transmitting beams in time depending on the traffic demand [4]. Therefore, the beam hopping system is expected to be a solution for increased HTS capacity with fewer satellites and gateways resources. Where, the aforementioned frequency and time duality is clarified in [9]. Since the satellite amplifies and forwards single carrier signal in beam hopping system, it does not need larger Output Back Off (OBO) than multi-carrier amplifier because of the lower Peak to Average Power Ratio (PAPR). As a result, beam hopping system is more efficient.

2.2. *Signalling aspects*

In contrast to the current HTS of four coloured (250MHz), beam hopping system transmits data by using whole user-link bandwidth (ex. 500MHz). However, the terminal need not to receive wideband signal continuously but only receives time-sliced PL-Frame specified in [8]. Furthermore, timing

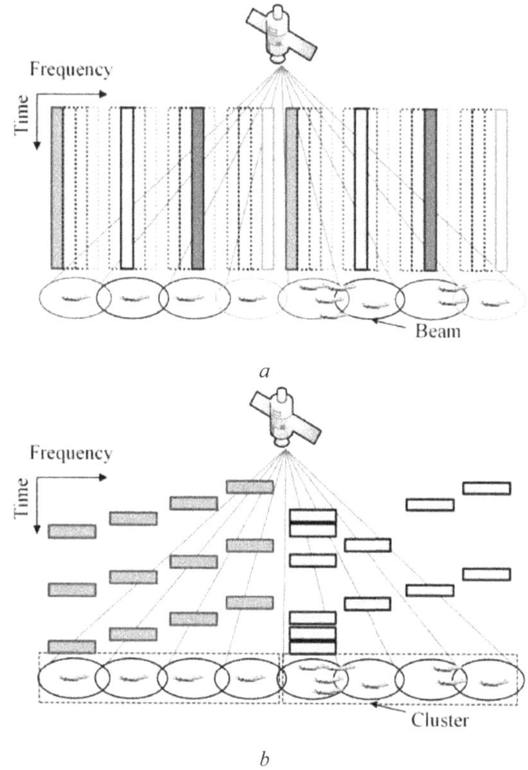

Fig. 2 (a) Static resource assignment in current HTS system, (b) Flexible resource assignment in beam hopping HTS system

synchronization is required among the satellite, gateways and terminals to relay signals at appropriate time slot. Since the coded frame length is fixed, modulated frame length (PLFRAME) varies depending on the modulation scheme and coding rate (MODCOD). Therefore, [8] specifies super frame whose length is fixed of 612,540 symbols. From the control point of view, novel signalling of Beam Hopping Time Plan (BHTP) which is similar to Terminal Burst Time Plan (TBTP) is determined for specification [10].

2.3. *Availability and a number of gateways*

In general, the HTS systems improve throughput as the number of user-link beams is increase, that is, it also needs more feeder-link beams to reuse bandwidth. If the user-link bandwidth is allocated for each of beams equally, uplink system bandwidth, B_U, can be formulated from uplink user-link bandwidth, W_U, downlink user-link bandwidth, W_D, number of beam, N_b, number of polarization, N_p, number of frequency reuse, N_f, and number of feeder link, N_g, as follows,

$$B_U = N_f W_U + \left\lceil \frac{N_b}{N_p N_g} \right\rceil W_D. \tag{2.4}$$

If we assume general Ka band HTS, $B_U = 2.5$GHz, W_U and $W_D = 250$MHz, $N_p = 2$, $N_f = 2$, $N_b = 60$, required number of gateways, N_g, is 4 from Eq. (2.4). However, it needs redundant gateways to operate four gateways simultaneously with keeping required availability under heavy rain circumstance. Where, system availability, p_t, is expressed as probability to operate more than N_g gateways out of number of redundant gateway, N_{gr}, plus N_g as follows,

$$p_t = \sum_{i=N_g}^{N_{gr}+N_g} \{{}_{N_{gr}+N_g}C_i\, p_1^i (1-p_1)^{N_{gr}+N_g-i}\}, \qquad (2.5)$$

where, p_l is availability of the each gateway. If we assume p_l = 99% and p_t = 99.9%, one redundant gateway is needed from Eq. (2.5). As a result, total number of required gateways ($N_{gr}+N_g$) is five. Where, rain attenuation, L_r, of availability, p_l, is formulated from the elevation angle, $\phi(t)$, effective slant path, L_E, area specific weighting factor, γ_R, as follows [12],

$$L_r = \gamma_R L_E \left(\frac{1-p_1}{0.01}\right)^{\alpha} \qquad (2.6)$$

$$\begin{aligned} \alpha = -(0.655 + 0.033\, ln(1-p_1) \\ - 0.045\, ln(\gamma_R L_E(t)) \\ - p_1(1-p_1)\, sin\, \phi(t)) \end{aligned} \qquad (2.7)$$

Fig. 3 shows rain attenuation performance with Ka band and Q band. If we assume an availability of 99% in Ka band, an availability of Q band decreases to 92% to consider same rain attenuation level with Ka band. Therefore, the number of redundant gateways is three and total number of required gateways is seven in Q band which is calculated based on Eq. (2.5) with p_l=92%, p_t=99.9%. As a result, Q band system requires 1.4 times as many gateways as that of Ka band. In other words, Q band system needs to improve efficiency 1.4 times higher.

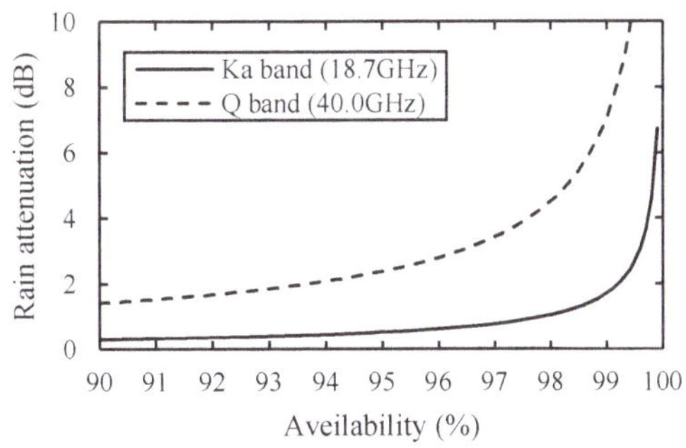

Fig. 3 Availability versus rain attenuation.

3. Beam hopping systems

3.1. Efficiency degradation of conventional method

As described in previous section, the conventional beam hopping system controls beam hopping time plan independently for every cluster. Therefore, as shown in Fig. 4, surplus or unmet resource allocation among the clusters may be caused when the traffic demands are unevenly distributed. As a result, frequency efficiency decreases in such a case.

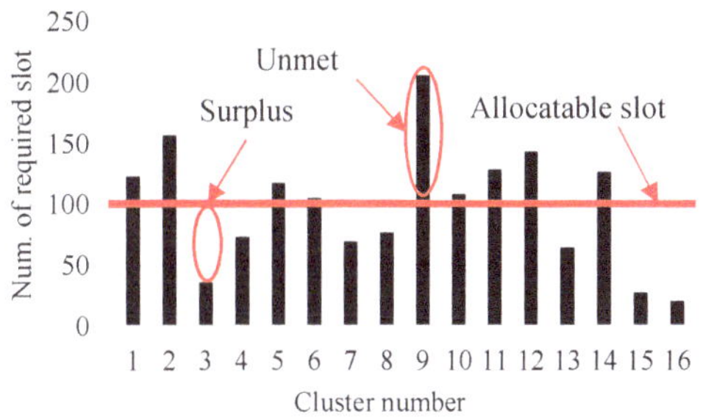

Fig. 4 An example of traffic variation and resource allocation of conventional beam hopping system.

3.2. Proposed overlapping clustering method

In order to solve aforementioned, this paper proposes overlapping clustering for beam hopping system. As shown in Fig. 5, the proposed method configures clustering groups to share each transmitting resources in the group. Thus, two or more beams can illuminate to desired areas in a group flexibly even if the traffic demands are unevenly distributed. To increase a number of clusters configuring a group is equivalent to increasing a number of samples to pick out from a population. Where, the standard error, σ_n, for n sample is formulated from the standard error of population, σ as follows.

$$\sigma_n = \frac{\sigma}{\sqrt{n}} \qquad (3.1)$$

Fig. 6 shows sampling error with varies number of cluster group calculated by Eq. (3.1). Where, we set standard error of 0.95 and four beams in a cluster. It shows that the slot assignment becomes efficient to enlarge clustering groups because of decreasing the standard error. However, improvement effect is limited at more than 4 cluster group because the reduction of sampling error becomes mild.

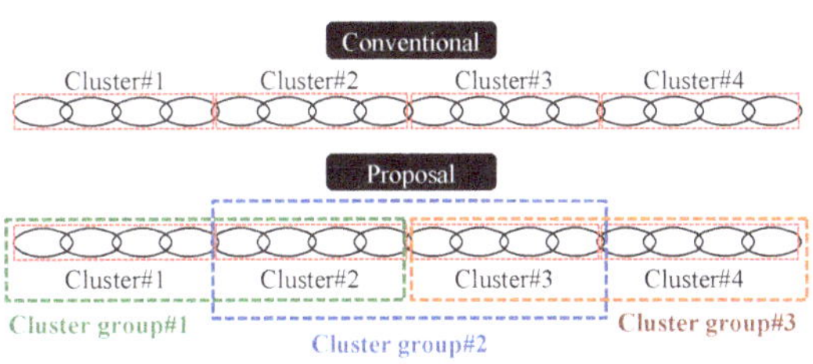

Fig. 5 Clustering configuration of proposal

In addition, since the conventional beam hopping system selects out of the predefined beam directions, the receiving SINR may be degraded due to the pointing loss if the receiving nodes are located in edge of beam coverage. As a result, the throughput becomes decreasing. By contrast, the proposed method decides beam direction so as to maximize system

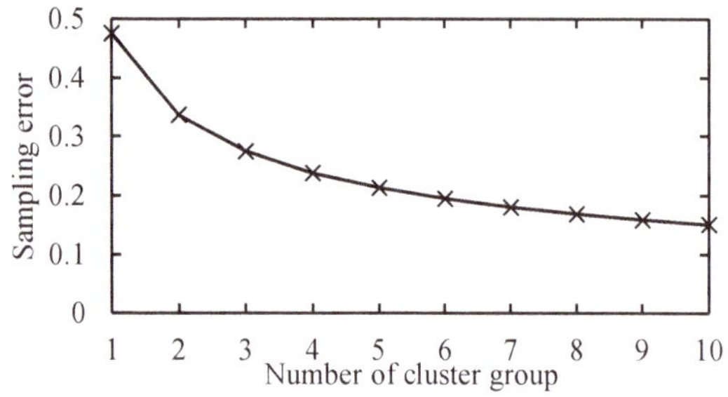

Fig. 6 Number of cluster group versus sampling error.

throughput based on nodes' position and inter-satellite-beam interference caused by satellite's antenna pattern.

The beam assignment algorithm of proposed overlapping beam hopping (OBH) is presented in Algorithm 1. In addition, schematic diagram of the proposed algorithm is shown in Fig. 7. First, initial beam direction, $B_0 = (b_0, b_1, \dots, b_{n-1})$, for n beams are selected from cluster groups which are predefined range of directivity. Similarly, initial position of candidate nodes in each of cluster groups, $D_0 = (d_0 \in DG_0, d_1 \in DG_1, \dots, d_{n-1} \in DG_{n-1})$, out of set of node position, $(DG_0, DG_1, \dots, DG_{n-1} \in D_{all})$, within an each of cluster groups is selected from whole of nodes' position, D_{all}. Where, an element, d_x, of D_0 is selected every scheduling timing from which the residual data ratio is the largest as shown in Eq. (3.2) so as to improve fairness among nodes. Second, beam direction, B, is decided temporary in an inner iteration so as to maximize the objective function, $f(b)$, utilizing mini-batch SGD as shown in Eq. (3.3). Where, objective function, $f(b)$, is formulated from generated data size, $z(n)$, reported by nodes every scheduling timing and estimated transmission data size, $e(q(n))$, at SINR, $q(n)$, of nth node as shown Eq. (3.4). That means SGD provides optimal solution to minimize sum of residual data size. Furthermore, $q(n)$ is formulated from SNR $Snr(b_n, d_n)$, at node positon, d_n and sum of SIR, $Sir(b_m, d_n)$, suffered from undesired beams, b_m, as shown in Eq. (3.5). Position set of candidate nodes, D_{i+1}, is updated after the inner iteration because initial position, D_0 is not always optimal. Where, D_{i+1} is updated to closest nodes from temporal beam position, B_i, as shown in Eq. (3.6).

Algorithm 1. Beam assignments for a slot:

Set initial parameter B_0, D_0, η
$i \leftarrow 0, j \leftarrow 0$
while num. of outer iteration, i, **do**
  while num. of inner iteration, j, **do**
    Select m sample of data $b \in B$ randomly
    for each $b = 0, 1, \dots, m-1$ **do**
      Update B_{j+1}utilizing eq.(3.3)
  Update D_{i+1} utilizing eq.(3.6)
If beam controlling mode = DBC **then**
  Adjust B to predefined beam position
return beam position, B, and the set of points, D, to be transmitted the data.

Finally, beam position, B, is updated depending on the satellite configuration. If the satellite has beam forming ability, continuous beam controlling (CBC) is adapted which updates optimal beam direction based on the above algorithm. In contrast, if the satellite simply switch beam direction, discrete beam controlling (DBC) is adapted which adjusts beam position, B, to the closest predefined candidate point. Since DBC misalign beam direction and location of traffic demand, SNR is slightly less than CBC caused by the antenna pointing loss. However, DBC can be adapted without beam forming.

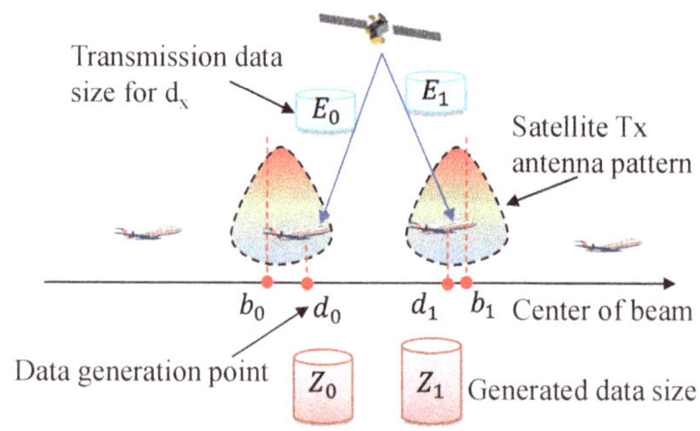

Fig. 7 Schematic diagram of the proposed algorithm.

$$d_x \in D_0 = max\left(y \in DG_x: 1 - \frac{\sum_{s=0}^{t} E(S(y,s))}{Z(y)}\right) \quad (3.2)$$

$$b_{n,j+1} = b_{n,j} - \eta \frac{\partial f}{\partial b}(b_{n,j}) \quad (3.3)$$

$$f(b) = \sum_{n=0}^{N-1} z(n) - e(q(n)) \quad (3.4)$$

$$q(n) = \left\{ Snr(b_n, d_n)^{-1} + \sum_{m \in \bar{n}}^{N-1} Sir(b_m, d_n)^{-1} \right\}^{-1} \quad (3.5)$$

$$d_x \in DG_x = min(y \in DG_x: \|b_{x,i} - d_y\|) \quad (3.6)$$

4. Performance evaluation

4.1. Simulation settings

In order to verify the effectiveness of the proposed beam control technique, we first observe difference of beam assignment strategy with an example. Then, we evaluate the throughput performance.

Table 1 shows simulation settings. The number of beams simultaneously transmitted is set to 8. In the conventional method, an each of beams points in range that corresponds to deploy 4 beams uniformly. The number of iteration for proposal set to 25 for inner loop, 2 for outer loop, and mini-batch size is set to 2 which are evaluated in advance so as to converge parameters. Consequently, the total number of iterations in a slot is 200. Scheduling for beam hopping and data generation is processed every frame, that is, the satellite hops their beams every slot based on the scheduling information decided at beginning of each frames. Data size is generated as multiplication of random numbers that follow Poisson distribution with mean, λ, from 10 to 60 and Normal distribution with average of 1 and standard deviation, σ, from 0.05 to 0.95. Furthermore, every 4 nodes are randomly deployed in same area as a cluster. In the simulation, we evaluate normalized throughput which is the ratio of generated data size to transmitted data size. Where, transmitted data size is equivalent to spectrum efficiency (bps/Hz) calculated from SINR and MODCOD table defined in [8]. In order to verify the effectiveness, we compare our proposal to Round Robin algorithm (RR) as a conventional method. Predefined candidate points for DBC are set to centre of beams which are deployed uniformly so that the edge of coverage (EOC) is 5dB down from the peak gain.

Table 1 Simulation settings

No.	Parameter	Value
1	Number of clusters	8
2	Number of candidate beams in a cluster	4
3	Carrier frequency	40GHz
4	Satellite Tx. Antenna diameter	2.0m
5	EOC gain	5.0dB
6	Number of outer iteration	2
7	Number of inner iteration	25
8	Frame structure	32 slot/frame
9	Traffic pattern	Poisson distribution ($\lambda = 10 \sim 60$) and Normal distribution ($\sigma = 0.1 \sim 0.9$)
10	Traffic location	Uniform distribution

4.2. *Evaluation results*

Fig. 8 show an example of beam assignment of RR and the proposed method. Where, dots shows beam direction and dashed line shows nodes' position. As shown in figures, RR sometimes points beams far from nodes, on the other hand, the propose method always points beams close to the nodes. Furthermore, if the some nodes are close to each other, the proposed method adjusts beam direction so as to minimize inter-beam interference.

Fig. 9 shows offered load versus normalized throughput performance where σ is set to 0.9. For the conventional RR algorithm, the value of normalized throughput is nearly same as that of the proposed method when λ=10, because there are enough slots to send generated data during scheduling timing. However, the performance of RR linearly decreases as λ increases. On the other hand, we can find the normalized throughput of our proposal nearly keeps 1 when λ is less than 30 and decreases linearly after λ is above 30. Therefore, we can find our proposal can achieve nearly 40% higher throughput than the conventional method whenever λ increases due to the improvement of time efficiency. In addition, OBH-CBC achieves 10% higher performance than OBH-DBC because it improves receiving SINR by fine beam control.

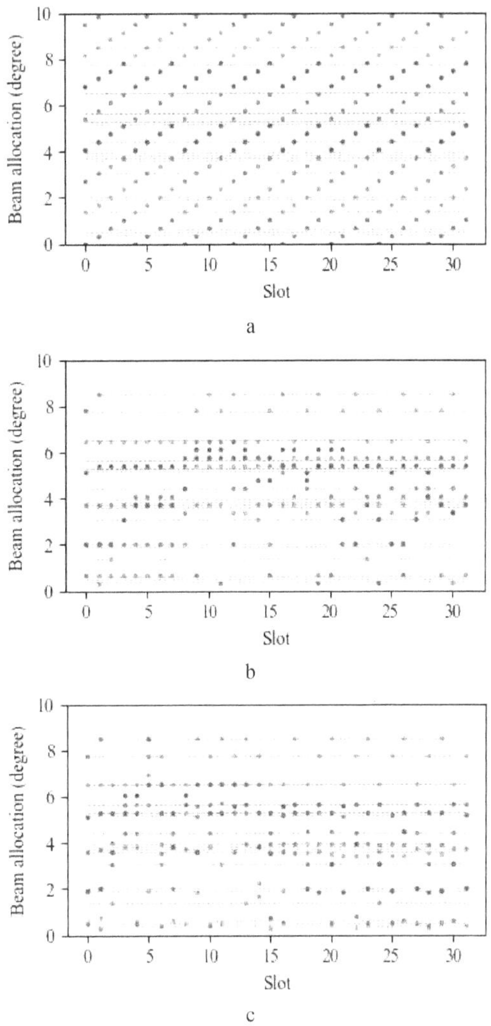

Fig. 8 An example of beam assignment of (a) RR, (b) OBH-DBC, (c) OBH-CBC.

Fig. 10 shows traffic variance versus normalized throughput performance where λ is set to 40. We can find all of the method gradually decrease as σ increase. However, the decreasing rate of our proposal is slightly better than conventional method. Because our proposal can avoid surplus or unmet resource allocation among the clusters to overlap clustering group even if the traffic demands are unevenly distributed.

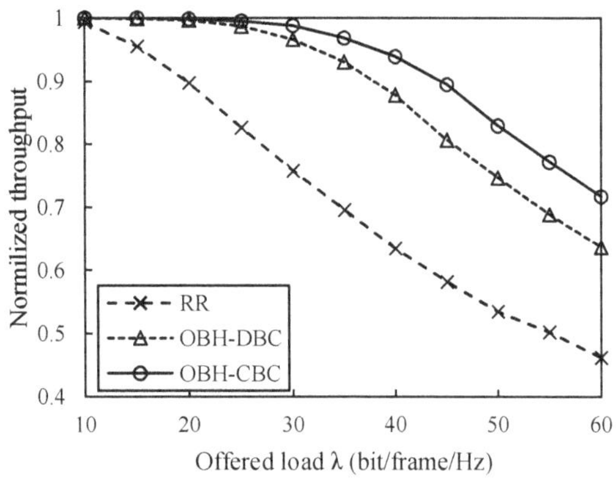

Fig. 9 Offered load versus throughput performance.

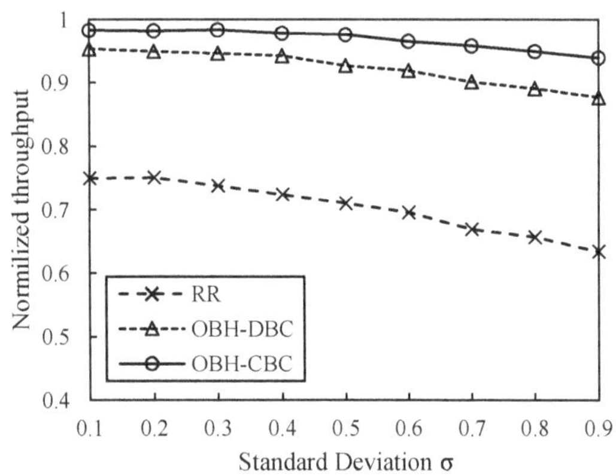

Fig. 10 Traffic variance versus throughput performance.

5. Conclusion

This paper proposed novel beam hopping technique which overlaps each of clusters to share transmitting resources among clusters. Moreover, this paper proposed optimal beam assignment algorithm using mini-batch stochastic gradient descent (SGD) to maximize system throughput considering satellite transmitting antenna pattern and traffic generated area. The simulation results verified its effectiveness.

References

[1] S. Gardner, "Satellite Broadband Enters the Mass Market: Now Everything is Different," MILCOM 2013 - 2013 IEEE Military Communications Conference, San Diego, CA, 2013, pp. 1075-1080.

[2] J.Nessel, J.Morse, M.Zemba, C.Riva, L.Luini, "Performance of the NASA beacon receiver for the Alphasat Aldo Paraboni TDP5 propagation experiment," 2015 IEEE Aerospace Conference, pp.1-8, Mar. 2015.

[3] Shigenori Tani, Katsuyuki Motoyoshi, Hiroyasu Sano, Atsushi Okamura, Hiroki Nishiyama, and Nei Kato, "Flexibility-Enhanced HTS System for Disaster Management: Responding to Communication Demand Explosion in a Disaster," IEEE Transactions on Emerging Topics in Computing (TETC). DOI : 10.1109/TETC.2017.2688078.

[4] J. Lizarraga, P. Angeletti, N. Alagha and M. Aloisio, "Flexibility performance in advanced Ka-band multibeam satellites," IEEE International Vacuum Electronics Conference, Monterey, CA, 2014, pp. 45-46.

[5] H. Fenech, S. Amos, A. Tomatis and V. Soumpholphakdy, "High throughput satellite systems: An analytical approach," in IEEE Transactions on Aerospace and Electronic Systems, vol. 51, no. 1, pp. 192-202, Jan. 2015.

[6] G. Cocco, T. D. Cola, M. Angelone and Z. Katona, "Radio resource management strategies for DVB-S2 systems operated with flexible satellite payloads," 2016 8th Advanced Satellite Multimedia Systems Conference and the 14th Signal Processing for Space Communications Workshop (ASMS/SPSC), pp. 1-8, Sept. 2016.

[7] H. Fenech, S. Amos and T. Waterfield, "The role of array antennas in commercial telecommunication satellites," 2016 10th European Conference on Antennas and Propagation (EuCAP), Davos, 2016, pp. 1-4.

[8] Second generation framing structure, channel coding and modulation systems for Broadcasting, Interactive Services, News Gathering and other broadband satellite applications; Part 2: DVB-S2 Extensions (DVB-S2X) ETSI EN 302 307-2 V1.1.1, Oct. 2014.

[9] J. Lei and M. Á. Vázquez-Castro, "Multibeam satellite frequency/time duality study and capacity optimization," in Journal of Communications and Networks, vol. 13, no. 5, pp. 472-480, Oct. 2011.

[10] Sunil Panthi, Dirk Breynaert, Christopher McLain, and Janet King. "Beam Hopping - a Flexible Satellite Communication System for Mobility", 35th AIAA International Communications Satellite Systems Conference, International Communications Satellite Systems Conferences (ICSSC), (AIAA 2017-5413)

[11] DVB-RCS2 Lower Layer Satellite Specification ETSI TS 101 545-2 V1.2.1, Apr. 2014.

[12] ITU-R P.618-11, "Propagation data and prediction methods required for the design of Earth-space telecommunication systems," Sept. 2013.

ADJACENT BEAMS RESOURCE SHARING TO SERVE HOT SPOTS: A RATE-SPLITTING APPROACH

Tomás Ramírez[1], Carlos Mosquera[2], Màrius Caus[3], Adriano Pastore[4], Nader Alagha[5], and Nele Noels[6]

[1,2] *AtlanTTic research center, University of Vigo, Galicia, Spain*
[3,4] *Centre Tecnològic de Telecomunicacions de Catalunya, Castelldefels, Barcelona, Spain*
[5] *European Space Agency Technical Research Center (ESTEC) , Noordwijk, The Netherlands*
[6] *DIGCOM research group, TELIN Department, Ghent University, Ghent, Belgium*

Abstract

This work addresses the uneven traffic demand scenario in multi-beam satellite systems, in which a hot-spot beam is surrounded by cold beams. After partitioning the hot-spot beam in different sectors, resource pulling from cold neighbouring beams is allowed following an aggressive frequency-reuse scheme. As a consequence, the level of the co-channel interference within the hot-spot beam increases. A scheme known as Non-Coherent Rate-Splitting (NCRS) is employed to cope with this interference, based on the exclusive use of magnitude channel state information at the transmitter (CSIT). The receiver complexity is increased with respect to full CSIT precoding schemes, which are considered for benchmarking purposes. Different NCRS strategies are analyzed and compared with several partial and full CSIT schemes. The proposed solution not only shows an improvement with respect to partial CSIT benchmarks, but also displays a competitive performance against full CSIT precoders.

1. Introduction

The ever growing demand of data throughput and the limited available bandwidth for satellite communications has created the necessity of more spectrally efficient techniques, which involves the adoption of aggressive frequency reuse schemes. If more aggressive reuse schemes such as 1-color or 2-color are considered, the performance of single user detection (SUD) can suffer due to the increased level of co-channel interference in the satellite forward link from the gateways to the end user terminals. (see Fig. 1). In consequence, additional complexity needs to be implemented at the transmit gateway, receive terminals, or both. Precoding is one of the most promising techniques in the forward link to cope with co-channel interference in multibeam satellites. As main drawbacks, it requires a permanent reporting of the channel state information (CSI) to the gateway, and synchronism across the different beams. In [1], a multi-user detector (MUD) approach is considered to avoid these drawbacks and to only work with partial CSI at the transmitter (CSIT), e.g. the channel strength knowledge, at the cost of increasing the receiver complexity. Under similar partial CSIT constraints, in [2] and [3] a rate-splitting approach named Non-Coherent Rate-Splitting (NCRS) was introduced to cope with the interference of neighboring beams. NCRS splits the messages into private and public components, and performs successive interference cancellation (SIC) at the receive terminals to extract messages that use the same time and frequency resources.

The uneven traffic scenario, with all the traffic demand concentrated in a single beam, has been recently studied in [1],[4] and [5]. This scennario is usually referred to as hot spot (HS). Assuming partial CSIT, the MUD technique from [1] presents a significant improvement with respect to the 4-color Frequency Division Multiplexing (FDM) baseline, where terminals perform single-user detection (SUD) and no resource sharing among beams is performed. However, if resource pulling from neighbouring beams is allowed [4], adjacent beam resource sharing (ABRS) presents significant improvement in the capacity in comparison to conventional multi-beam systems without resource sharing. As presented in [1], the ABSR with 3-color scheme even offers slight improvement over the MUD techniques without increasing the complexity of the receivers (using SUD) and without significant increase in the gateway uplink frequency (compared to 1 or 2-color schemes). On the other side, precoding techniques are analyzed in a Hot Spot (HS) scenario with full CSIT [5]. In addition to full CSIT, the performance of precoding is boosted by assuming a flexible payload and double polarization per beam. Diverse flavours of precoding are presented in [5] by following different optimization criteria. The so-called *performance precoding* presents the highest sum-rate after numerical optimization of the precoder coefficients, at the cost of introducing high imbalance in the rates assignment. A trade-off between fairness and sum-rate is also pursued in this reference by changing the optimization metric.

The current paper extends the work in [4], by applying rate splitting ideas to a HS scenario with the adoption of aggressive frequency reuse patterns. NCRS has already proved a significant improvement with respect to the 4-color FDM benchmark, achieving up to 20% for low signal-to-noise ratio (SNR) cases in a uniform traffic scenario [3]. Nevertheless, it is yet to be confirmed whether NCRS can present an improvement over the previous techniques for HS. In order to quantify the potential improvement, both partial and full CSIT benchmarks are considered. Caution is also required when comparing different schemes due to the varying degrees of complexity at the receivers. Some comments will be made on the required complexity at the receive terminals, which will be assumed to operate ideally without any implementation losses.

The rest of the paper is organized as follows. In Section II, the satellite system model is introduced. Next, NCRS is presented in Section III. After that, the simulation results are shown in Section IV and some conclusions are given in Section V.

Notation: Upper (lower) boldface letters denote matrices (vectors). $(.)^H$, $\mathbf{I}_N$ denote Hermitian transpose and $N \times N$ identity matrix, respectively. $\mathbb{E}[\cdot]$ is the expected value operator.

2. System Model

In this work we focus on the forward link of a multi-beam satellite communication system. More specifically, we focus on a Hot-Spot scenario in which a central beam is surrounded by six other beams which have none or negligible traffic. To keep the analysis simple, the second ring and outer rings beams are not considered. Assuming perfect synchronization, an ideal feeder link, a single feed per beam payload architecture, and neglecting non-linear effects in the satellite payload, the values received by M interfering users at a given time instant can be written as

$$\mathbf{y} = \mathbf{Hx} + \mathbf{w}, \tag{1}$$

where $\mathbf{y} \in \mathbb{C}^{M\times 1}$ is the vector of received signals given by $\mathbf{y} = [y_1[k] \quad y_2[k] \quad \ldots \quad y_M[k]]^T$, $\mathbf{H} \in \mathbb{C}^{M\times M}$ is the channel matrix, $\mathbf{x} \in \mathbb{C}^{M\times 1}$ is the vector of transmitted signals given by $\mathbf{x} = [x_1[k] \quad x_2[k] \quad \ldots \quad x_M[k]]^T$ and $\mathbf{w} \in \mathbb{C}^{M\times 1}$ is the noise vector given by $\mathbf{w} = [w_1[k] \quad w_2[k] \quad \ldots \quad w_M[k]]^T$. To keep the notation simple, the time index is dropped from now on. All the antenna feeds transmit with the same power, so that $\mathbb{E}\left[\mathbf{xx}^{\mathrm{H}}\right] = \frac{P}{M}\mathbf{I}_M$ for a total power P. The samples of the Additive White Gaussian Noise (AWGN) vector $\mathbf{w}$ are statistically independent, i.e. $\mathbb{E}\left[\mathbf{ww}^{\mathrm{H}}\right] = \frac{\sigma^2}{\mathrm{C}}\mathbf{I}_M$, where C is the number of frequency colors in the system.

The (q, m)th entry of the matrix $\mathbf{H}$ is given by:

$$h_{qm} = e^{j\phi_{qm}} \frac{\sqrt{G_R G_{qm}}}{4\pi d_q/\lambda}. \tag{2}$$

Here, the term G_R refers to the receive antenna gain and G_{qm} represents the transmit antenna gain from beam m to the q-th user terminal. As for the rest of the terms, λ is the carrier wavelength, ϕ_{qm} is the phase rotation introduced by the channel and d_q is the distance from the satellite to the q-th user.

Since the rate-splitting approach only uses magnitude information, we introduce also the link strength between the qth feed and the mth receiver:

$$\gamma_{C,qm} = \frac{\frac{P}{M}|h_{qm}|^2}{\sigma^2/\mathrm{C}}. \tag{3}$$

2.1 Hot-Spot scenario

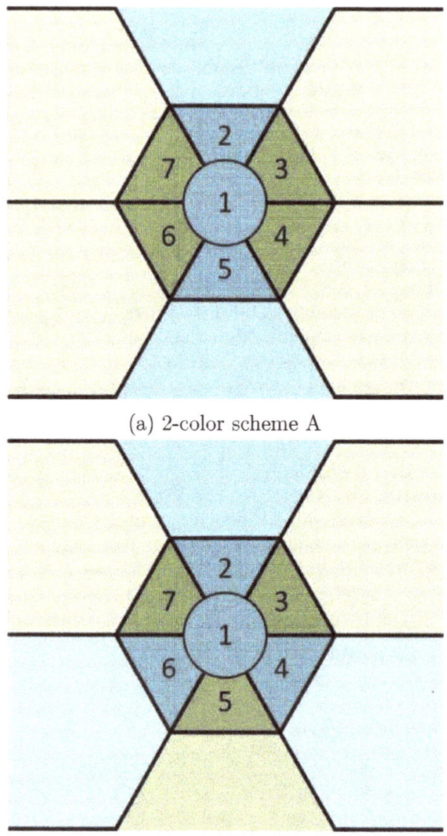

(a) 2-color scheme A

(b) 2-color scheme B

Fig. 1 Mapping of colors to the central beam for 2-color frequency reuse schemes.

The resources of the adjacent beams are shared with the hot spot beam in order to satisfy the traffic demand as in [4]. To this end, the hot spot beam is divided into seven sectors which are served by the respective beams. Figs. 1 and 2 sketch the boundaries of the sectors for different frequency schemes, with different colors denoting different frequency bands. Examples of 3-color and 4-color frequency allocations are given in Fig. 2 according to [1]. The radius of the inner zone is determined by a feed gain threshold, G_{th}, relative to the maximum feed gain of the beam*.

*The relative gain G_{th} is used for theoretical purposes. In an actual satellite system, it would be more practical to use instead

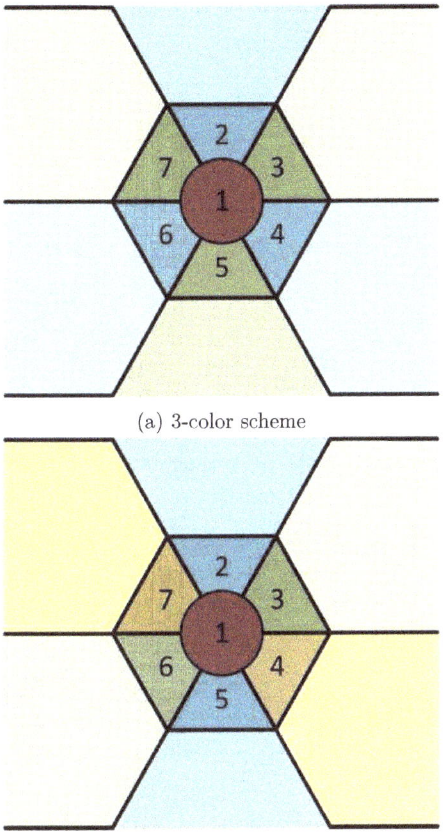

(a) 3-color scheme

(b) 4-color scheme

Fig. 2 Mapping of colors to the central beam for different frequency reuse schemes.

The more aggressive 2-color schemes from Fig. 1 gives rise to a much higher amount of interference; the lack of full CSIT at the gateway, with the only knowledge of the magnitude of the received signals from the different beams, makes us consider a non-orthogonal rate splitting scheme known as NCRS and presented in [2]. NCRS can operate with groups of two or more simultaneous users using the same frequency resources, and does not require tight synchronization requirements among the superimposed signals. A SIC strategy is employed at the receiver side to get the most out of the downlink flow. The beam grouping depicted in Fig. 1(a) is such that two pairs of beams (in green color) and a group of three beams (in blue color) are handled separately. In Fig. 1(b) NCRS is used to exploit the group of 4 beams (in blue color), with the users in the other three green zones performing SUD. As discussed in the next section, partial CSIT benchmarks will be also simulated for both 3-color and 4-color reuse schemes from Fig. 2 based on the ABRS approach of [4]. We will also consider full CSIT benchmarking for the full frequency reuse (FFR) scenario, in the form of precoding.

2.2 Satellite payload

The payload constraints are the same as in [1] and [4], where conventional payload structures for multi-beam satellites are assumed. For the sake of simplicity, perfect synchronization across beams is assumed in this work, although it is not required in the case of ABRS and NCRS [6]. However, in the latter case, the lack of synchronism demands more complexity from the receiver if applying joint decoding of the received signals.

3. Non-Coherent Rate-Splitting (NCRS)

Inspired by the Generalized Degrees of Freedom(GDoF) framework in [7], NCRS is presented in [2] and [3] to cope with the co-channel interference by using rate-splitting. The transmitters send two kind of messages: private and public. The former is decoded by and intended for one of the users and the latter is decoded by all the users. The amount of power allocated to each message is controlled by a variable λ_i, which needs to be optimized for a given quality of service (QoS) criterion. For a two user scenario the transmitted signals are expressed as

$$x_1 = \sqrt{\frac{P}{2}(1-\lambda_1)}x_{c_1} + \sqrt{\frac{P}{2}\lambda_1}x_{p_1} \tag{4}$$

$$x_2 = \sqrt{\frac{P}{2}(1-\lambda_2)}x_{c_2} + \sqrt{\frac{P}{2}\lambda_2}x_{p_2} \tag{5}$$

with $\mathbb{E}\begin{bmatrix}|x_1|^2 & x_1x_2^* \\ x_2x_1^* & |x_2|^2\end{bmatrix} = \frac{P}{2}\mathbf{I}$.

The public message is transmitted by sending two public messages x_{c_1} an x_{c_2} which can be jointly or successively decoded. At the receiver side, the public messages are decoded first by treating the private messages x_{p_1} and x_{p_2} as noise. After subtracting the public message, the private messages are decoded. In this paper, joint decoding (JD) of the public messages is assumed and thereby, the NCRS receiver involves JD and successive interference cancellation (SIC). The application of NCRS to more than two users increases substantially the complexity of the process to optimize the weights of the different public messages, and requires more cancellation stages at the receivers. As a suboptimal extension, we just consider a public message component x_{c_i} for each user. For the 3-user case, NCRS signals would read as

$$\begin{aligned} x_1 &= \sqrt{(1-\lambda_1)P}x_{c_1} + \sqrt{\lambda_1 P}x_{p_1} \qquad (6) \\ x_2 &= \sqrt{(1-\lambda_2)P}x_{c_2} + \sqrt{\lambda_2 P}x_{p_2} \\ x_3 &= \sqrt{(1-\lambda_3)P}x_{c_3} + \sqrt{\lambda_3 P}x_{p_3} \end{aligned}$$

signal to interference and noise (SINR) measurements at the receiver side which are fedback to the gateway.

whereas the corresponding 4-user NCRS signals are given by

$$\begin{aligned} x_1 &= \sqrt{(1-\lambda_1)P}x_{c_1} + \sqrt{\lambda_1 P}x_{p_1} \\ x_2 &= \sqrt{(1-\lambda_2)P}x_{c_2} + \sqrt{\lambda_2 P}x_{p_2} \\ x_3 &= \sqrt{(1-\lambda_3)P}x_{c_3} + \sqrt{\lambda_3 P}x_{p_3} \\ x_4 &= \sqrt{(1-\lambda_4)P}x_{c_4} + \sqrt{\lambda_4 P}x_{p_4}. \end{aligned} \tag{7}$$

Note that the signals indexes in (7) and (8) are expressed in a general way, and do not hold any relation to the sector indexes in Figs. 1 and 2.

Therefore, for an N-user NCRS scenario, $N+1$ rates have to be selected, splitted as N private rates and one public rate. If W denotes the available bandwidth, the ith user private rate is given by

$$R_{p_i} = \frac{W}{2}\log_2\left(1 + \frac{\lambda_i \gamma_{2,ii}}{1 + \sum_{\substack{m=1 \\ m\neq i}}^{N} \lambda_m \gamma_{2,im}}\right), \tag{8}$$

whereas the rate of the public message at the ith receiver reads as

$$R_{c_i} = \frac{W}{2}\log_2\left(1 + \frac{\sum_{m=1}^{N}(1-\lambda_m)\gamma_{2,im}}{1 + \sum_{m=1}^{N} \lambda_m \gamma_{2,im}}\right). \tag{9}$$

The public message is to be decoded by all users, so that its rate is given by

$$R_c = \min_j R_{c_j}, \tag{10}$$

Since the public message is decoded by all the users, it can be made of the aggregation of messages addressed to the different users. There is not any restriction in how to share the public information among users. The user rates are given

$$R_m = R_{p_m} + \beta_m R_c \tag{11}$$

where β_m is the portion of the public message which is assigned to the mth user and $\sum_{m=1}^{N} \beta_m = 1$. In this paper, we are going to assume that the criterion of the β_m weights is to assign the portions in such a way that user rates are as evenly distributed as possible.

3.1 Rate optimization

N optimization problems need to be solved, one for each user setting the minimum rate of the public message in (10). From this relation, we have that $R_c = R_{c_i}$ if, for $j = \{1, 2 ..., N\}$ and $j \neq i$,

$$L_{0ij} + \sum_{m=1}^{N} L_{mij}\lambda_m \geq 0 \tag{12}$$

with

$$\alpha_i = 1 + \sum_{s=1}^{N} \gamma_{2,is} \tag{13}$$

$$L_{0ij} = \frac{1}{\alpha_i} - \frac{1}{\alpha_j} \tag{14}$$

$$L_{mij} = \frac{\gamma_{2,im}}{\alpha_i} - \frac{\gamma_{2,jm}}{\alpha_j}. \tag{15}$$

With this, the N non-convex sub-problems P_i which need to be solved to maximize the sum-rate are expressed as

$$\begin{aligned} (P_i)\quad & \underset{0\leq\lambda_1,\dots,\lambda_N\leq 1}{\text{argmax}} \quad R_{c_i} + \sum_{m=1}^{N} R_{p_m} \\ \text{s.t.}\quad & L_{0ij} + \sum_{m=1}^{N} L_{mij}\lambda_m \geq 0 \\ & j = 1, 2, \dots, N, j \neq i. \end{aligned} \tag{16}$$

This sum-rate optimization cannot guarantee a fair assignation of rates. In order to prevent some users from being drastically underserved, the harmonic mean can be used as optimization metric:

$$\begin{aligned} (P_i)\quad & \underset{0\leq\lambda_1,\dots,\lambda_N\leq 1}{\text{argmax}} \quad \frac{N}{\sum_{m=1}^{N} \frac{1}{R_m}} \\ \text{s.t.}\quad & L_{0ij} + \sum_{m=1}^{N} L_{mij}\lambda_m \geq 0 \\ & j = 1, 2, \dots, N, j \neq i. \end{aligned} \tag{17}$$

For both optimization problems, the different maximization sub-problems can be pursued by applying a sequential quadratic programming method [8]. The optimal solution will be the best among the obtained solutions.

4. Competitive techniques

In order to measure the potential improvement of NCRS in a HS scenario, other techniques, either using full or partial CSIT, will be also implemented for benchmarking purposes.

4.1 Partial CSIT

In [4], ABRS has shown a potential improvement with respect to the MUD technique in [1], requiring also a simpler receiver. Therefore, ABRS is chosen as a partial CSIT benchmark and will be simulated for the 3-color and 4-color schemes. The main idea of ABRS is to pull the adjacent cold beam resources to cope with the traffic demand in the HS beam. Thus, the system capacity can be increased without requiring any additional complexity at the satellite or the receiver with respect to traditional FDM. Both ABRS and NCRS make use of the magnitude CSIT with some differences on how this information is reported; as detailed earlier, NCRS operates with the knowledge of the disaggregated power values (3) in the form of SNR and interference to noise ratios (INR), whereas ABRS only needs the signal to interference and noise ratio (SINR) at the receiver. The corresponding ABRS rates for the ith user in the 3-color and 4-color cases are given by

$$R_i = \frac{W}{3}\log_2\left(1 + \frac{\gamma_{3,ii}}{1 + \sum\limits_{m\in\mathcal{S}_i}^{N}\gamma_{3,im}}\right), \tag{18}$$

and

$$R_i = \frac{W}{4}\log_2\left(1 + \frac{\gamma_{4,ii}}{1 + \sum\limits_{m\in\mathcal{G}_i}^{N}\gamma_{4,im}}\right), \tag{19}$$

respectively, where $\mathcal{S}_i$ and $\mathcal{G}_i$ are the subset of beam indexes which are treated as noise by the ith receiver at each case.

4.2 Full CSIT

In [5], different precoders are analyzed in a full frequency reuse HS scenario. For comparison purposes, the performance precoding in [5] will be simulated. The input-output relation is expressed as

$$\mathbf{y} = \mathbf{HFx} + \mathbf{w}, \tag{20}$$

with $\mathbf{Fx}$ the precoded symbols. The precoding matrix $\mathbf{F}$ is obtained as the solution of the following non-convex optimization problem:

$$\begin{aligned} \mathbf{F}_{opt} = \underset{F}{\operatorname{argmax}} \quad & \sum_{m=1}^{N} R_m \\ \text{s.t.} \quad & \|\boldsymbol{F}_i\|^2 \leq \tfrac{P}{M} \\ & R_m = W\log_2\left(1 + \frac{|H_m f_m|^2}{\sigma^2 + \sum\limits_{l\neq m}|H_l f_l|^2}\right) \end{aligned} \tag{21}$$

with $\mathbf{F}_i$ and $\boldsymbol{f}_i$ the ith row and the ith column of the matrix $\mathbf{F}$, respectively, and $\mathbf{H}_i$ the ith row of the channel matrix $\mathbf{H}$.

In addition, the closed form MMSE precoder will be also simulated. It reads as

$$\mathbf{F} = \sqrt{\nu}\mathbf{H}^{\mathrm{H}}\left(\mathbf{H}\mathbf{H}^{\mathrm{H}} + \frac{M}{P}\sigma^2\mathbf{I}\right)^{-1} \tag{22}$$

where ν is a scalar such that the power constraint per feed is satisfied, and it is expressed as

$$\nu = \frac{P/M}{\max\left(\operatorname{diag}\left(\mathbf{H}^{\mathrm{H}}(\mathbf{H}\mathbf{H}^{\mathrm{H}})^{-2}\mathbf{H}\right)\right)}. \tag{23}$$

5. Numerical Results

The performance of the considered techniques has been tested for the different color schemes in Figs. 1 and 2, and the system parameters included in Table 1. The simulations are parameterized by G_{th}[†], the radius of the inner zone. For each different value of G_{th}, 5,000 Monte Carlo realizations have been run, with uniform distribution of users in each HS section. The total average spectral efficiency is displayed in Fig. 3 for all techniques. Please note that perfect CSI is being assumed for the precoding techniques, and perfect cancellation at the receive terminals when applying SIC in the case of NCRS.

Table 1 System Parameters.

Satellite foward link	
Diagram pattern	Provided by ESA [1]
Number of beams	7
Number of feeds	7
Feed synchronization	Perfect synchronization
Frequency band [GHz]	20
EIRP/beam	67 dBW
Far Field Noise Power Ratio (NPR)	19
Receiver Parameters	
Receiver antenna efficiency	0.65
Receiver antenna diameter	0.6 m
Receiver cloud noise temperature	280°K
Receiver terminal noise temperature	310°K
Receiver ground noise temperature	45°K
LNB Noise Figure	2 dB
Interference cancellation	Ideal cancellation

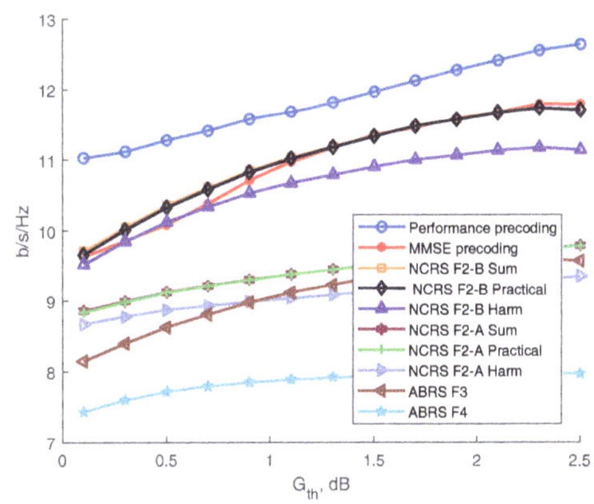

Fig. 3 Aggregated spectral efficiency versus gain threshold.

As expected, the maximum rate is achieved after the optimization of the precoding coefficients in (21), by the

[†]Note that for G_{th}=0 the inner zone would collapse into a point.

so-called *performance precoder*. MMSE performance compares with that of NCRS-B, whereas ABRS under three colors is similar to NCRS-A. Interestingly, the optimization of $\{\lambda_i\}, i = 1,2,3,4$ in (16) -to maximize the overall throughput of the group of 3 and 4 beams in Fig. 1- is such that most of the content of the message sent from the central beam should be public, whereas most of the content of the messages sent from the three green peripheral beams should be private. Therefore, some complexity can be saved by making a practical design with $(\lambda_1, \lambda_2, \lambda_3) = (0,1,1)$ and $(\lambda_1, \lambda_2, \lambda_3, \lambda_4) = (0,1,1,1)$ for the 3-user and 4-user cases, respectively. This operation point, labeled as "Practical" in Fig. 3, suffers only from a marginal degradation, within the range of 0-2%. The corresponding signals for the group of 3 beams (2-color scheme B) are written now as

$$\begin{aligned} x_1 &= \sqrt{P} x_{c_1} \\ x_2 &= \sqrt{P} x_{p_2} \\ x_3 &= \sqrt{P} x_{p_3} \end{aligned} \tag{24}$$

whereas for the group of 4 beams (2-color Scheme A) are written as

$$\begin{aligned} x_1 &= \sqrt{P} x_{c_1} \\ x_2 &= \sqrt{P} x_{p_2} \\ x_3 &= \sqrt{P} x_{p_3} \\ x_4 &= \sqrt{P} x_{p_4} \end{aligned} \tag{25}$$

where x_1 is the transmitted signal by the central beam. Note that this practical design avoids the optimization step, and reduces the complexity of the receiver: JD is not needed, and only a one SIC stage is applied. On the other side, some flexibility in the allocation of the rates is lost as the λ_i values are fixed.

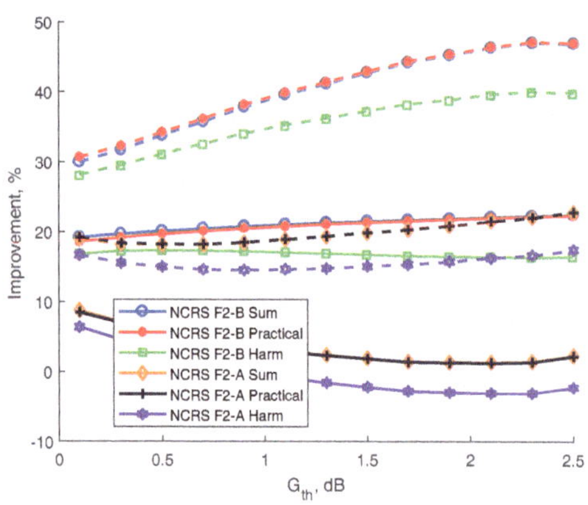

Fig. 4 Relative sum-rate performance between non-coherent schemes (NCRS) and partial CSIT benchmarks. Solid curves report the improvement over 3-color ABRS. Dashed lines report the improvement over 4-color ABRS.

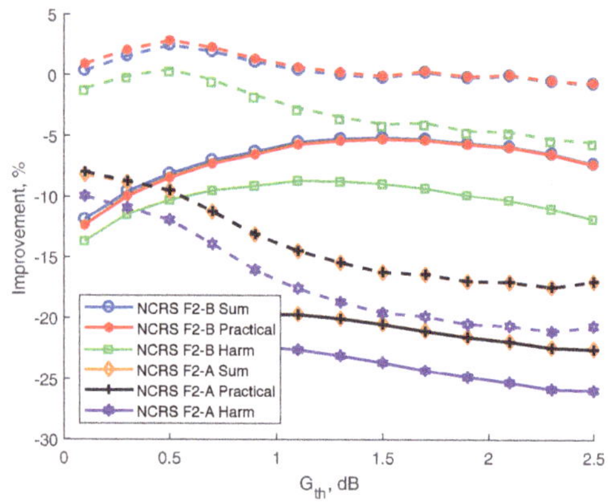

Fig. 5 Relative sum-rate performance between non-coherent schemes (NCRS) and full CSIT precoding. Dashed curves report the improvement over MMSE precoding. Solid lines report the improvement over performance precoding.

A close look at the comparison between partial CSIT schemes is depicted in Fig. 4. The improvement is significant in the case of the 2-color scheme B. However, the more modest improvement of the 2-color scheme A can be more significant from a satellite payload complexity point of view. In [3], the 2-color scheme A was simulated in a uniform traffic distribution scenario, with an improvement over the traditional 4-FDM benchmark. Without requiring any change in the frequency of the beams, NCRS is able to provide some improvement for both the uniform and uneven traffic scenario. This might not be the case for the 2-color scheme B, with more adjacent beams with the same color and, in consequence, higher co-channel interference.

NCRS is also competitive when compared with precoding at the gateway, despite the use of a lower amount of CSI. Fig. 5 shows how the *performance precoder* [5] offers the best aggregate sum-rate with some margin over the others techniques, whereas NCRS and the MMSE precoder perform quite similarly. The design of the MMSE precoder is simple given the availability of a closed form expression (22). The *performance precoding* presents high complexity since an $N \times N$ matrix needs to be found after a non-convex optimization problem. In the case of NCRS the maximum number of elements to be optimized is N. On the other side, NCRS demands more complexity from the receive terminals as compared with the benchmark receivers.

In addition to the overall rate, how the bit rate is allocated to the different users is especially relevant. We have used the Jain's fairness index [9] to compare the rate allocation results of the different schemes. The results are shown in Fig. 6 for different values of G_{th}. The use of full CSIT and a more complex optimization process, by employing the performance precoding technique in [5], can

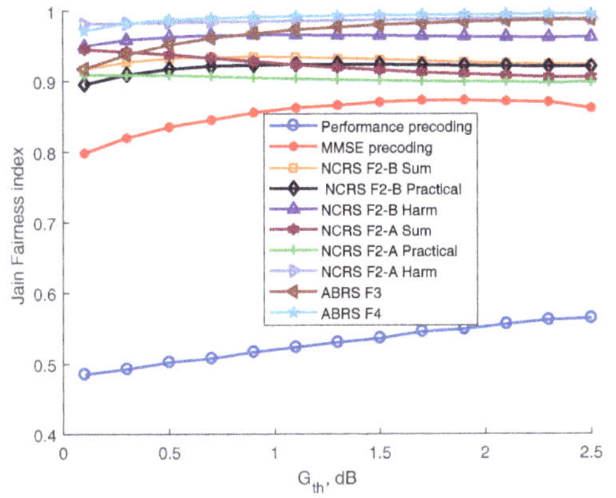

Fig. 6 Jain index versus gain threshold.

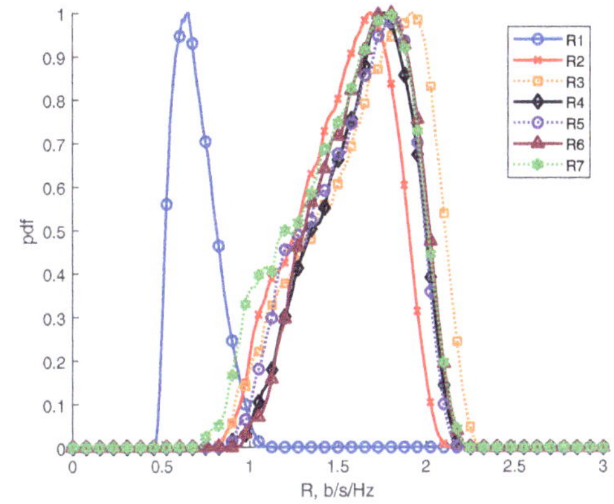

(a) 2-color B. NCRS with sum-rate optimization.

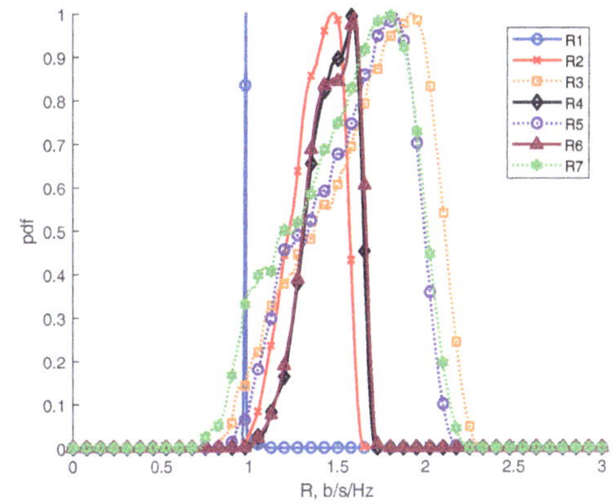

(b) 2-color B. NCRS with harmonic mean optimization.

Fig. 7 Different rate probability distribution functions for 2-color NCRS B. $G_{th} = 0.5$.

achieve a larger sum rate, at the cost of creating strong differences in the allocated rates. The required complex numerical optimization to obtain the optimal precoder in [5] can be alleviated by using the MMSE precoder, which is also more balanced when assigning rates, although still outperformed by ABRS and NCRS in terms of fairness.

The use of the harmonic mean criterion (17) serves to improve the fairness in NCRS, as concluded from Fig. 6, for a moderate degradation of the overall sum-rate (Fig. 3). This is further illustrated in Fig. 5, which displays the probability density function (pdf) of the rates for different NCRS optimization criteria, where R_i represents the rate of users located at the ith sector from Fig 1. Fig. 5 clearly depicts how the harmonic mean criterion guarantees a more fair allocation of user rates. This is especially significant for the user served at the central area of the HS, labeled as R1. Thus, by changing the optimization metric in NCRS, different sets of weights λ_i are obtained.

Overall, the rate-splitting approach of NCRS outperforms the partial CSIT benchmarks and, surprisingly, can match and even surpass the MMSE precoder in some cases. On the other side, performance precoding presents the highest aggregated sum-rate as a result of the numerical maximization of the overall rate with full CSIT. The comparison between the different techniques is summarized in Table 2.

6. Conclusions

In this paper, a rate-splitting approach has been successfully tested in a Hot-Spot scenario. For this purpose, the work in [2] and [3] has been extended to embrace the Hot-Spot case analyzed in [4]. Since NCRS has been initially designed for a two user case, a simplified scaled version for three and four users has been introduced. NCRS is able to provide a modest improvement over the partial CSIT benchmarks [4], around 20% and 40% with respect to the 3-color and 4-color ABRS, respectively, at the cost of increasing the receiver complexity. Somewhat surprisingly, NCRS achieves competitive rates with respect to the full CSIT precoding benchmarks. The practical MMSE precoder shows a close performance with respect to NCRS, despite the different CSIT requirements. However, if the precoder is designed to maximize the sum rate [5], NCRS cannot match the performance. Nonetheless, this precoder assigns quite unevenly the rates among the users. Alternatively, when it comes to the fairness, NCRS seems to be more amenable to a fair allocation of rates than precoding, specially when the harmonic mean is the optimization criterion. It should be remarked that the capacity analysis in this work does not involve any modulation and coding nor practical implementation assumptions, which will undoubtedly impact the results in the system analysis. In this regard, 3-color ABRS seems to pose a good trade-off between performance and implementation simplicity.

‡Even though strict synchronism is not required, the time misalignment among beams will require more complexity at the receive

Table 2 Main features of the compared techniques

Color scheme	Technique	Sum-Rate	Fairness	CSIT requirement	Receiver complexity	Optimization complexity	Beam synchronization
4-colors	ABRS	Poor	Good	Partial (SINR)	Low	None	Not required
3-colors	ABRS	Fair	Good	Partial (SINR)	Low	None	Not required
2-colors	NCRS-A	Fair	Good	Partial (SNR,INR)	High	Fair	Not required ‡
2-colors	NCRS-B	Good	Good	Partial (SNR,INR)	High	Fair	Not required
FFR	MMSE precoding	Good	Fair	Full	Low	None	Required
FFR	*Performance* precoding	Very good	Poor	Full	Low	High	Required

Acknowledgements

This work has been supported by European Space Agency funded activity SatNEx IV CoO2-Part 1 WI 3 "Non-Orthogonal Superposition Techniques for Multi-Beam Satellite Networks". The views of the authors of this paper do not reflect the views of ESA. This work was partially funded by the Agencia Estatal de Investigacion (Spain) and the European Regional Development Fund (ERDF) through the projects TERESA (TEC2017-90093-C3-1-R) and MYRADA (TEC2016-75103-C2-2-R). Funded also by the Catalan Government under project 2017-SGR-1479, and Galician Government under project ED431C 2017/53.

References

[1] R.D. Gaudenzi, N. Alagha, M. Angelone, and G. Gallinaro. 'Exploiting Code division Multiplexing with Decentralized Multiuser Detection in the Satellite Multibeam Forward Link '. *International Journal of Satellite Communications and Networking*, 36(3):239–276.

[2] M. Caus, A. Pastore, T. Ramírez, C. Mosquera, and N. Noels. 'Exploratory Analysis of Superposition Coding and Rate-Splitting for Multibeam Satellite Systems'. In *15th International Symposium on Wireless Communications Systems*, Aug 2018.

[3] T. Ramírez, C. Mosquera, M. Caus, A. Pastore, M. Navarro, and N. Noels. 'Message-Splitting for Interference Cancellation in Multibeam Satellite Systems'. In *9th Advanced Satellite Multimedia Systems Conference (ASMS) and 15th Signal Processing for Space Communications Workshop (SPSC)*, Berlin, Germany, 2018.

[4] N. Alagha. 'Adjacent Beams Resource Sharing to Serve Hot Spots'. In *35th AIAA International Communications Satellite Systems Conference, International Communications Satellite Systems Conferences (ICSSC)*, Oct 2017.

[5] G. Taricco and A. Ginesi. Precoding for flexible high throughput satellites: Hot-spot scenario. *IEEE Transactions on Broadcasting*, pages 1–8, 2018.

[6] N. Noels, M. Moeneclaey, T. Ramírez, C. Mosquera, M. Caus, and Adriano Pastore. 'Symbol-asynchronous transmission in multibeam satellite user down-link: rate regions for novel superposition coding schemes'. In *2018 6th IEEE Global Conference on Signal and Information Processing*, Anaheim, USA, 2018.

[7] A. Gholami Davoodi and S. A. Jafar. 'Transmitter Cooperation Under Finite Precision CSIT: A GDoF Perspective'. *IEEE Transactions on Information Theory*, 63(9):6020–6030, Sept 2017.

[8] J. Nocedal and S. J. Wright. *Numerical Optimization*. Springer, New York, NY, USA, second edition, 2006.

[9] Raj Jain, Dah-Ming Chiu, and W. Hawe. 'A quantitative measure of fairness and discrimination for resource allocation in shared computer systems'. tech. rep., Digital Equipment Corporation, DEC-TR-301, 1984.

terminals [6]. Alternatively, the performance of NCRS can be reduced [3].

Section 2 – Cognitive Communications and Propagation Channel Optimization

Time Correlation Used to Improve Time Diversity Gain of Rainfall Prediction

Peeramed Chodkaveekityada[1]
Faculty of Engineering, King Mongkut's Institute of Technology Ladkrabang, Bangkok, 10520, Thailand

Hajime Fukuchi[2]
Tokyo Metropolitan University, Tokyo, 191-0065, Japan

Correlation of the rainfall rate with time depends on geography, environment, and the type of rainfall. In this letter, we consider the use of such a correlation to improve the time diversity gain of a system that predicts rainfall over Japan. From a rain may with 4 years of data, we selected 43 locations throughout mainland Japan, and from these data, we derived the cumulative time percentage, the diversity gain from a time diversity method using 10 to 120 min delays, and the correlation from a time series of rainfall. This information is used to improve the prediction of rainfall. Root-mean-square errors show an improvement in the diversity gain.

Nomenclature

G_{t_d} = diversity gain
R_{CDF} = rainfall rate
β = estimated parameters
ρ_{t_d} = correlation value
$\tau_{\rho 0.5}$ = approximation coefficient of diversity gain

1. Introduction

Time diversity [1-3] has been investigated as a way to reduce rain attenuation of non-real-time satellite communications, with the aim of improving future high-throughput satellite communications. It has been demonstrated that this approach is highly effective in obtaining a power margin that is higher than that obtained using other methods. Time diversity techniques are based on the repeated transmission of data, and this reduces the effects of rain. These require a ground receiver system with large-capacity memory, since it will sometimes be necessary to store information until reception is complete.

1

Lecturer, Telecommunication Engineering Department, peeramed.ch@kmitl.ac.th, Not a Member.

2

Professor, Department of Aerospace Engineering, fuku@tmu.ac.jp, Senior Member.

Rain attenuation is an important problem in satellite communication systems, especially for frequencies above the Ku-band. For this reason, it is important to have information about rainfall rates. Data from rain gauges are reliable, but can only provide information for localized conditions. Rain radar data are valuable, since they allow the incorporation of information over a wider area [4,5].

In a previous study [6], we used rain radar data to develop a time diversity method based on 15 locations on mainland Japan. The results show that the behaviour of the rainfall was different at each of the locations considered; for example, even if two locations had the same average rainfall rates for the 0.1% or 0.01% cumulative times, there would be differences in the time diversity gains. Because of this, the time correlation plays a significant role in causing different rain behavior.

2. Data Description

A rain radar map of Japan is produced by the Japan Meteorological Agency. The rain radar data are added to data from a rain gauge network of more than 1300 stations in order to increase the precision of the estimated rainfall rate. The mesh size of the radar is approximately 1 km × 1 km, and at 5 min intervals. We considered data spanning 4 years from July 2009 to June 2013, and we evaluated the time correlation. We used this to add diversity gain and thus improve the formula for predicting rain attenuation for mitigation technologies, especially those that use a time diversity method. In addition, we selected 43 points throughout mainland Japan (see Fig. 1) and derived the time correlation at each observation point; this was then used to express the behaviour of rainfall for that location.

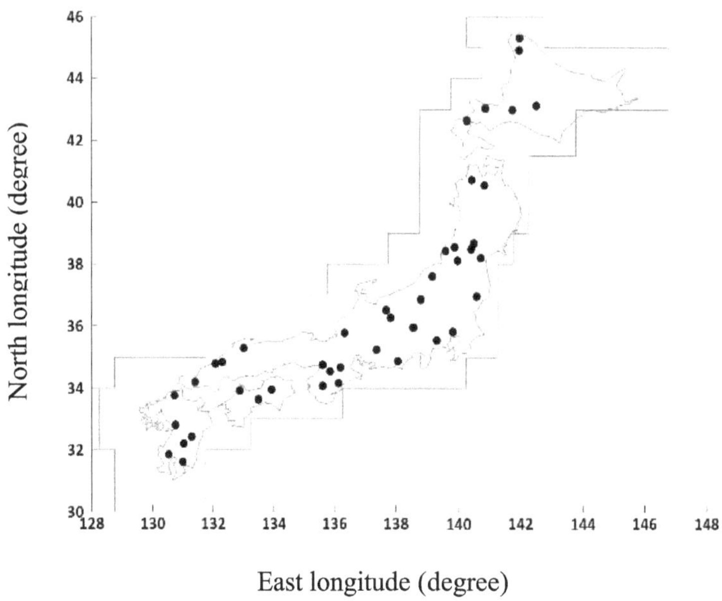

Figure 1. Forty-three observation locations throughout mainland Japan.

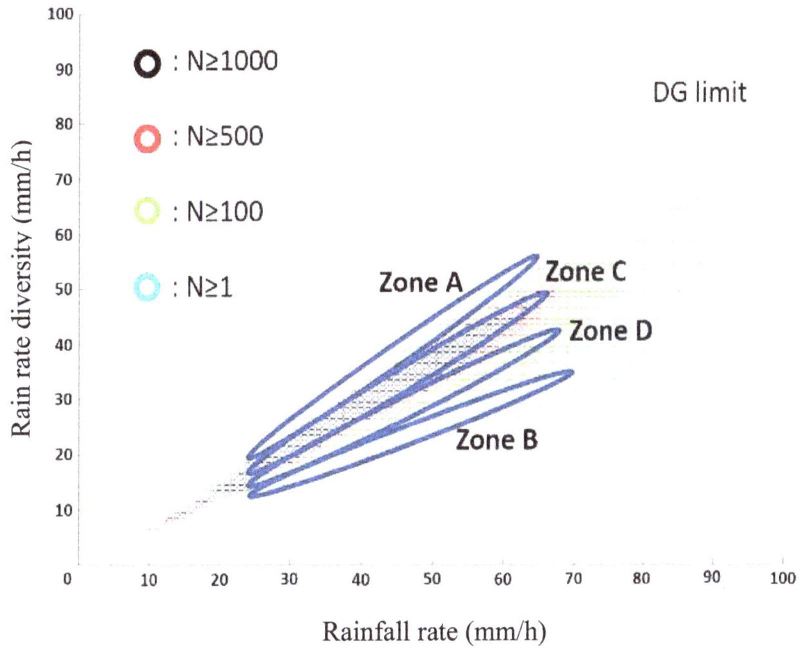

Figure 2. Scatter plot of all grid locations from rain radar map at 0.01% of cumulative time percentage with 120 min delay.

3. Methodology and Results

As mentioned above, we derived the time diversity property from rain radar data and found some interesting points that can be used to improve the predicted diversity gain. Fig. 2 shows a scatter plot of the rain rate diversity gain versus the rainfall rate for a time delay of 120 min. The colour of each point indicates the number of locations, N, corresponding to that point. It can be seen that when the time delay is large, the slope closely approaches that of the diversity gain line, and when the time delay is short, the slope is less.

To consider the behaviour rainfall throughout Japan, we separated it into 4 zones, as shown in Fig. 2. Zone A represents a few locations for which the data are close to the limit, Zone B represents a few locations for which the data are far from the limit, Zone C represents locations for which the prediction was improved by the time diversity method, and Zone D represents the remaining locations (that is, those between Zones B and C). We started with 15 locations and found that most of these were located in Zone C; we then randomly selected an additional 28 locations in order to evaluate the rainfall behaviour in all zones.

Fig. 3 shows an example of cumulative time percentage for locations A06 and B04, which were 2 of the 43 observations for which the time delay was either 0 or 120 min. When there was no time delay, the average rainfall rate for A06 was somewhat lower than it was for B04 when the cumulative time exceeded 0.01%, but when the cumulative time was only 0.001%, the rainfall rate was greater. On the other hand, when we applied the time diversity method with a delay of 120 min, the rainfall rate at B04 was greater than that at A06. This shows that the time correlation of the rainfall rate for each area influences the predicted diversity gain.

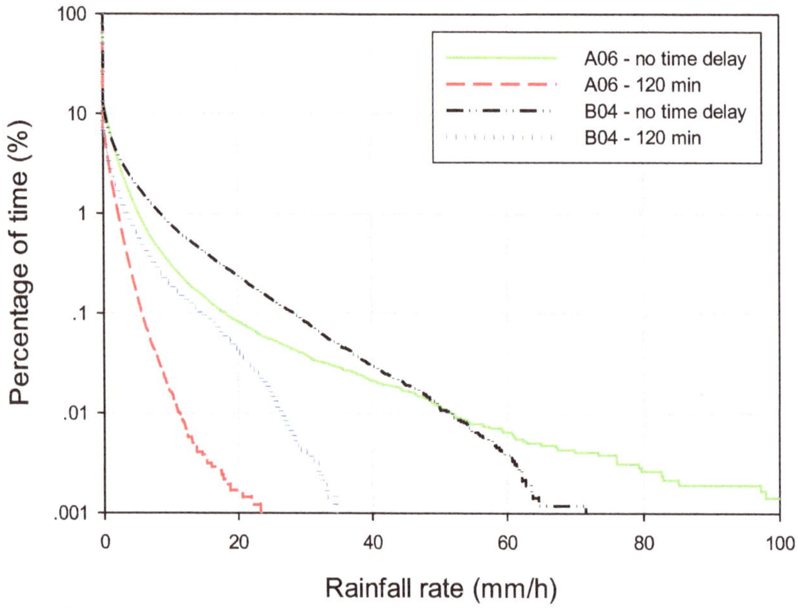

Figure 3. Example of cumulative time percentage of rainfall rate at A06 and B04, for time delays of 0 and 120 min.

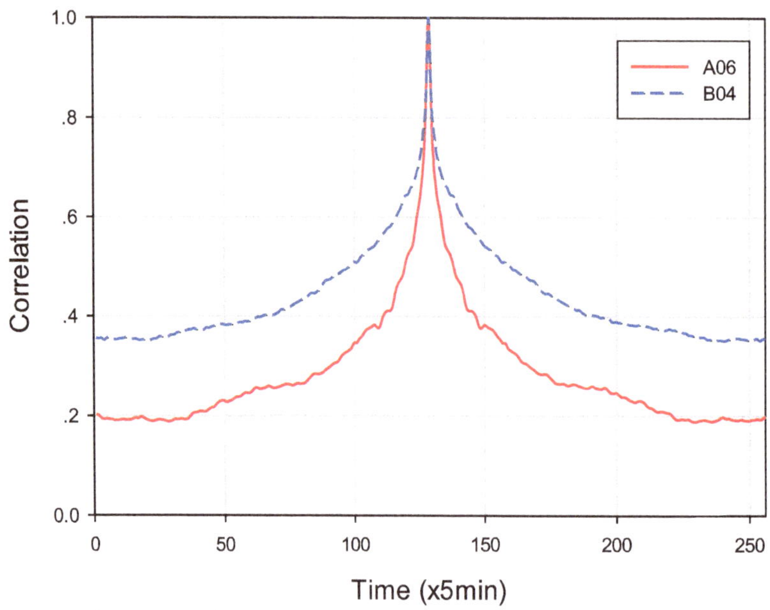

Figure 4. Example of time correlation of A06 and B04.

To analyse the time correlation factor, we collected time series of 256 points at 5 min intervals. Each set contained rainfall data for at least 128 points, and we set the threshold for collecting the rainfall rate to 10 mm/h. We obtained the fast Fourier transform of the rainfall data and the Wiener-Khinchin relation of the time correlation; this is shown in Fig. 4 for A06 and B04. The results show clearly that the rain behaviour is different in these two locations.

4. Improvement of Diversity Gain Formulas

For the time correlation results in Fig. 4, we considered the use of two parameters to improve the diversity gain prediction; these were the time and the correlation. The time parameter was defined to be the duration of time (min) from over which the correlation value of ρ decreased to 0.5, and the correlation was evaluated for each of the time delays. The relationship formulas are as follows:

$$G_{t_d} = \beta_0 + \beta_1 R_{CDF} \tag{1}$$

$$G_{t_d} = \beta_0 + \beta_1 R_{CDF} + \beta_2 \rho_{t_d} \tag{2}$$

$$G_{t_d} = \beta_0 + \beta_1 R_{CDF} + \beta_2 \tau_{\rho 0.5} \tag{3}$$

where G_{t_d} is the diversity gain in mm/h; R_{CDF} is the rainfall rate in mm/h at a particular point in the cumulative distribution function; β_0 , β_1 and β_2 are estimated parameters; ρ_{t_d} is the correlation value at a particular point with a given time delay; and $\tau_{\rho 0.5}$ is the time in minutes at which the correlation equals 0.5.

Note that (1) is the conventional relationship between diversity gain and rainfall rate; (2) is our proposed relationship between diversity gain, rainfall rate, and correlation value for a given time delay; and (3) is our proposed relationship between diversity gain, rainfall rate, and time at which the correlation value equals 0.5. We estimated the parameters from the data for the 43 observation points, and the resulting improvement is evaluated by the root-mean-square errors, which are listed in Table 1. Note that the relationship in (3) has the smallest error when the time is considered

Table 1. Root mean square error of the prediction formulas (1) to (3).

CDF	Time delays (t_d)	Eq. (1)	Eq. (2)	Eq. (3)
0.1%	60 min	1.1504	0.8973	0.8206
	120 min	1.1781	0.9151	0.8148
0.01%	60 min	4.1606	3.1952	3.0133
	120 min	4.6397	3.4646	3.3836

5. Conclusion

We have presented an improved formula for predicting the diversity gain for use in a time diversity method; this is based on the time correlation derived from time series of radar data for rainfall over 4 years in Japan. The results show that the use of the time correlation of the rainfall rate significantly improves the methods for mitigating rain attenuation; it does this by boosting the power margin for each area. We note that increasing the diversity of sites or using an adaptive power control method could also improve the time correlation factor.

References

[1] H. Fukuchi and T. Nakayama, "Quantitative evaluation of time diversity as a novel attenuation mitigation technology for future high speed satellite communication," IEICE Trans. Commun., vol.E.87-B, no. 8, 2004, pp. 2119-2123.

[2] C. Capsoni, M. D. Amico and R. Nebuloni, "Performance of time diversity satellite communication systems investigated through radar simulation," Proc. European Con. Antennas Propagat., 2007.

[3] P. Chodkaveekityada and H. Fukuchi, "Time diversity evaluation for attenuation mitigation method using attenuation data in Thailand and Japan," Int. J. Satell. Commun. Network, 2016, doi: 10.1002/sat.1184.

[4] J. X. Yeo, Y. H. Lee and J. T. Ong, "Performance of site diversity investigated through radar derived results," IEEE Trans. Antennas Propagat., vol. 59, no. 10, 2011, pp. 3890-3898.

[5] C. Capsoni, M. D. Amico and R. Nebuloni, "Radar simulation and physical modelling of time diversity satellite systems," Radio Sci., vol. 44, RS4009, 2009, doi: 10.1029/2009RS004142.

[6] P. Chodkaveekityada and H. Fukuchi, "Prediction model of time diversity using Japan rain radar data," Int. J. Satell. Commun. Network, 2016, doi: 10.1002/sat.1182.

PROTECTION OF THE MOBILE STATION FROM THE INTERFERENCE BY MARITIME EARTH STATION IN MOTION IN THE 28 GHZ BAND

Daesub Oh, Jongmin Park

Radio & Satellite Research Division, ETRI, Daejeon, Republic of Korea
trap@etri.re.kr

Keywords: ESIM, FSS, MS, Satellite system, Frequency sharing

Abstract

The Earth Stations in Motion (ESIM) system are one of the most modern satellite systems, providing high-speed and high-capacity communication with mobility. ESIM in the Ka-band should be operated under the condition ensuring the protection of existing services in the same frequency band. Therefore compatibility studies of ESIM with incumbent services in the same frequency band should be conducted before authorization, licence or operation of EISM. This paper provides a methodology to assess the interference from maritime ESIM to an MS stations in the 28 GHz frequency band. Taking into account a point-to-point interference scenario, the simulation results show the separation distances between maritime ESIM and MS station for protecting the MS stations. We considered the heights of station, elevation angle and time percentage in calculating the interference to reflect various factors affecting the interference level. The separation distance of up to 155 km would be required to ensure the protection of MS receiver taking the results of simulation into consideration.

1 Introduction

In modern satellite communications, Earth Stations in Motion (ESIM) are the one of the most modern satellite technological developments and were designed for use on aircraft, ships, and land vehicles. ESIM are small sized terminal with high-precision tracking capabilities associated with state-of-the-art Ka-band satellites providing high-power multiple spot beam coverage, allowing transmission rates on the order of 10-50 Mbits/s [1]. Owing to these characteristics, ESIM can be operated in the FSS frequency band although ESIM are a kind of mobile earth stations which should be operated in the MSS frequency bands.

Some satellite operators have a plan for ESIM system to transmit the signal in the 27.5 – 29.5 GHz frequency bands in which the terrestrial services including mobile service are also allocated. Therefore, we expect that the receiving mobile stations will suffer from the radio interference caused by the transmit ESIM system in the same frequency band. Since the ESIM are subject to be operated on a non-interference and non-protection principle with respect to other existing services allocated in these bands, appropriate protection criteria should be established between ESIM system and receiving MS stations as a victim operating in the co-channel frequency band.

There are three types of ESIM: aeronautical (A-ESIM), maritime (M-ESIM) and land (L-ESIM), depending on which vehicle they are installed. Each type of ESIM would require different regulatory and/or technical measures to protect existing services [2]. In the International Telecommunication Union Radiocommunication Sector (ITU-R), the operational, technical and regulatory measures for ESIM operation in 27.5 – 29.5 GHz frequency band have been discussed including sharing studies with respect to the existing services under the WRC-19 agenda item 1.5 [3].

This paper conducts a sharing study between transmitting M-ESIM and MS victim station by means of the computation the interference levels and analyses the separation distances not to exceed the unacceptable interference. Taking into account the results, this paper proposes separation distances as a protection criteria with respect to the victim MS station from transmitting M-ESIM.

2. Interference Scenario

The interference scenario between M-ESIM and mobile service (MS) stations which consist of base station (BS) and user equipment (UE) is shown in Figure 1. It is assumed that the satellite (space station), M-ESIM and mobile stations are located on line (at same longitude), therefore the azimuth direction of M-ESIM transmission communicating GSO satellite is toward mobile stations. The azimuth direction of the BS and UE antenna of MS is also toward the M-ESIM with maximum antenna gain taking into account the worst case interference scenario.

When M-ESIM move toward MS station, the azimuth direction of M-ESIM antenna is retained to MS station with fixed elevation angle. On moving of M-ESIM, the interference levels are calculated at each position, and the separation distances are determined by comparing the interference levels with acceptable interference level based on the I/N = - 6 dB.

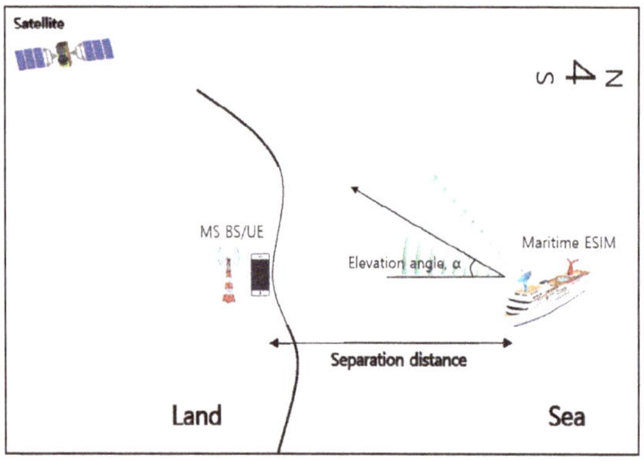

Figure 1. Interference scenario

As a propagation loss mechanism in the interference scenario, we adopt the Recommendation ITU-R P. 452 model with gaseous attenuation (A_g) taking account of sea path between interferer and interfered stations [4]. The path loss between the M-ESIM system and MS station with frequency f (GHz) and distance r (km) is expressed as follows.

$$P_L(r) = 92.5 + 20\,log(f) + 20\,log(r) + A_g \quad (1)$$

The gaseous attenuation consists of dry-air and water-vapour attenuations is considered based on the Recommendation ITU-R P.676 [5]. Assuming that the water-vapour density is 7.5 g/m^3, the frequency is 28.5 GHz and the height of the MS station is 50 m, A_g becomes

$$A_g \approx 0.107 \times r\,. \quad (2)$$

3 System Characteristics

In general, M-ESIM communicate with the GSO satellite station with elevation angle, so off-axis emission to the horizon would make the radio interference to the terrestrial MS station. In the ITU-R, the off-axis e.i.r.p. density mask in 40 kHz for ESIM has been suggested as a transmitting ESIM's characteristics for sharing studies between ESIM and incumbent services operating in the 28 GHz frequency band [6]. The shape of this off-axis e.i.r.p. density mask is shown in the Figure 2 and the Table 1 provides formula of this for assessing the interference level from M-ESIM to receiving MS station.

In this interference scenario, the antenna height of M-ESIM is set 40 m above sea level and the M-ESIM is operated with center frequency of 28.5 GHz and frequency bandwidth of 100 MHz.

With respect to the MS station as interfered station, both BS and UE are considered respectively. MS station would be located at the edge of land or shore shown in Figure 1, so there is no obstacle in the radio path between M-ESIM and MS station. The each antenna height of BS and UE is assumed 20 m and 1.5 m above terrain height respectively. The MS station is operated with the same center frequency band and bandwidth of M-ESIM. The system characteristics of MS system are shown in Table 2 [7].

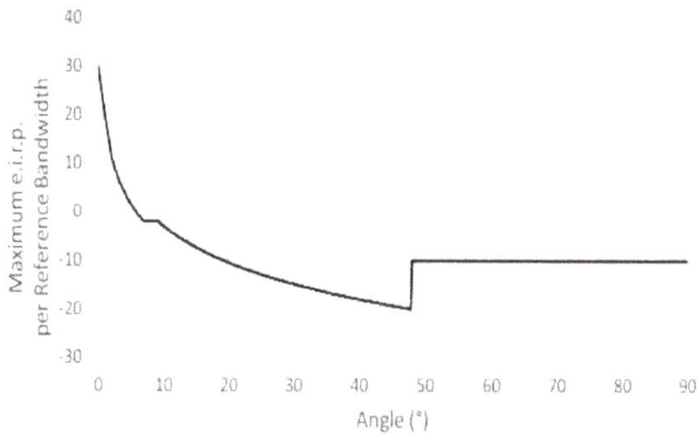

Figure 2. Off-Axis e.i.r.p. density mask of M-ESIM

Table 1. The formula of off-axis e.i.r.p. density of M-ESIM

Angle θ	Maximum e.i.r.p. per 40 kHz
$2° \leq \theta \leq 7°$	$(19 - 25\log\theta)$ dB (W/40 kHz)
$7° < \theta \leq 9.2°$	−2 dB (W/40 kHz)
$9.2° < \theta \leq 48°$	$(22 - 25\log\theta)$ dB (W/40 kHz)
$48° < \theta \leq 180°$	−10 dB (W/40 kHz)

Table 2. The system characteristics of MS system

Characteristics	Mobile system	
	BS	UE
Frequency range (GHz)	27.5-29.5	
Receiver bandwidth (MHz)	100	
Antenna pattern type	Directional	
Antenna polarization	Linear	
Peak antenna gain (dBi)	29	20
Antenna height (m)	20	1.5
Receiver noise figure (dB)	6	6
Protection Criterion (I/N) (dB)	-6	
Base station antenna downtilt (degrees)	10	

4 Results

In the interference scenario between M-ESIM and MS station described in previous section, the interference levels at the MS station is calculated taking into account the mobility of M-ESIM. In order to consider various factors which can affect the interference level and separation distances as

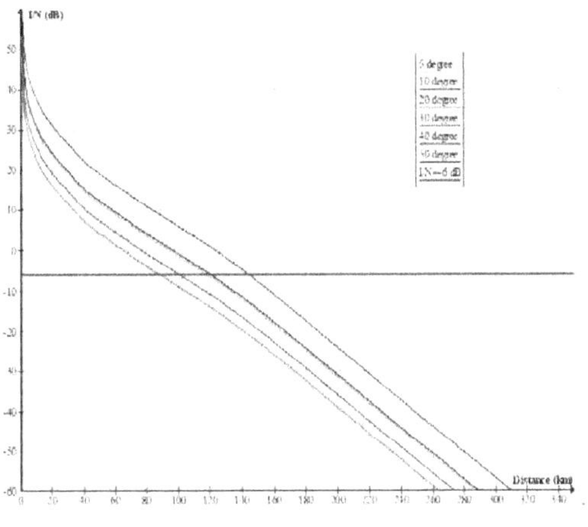

(a)

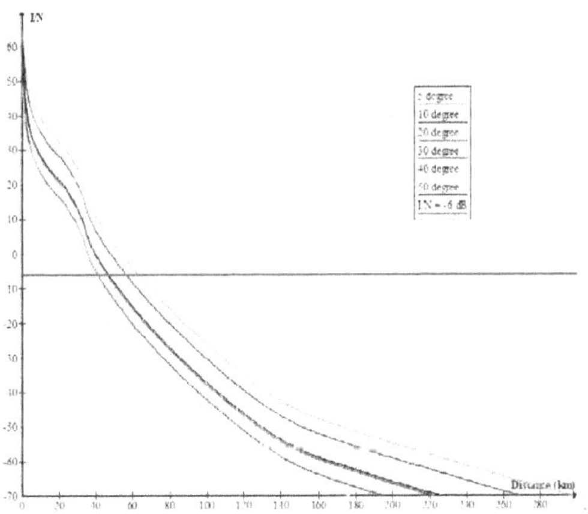

(b)

Figure 3. I/N results for base station of MS

(a) p = 1 %

(b) p = 20%

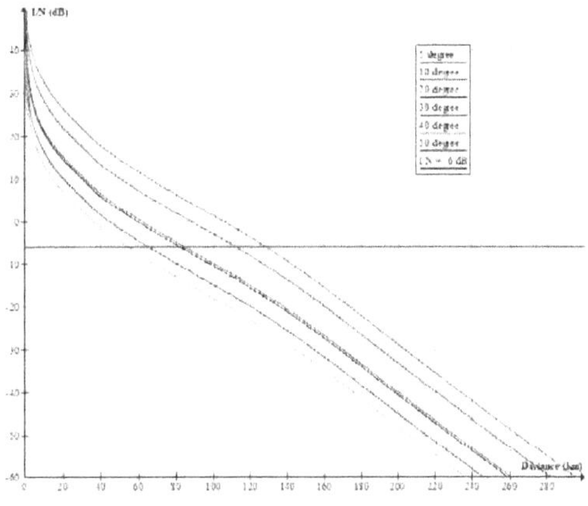

(a)

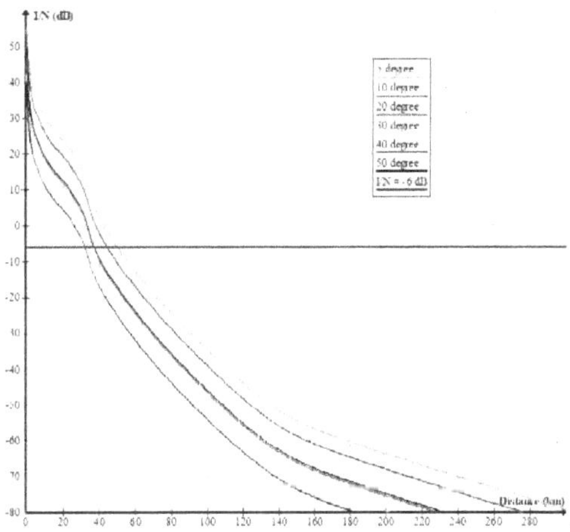

(b)

Figure 4. I/N results for user equipment of MS

(a) p = 1 %

(b) p = 20%

protection criteria, the different location and elevation angle of M-ESIM, two terrain heights of victim stations and time percentage of exceeding the interference level are applied. The time percentage of exceeding the interference level, *p*, is considered in the calculation. We look over the simulation results for 1%, 10% and 20% of time percentage respectively, the time percentage of exceeding the interference level is also dominant factor to determine the interference level and separation distance for protecting the victim station.

Figure 3 shows the results of I/N values according to the distances between BS of MS and M-ESIM and elevation angle. The I/N values are calculated taking into account the time percentage of 1% (short term) and 20% (long term) at the terrain height of 50 m. Since time percentage implies that the probability of unacceptable interference is more than a certain level, the lower time percentage requires the more stringent protection measures for victim receiver. The results between UE of MS and M-ESIM carried out in the same manner as the case of BS and M-ESIM are shown in Figure 4.

In these graphs of the Figure 3 and 4, the interference (I/N) toward MS station would be increased at the low elevation angle of M-ESIM antenna because low elevation angle has high e.i.r.p. density level form the Table 1. We expect easily that long distance between two stations can reduce the interference level due to the propagation loss. At the I/N = - 6 dB as protection criterion, the separation distances are determined to ensure the MS operation. Depending on the system or environment parameters such as elevation angle, time percentage and location of station, the distances are varied. The table 3 and 4 summarize the separation distance to meet the I/N = -6 dB taking into account several different parameters. In the worst case, the required separation distance is up to 155 km between two stations.

In order to ensure the protection with respect to the MS stations, the separation distance as protection criterion needs to be determined taking into account the interference environment and results of this study.

Table 3. Separation distances to BS of MS

I/N = -6 dB	Elevation angle of ESIM (degree)	Separation distance (km) Time percentage 1 %	10 %	20 %
BS (Terrain height 50m)	5	152	104	83
	10	139	95	78
	20	114	86	70
	30	98	85	66
	40	86	79	62
	48 - 90	116	86	70
BS (Terrain height 100m)	5	155	110	91
	10	140	104	86
	20	115	101	78
	30	100	92	73
	40	90	80	70
	48 – 90	117	101	79

Table 4. Separation distances to UE of MS

I/N = -6 dB	Elevation angle of ESIM (degree)	Separation distance (km) Time percentage 1 %	10 %	20 %
UE (Terrain height 50m)	5	127	87	70
	10	111	79	65
	20	83	74	58
	30	67	58	53
	40	56	48	45
	48 - 90	85	76	58
UE (Terrain height 100m)	5	125	98	78
	10	109	96	74
	20	84	74	67
	30	67	58	56
	40	56	48	46
	48 - 90	86	76	67

5 Conclusions

This paper proposed the separation distances to protect MS stations from the interference of M-ESIM in 28 GHz frequency band. This study results are very important to determine the operational requirement of M-ESIM system because the M-ESIM system should be operated under the condition that is to ensuring the protection of the existing other radio station including MS station. It means that the ESIM operators or administrations which allow its operation in their territory should keep the separation distance unless there is an agreement with affected administrations.

As we can see the study results, the separation distances would be changed according to an assumption of interference scenario, location of station and time percentage of exceeding the interference level. We expect that the long separation distance would reduce the coverage of ESIM operation whereas, the short separation distance would increase the possibility of interference to victim station.

Therefore, the appropriate assumption including time percentage should be considered for frequency sharing between M-ESIM and terrestrial mobile system.

Acknowledgements

This work was supported by Institute for Information & communications Technology Promotion (IITP) grant funded by the Korea government (MSIP) (2016-0-00026, Development on spectrum assessment technique for utilizing the national satellite frequency resources)

References

[1] http://www.inmarsat.com/blog/earth-stations-in-motion-in-fss-kaband.

[2] Dae-Sub Oh, Young-Houng Kang, "Effective Interference Assessment from Earth Station in Motion to Fixed Station in the 28 GHz Band," Proceedings of APMC, Kuala Lumpur, 2017

[3] International Telecommunication Union, "Agenda for the 2019 World Radiocommunication Conference," Resolution 809, Nov. 2015

[4] International Telecommunication Union, "Prediction procedure for the evaluation of interference between stations on the surface of the Earth at frequencies above about 0.1 GHz," Recommendation ITU-R P. 452, July 2015.

[5] International Telecommunication Union, "Attenuation by atmospheric gases," Recommendation ITU-R P. 672, Sep. 2016.

[6] International Telecommunication Union, "Use of the frequency bands 19.7-20.2 GHz and 29.5-30 GHz by earth stations in motion communicating with geostationary space stations in the fixed-satellite service," Resolution 156, Nov. 2015

[7] International Telecommunication Union, "Sharing and compatibility between earth stations in motion operating with geostationary FSS networks and current and planned stations of the MS in the frequency band 27.5-29.5 GHz," Annex 14 to Document 4A/826, Aug. 2018

RATELESS CODES FOR SATELLITE SYSTEMS OVER RAIN FADING CHANNELS

***Satya Chan*[1], *Meixiang Zhang*[2], *Daesub Oh*[3], *Sooyoung Kim*[1*]**

[1]*Div. of Electronics Eng., Chonbuk National University, 567 Baekje-daero, Jeonju, Korea*
[2]*College of Information Engineering, Yangzhou Univ., 196 Huayang West road, Yangzhou, China*
[3]*Radio & Satellite Research Division, ETRI, 218 Gajeong-ro, Daejon, Korea*
**sookim@jbnu.ac.kr*

Keywords: rateless codes, satellite broadcasting, rain fading, adaptive transmission

Abstract

Modern satellite systems utilize high frequency bands usually above 15 GHz to provide high speed transmission services, and adopt adaptive modulation and coding (AMC) schemes in order to countermeasure serious rain fading occurred in the utilized frequency bands. This paper introduces an efficient scheme to countermeasure rain fading by using rateless codes. The proposed scheme utilizes existing AMC scheme with the LDPC codes, and thus it can be easily combined with existing standards. The error rate and spectral efficiency performance simulation results demonstrate performance enhancement over a satellite rain fading channel, with almost the same code rate as the existing LDPC codes.

1. Introduction

Low-density parity check (LDPC) code is a forward error correction (FEC) coding scheme, and it is known to provide a capacity limiting performance by virtue of soft iterative decoding process at the receiver. For this reason, it was defined in many modern communication and broadcasting standards, including digital video broadcasting via satellite – 2nd generation (DVB-S2) and its extension (DVB-S2X) [1]. The DVB-S2 and DVB-S2X define two kinds of FEC frames which are the normal frame with a length of 64800 bits and short frame with a length of 16200 bits, and they also define adaptive coding and modulation (ACM) scheme to counteract time-varying channel conditions. Eleven different code rates for the normal frame and ten rates for the short frame are defined for ACM operation [2]. Even though, multiple code rates of the LDPC codes can be utilized in the ACM scheme, LDPC codes do not have their own rate compatibilities. This makes it very difficult for LDPC codes to be applied to incremental redundancy applications, such as hybrid automatic repeat request (HARQ) [3].

Luby transform codes (LT codes) are the first class of practical fountain codes that are near-optimal erasure correcting codes. LT codes are rateless because the encoding algorithm can in principle produce an infinite number of message packets [4]. By taking advantages of its unlimited code rate generation capability, previous studies investigated application of rateless codes an effective retransmission means to counteract channel fading [3][5-7]. A study in [5] proved that rateless codes could be used as an effective means for incremental redundancies, and the simulation results showed that the rateless-coded HARQ outperformed at low signal-to-noise ratio (SNR) compared to the conventional AMC with HARQ. As an example of rateless-coded HARQ, our previous study proposed a HARQ scheme with rateless LDPC codes, with an application example for satellite broadcasting systems [3]. Theoretical analysis showed that combining existing AMC with HARQ with rateless codes could achieve quasi error-free condition [6]. More specific study results showed that adding HARQ on top of AMC might be counter-productive in the high average SNR regime for fast fading channels, while HARQ was useful for slow fading channels [7].

Motivated by these results, we propose to apply rateless codes in combination with the existing LDPC codes to compensate channel fading. Main target to compensate in satellite channel should be slow-fading, and we should consider very low SNR condition in a severely rain faded situation. Therefore, HARQ with rateless codes can be a powerful means to counteract rain fading in satellite system. As an extension of our previous works in [3], the proposed scheme achieves rate-compatibilities of the LDPC codes by concatenating the existing LDPC codes with LT codes. In such a way, efficient joint-soft-iterative decoding algorithm can be applied at the receiving end, and high performance gain can be achieved.

After this introduction, this paper is organized as follows. Section 2 describes the effect of rain fading in satellite system and operational principles of the conventional satellite system with ACM to counteract channel fading. Section 3 presents the proposed schemes. After describing the operational principles of the HARQ scheme with rateless codes, the performance enhancement mechanism with joint-soft-iterative decoding algorithm is presented. Section 4 demonstrates the simulation results over a rain fading channel, with description of simulation model. Finally conclusions are drawn in Section 5.

2. Satellite system with ACM

ACM scheme is mandatory for the satellite systems operating at higher frequency bands over 10 GHz, in order to suitably counteract severe rain fading. A system with ACM employs spectrally efficient transmission schemes in good channel conditions and switches them to power-efficient schemes in bad channel conditions [8]. In this way, the system is supposed to allocate an ACM mode with lower code rate and lower-order modulation scheme to a faded user. The system with ACM requires to investigate the history of the received signal quality and to predict it for the optimum allocation of the resources.

The control mechanism consists of estimation of fading, prediction of fading, and mode selection. The prediction, on the receiving side, must consider the round trip delay over a satellite link. The updating interval of the mode allocation can be no lower than two times of round-trip delay. Mode selection adaptively allocates suitable transmission schemes, such as coding and modulation, to both the transmission and receiving sides. Additionally, a power margin may be adaptively applied to compensate for any error that may occur due to the estimation, prediction, or mode selection errors [9].

In the prediction algorithm, discrete-time low-pass filtering (LPF) is usually used to eliminate scintillation effect because rain fading variation by time due to scintillation is much faster than the response speed of an adaptive system. The predicted signal level, $\hat{y}_{t+p}$, after a prediction time, p, using the estimated signal levels can be represented as follows, using the following generic linear regressive filtering:

$$\tilde{y}_{t+p} = \sum_{i=0}^{n-1} w_{i,t} y_{t-i}, \tag{1}$$

where n is the observation time, $w_{i,t}$ is the weighting factor, and y_t is the estimated signal level at time t. One of the simplest implementations of (1) uses two constant weight values for two end points of the observation period, $w_{0,t} = p/(n-1)+1$ and $w_{n-1,t} = -p/(n-1)$ and assumes that future variation of the signal level will remain the same as the previous variation [9]. A simple prediction algorithm can be implemented by using a simple first-order low pass filter (LPF), i.e., when n=2 in (1), as follows:

$$\hat{y}_{t+1} = 2y_t - y_{t-1}. \tag{2}$$

The mean-error correction with a marginal value can be used to compensate for prediction error. In addition, a resource allocation algorithm is required to select an appropriate transmission scheme with the best spectral efficiency and performance based on the predicted signal level. Figure 1 is the block diagram showing operational principles of the ACM system to compensate rain fading [8].

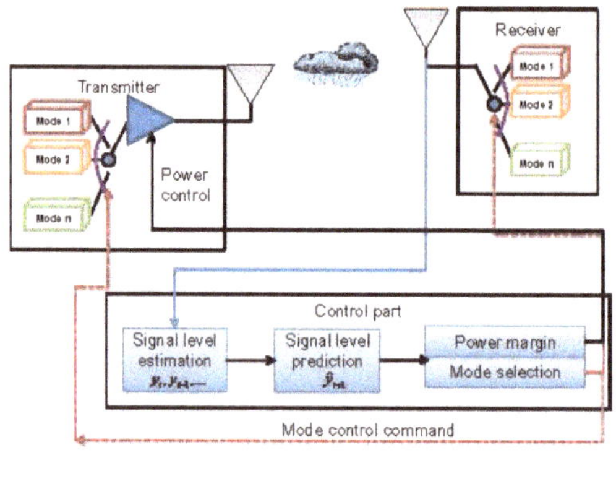

Fig. 1 Operation of ACM system compensating rain fading

3. Proposed scheme

3.1 Operational principle

Even though data transmission with ACM can be effectively countermeasure the rain fading, prediction and allocation error may cause performance degradation. In this case, utilization of rateless codes can be an effective solution. Having the LDPC codes used in the ACM operation as it is, we can generate additional parities by considering the codeword of the LDPC codes as a systematic part of the LT codes. Figure 2 shows operational principles of the proposed system.

Referring to Fig. 2, the system allocates suitable ACM mode based on the predicted signal level. For the given ACM mode, the LDPC encoder generates the codeword **U**, with the information word, **u**. If decoding of the received information succeeds, the estimated code rate of the LT code is subjected to be 1 at the next transmission flow, indicating that no LT parity transmission is required. Next information is then transmitted based on the next predicted signal level.

On the other hand, if the decoding at the receiver fails, the controller selects the proper length of LT parity, i.e., a code rate of the LT code, R to be transmitted at the next retransmission time. Using the codeword generated from the LDPC encoder as the information word, **U**, the parity words with the determined length are generated. Therefore, the system allocates adaptive parity lengths depending on the channel condition. When the generated parities for the LT code is transmitted, the same modulation scheme as used for the original information is kept even if the channel quality is changed. In the case of severely faded condition, length of parity words can be longer than that of **U**. In Section 4, we detail the simulation model we used and the performance simulation results.

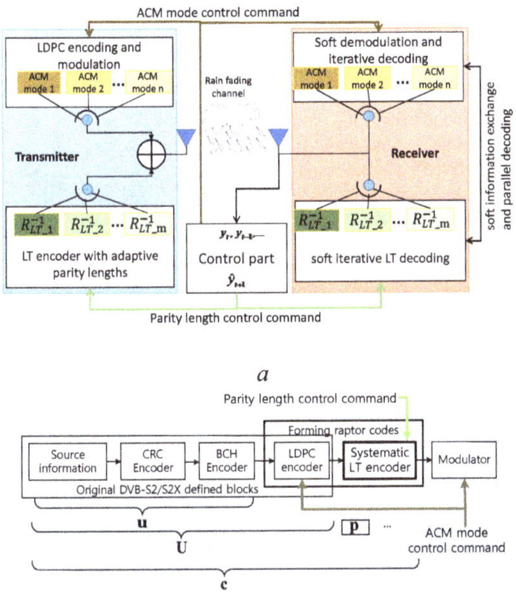

b

Fig. 2 Operational principle of the proposed HARQ with rateless codes in combination with ACM (a) Block diagram of the system (b) Signal formation at the transmitter

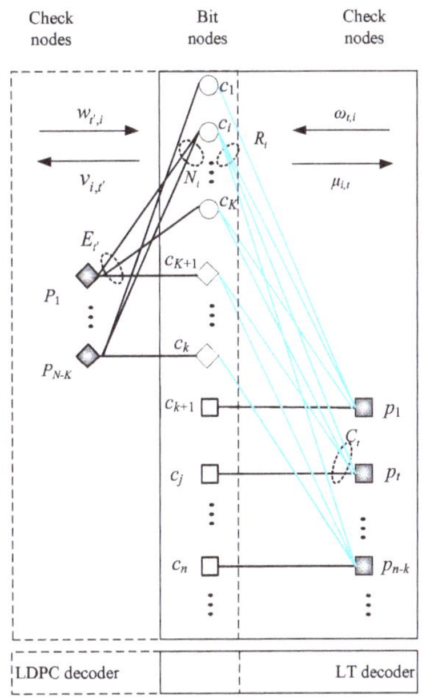

Fig. 3 Modified Tanner graph for joint soft iterative decoding

3.2 Joint soft-iterative decoding

With the retransmission scheme using the proposed HARQ schemes, we note that the utilized code can be considered a raptor code which is a concatenated code with a LDPC and LT code. It is also noted that both of the codes are modelled with a Tanner graph, and thus the same soft iterative decoding algorithm based on belief propagation can be used. In this way, we can further improve the decoding performance by utilizing joint soft iterative decoding algorithm.

For the efficiency of decoding, we adopt the modified Tanner graph approach proposed in [10], where the Tanner graph of both component codes are represented in terms of bit and check nodes as shown in Fig. 3. In this way, the same decoding algorithm can be used not only at the LDPC decoding but also at the LT decoding processes at the receiver. In addition, bit nodes of two component codes are shared, and thus the decoding process can be performed in parallel by utilizing the shared bit nodes. This means that the joint iterative decoder hardly require additional decoding time due to the added parities, but only requires additional computations to enhance the performance. .

With the Tanner graph in Fig. 3, we can utilize three iterative loops with soft-input and soft-output, including iterative loops inside LT and LDPC decoders, and iterative loops between two decoders. We note that, after an initial iteration inside the LT decoder, not only the LT decoder but also LDPC decoder can be activated. In this case, the LT decoder continues its iterative process, while the LDPC decoder can start the decoding by using the soft information provided by the LT decoder. By this way, two decoders can be processed in parallel, and neither additional hardware processors nor memories are required.

With this basic concept, the proposed parallel decoding method is explained as follows. Referring to Fig. 3, first, soft input vector for to the LT decoder, $\hat{\mathbf{r}}$, is estimated by using a soft demodulator. The decoding process is first started by initialization of the soft input value on the ith bit node, λ_i with the *a priori* channel information, i.e., $\lambda_i = r_i$. With these initial values, the LT decoder estimates the log-likelihood ratio (LLR) values from the ith bit to the tth check nodes in the LT decoder as follows:

$$\mu_{i,t} = \lambda_i, \quad t \in R_i, \tag{3}$$

where R_i is the set of the check node indices connected with the ith bit node. Afterwards, the LT decoder updates LLR values between the bit and check nodes alternatively, i.e., LLR value from the tth check node to the ith bit node, $\omega_{t,i}$ is updated as follows:

$$\omega_{t,i} = 2\tan^{-1}\left(\prod_{s\in C_t, s\neq i} \tanh\frac{\mu_{s,t}}{2}\right), \quad (4)$$

where C_t is the index set of the bit nodes connected with check node, t. With this, $\mu_{i,t}$ can be re-estimated by,

$$\mu_{i,t} = \lambda_i + \sum_{x\in R_i, x\neq t} \omega_{x,i}, \quad (5)$$

If this initial LT decoding process is finished, then λ_i is updated to produce soft output for c_i, as follows:

$$\lambda_i = \sum_{t\in R_i} (\omega_{t,i} + \mu_{i,t}), \quad (6)$$

In parallel to this LT soft iterative decoding processes with (5) and (6), similar soft iterative LDPC decoding processes are performed with the following initialization. The LLR values from the ith bit node to the t'th check node is estimated as follows:

$$v_{i,t'} = \lambda_i, \quad t' \in N_i \quad (7)$$

where N_i is the index set of the check nodes connected with information node i.

With this initialization, the LLR values between the bit and check nodes alternately updated. The LLR value from the t'th check node to the the ith bit node is estimated by:

$$\omega_{t',i} = 2\tan^{-1}\left(\prod_{s\in E_{t'}, s\neq i} \tanh\frac{v_{s,t'}}{2}\right), \quad (8)$$

where $E_{t'}$ is the index set of the bit nodes connected with the check node t', and then re-estimation of the LLR values from the ith bit node to the t'th check node is performed by:

$$v_{i,t'} = \lambda_i + \sum_{x\in N_i, x\neq t'} \omega_{x,i}, \quad (9)$$

The final soft output information is estimated by integrating the decoded results from both of the LDPC and LT decoding processes as follows:

$$\lambda_i = \sum_{t\in R_i} (\omega_{t,i} + \mu_{i,t}) + \sum_{t'\in N_i} (\omega_{t',i} + v_{i,t'}). \quad (10)$$

If the parity check equation by using the hard decision value of λ_i for the LDPC code is satisfied, then the decoding process will be terminated. Otherwise, both of the LT and LDPC decoders will repeat the above process in parallel until they reach the maximum number of iterations. In this case, LT decoder repeats (3) to (5) by using λ_i in (10) in parallel to LDPC decoding of (7) to (9).

4. Simulation results

4.1 Simulation model

We simulated the performance of the conventional ACM and the proposed methods over a synthesized satellite rain fading channel at Ka frequency band [11]. By considering the round trip delay of geo-stationary orbit satellite system, the fading value is generated with an interval of 1 second, and it is assumed that the switching operation is made with the same time interval.

For the operation of ACM, we utilized the short frame sized DVB-S2 LDPC codes with various code rates and modulation schemes of binary phase shift keying (BPSK), quadrature phase shift keying (QPSK), 16 amplitude and phase shift keying (APSK), and 32APSK. We set the target bit error rate (BER) performance to be achieved to 10^{-5}. We note this target BER value is just for simulation purpose not for the practical operation. It is also noted that there is no more than 0.1 dB differences in SNR from BER value of 10^{-5} to 10^{-10} due to very steep slope of BER performance of the LDPC codes. For the rateless codes, we utilized LT codes with Robust Soliton distribution [4].

The dynamic ranges of the employed modulation and coding scheme is 24 dB, and thus in the simulation we limit the minimum fading depth to -25 dB. Table 1 shows the modulation and coding schemes utilized in the simulation, and Figure 4 shows bit error rate (BER) simulation results over an additive white Gaussian noise (AWGN) channel, where E_s/N_0 is symbol energy to noise (needs modification)

Table 1 ACM modes used in the simulation

Modulation Scheme	Code Rate	E_s/N_0 (dB) @BER=10^{-5}
BPSK	1/5	-5.7
BPSK	1/3	-3.45
BPSK	4/9	-2.1
BPSK	1/2	-1.3
QPSK	1/3	-0.5
QPSK	4/9	0.95
QPSK	1/2	1.75
QPSK	2/3	3.5
QPSK	11/15	4.35
QPSK	37/45	5.42
QPSK	8/9	6.5
16APSK	1/2	8.5
16APSK	2/3	10.1
16APSK	11/15	11.15
16APSK	37/45	12.2
16APSK	8/9	13.3
32APSK	11/15	15.87
32APSK	7/9	16.44
32APSK	37/45	17.05
32APSK	8/9	18.3

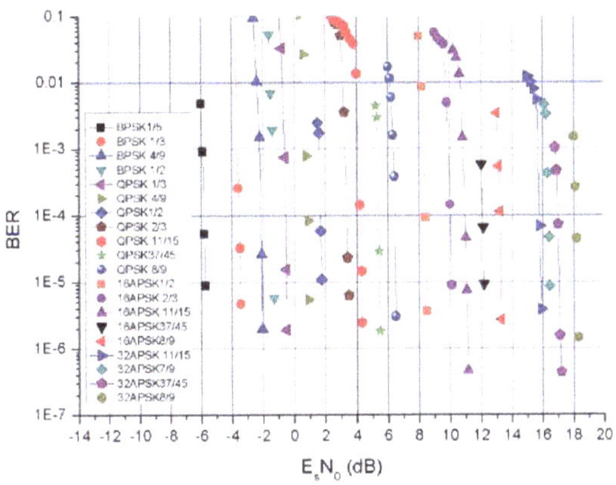

Fig. 4 BER performance of DVB-S2 modulation and coding schemes over an AWGN channel

We assumed that the switching operation is made with an interval of one second, with various margin values. Depending on the bit rates, the amount of data transmitted during one second will vary. Nevertheless, not only the generated fading value but also the data transmitted with the generated fading value can represent expected values in statistical sense. With this simulation model, the average throughput was estimated as follows:

$$\bar{\eta} = \frac{\sum_{i=1}^{T} \eta_i}{T}. \tag{11}$$

where η_i is expected value of the throughput estimated at the ith one second interval, and it is estimated by the ratio of the number of successfully decoded information bits to the total number of transmitted symbols. T is the total simulation time, i.e., T seconds. The expected value of throughput can be estimated depending on what kind of information was carried-over. Table 2 shows how η_i is estimated, where η_{ACM} is the code rate of the LDPC code multiplied by the number of bits per symbol.

Table 2 Estimation of η_i depending on the transmission modes

η_i	Transmitted condition during 1 second
0	Decoding fails or only LT parities are transmitted
η_{ACM}	Decoding succeeds without LT parities
$(1-\tau)\eta_{ACM}$	Decoding succeeds and LT parities of the previous codeword are transmitted together.

Table 3 shows the required E_s/N_0 (dB) to achieve BER of 10^{-5} by using ACM mode operation in combination with rateless codes. The first column indicates various ACM modes, while the first row indicates the inverse code rate of LT, R^{-1}. For example, BPSK modulation scheme using LDPC codes of rate = 1/3 concatenates with LT codes of R^{-1}= 1.2 requires E_s/N_0 = -3.75 dB to obtain BER of 10^{-5}. We use this table to allocate LT code rate adaptively when LDPC decoder fails.

Table 3 Required E_s/N_0 (dB) @BER=10^{-5} with rateless codes

R^{-1} (LT)	1	1.2	...	2.8	3
ACM mode					
BPSK1/5	-5.7	-5.85	...	-6.66	-6.52
BPSK1/3	-3.45	-3.75	...	-5.8	-5.23
BPSK4/9	-2.1	-2.48	...	-4.35	-4.55
BPSK1/2	-1.3	-1.62	...	-3.92	-4.15
QPSK1/3	-0.5	-0.72	...	-2.03	-2.23
QPSK4/9	0.95	0.57	...	-1.31	-1.5
QPSK1/2	1.75	1.33	...	-0.92	-1.15
QPSK2/3	3.5	2.88	...	0.13	-0.12
QPSK11/15	4.35	3.72	...	0.67	0.45
QPSK37/45	5.42	4.8	...	1.38	1.13
QPSK8/9	6.5	5.74	...	1.73	1.38
16APSK1/2	8.5	8.08	...	5.98	5.74
16APSK2/3	10.1	9.65	...	6.95	6.7
16APSK11/15	11.15	10.48	...	7.35	7.15
16APSK37/45	12.2	11.5	...	8.13	7.88
16APSK8/9	13.3	12.48	...	8.43	8.13
32APSK11/15	15.87	15.25	...	12.2	11.9
32APSK7/9	16.44	15.82	...	12.44	12.25
32APSK37/45	17.05	16.35	...	12.9	12.6
32APSK8/9	18.3	17.4	...	13.2	12.92

4.2 Performance comparison with simulation results

In the simulation, E_s/N_0 was set to 18.3 dB with which BER performance of 10^{-5} can be achieved by the ACM mode with the best spectral efficiency and lowest power efficiency, that is with 32APSK and the LDPC code with a rate of 8/9. At the simulation of the proposed methods, we set the maximum number of iterations at the decoder to 20, 40, and 3 for LT decoding, LDPC decoding, and joint iterative decoding, respectively. We limit the maximum inverse code rate of LT code to 3, and because of the adaptive code rate we used retransmission is made only one time.

The BER and frame error rate (FER) simulation results along with $\bar{\eta}$ are shown in Table 4. The results in Table 4 show that the proposed method shows better performance across all the performance parameters compared to the conventional ACM technique. It is noted that the proposed method utilizing the rateless codes provides better performance than the conventional ACM method with power margin of 2 dB. Table 5 shows the occupancy rate, γ of the utilized R^{-1}, where γ represents the portion of corresponding R^{-1} among total transmitted frames during the simulations. Then, average inverse code rate $E[R^{-1}] = \sum_{\gamma} \gamma R^{-1}$ is estimated to be 1.02 using the values in Table 5. This result indicates 2% increase in parity length provides better performance than the conventional ACM with power margin of 2 dB.

Table 4 Comparison of performance simulation results

Method (power margin in dB)	BER	FER	$\overline{\eta}$
ACM (0)	7.0×10^{-4}	2.4×10^{-2}	3.5
ACM (1)	1.2×10^{-4}	4.4×10^{-3}	3.5
ACM (2)	3.5×10^{-5}	2.2×10^{-3}	3.4
Proposed (0)	8.0×10^{-6}	1.1×10^{-4}	3.8

Table 5 Occupancy rate, γ of R^{-1} for LT codes

R^{-1}	γ
1	9.3×10^{-1}
1.2	5.0×10^{-2}
1.4	1.7×10^{-2}
1.6	4.5×10^{-3}
1.8	1.3×10^{-3}
2.0	4.3×10^{-4}
2.2	1.4×10^{-4}
2.4	8.2×10^{-5}
2.6	8.6×10^{-5}
2.8	3.0×10^{-5}
3.0	8.6×10^{-5}

5. Conclusion

We proposed a method of utilizing rateless codes in combination with ACM scheme for efficient providing of satellite broadcasting services. The proposed method uses conventional LDPC codes and APSK modulation schemes defined in the DVB-S2 system as the ACM scheme. From the simulation results demonstrated in this paper, the proposed scheme can be effectively used to improve the service quality of satellite broadcasting services. In the future, we will study on the performance improvement techniques on the proposed schemes. For example, the LT parities may be transmitted with a higher-order modulation scheme in order to further increase spectral efficiency. In addition, it is worth to investigate the effect of increasing retransmission times.

Acknowledgements

This work was supported by Institute for Information & communications Technology Promotion(IITP) grant funded by the Korea government(MSIP) (2018-0-01470, Standardization for spectrum resources based on the interference assessment between satellite system and 5G system), Basic Science Research Program through the National Research Foundation of Korea(NRF) funded by the Ministry of Education(2017R1D1A1B03027939), and the National Natural Science Foundation of China (Grant No. 61601403).

References

[1] ETSI EN 302 307-1 V1.4.1: 'Digital Video Broadcasting (DVB);Second generation framing structure, channel coding and modulation systems for Broadcasting, Interactive Services, News Gathering and other broadband satellite applications; Part 1: DVB-S2'. 2014, pp. 1–80

[2] Chandra, M., Harihara, S., Adiga, B., Balamuralidhar, P., Subramanian, P.: 'Effect of Check Node Processing on the Performance of Message Passing Algorithm in the Context of LDPC Decoding for DVB-S2'. the Fifth International Conference on Information, Communications and Signal Processing, Bangkok, Thailand, 2005, pp. 1369–1373

[3] Zhang, M., Li, C., Kim, S.: 'Efficient utilization of rateless LDPC codes for satellite broadcasting services'. Int. Conf. Wireless Communications & Signal Processing (WCSP 2016), Yangzhou, China, Oct 2016, pp. 1–5

[4] Luby, M.: 'LT Codes'. Proc. The 43rd Annual IEEE Symposium on Foundations of Computer Science, Canada, 2002, pp. 271–280

[5] Lee, C., Gao, W.: 'Rateless-Coded Hybrid ARQ'. Int. Conf. on Information, Communications & Signal Processing, Singapore, Dec. 2007, pp. 1–5

[6] Tan, G., Ma, S., Jiang, D., Li, Y., Zhang, L.: 'Towards Optimum Hybrid ARQ with Rateless Codes for Real-time Wireless Multicast'. 2012 IEEE Wireless Communications and Networking Conf (WCNC), Shanghai, China, 2012, pp. 1953–1957

[7] Sassioui, R., Jabi, M., Szczecinski, L., Le, L.B., Benjillali, M., Pelletier, B.: 'HARQ and AMC: Friends or Foes?'. IEEE Transactions on Communications, 2017, 65, (2), pp. 635–650

[8] Recommendation ITU-R S.2099: 'Allowable short-term error performance for a satellite hypothetical reference digital path'. 2017

[9] Shin, S. K., Lim, K., Choi, K., Kang, K.: 'Rain Attenuation and Doppler Shift Compensation for Satellite Communications'. ETRI Journal, 2002, 24, (1), pp. 31–42

[10] Zhang, M., Kim, S., Chang, J. Y., Kim, W. Y.: 'A New Soft Decoding Method for Systematic Raptor Codes'. IET Communications, Nov. 2015, 9, (16), pp. 1933–1940

[11] Zhang, M., Kim, S.: 'A Statistical Approach for Dynamic Rain Attenuation Model'. 29th AIAA International Communications Satellite Systems Conf. (ICSSC-2011), Nara, Japan, 28 November–1 December 2011

Channel States Information based Spectrum Sensing Algorithm in Satellite Cognitive Communication Networks

Zhang Weizhong; Yang Mingchuan; Guo Qing*

Communication Research Center, Harbin Institute of Technology, Harbin, China
email address of corresponding author: mcyang@hit.edu.cn

Keywords: Satellite Communication, Cognitive Radio, Spectrum Sensing, Channel States Information

Abstract: This paper proposes a spectrum sensing algorithm based on channel states information to solve the problem of the instantaneous power drop of the signal in the satellite cognitive network. The proposed algorithm adds an update mechanism in the traditional energy detection process and maps the current channel state statistic by using the channel statistics of the previous detections. This algorithm is studied in the satellite cognitive network scenario, in which GEO satellite network is primary user, LEO satellite network is secondary user. Moreover, a spectrum sensing model is built to help study this algorithm. Simulation results show that the proposed algorithm significantly improves the efficiency of spectrum sensing and solves the problem that the power of the signal to be detected is constantly changing due to shadow effect.

1 Introduction

Satellite communication is developing rapidly, but the existing satellite communication networks have the problem of shortage of spectrum resources. Limited spectrum resources can no longer satisfy the increasing needs for satellite communication services [1]. However, relevant survey data shows that the actual utilization rate of the allocated spectrum resources is less than 50% in the satellite communication systems, and sometimes it is even lower than 5% [2]. Because of the scarcity of spectrum resources and low utilization of spectrum in satellite communication systems, CR (Cognitive Radio) technology should be used in satellite communication networks to improve spectrum utilization.

In 1999, Dr.Mitola first proposed the concept of Cognitive Radio in order to solve the problem of increasingly tight spectrum resources and low utilization of allocated spectrum resources [3]. CR technology means that cognitive users acquire spatial environment spectrum information through spectrum sensing, and dynamically adjust the transmission parameters according to the test results, so that spectrum resources can be effectively shared without affecting the normal service transmission of primary users. One of the key technologies of CR is spectrum sensing, which is the basis for constructing satellite cognitive networks. Spectrum sensing means that secondary users collect spectrum information in the spatial networks and then adopt a certain decision mechanism to judge the spectrum status and find available spectrum resources [4]. The performance of spectrum sensing reflects secondary users' ability to find free spectrum.

In 2013, the European Union's Seventh Scientific and Technological Framework Program launched the Cognitive Radio for Satellite Communication (CoRaSat) project [5], which aims to enhance the spectrum utilization of satellite networks and ground networks by using cognitive radio technology. CoRaSat analyses the different application scenarios of cognitive satellite communication in S, C, Ku, Ka bands, and summarizes the challenges faced by cognitive satellite communication systems from four aspects: commercial market, management control, institutional standards and technical theories.

For satellite cognitive networks, the characteristics of dynamic topology, fast node movement and time-varying channel fading make the spectrum environment that cognitive users are facing more complex, and the signal strength of primary users received by secondary users in different time and regions is also changing constantly. As a result, the applicability of spectrum sensing based on static decision mechanism in satellite cognitive networks is greatly reduced [6]. Therefore, in view of the time-varying spectrum environment of satellite cognitive network, a spectrum sensing method based on dynamic decision is proposed to realize the adaptive optimization of spectrum state decision parameters of secondary users with the spatial environment, which improves the spectrum sensing efficiency [7].

According to [8], when applying spectrum sensing to satellite networks, it's necessary to solve some problems, for example, the power of the signal to be detected is also constantly changing due to shadow effect, and it leads to a sharp drop in the performance of the spectrum sensing. In order to solve the above problems, this paper proposes a spectrum sensing algorithm based on the channel states information, which adds an update mechanism in the traditional energy detection process, and it maps the current channel state statistic by using the channel statistics of the previous detections.

The rest of this paper is organized as follows. In Section II, the system model and the cognitive scenario are described. In Section III, channel states information based spectrum sensing algorithm in satellite cognitive communication networks is proposed. Algorithm performance verification and discussion is provided in Section IV, and concluding remarks are offered in Section V.

2 System Model And Cognitive Scenario

2.1 Cognitive Scenario

This paper studies a cognitive scenario, in which GEO satellite network is primary user, LEO satellite network is secondary user. LEO satellite network uses channel states information based spectrum sensing algorithm to detect available spectrum of GEO satellite network. GEO-LEO cognitive scenario has the advantages of wide coverage, small number of users, and simple network structure. The system model is shown in Figure 1.

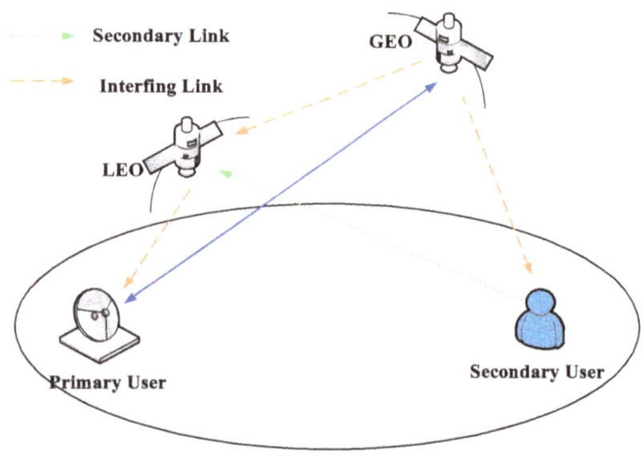

Figure 1 GEO-LEO cognitive scenario

In the above cognitive scenario, due to the continuous movement of LEO satellites and the time-varying fading of channels, the following issues need to be considered in the application of existing spectrum sensing methods in satellite cognitive wireless networks: For cognitive LEO satellites, due to the dynamic changes of signal strength and link fading characteristics of primary users, the signal strength received by LEO satellites from primary users of GEO satellites is also changing, cognitive LEO satellites not only need to quickly and accurately obtain the current temporal and spatial spectrum state, but also can make full use of the historical statistical information of spectrum state to further enhance the reliability and validity of spectrum decision results. To solve these problems, this paper will fully consider the LEO cognitive satellite motion caused by the dynamic changes of the sensing environment, proposed channel states information based spectrum sensing algorithm to improve the accuracy of decision in dynamic spectrum environment.

2.2 System Model

Spectrum sensing is a key part of the construction of satellite cognitive wireless networks. It can be considered as a binary hypothesis to judge whether a signal exists or not: H_0 indicates that the primary user does not exist (the frequency band is idle); H_1 indicates that the primary user exists (the frequency band is occupied).

$$y(n)=\begin{cases} w(n) & n=1,\cdots N \quad H_0 \\ s(n)+w(n) & n=1,\cdots,N \quad H_1 \end{cases} \tag{1}$$

Where $s(n)$ and $w(n)$ are the sampled signal and interference noise samples received respectively, N is the number of samples within the sensing time t.

Probability of Detection (P_d) and Probability of False-alarm (P_f) are two main indexes to measure spectrum sensing performance. Detection probability refers to the primary user exists and the judgement result of the secondary user is H_1, that is, $P_d = P(H_1 \mid H_1)$. False alarm probability refers to the primary user does not exist and the judgement result of the secondary user is H_1, that is, $P_f = P(H_1 / H_0)$. Obviously, the larger the P_d , the less interference the secondary user has to the primary user, and the larger the P_f , the greater the probability that the secondary user will miss the available spectrum resources. Therefore, the influence of spectrum sensing parameters on the performance of the algorithm can be measured by the receiver operating characteristic curve (ROC) corresponding to P_d and P_f .

3 Channel States Information Based Spectrum Sensing Algorithm

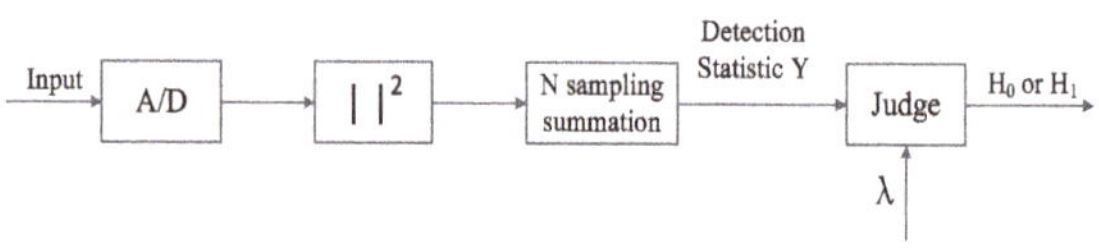

Figure 2 Principle of algorithm

The principle of the algorithm is to accumulate energy in a certain frequency band to obtain the energy detection statistic Y. If the value of Y is higher than the selected threshold lambda, the signal exists; otherwise, only noise signal exists. We assume that the interference noise is Gaussian white noise $w(t)$ and its amplitude obeys the normal distribution of mean zero and variance σ_w^2 , and then assume that the amplitude of the primary user's signal $s(t)$ obeys the normal distribution of mean zero and variance σ_s^2 , finally, we assume that the noise sampling values are independent and identically distributed, and are independent of the sampling values of the signals. We sampled $w(t)$ at N points and regarded the sampling result $w(1),\cdots,w(N)$ as a random variable. Obviously, these N random variables will obey the normal distribution of mean zero and variance σ_w^2. Similarly, because the amplitude of $w(t)+s(t)$ obeys the normal distribution of mean zero and variance $\sigma_w^2+\sigma_s^2$, the N random variables of $w(1)+s(1),\cdots, w(N)+s(N)$ obtained by N-point sampling also obey the normal distribution of mean zero and variance $\sigma_w^2+\sigma_s^2$ .The statistic Y obtained after sampling and summation can be expressed as:

$$Y=\sum_{n=1}^{N}y(n)^2 \quad (2)$$

According to the above assumption, the probability density function of noise signal $w(n)$ can be expressed as:

$$f(w_n)=\frac{1}{\sqrt{2\pi\sigma_w^2}}exp\left(\frac{-w_n^2}{2\sigma_w^2}\right) \quad (3)$$

Under the condition of H_0, Y obeys the non-central chi-square distribution with degree of freedom. For large N, the probability density function of chi-square distribution can be approximately Gaussian distribution, mean $E(Y\mid H_0)=N\sigma_w^2$ and variance $D(Y\mid H_0)=2N\sigma_w^4$ by using the central limit theorem. Its probability density function can be written as:

$$f_Y(y)=\frac{1}{\sqrt{4N\pi\sigma_w^4}}exp\left(\frac{-\left(y-N\sigma_w^2\right)^2}{4N\sigma_w^4}\right) \quad (4)$$

Through calculation, we can get the expression of false alarm probability as follows:

$$P_f^{ED}=P(Y\ge\lambda\mid H_0)=\int_{\lambda}^{\infty}f_Y(y)dy=Q\left(\frac{\lambda-N\sigma_w^2}{\sqrt{2N}\sigma_w^2}\right) \quad (5)$$

where $Q(x)=\int_{x}^{\infty}\frac{1}{\sqrt{2\pi}}\exp\left(-\frac{1}{2}t^2\right)dt$.

Under the condition of H_1, the probability density function can be written as:

$$f_Y(y)=\frac{1}{\sqrt{4N\pi\left(\sigma_w^2+\sigma_s^2\right)^2}}exp\left(\frac{-\left(y-N(\sigma_w^2+\sigma_s^2)\right)^2}{4N\left(\sigma_w^2+\sigma_s^2\right)^2}\right) \quad (6)$$

Through calculation, we can get the expression of detection probability as follows:

$$P_d^{ED}=P(Y\ge\lambda\mid H_1)=\int_{\lambda}^{\infty}f_Y(y)dy=Q\left(\frac{\lambda-N(\sigma_w^2+\sigma_s^2)}{\sqrt{2N}(\sigma_w^2+\sigma_s^2)}\right) \quad (7)$$

Simulation analysis of receiver operating characteristic should be carried out to verify the performance of the algorithm. The simulation conditions are as follows: The number of sampling points is 100, 500, 1000 ($N=100,500,1000$), SNR=-10dB. The simulation results are as follows:

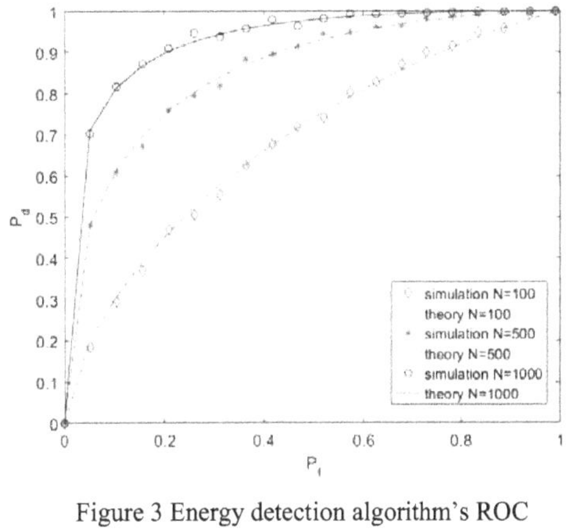

Figure 3 Energy detection algorithm's ROC

From Figure 3, it can be seen that the scatter plot obtained by simulation coincides with the ROC obtained by theoretical solution, which proves the correctness of the theory. We can see from the simulation results that under the same false alarm probability, the greater the sampling point N, the greater the detection probability. It shows that the detection performance of the energy detection algorithm is improved with the increase of the sampling point N.

Channel states information based spectrum sensing algorithm in satellite cognitive communication networks adds an update mechanism in the traditional energy detection process. It maps the current channel state statistic by using the channel statistics of the previous L detections, and we suppose the previous L detections are $Y_i=\{Y_1,Y_2,\dots Y_L\}$. If the current channel state is simply reflected by the average test quantity of the previous L channel states, it will lead to excessive influence of a certain channel state on the current state. The past statistics of the channel are fully considered here: the closer to the current judgment moment, the more accurate it is to reflect the current state of the channel. Consider the previous L detection statistics, $Y_i, i=1,2,\dots,L$,(Y_L represents the last one in the previous L detections, that is, the state before the current state) each past statistic multiplied by an attenuation function $f(k)$, which can be described as follows:

$$f(k)=e^{-\eta k} \quad (8)$$

where η stands for the forgetting factor, and the value of forgetting factor depends on the transition probability (θ) of the primary users' channel state. This paper will make the forgetting factor $\eta=\log\left(1/\left(1-\theta\right)\right)$ .Normalize the attenuation function $f(k)$,we can get $\tilde{f}(k)$ as follows:

$$\tilde{f}(k)=\frac{f(k)}{\sum_{j=1}^{L}f(j)}=\frac{e^{-\eta k}}{\sum_{j=1}^{L}e^{-\eta j}} \quad (9)$$

After introducing the attenuation function $\tilde{f}(k)$, The current statistic $\tilde{Y}$ $\tilde{Y}$can be described as follows:

$$\tilde{Y}=\sum_{i=1}^{L} Y_i \tilde{f}(L-i+1)=\sum_{i=1}^{L}\frac{Y_i e^{-\eta(L-i+1)}}{\sum_{j=1}^{L} e^{-\eta j}} \quad (10)$$

If the value of the current channel statistic Y is lower than the decision threshold λ, we should make an extra detection based on $\tilde{Y}$, if $Y<\lambda$, but $\tilde{Y}>\lambda$, it means that the primary users does exist and the power of the signal is low at this moment, so we can get the conclusion that the channel is in an occupied state. On the other hand, if $Y<\lambda$,and $\tilde{Y}<\lambda$, we can get the conclusion that the channel is in an idle state.

Since the test statistic Y_i is independent of each other and obeys a normal distribution when the number of samples is large enough, then $\tilde{Y}$ is the sum of L independent Gaussian distribution random variables, which also obey the normal distribution.

$$\tilde{Y}\sim N\left(\tilde{\mu},\tilde{\sigma}^2\right) \quad (11)$$

Where

$$\begin{aligned}\tilde{\mu} &= \sum_{k=1}^{L}\tilde{f}(L+1-k)Y_c(k)N\left(\sigma_s^2+\sigma_w^2\right)\\ &+\sum_{k=1}^{L}\tilde{f}(L+1-k)\left(1-Y_c(k)\right)N\sigma_w^2\end{aligned} \quad (12)$$

$$\begin{aligned}\tilde{\sigma}^2 &= \sum_{k=1}^{L}\tilde{f}^2(L+1-k)Y_c(k)2N\left(\sigma_s^2+\sigma_w^2\right)^2\\ &+\sum_{k=1}^{L}\tilde{f}^2(L+1-k)\left(1-Y_c(k)\right)2N\sigma_w^4\end{aligned} \quad (13)$$

Where $Y_c(k)=\{Y_c(1),Y_c(2),\ldots Y_c(L)\}$ is the state vector of the former L channels, $l\in[0,L]$ is the number of times that the primary users of GEO satellite actually exists in the previous L tests, that is, the number of 1 in vector Y_c.We can calculate the expression of detection probability and false-alarm probability as follows:

$$\begin{aligned}P_d &= P(Y>\lambda\mid H_1)+P\left(Y\le\lambda,\tilde{Y}>\lambda\mid H_1\right)\\ &= P(Y>\lambda\mid H_1)+P(Y\le\lambda\mid H_1)P\left(\tilde{Y}>\lambda\mid H_1\right)\\ &= P_d^{ED}+\left(1-P_d^{ED}\right)Q\left((\lambda-\tilde{\mu})/\tilde{\sigma}\right)\end{aligned} \quad (14)$$

$$\begin{aligned}P_f &= P(Y>\lambda\mid H_0)+P\left(Y\le\lambda,\tilde{Y}>\lambda\mid H_0\right)\\ &= P(Y>\lambda\mid H_0)+P(Y\le\lambda\mid H_0)P\left(\tilde{Y}>\lambda\mid H_0\right)\\ &= P_f^{ED}+\left(1-P_f^{ED}\right)Q\left((\lambda-\tilde{\mu})/\tilde{\sigma}\right)\end{aligned} \quad (15)$$

Because P_d^{ED}, P_f^{ED} and $Q(\cdot)\in[0,1]$, $P_d^{ED}\le P_d^{ED-PCS}\le 1$, $P_f^{ED}\le P_f^{ED-PCS}\le 1$. It is obvious that the channel states information based spectrum sensing algorithm improves the detection probability at the expense of false alarm probability.

Simulation analysis of receiver operating characteristic should be carried out to verify the performance of the algorithm. The simulation conditions are as follows: The number of sampling points is 1000 (N=1000), SNR=-10dB, L=3. The simulation results are as follows:

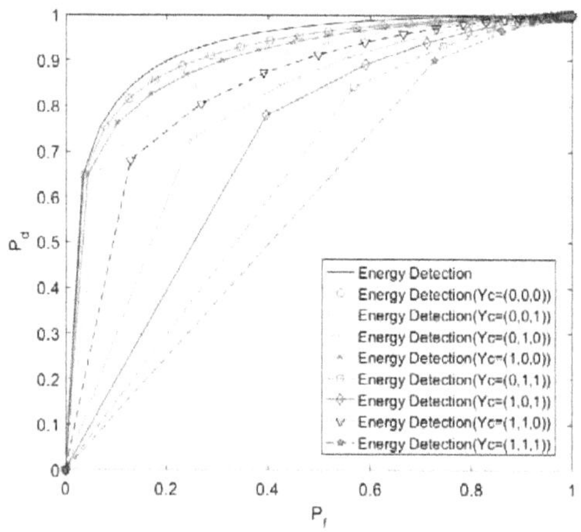

Figure 4 Channel states information based spectrum sensing algorithm's ROC

It can be drawn from the simulation results that the performance of Channel States Information based Spectrum Sensing Algorithm is worse when the channel state is 1 and the time is closer to the current time. Therefore, it can be concluded that the channel detection state at the last one in the previous L detections has a more important impact on the current channel state decision. So we revised the algorithm as follows: When $Y<\lambda$ and $\tilde{Y}>\lambda$, if $Y_L>\lambda$, it means that $Y<\lambda$ is caused by the decrease of the instantaneous energy of the primary users' signal of GEO satellite, and the detection result is H_1. If $Y_L<\lambda$, it means that $Y<\lambda$ is due to the release of GEO satellite primary users' channel, and the detection result is H_0. The detection probability and false alarm probability of the modified algorithm are as follows:

$$\begin{aligned}P_d^M &= P(Y>\lambda\mid H_1)+P(Y\le\lambda\mid H_1)\cdot\\ &\quad P\left(\tilde{Y}>\lambda\mid H_1\right)\cdot P\left(Y_L>\lambda\mid H_1\right)\\ &= P_d^{ED}+\left(1-P_d^{ED}\right)Q\left((\lambda-\tilde{\mu})/\tilde{\sigma}\right)P_d^{ED}\end{aligned} \quad (16)$$

$$\begin{aligned}P_f^M &= P(Y>\lambda\mid H_0)+P(Y\le\lambda\mid H_0)\cdot\\ &\quad P\left(\tilde{Y}>\lambda\mid H_0\right)\cdot P\left(Y_L>\lambda\mid H_0\right)\\ &= P_f^{ED}+\left(1-P_f^{ED}\right)Q\left((\lambda-\tilde{\mu})/\tilde{\sigma}\right)P_f^{ED}\end{aligned} \quad (17)$$

4 Algorithm Performance Verification And Discussion

Simulation analysis of receiver operating characteristic should be carried out to verify the performance of the algorithm. The simulation conditions are as follows: The number of sampling points is 1000 (N=1000), SNR=-10dB,

Figure 5 shows the ROC of the modified algorithm under different conditions (the number of L). Figure 6 shows the ROC of the modified algorithm under different Y_c conditions.

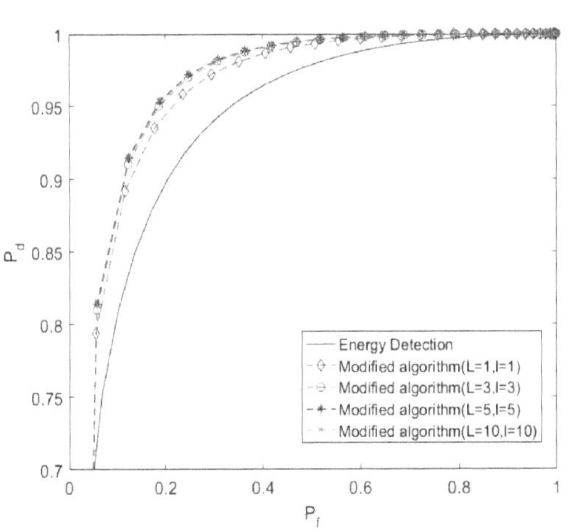

Figure 5 Modified algorithm's ROC ($l = L$)

Figure 5 shows that the detection performance of the modified algorithm increases slightly with the increase of L. This is due to the fact that $\bar{Y}$ calculated by historical channel states can more accurately reflect the real average energy of the signal and improve the accurate decision rate of the signal state to a certain extent. However, when the number of L increases to a certain extent, the energy value of the average signal can be accurately estimated, which makes further increase of L can not bring about further improvement of detection performance.

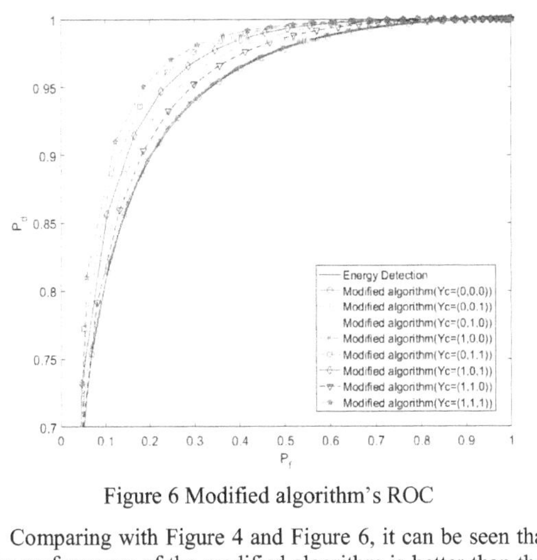

Figure 6 Modified algorithm's ROC

Comparing with Figure 4 and Figure 6, it can be seen that the performance of the modified algorithm is better than that of the energy detection algorithm.

5 Conclusion

This paper proposes a spectrum sensing method based on channel state information in the GEO-LEO cognitive scenario. A channel state sensing model is established to realize the sensing of the current channel state, on this basis, an additional sensing strategy based on the previous test results is added to modify the model to weaken the impact of the parameters in the model. Through simulation experiments, we can get the following conclusions: The performance of the modified channel states information based spectrum sensing algorithm is better than that of the energy detection algorithm, and the modified algorithm solves the problem that the power of the signal to be detected is also constantly changing due to shadow effect, moreover, the modified algorithm enhances the reliability and effectiveness of spectrum sensing results.

Acknowledgements

The paper is sponsored by National Natural Science Foundation of China (No. 91538104; No. 91438205).

References

[1] Xiao L L, Liang X J, Li X. Development and Application of Satellite Mobile Communication System [J]. Communication Technology, 2017, 50 (6): 1093-1100.

[2] Mchenry M. NSF Spectrum Occupancy Measurements[J]. 2005.

[3] Mitola J, Gerald Q, Maguire J. Cognitive radio: making software radios more personal[J]. IEEE Personal Communications, 1999, 6(4): 13-18.

[4] Axell E, Leus G, Larsson E G, et al. Spectrum sensing for cognitive radio: state-of-the-art and recent advances[J]. IEEE Signal Processing Magazine, 2012, 29(3): 101-116.

[5] Konstantinos L, Schlueter G, Krause J, et al. Cognitive radio scenarios for satellite communications: The CoRaSat approach[C]. IEEE International Conference on Future Network and Mobile Summit. 2013: 1-10.

[6] Wang X. The Application of Cognitive Radio Technology in Satellite Communication [J]. Wireless Interconnection Technology, 2017 (7): 3-4.

[7] Chatzinotas S, Evans B, et al. Cognitive approaches to enhance spectrum availability for satellite systems [J]. International Journal of Satellite Communications & Networking, 2016, 11(7): 1-36.

[8] Sina M, Symeon Chatzinotas, "Application of cognitive radio techniques to satellite communication", IEEE Communications Magazine, pp. 24-29, 2015.

WIDEBAND NONLINEARITIES CORRECTION IN DIGITAL PAYLOADS CHANNELS WITH PARALLEL ARCHITECTURES

Guglielmo Lulli [1], *Pietro Monsurrò* [2], *Felice Rosato* [1], *Giuseppe Tomasicchio* [1*], *Pasquale Tommasino* [2], *Alessandro Trifiletti* [2]

[1] *Thales Alenia Space Italia; Via Saccomuro, 24 Rome, Italy*
[2] *Dept. of Information, Electronics and Telecommunications Engineering, Sapienza, University of Rome, Via Eudossiana - 00184 Rome, Italy*
**giuseppe.tomasicchio@thalesaleniaspace.com*

Keywords: WIDEBAND NONLINEARITIES, VOLTERRA MODELS, PARALLEL ARCHITECTURES, MANY-CORE DSP

Abstract

The problem domain related to the correction of wideband non-linearities on satellite receiving or transmitting chains is more and more increasing due to the need of satellite missions with high throughput data handling, high precision accuracy and high energy efficiency.

Typical non-linear distortions could be introduced by different active on-board devices in the RX/TX chains such as LNA, mixers, active filters, A/D and D/A converters and high power amplifiers (in low back-off operations).

Modeling wideband distortions requires the introduction of memory effects, to account for more complex distortion mechanisms than static polynomial nonlinear models. In multi-channel (beam-forming) receivers/transmitters, especially if wideband, the data rate increases with the number and bandwidth of channels, and so does the need for hardware resources. Generalized models based on Volterra kernels are hardware-consuming, especially when applied for wideband systems with high data rates and potentially long memory effects.

In this paper the architecture of a Volterra filter, adapted in order to exploit high parallelism in a possible target space hardware, is presented, taking into account the design constraint to keep digital complexity to a minimum by using a priori (restricted models) or a posteriori (pruning) techniques for complexity reduction and low power exploitation.

Finally, the implementation of this filter performed on a many-core processors technology, in order to achieve high throughput performances on beam-forming and wideband systems, is evaluated in terms of computational cost and I/O, taking into account data rates, memory requirements, data dependencies, and raw processing power.

1. Introduction

Next generation SatCom payloads envisage an ever growing digital domain section that enables higher flexibility with respect to communication protocols and Digital Signal Processing (DSP) functions thanks to hardware reconfiguration capabilities. It is possible to extend these advantages even for the antenna section employing Direct Radiating Antenna (DRA) arrays and DBFN processing capable to realize programmable spotted coverage and adaptive interference robustness. The figures of merit of a phased array satcom antenna such as directivity and process gain are highly dependent on the channels' frequency-selective responses and on frequency responses mismatch between different channels giving rise to linear distortions. At the receiving side, the presence of receiver nonlinearities, typically higher in low power RF circuitry, will also impact the spatial resolution and the dynamic range of the system. At the transmitting side, nonlinear distortions are usually more pronounced due to the high power amplifiers working in low back-off. To compensate these errors, due to RF chain non-idealities, many solutions have been proposed so far relying on digital signal processing using nonlinear models.

The most common solutions are based on the Volterra series, very important for its general approximation capability of nonlinear systems. Volterra model and its subsets are widespread in pre-distortion applications to enhance the efficiency of power amplifiers in transmission chains. Post-distortion applications include calibration techniques applied to receivers [1], Sample and Hold Amplifier (SHA) stages [2] and to A/D converters [3], even if in some cases additional specific models or simplifications of the Volterra model could be needed [4], [10].

1.1. Parallel HW implementation

On the basis of the hardware implementation of different systems and the features of the processed signal, different kinds of architectures could be required to implement the filtering algorithm. Therefore, an interest is raising in flexible filtering stages able to adapt to different operating scenarios.

With Moore's law no longer working for single-core CPUs, the trend in high performance computing is to exploit parallel architectures which perform several (from a couple to thousands) operations at the same time. To the extent that the underlying algorithm can be parallelized, i.e., it performs many independent operations that do not rely on the output of previous instructions, and possesses sufficient data locality to exploit the bandwidth of wide (for instance, 64 to 512 bits) digital buses.

In the framework of flexible / reconfigurable payloads, DSP algorithms can exploit parallelism of programmable space graded platforms based on a mixed many-cores CPU – FPGA architecture [5]. Next generation many-cores DSP, such as RC-64 [6], have many instances of cores (10-1000), which are simpler than the cores in conventional CPUs and can scale parallelism up by more than an order of magnitude respect to them. FPGAs can be configured to perform parallel computations at the hardware level, and are usually programmed in VHDL, though recently support for OpenCL has been introduced, making FPGA development somewhat more similar to that of CPUs and GPUs.

The efficiency in the use of parallel architectures is strongly dependent on the type of algorithm being performed: in FIR digital filters, and FIR filter banks (including the DFT algorithm), efficiency is usually high. On the other hand, for IIR and adaptive filters, which make use of feedback loops, efficiency is much lower. Besides the computing power, other limitations arise from bandwidth limitations in the bus connecting the device and its RAM, and the acceleration board and the host computer. Finally, memory limitations may be an issue, if for instance tens or hundreds of GB of data are to be stored in RAM.

In the following sections the implementation issues on parallel architecture of a Volterra filter, adapted in order to exploit high parallelism in a possible target space hardware, is presented and evaluated with reference to a many-cores DSP HW architecture.

2. Methodology

In this work it is considered a full Volterra model with kernels of order $k \in O \in \mathbb{N}$, with lags L_k for the kernel of order k [7]. Each kernel is defined as:

$$y_k[n] = \sum_{i_1=0}^{L_k} \sum_{i_2=i_1}^{L_k} \dots \sum_{i_k=i_{k-1}}^{L_k} h_{i_1 i_2 \dots i_k} x[n-i_1]x[n-i_2]\dots x[n-i_k]$$

The full Volterra model is expressed as:

$$y[n] = \sum_{k \in O} y_k[n]$$

These equations show no dependencies over n, so that it is possible to parallelize the computation of the Volterra model by computing multiple values of the Volterra model at the same time.

It is advisable to first compute the primary functions:

$$x^k_{i_2,\dots i_k}[n] \equiv x[n]x[n-i_2]\dots x[n-i_k]$$

And then the kernels:

$$y_k[n] = \sum_{i_1=0}^{L_k} \sum_{i_2=i_1}^{L_k} \dots \sum_{i_k=i_{k-1}}^{L_k} h_{i_1 i_2 \dots i_k} x^k_{i_2,\dots i_k}[n-i_1]$$

In fact there are $\frac{(L_k+k)!}{L_k!\,k!}$ terms in each kernel of order k, but only $\frac{(L_k+k-1)!}{L_k!\,(k-1)!}$ primary functions. Furthermore, the primary functions of order k can be computed starting from the primary functions of order $k-1$, i.e., the primary function of order 3 $x_n x_{n-1} x_{n-2}$ can be computed from the primary function of order 2 $x_n x_{n-1}$.

Because all operations are independent for different values of n, it is possible to parallelize ad libitum the computation of the primary functions, the kernels, and the final Volterra model, if a batch of N, with $n = kN, kN+1+\dots+kN+N-1$ terms are computed together.

In the end, there always are $\frac{(L_k+k)!}{L_k!\,k!}$ operations required per sample to compute $h_{i_1 i_2 \dots i_k} x^k_{i_2,\dots i_k}[n-i_1]$, plus all the operations required for the $\frac{(L_k+k-1)!}{L_k!\,(k-1)!}$ primary functions. The number of operations per sample for each order k is then $(L_k+2k)\frac{(L_k+k-1)!}{L_k!\,k!}$. For complex Volterra models, this number must be multiplied by 4. Typically, the lag used for the linear kernel is larger than the ones used for the nonlinear kernels.

Memory requirements are given by the need to store the values of all the filter coefficients $h_{i_1 i_2 \dots i_k}$ and the N values of the input frame $x[kN], \dots, x[kN+N-1]$. Furthermore, the primary functions should be stored in memory, and there are $\frac{(L_k+k-1)!}{L_k!\,(k-1)!}N$ such values for each kernel of order k to be stored. The latter requirement constrains memory usage, so that it is likely that only hardware with large memory resources, e.g. the 4MB shared memory in RC64 [6], can store the primary functions in memory. Because N_{bit} of memory are necessary to store each real data, and data are complex, the total memory occupation is $2N_{bit}$ times the number of data to be stored. The Volterra model parameters $h_{i_1 \dots i_k}$ may be stored in the internal memory for faster access: for instance, the Constant memory in GPU architectures. A summary of the algorithm details is reported in Table 1.

Table 1 Summary of Volterra algorithm performances

Cost	Memory
$f_S \sum_{k \geq 2} (L_k+2k)\frac{(L_k+k-1)!}{L_k!\,k!}$	$\left(1 + \sum_{k \geq 2} \frac{(L_k+k-1)!}{L_k!\,(k-1)!}\right) N N_{bit}$

Other models, like Hammerstein or Wiener (or Hammerstein-Wiener) models can be devised, which are simpler, i.e., require fewer operations per sample and use fewer parameters. However, these restricted models are usually less accurate than the more general ones [2-3]. Fairly accurate models [2-3] can be obtained using the first-order derivative of the signal multiplied by some polynomial of the same signal, but the resulting models are still less accurate than Volterra models.

A posteriori pruning of the coefficients may be a more promising avenue of investigation [2-3]: pruned models force to zero as many parameters as possible, to obtain simpler (though slightly less accurate) models, and an attempt is made to minimize complexity without affecting linearity significantly: a significant complexity reduction of a factor two or three can be often obtained without hampering accuracy.

All the nonlinear models discussed above are generalizations of linear FIR filters, hence they are easy to implement on parallel hardware, if a batch of multiple consecutive input samples is processed concurrently. Multiple independent operations performed on many input samples also help masking the pipelining of consecutive operations on data with dependencies, allowing high levels of computational efficiency at the expense of additional latency.

3. Many-Core DSP Architecture

A new trend already under study as flexible and reconfigurable alternative to solutions based on ASIC/FPGA, is to use many-core DSP processor technologies to execute parallel digital signal processing in order to maintain and extend the degree of on-board flexibility, while respecting the need to reduce requirements in terms of costs, power and mass. Additionally, a many-core DSP processor can help to cope with some evolutionary requirements such as: (i) fast memory access; (ii) high I/O throughput to support full bandwidth processing up to 500 MHz; (iii) high degree of parallelization and computation efficiency (FFT and filter routines, encoders/decoders HW accelerators, PSK modulators/demodulators). Indeed, there are main issues not only concerned with the HW computational power, but also with memory access efficiency.

The main applications of such architecture can be for Dual Use Satellite System where the traffic demands are varying and where also specific requirements concerning anti-jamming/anti-interference are present. Furthermore, system flexibility could be extended also to support future application to commercial systems.

A recent application of many-core DSP HW technology for Satellite systems is shown in Figure 1 and is based on the Ramon Chips RC64 device [6]. Figure 1 depicts the RC64 Many-Core architecture with 64 DSP cores, HW accelerators and multiple DMA controllers of I/O interfaces accessing the multibank shared memory through a logarithmic network. The hardware scheduler dispatches fine grain tasks to cores, accelerators and I/O interfaces towards other external systems such as ADCs, DACs and possible control external systems and memories.

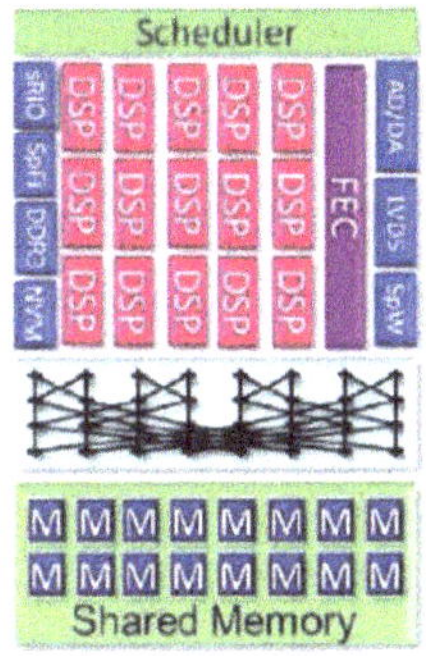

Figure 1 RC64 Many-Core Architecture with 64 DSP cores

The RC64 Many-Cores DSP comprises many DSP cores connected to a single address space multi-bank memory through a dedicated network. The RC64 employs a Multistage Interconnection Network offering N-to-M high-bandwidth connectivity for N processors and M memory banks. The memory is organized as a single shared local store. There are no cache coherency issues thanks to a single memory address space that is shared by all cores and thanks to the unique Task Oriented Programming (TOP) model. Task distribution is controlled by a hardware scheduler and high-capacity network connecting it to all cores.

RC64 operates with 64 DSP cores at 200MHz and achieve 102 Giga 16-bit fixed-point operations per second (102 GOPS). When single-precision floating point operations are needed, a rate of 25 GFLOPS is achievable. RC64 supports up to 120 Gbit/sec input and output rates over 12 high speed serial links. It can also connect to space-grade ADCs and DACs at a combined data rate of close to 40 Gbit/sec and to DDR2/DDR3 memories at up to 25 Gbit/sec [11].

The RC64 architecture has a large shared memory of 4MB per chip, with low latency and very large bandwidth where the DSP cores can perform over independent operations, whereas in other architectures such as on the GPUs all the cores in a streaming multiprocessor needs to perform the same operation on multiple data (each streaming multiprocessor contains tens or hundreds of single cores). This approach allows a much more flexible computing architecture for space technology, where there are fewer constraints on coalesced memory accesses, code divergences, etc. The limited amount of shared memory makes it harder to store a large amount of data, but the ability to perform different operations at each core allows the use of shorter blocks of N samples, thus also reducing the latency of the processing.

It is advisable to use a fixed-point (16-bit) representation for algorithms on RC64 because the DSP cores can perform n. 4 fixed-point MACs, but only n. 1 floating-point MAC operation. Furthermore, the vector VLIW architecture of each core, with n. 4 fixed-point MACs, needs to be exploited for optimized processing.

4. Software defined Volterra filter design

In real time filtering applications input data is typically available in blocks with the size of the input buffer. FIR filtering in block (or frame) based processing requires a proper organization of the input and output data to take into account the memory of the linear convolution operator. Volterra filters belong to the class of Linear-In-the-Parameter (LIP) models and are thus synthesizable as FIR filters. From a software point of view, such a filtering algorithm can be implemented using the scheme in Fig. 2.

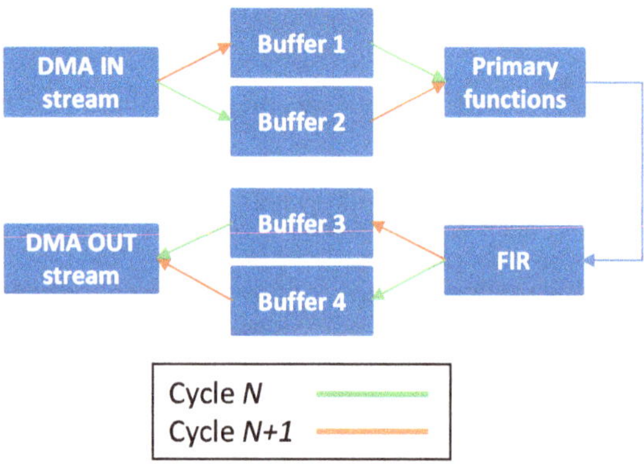

Figure 2 Volterra filter implementation scheme

This kind of implementation introduces an additional latency with respect to a simple FIR filter due to the computation of the primary functions. The double buffer mechanism is represented using the green and orange arrows: while the algorithm processes the data in the buffer 1 and writes the results in buffer 4, buffer 2 stores new input data samples and output data flows from buffer 3. Buffers switch at each cycle that processes a frame of N input data.

An efficient implementation of the FIR function on the RC64 architecture, [6], must exploit parallelism between cores and within each core. Given the number of four 16 bits fixed point MACs per core, a straightforward approach consists in distributing 4 multiplications between input samples and filter kernels on each core. The number of core instances needed to compute all the operations for each data block is given by:

$$\mathrm{N_c} = \left\lceil \frac{N_{ops}}{4} \right\rceil$$

Using this software defined approach we can assess the computational cost of some Volterra models adopted in the literature [2,3] with different kernels and lags. The identification techniques are out of the scope of this paper. See [1] for further information.

In [2] two different lag configurations of a truncated Volterra model are adopted to compensate the distortions of a sample-and-hold amplifier, with 83 and 108 parameters respectively. These models are then reduced using backward pruning, obtaining a better linearity performance with fewer parameters with respect to the full models. Considering a sampling frequency of 50MHz, the computational cost of the RC64 implementation of the pruned model is 3GOPS.

In [3] Volterra models with higher complexity are used to compensate for the dynamic nonlinearities of a pipeline ADC. The full model has 162 parameters and its RC64 implementation requires 32 GOPS when data sampling frequency is 125MHz. Pruning improves linearity, initially, and reduces model complexity by a factor of about 2.

The hardware resources of a single RC64 board suffice to realize a multichannel implementation (MIMO) of the Volterra filters described above. Table 2 summarizes the models' structures and the computational cost of the examples shown. Its last column contains the maximum channel number of a MIMO implementation sustainable on a single RC64 board. Volterra models have only odd kernels, whose lags are reported as L_1, L_3, L_5, etc.

Table 2 Summary of algorithm implementation on RC64 many-cores parallel hw.

Model	f_S	Cost	N_{ch}
[2] 15,4,2,2	50 MHz	8.9 GOPS	≤ 11
[2] Pruned	50 MHz	3 GOPS	≤ 34
[3] 30,4,2,2,1,1,1,0,0,0	125 MHz	32 GOPS	≤ 3
[3] Pruned	125 MHz	16 GOPS	≤ 6

5. Conclusions and Future Works

The architecture of a Volterra model for the correction of satellite payload elements wideband non-linearities, while exploiting high parallelism in a possible target space-grade hardware, has been presented.

Even though the generalized models based on Volterra kernels are complex from the computational view point, especially when applied for wideband systems with high data rates and potentially long memory effects, some new space-grade HW architectures are coming which offer advanced technology, such as many-cores DSP, able to support HW consuming algorithms exploiting HW parallelism.

In this paper the implementation of the Volterra filters on the Ramon Chips RC64 many-cores DSP technology, have been studied in terms of computational cost and I/O, taking into account data rates, memory requirements, data dependencies, and raw processing power.

On the basis of this feasibility evaluation and the achieved results, the future activities will concern the application of the Volterra algorithms on a breadboard HW system based on RC64 chips in order to analyse in details the performances and benefits of the implementation on a target space-grade parallel hardware.

References

[1] F. Rosato, "Nonlinear models and algorithms for RF systems digital calibration", PhD Thesis, Sapienza University, 2018.

[2] F. Centurelli, P. Monsurrò, F. Rosato, D. Ruscio, A. Trifiletti, "Calibrating sample & hold stages with pruned Volterra kernels", Electronics Letters, 2015

[3] F. Centurelli, P. Monsurrò, F. Rosato, D. Ruscio, A. Trifiletti, "Calibration of pipeline ADC with pruned Volterra kernels", Electronics Letters, 2016

[4] F. Centurelli, P. Monsurrò, A. Trifiletti, "Behavioral Modeling for Calibration of Pipeline Analog-To-Digital Converters", IEEE Trans. Circuits Syst. I, Reg. Papers, vol. 57, no. 6, pp. 1255–1264, 2010.

[5] G. Lulli, F. Iacomacci, G. Losquadro, G. Tomasicchio, "Software Radio for OBP Systems Implementations: Architecture and Technology for a Reconfigurable Platform", 19th Ka and Broadband Communications Conference, 2013

[6] R.Ginosar, P. Aviely, F. Lange, T. Israeli, "RC64: High Performance Rad-Hard Manycore," ESA DSP Day 2016, AMICSA&DSP 2016, Gothenburg, Sweden, June 2016.

[7] V.J. Mathews, G. Sicuranza, "Polynomial Signal Processing", Wiley, 2000.

[8] P. Cornfield, A. Bishop, R. Masterton, S. Weinberg, "A Generic On-Board Digital Processor suitable for Multiple Missions", Proceedings of the 2nd ESA Workshop on Advanced Flexible Telecom Payloads, Noordwijk, The Netherlands, 17-19 April 2012.

[9] G. Tomasicchio, G. Lulli, P. Monsurrò, F. Rosato, P.Tommasino, A. Trifiletti, "Models for Wideband Nonlinearities in Satcom Payloads Receiver Channels and their Parallelism", 23rd Ka Band and Broadband Communications Conference, Trieste, October 2017.

[10] D. Goberman, D. Rainish, A. Freedman, Power Amplifier Non Linearity Effects Mitigation In SDR Asic, 21st Ka Band and Broadband Communications Conference, Cleveland, October, 2016.

[11] T. M. Lovelly, A. D. George, "Comparative Analysis of Present and Future Space-Grade Processors with Device Metrics", Journal of Aerospace Information Systems, Vol. 14, No. 3, March 2017.

Section 3 – Flexible High Throughput Satellite Systems and Interference Mitigation Techniques

Modifications to Multi-beam Systems for DRRM

***Liping Ai*[1] *and Harry C. Shaw*[2]**

[1]*Independent Consultants, liping.ai11111@gmail.com*

[2]*Staff Engineer, Telecommunications and Technologies Branch, NASA/GSFC, Greenbelt, MD USA 20771*

Keywords: disaster relief and rescue mission (DRRM), on-board processing (OBP), discrete Fourier transform filter bank (DFTFB), frequency response mask (FRM) technique

Abstract

An initial study of the emergency communications services offered by current multi-beam satellite systems for disaster relief and rescue mission (DRRM) effort has been conducted within the frame work of DVB and LTE/5G in ETSI. A multi-beam communications satellite system with its unique coverage and beam hopping advantage can be adapted to provide DRRM communications services when and where the other communications services are absent. A proposal of using partial multi-beam system resources for the DRRM is made and a sample design of the digital analysis filter bank for the modified satellite transponder is presented and partially optimized.

1 Introduction

Natural and man-made disasters cause human casualties and distress and damages to properties. Wild fire, transportation accident, tornado, earth quake, and hurricane are the major such disasters that result in the loss of human lives. Various disaster relief and rescue mission (DRRM) efforts have been made including the communications services made available to save lives and lessen the damages in disasters [1-21, 33]. Despite these efforts, communications services have shown deficiencies in recent hurricane rescue missions [19]. Therefore, complementary communications satellite services and systems are sought and described in this paper taking the coverage advantage of the communications satellite to relay calls in such scenarios.

Complementary to the existing DRRM services offered, the satellite communications service will aid in saving lives and speeding up recovery by supporting power restoration, flood drainage, water restoration, foods and drink delivery and communications network restoration. Therefore, it is envisioned that the services will provide telephone calls, text messages and GPS services between handhelds. Currently, the personal mobile communication services via satellite systems offering voice calls use wide beams and are only affordable to professionals [22]. Most of the high-throughput-satellite (HTS) systems in wide use today provide broadcast and Internet services to fixed stations (FS) and mobile stations (MS). It is with the HTS systems, we will divert part of the system's resources during the time of a regional disaster to aid in speeding up the disaster recovery. To serve this market, modifications will be made to the various parts of the multi-beam system. This paper presents several considerations involved in such modifications with a focus in digital channelization design needed for transponder modification.

Based on our initial study, the most frequent disasters in the US can be covered by one to three Ka-band satellite spot beams [1-9, 15, 19 and 33]. For massive hurricanes, up to 13 beams may be necessary considering the paths of the recent hurricanes [14-15]. The design in present paper is focused on but not limited to the former. The proposed DRRM communications services can also aid in the massive hurricane aftermath with critical infrastructure recovery including water and power restoration. A trade-off of time versus coverage area can be made.

In the following, we present the current HTS system architecture, the modification to the HTS system architecture for DRRM communications services, the satellite transponder modification design and a sample digital analysis filter bank proto-type filter design.

2 Current HTS System Architecture

A HTS system typically consists of a satellite in the geostationary orbit, a few gateways for forward link broadcast or broadband information transmission and many FS and MS users in the frequency and polarization (color) reused spot beams shown by repeated numbers in Figure 1. It most often operates in Ka-band. The satellite spot beam coverage area is typically divided into several service and market regions illustrated by the colored clusters of spot beams in Figure 1. Each gateway is assigned to serve either a service or a market region covered by the same colored spot beams. One or two gateways control the configuration and operation of the system forming a star system topology. In Figure 1, it is the purple gateway taking the control gateway role. Forward link is from a gateway via satellite to a user terminal (UT) while the reverse link is in the opposite direction. In the forward link, the satellite receives the RF signals from gateways via feeder links and broadcasts or multicasts the wideband information to FS and MS users in spot beams. The channelization and routing in the transponders is typically hardwired for the service or market regions. The purple gateway serves the purple spots whereas the green gateway serves the green spot beams and etc.. This architecture is supported by DVB-S2X [23-24]. Channels and channel capacities are allocated according to user-operator agreement and system operational conditions.

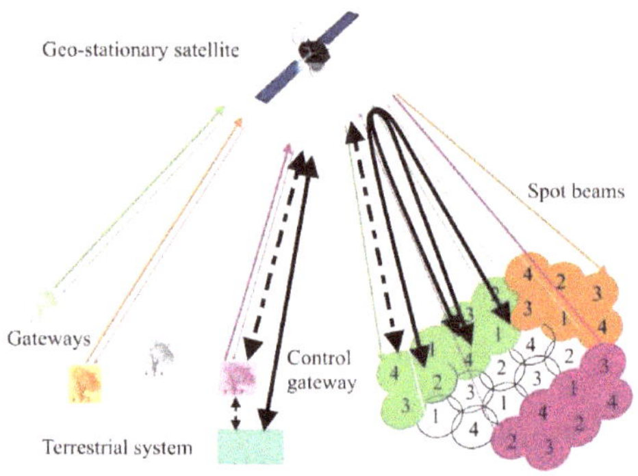

Figure 1 HTS system architecture with modifications for DRRM communications services

3 HTS System for DRRM Communications

The HTS system architecture modification is made as marked by the broad black lines and the blue box in Figure 1. The solid lines indicate the signal paths. The dashed lines are for signalling messages. The blue box symbolizes the terrestrial wireless network. Our modifications to the HTS systems include

1. Identifying three or more transponders and providing additional hardware and software to them for digital channelization
2. Allocating a portion of the existing control gateway resources or adding a new gateway for the control function of emergency service
3. Adding the RF adaptors to the LTE/5G smart phones in wide use today for the DRRM communications service users or making new dedicated handhelds per DVB-SH

The functionalities envisioned for the modified control gateway are to switch one to three spot beams attached to the modified transponders to illuminate the disaster affected areas with the beam centres tracking the disaster developments, switch the broadcast and broadband transponder channelization to DRRM service digital channelization and provide signaling for the DRRM communications services; either LTE/5G or DVB-SH signaling depending on the standards with which the handhelds are made to support. For modified smart phones, the terrestrial wireless network standard LTE/5G signalling messages are exchanged between the control gateway and the handhelds via satellite and controlled by the control gateway.

One to three satellite transponders are specially designed. Each of them is added with one or more digital channelization blocks and baseband processing blocks depending on the original transponder architecture. OBP is adopted for the DRRM communications services and performs the digital channelization and baseband processing that route the spot-beam loop-back calls in the disaster affected area. The switch over from broadcast service to the DRRM service is software configurable at the control gateway. Beam hopping technique is well established. The special transponder hardware also includes a routing to terrestrial network connection capability to a recipient outside of the disaster affected area via gateway terrestrial connection capability also shown in Figure 1.

To minimize cost and simplify adoption, adaptors can be designed for the smartphones in wide use today. The adaptor replaces the RF front-end of the smart phone, includes a patch Ka-band antenna and a RF portion that consists of filters and amplifiers in Ka-band frequency. It also performs down-conversion to the IF band of the smart phones.

Three services will be provided to each user, caller GPS position tracking, telephone calls and text messages. The GPS caller position tracking provides three parameters of longitude, latitude and altitude of the smart phone when enabled. To mitigate the rain fades, resource block (RB) aggregation can be used in addition to the error mitigation techniques used for terrestrial multi-path fading channel condition.

4 Transponder Modification

This paper focuses on the study of transponder modification required for the DRRM communications services with LTE/5G smart phones as user terminals (UTs). Figure 2 shows both the original transponder for reverse channel access by DVB-RCS and the modified transponder for the DRRM communications services by LTE. The DVB-RCS typically operates in bent-pipe mode with amplification, filtering and frequency translation functionalities which are in the analog RF and IF blocks of Figure 2. The DRRM communications transponder for LTE UTs is digital at and after IF and it requires mesh switching of calls on the beam-to-beam basis in the service coverage area as shown in the bottom figure of Figure 2. This requires the transponder to operate in transparent or OBP mode into digital baseband.

The original reverse channel satellite receiver RF portion can be reused in the loop-back DRRM communications for transmission to the same or neighbouring beams covering the disaster affected area also shown in Figure 1-2. The reverse channel for DRRM communications operation with modified smart phone as the user terminal (UT) adopts single-carrier frequency division multiple access (SC-FDMA) seen by the satellite, each call is modulated by a unique carrier and calls modulated by consecutive carriers form the wide band signal at the satellite receiver RF front-end. Because most of the calls originate and end in the disaster affected areas, the calls are simple loop-back broadcast calls to the same beams or the neighbouring 1 to 2 beams. A switch between the analysis digital filter bank and the synthesis digital filter bank in the transponder performs this task [28-31]. Moreover, the switch routes the calls destined to smart phones in areas not affected by disaster via the control gateway with terrestrial wireless network connection. This is shown by the various parts of the HTS system connected by the broad black lines in Figure 1.

The disaster areas considered in this study are assumed to be less than 700 km in diameters and can be covered by one to three spot beams [1-9, 15, 19, and 33]. We use the US system model in [32]. The communications services are used by the professional DRRM workers. Depending on the scale and type of the disaster, communications services may be required for a few hours to tens of days to up to months. Personal communications between individuals are provided in all DRRM scenarios in the first phase of the system modification design with caller's position sent together with the telephony and/or text message calls to the call recipients. In the second phase, all calls will be recorded by the control gateway and displayed at the rescue mission centers or sheriff's headquarters real time. The GPS service can be rendered with an application to the current smart phone handhelds. Because current geostationary spot beam systems are in Ka-band, the DRRM emergency communications service will be provided in this band. Two standard LTE channel bandwidths of 200 KHz and 1.4 MHz are considered for the clear sky line of sight channel conditions in typical fire, earth quake and transportation accidents scenario as well as in storm, tornado and hurricane channel conditions. The system will operate mainly after the intense storm period in areas the other communications systems are knocked out of services and rain storms have subsided. With channel aggregation and the established adaptive coding and modulation (ACM) fading mitigation gains [20-21, 23-24, 28], the 200 KHz channels can provide basic telephony or text message calls while the 1.4 MHz channels are made available to sustain the heavy rain fading conditions and provide value added services optionally.

Our digital channelizer analysis part consists of the band-pass filter (BPF) blocks and the down-sampler (DS) blocks shown in Figure 2(b). Its design for the DRRM service is based on the typical HTS system transponder of 500 MHz and a model system of about 180 km radius average beams [24 and 32]. It can be shown that 256 (16x16) 1.4 MHz or 1280 (16x16x5) 200 KHz calls can be supported by a simple design of the digital channelizer of a transponder. It can also be shown that this is sufficient for the DRRM workers to coordinate in the disaster recovery efforts in the worst case scenario hurricane affected area.

The use of the satellite DRRM communications capability starts when the terrestrial system fails during or after the disaster. The NOAA beacon rescue system is the first line emergency communications system useful for life saving efforts [11-12]. The NASA imagery services will uncover the most affected areas and other useful DRRM information [33]. The worst case scenarios are hurricanes for Ka-band. Typically the rain fall subsides after the hurricane but power, water and communications systems are out of work. The satellite communications system for DRRM proposed in this paper can aid in coordination among key individuals and organizations to speed up the recovery and potentially reduce the cost of the recovery efforts.

5 Sample Digital Analysis Filter Bank Design

Digital channelization consists of two digital filter banks and a switch in three stages [26]. The analysis filters extract the user channels from the wide-band input signal after the analog to digital conversion at the IF block of the receiver at satellite. The switch routes the user sub-channel signals to the proper beams according to their destination address in baseband processing and the synthesis filter bank recombines the user sub-channels into wide-band signals each for its destined beam [26-27]. Figure 2(a) illustrates the original 500 MHz transponder that performs frequency translation and a hardwired switching with a wideband filter optionally. The digital front-end added to the transponder starting from IF block digitizes IF waveform and enters the digital channelization three stages shown in Figure 2(b). Instead of going to a gateway, the digitally channelized narrow band user signals riding on the recombined wide-band signal reaches the destined recipient via broadcast to the beam directly. This is achieved via the switch between the analysis filter bank and synthesis filter bank [26].

Because 200 KHz and optionally 1.4MHz are small percentages of a satellite wide-band 500 MHz signal [24], it requires the extraction of a large number of sub-channels of uniform bandwidths from the satellite transponder wide-band signal. Considering the computation consumption is limited on-board satellite, we choose a pipelined two to three stage DFT filter bank (DFTFB) architecture shown in Figure 3 for its simplicity and feasibility in our filter design [25-27]. With DFTFB, we have the *k*th sub-band transfer function as:

$$
\begin{aligned}
H_k(z) &= \sum_{n=0}^{L-1} P(n) e^{-j2\pi (kn)/M} z^{-n} \\
&= \sum_{l=0}^{M-1} e^{-\frac{j2\pi kl}{M}} z^{-l} \sum_{m=-\infty}^{\infty} P_l(m) z^{-mM} \\
&= \sum_{l=0}^{M-1} z^{-l} P_l(z^M) e^{-j2\pi kl/M} \qquad (1)
\end{aligned}
$$

where the complex exponential designates the modulation carriers of the sub-channels whereas P_l is the proto-type low-pass filter.

Clearly, the 256 1.4 MHz digital filters are reused for the 200 KHz channelization in this architecture and 16 sub-channels are used in the first two stages for DFT computation savings via FFT. We design the digital analysis filters as a filter bank in this paper only. A slight modification can render the synthesis filter bank. At the center of the DFTFB design is the proto-type filter design. This is typically a low-pass FIR filter with perfect linear phase. Because of the primary applications used in the DRRM communications system, the specifications for each of the 200 KHz filters are chosen to be transmission bandwidth f_p=180KHz, channel bandwidth f_s= 200KHz,, pass-band variation $\varepsilon_p \leq 0.1dB$, and stop-band attenuation $\delta_s \geq 60\ dB$, whereas for 1.4 MHz, it is f_p=1.08 MHz, f_S=1.4 MHz, $\varepsilon_p \leq 0.1$dB and $\delta_s \geq 60$dB respectively. The sub-band filters of the analysis filter bank are derived from the proto-type low-pass filter by modulation as shown by Equation (1) [25-27, 34].

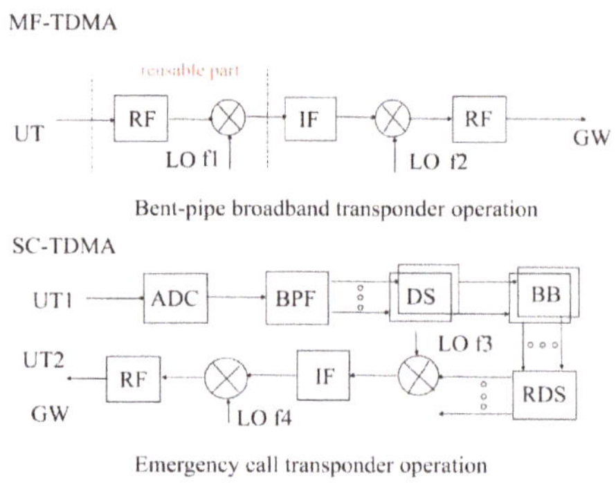

Figure 2 HTS transponder and its modification

According to the overall specification, the specification for each stage is allocated as shown in Table 1. Many low-pass FIR filter design techniques exist. We choose equal-ripple in our proto-type FIR filter design for its superior pass-band ripple and stop-band attenuation performance. This results in the filter lengths L in column 6 and the number of multiplication operations N with the DFTFB structure using FFT where possible. The first column of Table 1 shows the stage number. Figures 5-7 show the frequency responses of the filters in Table 1

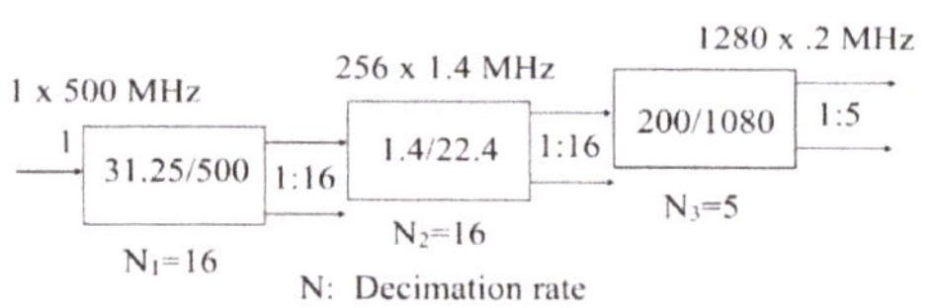

Figure 3 Two to three stage analysis DFT filter bank

Table 1 Proto-type filter designs

	f_p (M)	f_s(M)	δ_p (dB)	δ_s (dB)	L	N*
1	22.4	31.25	0.03	20	208	17
2	1.08	1.4	0.03	60	446	33.06
3	0.18	0.2	0.04	60	329	75.8

*accumulated number per channel

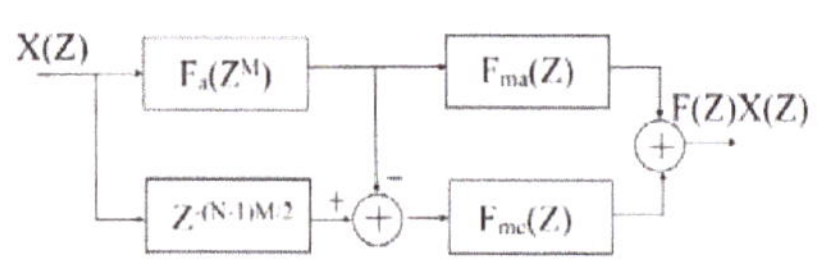

Figure 4 FRM filter structure

Table 2 FRM design of stage 3 proto-type filter

	(δ_p, G_s) dB	(f_p , f_s)**	L	N*
F	(0.04, 60)	(1/6, 0.18518519)	246	<20
F_a	(0.1, 19)	(1/6, 0.2963)	23	-
F_{ma}	(0.02, 60)	(1/6, 0.2434)	85	-
F_{mc}	(0.02, 60)	(0.119, 0.2963)	37	-

*multiplications per sub-channel

**normalized frequency

Optimization is made for the proto-type filter at stage three using frequency-response mask (FRM) filter design technique. A shorter filter length is obtained by FRM technique because a closer examination shows the narrow transition band at about 10% of the bandwidth of the desired filter [27, 34]. Briefly, the FRM filter assumes the architecture of Figure 4 and Equation (2). Table 2 shows the one conservative design specification allocation and design result using FRM technique for the stage three filter bank proto-type low pass filter, where

$$F(z) = F_a(z^M)F_{ma}(z) + F_c(z^M)F_{mc} \tag{2}$$

A comparison of the filter length L of F in Table 2 with the L of stage 3 of Table 1 shows the filter length reduction which reduces the number of multiplication operations.

The current simple 16 x 16 x 5 design wastes much of the 500 MHz bandwidth. Therefore alternatives such as multiple 125 MHz wide-band with solid state power (SSP) amplifiers can be considered. On the other hand, because the first stage does not yield real desired sub-channels, the stop-band attenuation of it can be left unspecified. This leaves room for design optimization. It is also possible to divide the 500 MHz into 20 to 22 blocks to increase the total number of sub-channels. This will tighten the transition band bandwidth of the filters at stage one and need prime number FFT algorithms for DFT computation reduction. FRM filters can be used once again to design the sharp transition band proto-type filter for less computation load.

Other optimizations including specification reallocation to sub-channels, maximization of sub-channel numbers and arithmetic operation exploitation for the minimum number of operations per sample per channel are possible to render the desired design objectives of the analysis filter bank.

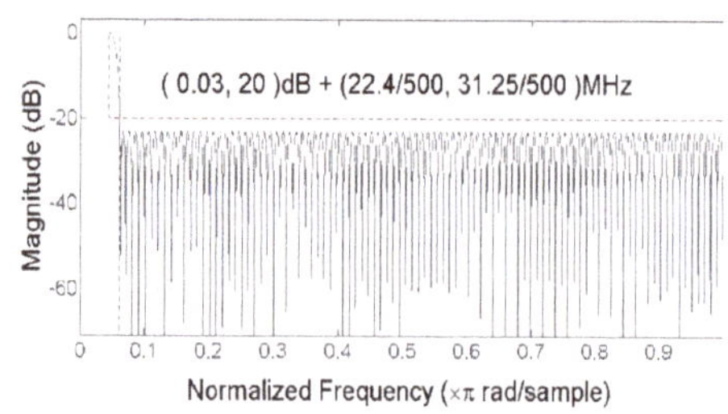

Figure 5 stage one proto-type filter

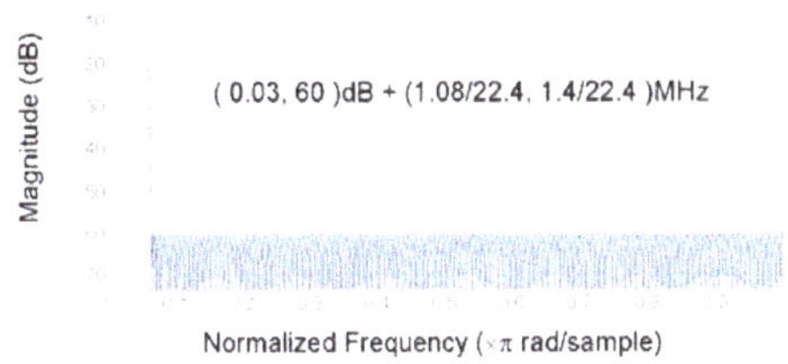

Figure 6 stage two proto-type filter

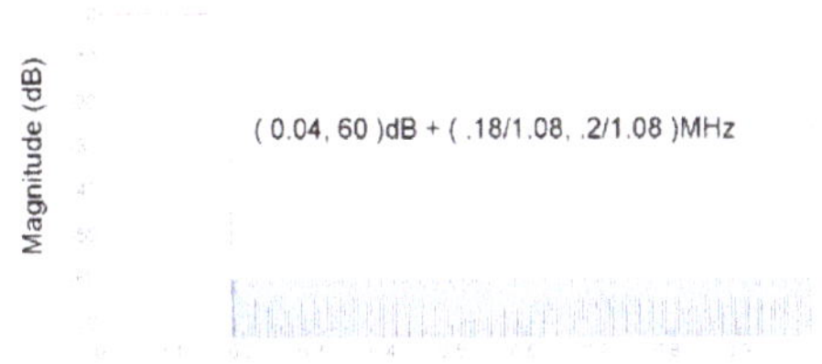

Figure 7 stage three proto-type filter

6 Conclusions

This paper proposes a HTS system modified for use in the DRRM efforts to speed up the disaster relief and aftermath recovery. It presents a sample analysis filter bank proto-type filter design used in the DFTFB architecture for the OBP transponder working with modified LTE smart phones. We choose the DFT filter bank architecture for its simplicity. Design principles will remain the same for DVB_SH handhelds or 5G UTs with different numeric values of the specifications of the proto-type filters. Superior equal-ripple FIR filters are used as the proto-type filters for the DFTFB. It is worth noting that further channelization after 200 KHz to symbol level will take place before baseband processing and switching. The paper also shows the optimization possibilities with or without the additional computational load. The simple performance assessment is the number of multiplication operations per sub-channel per sample. It shows that the computation complexity is moderate and affordable. We can also infer that the total delay of the OBP digital channelization is not the limiting factor. Total latency, computation complexity, and storage requirement can best be assessed by simulation.

Future work concerning transponder design would be switch design and end-to-end optimization of the digital channelizers. Signaling modification design and smart phone RF adaptor design will follow before the simulation of the whole system operation for performance evaluation. Certainly, the study of DRRM communications services proposed in this paper can be incorporated into the satellite 5G system design architecturally.

References

Websites:

[1] 'Wikipedia, 2017 Atlantic hurricane reason' https://en.wikipedia.org/wiki/2017Atlantic_hurricane_season

[2] 'CISION PRWeb: 2018 Hurricane Prediction – Strongest Cycle in 70 Years' http://www.prweb.com/releases/2018/01/prweb1509592.htm

[3] John Lilington, "Comparison of wideband Channelization Architectures", TechOnLine (TOL). www.edn.com/Pdf/viewPdf?/contentitemid = 4133641

[4] 'Wikipedia, 2017 California wildfires' https://en.wikipedia.org/wiki/2017_California_wildfires

[5]'Atlantic: Wildfires Rages Across the American West' https://www.theatlantic.com/photo/2017/09/wildfires-rage-across-the-american-west/538977/

[6] 'CA.gov: California Statewide Fire Map' http://www.fire.ca.gov/general/firemaps

[7] 'National Transportation Safety Board: NTSB 2017-2018 Most Wanted List of Transportation Safety Improvments' https://www.ntsb.gov/safety/mwl/Documents/2017-18/MWL-Brochure2017-18.pdf

[8] 'List of Earthquakes in 2017' https://en.wikipedia.org/wiki/List_of_earthquakes_in_2017

[9] 'List of Earthquakes in 2016' https://en.wikipedia.org/wiki/List_of_earthquakes_in_2016

[10] 'FMCSA: Large Truck and Bus Crash Facts 2015' https://www.fmcsa.dot.gov/safety/data-and-statistics/large-truck-and-bus-crash-facts-2015

[11] 'NOAA: Search and Rescue Satellite Aided Tracking' http://www.sarsat.noaa.gov/

[12] 'Wikipedia: International Cospas-Sarsat Programme' https://en.wikipedia.org/wiki/International_Cospas-Sarsat_Programme#GEOSAR

[13] 'INMARSAT: GMDSS and safety information' https://www.inmarsat.com/services/safety/gmdss-safety-information/

[15] 'Harvey Response: NASA Lends Space-Based Eyes to Recovery' https://www.space.com/38007-nasa-aids-harvey-relief-effort.html

[16] 'FirstNet: Public Safety Network' https://firstnet.gov/

[17] 'FirstNet: Devices are Ready for Public Safety's Use on FirstNet' https://firstnet.gov/newsroom/blog/devices-are-ready-public-safetys-use-firstnet

[18] 'NIST: Process Document for the NIST List of Certified Devices' https://www.nist.gov/ctl/pscr/process-document-nist-list-certified-devices

[19]'NTIA: How NTIA Ensures Responders Can Communicate During Hurricane Recovery Efforts' https://www.ntia.doc.gov/blog/2017/how-ntia-ensures-responders-can-communicate-during-hurricane-recovery-efforts

[22] 'THURAYA: Network Coverage' http://www.thuraya.com/network-coverage

[29] '3GPP: LTE'
http://www.3gpp.org/technologies/keywords-acronyms/98-lte
[33] NASA Disaster Response: A very Tough Month for Natural Disaster'
https://blogs.nasa.gov/disaster-response/tag/irma/

Standards*:*

[20] ETSI TS 103 260-1, "Satellite Earth Stations and Systems (SES); Reference scenario for the deployment of emergency communications; Part 1: Earthquake Satellite Earth Stations and Systems (SES);" ETSI V 1.1.1 2015-05
[21] ETSI TS 103 260-1, "Reference scenario for the deployment of emergency communications; Part 2: Mass casualty incident in public transportation;" ETSI V 1.1.1 2015-05
[23] Digital Video Broadcasting (DVB), "Digital Video Broadcasting (DVB) Implementation guidelines for the second generation system for Broadcasting, Interactive Services, News Gathering and other broadband satellite applications; P01.art 2 - S2 Extensions (DVB-S2X),'' Digital Video Broadcasting (DVB), DVB Document A1712, March 2015
[24] ETSI EUROPEAN STANDARD "Digital Video Broadcasting (DVB); Second generation framing structure, Channel coding and modulation systems for Broadcasting, Interactive services, News gathering and other broadband satellite applications; Part 2: DVB-S2 Extension (DVB-S2X)", ETSI EN 302 307-1 V1.4.1, 2014-07
[30] ETSI, "LTE; Evolved Universal Terrestrial Radio Access (E-UTRA); Physical channels and modulation (3GPP TS 36.211 version 10.0.0 Release 10)", ETSI TS 136 211 V10.0.0 (2011-01) ETSI,
[31] ETSI, "User Requirement Specification; Mission Critical Broadband Communication Requirements", ETSI TR 102 022- V1.1.1 (2012-08)

Journal articles

[25] Junpeng Hu *et al.* "Dynamic Digital Channelizer Based on Spectrum Sensing", PLOS, August 26, 2015; http://journals.plos.org/plosone/article?id=10.1371/journal.pone.0136349
[26] R. Mahesh *et al.* "Filter Bank Channelizers for Multi-Standard Software Defined Radio Receivers", Journal of Signal Processing Systems, Feb. 2011, Vol. 62, pp157-171
[28] Stefan Parkvall *et al,* NR: The New 5G Radio Access Technology", IEEE Communications Standards Magazine, December 2017
[34] Yong Ching Lim, "Frequency-Response Masking Approach for the Synthesis of Sharp Linear Phase Digital Filters", IEEE Trans. On Circuits and Systems, Vol. CAS-33, No. 4, April 1986

Conference papers

[14] Tim Hentschel, "Channelization for software defined base-stations", Annales des Telecommunications, ISSN 0003-4347, Vol. 57, no. 5-6, May/June 2002, pp.386-420 https://mns.ifn.et.tu-dresden.de/Lists/nPublications/Attachments/683/Hentschel_T_Annales_2002.pdf
[27] R. Mahesh and A.P.Vinod,"Coefficient Decimation Approach for Realizing Reconfigurable Finite Impulse Response Filters", ISCAS 2008. IEEE International Symposium on Circuits and Systems, 2008
[32] Liping Ai, "A Ka-band Spot Beam High Throughput Satellite System Performance Prediction Procedure", 23rd AIAA ICSSC, October, 2017

Adaptive Resources Allocation for Flexible Payload enabling VHTS systems: Methodology and Architecture

Flor G. Ortíz-Gómez[1*], Ramón Martínez Rodríguez-Osorio[1], Miguel Salas-Natera[1], Salvador Landeros-Ayala[2]

[1]Universidad Politécnica de Madrid, ETSI de Telecomunicación, Madrid, Spain
[2]Universidad Nacional Autónoma de México, Facultad de Ingeniería en Telecomunicaciones, Mexico City, Mexico
**fdg.ortiz@upm.es*

Keywords: ADAPTIVE RESOURCES, HTS/VHTS SYSTEMS, FLEXIBLE PAYLOAD, COVERAGE OPTIMIZATION, REGULAR AND IRREGULAR BEAM SIZE

Abstract

Very High Throughput Satellites (VHTS) are optimized satellite systems that will meet the demands on increasing data traffic. VHTS outperforms the capacity of traditional systems that provide FSS and MSS (Fixed and Mobile Satellite Services, respectively) which use contoured regional footprints. VHTS aims at achieving 1 Terabit/s per satellite in the near future, and are based on multi-spot coverage with frequency and polarization reuse schemes using larger bandwidths in the feeder link in the frequency Q/V bands or by optic links. These systems provide higher satellite links capacity with a reduced cost per Gpbs-in-orbit. Since traffic demand is not always the same in the service area and it may change over time, new flexible payload concepts are required. In this sense, this contribution presents the study of a flexible payload concept, the methodology for the adaptive algorithm to be used during the on-line or off-line processes for optimal resources allocation and the architecture for adaptive resources allocation as well with a set of preliminary results.

1 Introduction

In the near future (2020-2025), it is expected that satellite communication systems provide capacities close to 1 Terabit/s [1]. One objective is to prepare Europe for the deployment of the next generation of broadband infrastructure, including fixed and mobile Internet access (5G) [1]. An intermediate step will be the KONNECT VHTS by EUTELSAT, expected to be launched in 2021, it will bring 500 Gbps of Ka-Band capacity over Europe [2]. Therefore, to meet the requirement on capacity increment there is a trend towards on the use of larger, more powerful GEO satellites, and/or flexible payloads.

VHTS systems are proposed to use Q/V bands [1] or optical links [3, 4] for the feeder link. The flexible payload will be a step forward approach to meet the changing traffic demands [1, 5]. Furthermore, VHTS has been appointed by users, the satellite operators and the manufacturers to play a key role in future 5G networks to complement the terrestrial networks. Besides, VHTS satellite systems assume uniform traffic requested per beam, but it is anticipated that future traffic demands will be very uneven in the service area. Authors in [6] explain that VHTS systems, uniform traffic capacity distribution causes some beams to be overprovisioned while others may be saturated. This results that actual HTS systems do not fully use their capacity advantage which will be more inefficient for VHTS systems. Thus, the design of new generation of satellite payloads, capable of flexible resources allocation, is clearly required. For this, new proposals for flexible payloads systems not only need the analysis of novel efficient cost function to optimize resources allocation, but also need the analysis of the payload architecture and its adaptive allocation methods for new traffic demand scenarios [7] and its mentioned changes during operation within the optimizations or models used.

Furthermore, it is necessary first to understand the relationship between the number of beams and the co-channel interference in order to propose an approach for the solution of the coverage versus traffic demand changes, and so to minimize the cost of the entire operation meanwhile the channel capacity is increased [7]. Since the satellite feeder link has a key role for the increase of the capacity offered by the satellite, it is important also to study the different techniques to improve its bandwidth provided to the satellite. In this sense, from the point of view of the operators, new flexible payload solutions are a technological challenge for manufacturers but needed to guarantee the operations sustainability of the future satellite communication systems.

More promising proposal for flexible payload solutions are discussed from the state of the art. In this sense, authors in [6] have proposed a solution for non-uniform traffic demand applying a non-uniform bandwidth distribution and designing the non-uniform beam sizes, this optimizes the capacity distribution of the beams with respect to the traffic demand but is not a flexible solution to when traffic demand change, limiting the use of the capacity since the solution is not adaptive.

As a solution for the above mentioned limitation, authors in [5] addressed the problem of the allocation of the multi-beam satellite networks systems, by proposing a cost function to meet the requested traffic demand across the beams allocated within the coverage region using radio resource management (bandwidth and power per beam) on SDR (Software Defined per Radio). However, a payload architecture is not proposed, so it is not mentioned the technological implications that could be had when using this methodology.

On the other hand, also in the state of the art, the joint power and carrier allocation for the multibeam satellite downlink with individual CINR constraints is studied [8] and [9]. However, the technological implications that the payload might require are not explained either.

Nowadays, multi-spot systems usually make use of Single Feed per Beam (SFB) when no adaptive beamwidth or beam steering capability are required. Using SFB, the VTHS systems have had an important evolution of their capacity [10], that has been increased from 10 Gbps of the Anik F2, to 90 Gbps of the Ka-SAT, up to the 500 Gbps of the proposed KONNECT [2]. For this, significant effort has been made in the research and development of SFB systems such those presented in [11-19]. In this sense, a manufacturing and testing of Ka-band communication payload using SFB [11]. In [12], the recent Korean R&D in Satellite Communications using SFB for multibeam satellite is presented. Furthermore, multibeam reflector antenna fed by few elements for communication satellite is presented in [13]. In [14, 15] , authors present designs of the Ka-band communication payload for satellite communication using SFB. In this sense, advanced antenna technologies for a Broadband Ka-Band communication satellite are explained in [16, 17, 18]. Other important contributions for satellite payload is the multibeam deployable antennas for Ka, C, and S frequency bands presented in [19].

On the other hand, the use of Multiple Feed per Beam (MFB) became as part of the solution to scenarios with significant changes on the traffic demand pattern over the service area. Due to the relationship between the system capacity, the MFB antenna design and the payload architecture, MFB might be included into the approach proposed here for the flexible payload system optimization. From the recent effort in the development of more flexible on-board antenna systems it is interesting to highlight the work presented in [20] of a new concepts for active MFB antennas with shared power amplifiers for HTS systems. In [21], it is mentioned that can be realized by multiple-feed-per-beam antennas. Authors in [22] explain a multibeam antenna with very high G/T for ka-band multimedia via satellite missions. Authors in [23] explain that multi-spot beam antennas are a key element for the frequency reuse scheme and single feed per beam and multiple feeds per beam, and their application were discussed.

As discussed above, there are different limitations in the state of the art about the proposed methodologies for non-uniform allocation of resources in each beam. In this sense, this work proposes a methodology for optimal allocation of resources depending the traffic demand that changes over time, the preliminary results help to propose a flexible payload architecture.

This paper is organized as follows. Section two presents the problem definition. Section three explain the methodology for the adaptive algorithm. In section four preliminary results are discussed for the study cases. Section five explains the concepts for payload architecture. Finally, section six presents the conclusions of this work.

2 Problem Definition

The optimization starts taking into account the input parameter such as the service area, the change on the traffic demand distribution, the capacity of the satellite system, the limitation of the number of transponders, the limitation of the on-board antenna geometry, user station specifications, and other parameters. Then, the aggregated CIR and CINR as well as the required EIRP parameters are computed over the full service area under optimization in a latitude and longitude grid.

The allocation of the satellite resources is under discussion since the traffic demand in being growing and the cost of the served bit per second have to be reduced. In this sense, for HTS or VHTS multi-beam missions, the optimization of the traffic changes over time for each beam becomes a key factor. In a flexible multibeam satellite system using an on-board antenna with beamforming and adaptive beamwidth capability, the system may generate N_b beams over the coverage area over time that maximize the system efficiency. In this sense, the actual capacity $C_i(t)$ offered towards the i-*th* beam may by optimized with respect to the actual required capacity $R_i(t)$ to be offered in the i-*th* beam. This means that the beamwidth or beam-pattern as well as the multi-spot distribution may be changed to meet the new traffic demand distribution over time.

The offered capacity (C_i) towards the i-*th* beam depends of Carrier to Interference plus Noise ratio ($CINR(t)$), The $CINR(t)$ of a point inside i-*th* beam depends mainly of Equivalent Isotropic Radiated Power ($EIRP(t)$) and the interference at t. This work is focussed to minimize the normalised mean square error to match the traffic demand with the available capacity of the system. If we assume that $R_i(t)$ is the demanded traffic for the i-*th* beam at the instant t, it can be expressed as the following optimization problem:

$$F^{min}_{\mathrm{CINR}_i(t)} \rightarrow F = \frac{1}{N_b} \sum_{i=1}^{N_b} |C_i(\mathrm{CINR}_i(t)) - R_i(t)|^{w_i} \tag{1}$$

$$w_i = \begin{cases} wc_1 & C_i(CINR_i(t)) \geq R_i(t) \\ wc_2 & C_i(CINR_i(t)) < R_i(t) \end{cases}$$

where

$$CINR_i(t) = E[CINR(lat, lon, t)] \tag{2}$$

Where, w_i is the weight of the *i*-th beam. If $C_i(CINR_i(t))$ is minor than $R_i(t)$, the system conditions are worse, for this reason wc_2 must be greater than wc_1. Furthermore, the pair (*lat, lon*) is the geographic position and $CINR_i(t)$ is the *CINR* expected value over the area inside *i*-th beam at *t*. The calculation of the *CINR* for each geographical position is defined in [7], taking to account the gain and EIRP corresponding to each latitude and longitude of the service area.

On the other hand, the requested traffic for the ith-beam R_i (in bps) is defined as:

$$R_i(t) = r_i(t)\, Ab_i \tag{3}$$

$$r_i(t) = E[r(x, t)] \tag{4}$$

x is defined as a specific geographic area of one square kilometer, r as the required throughput density per square kilometer (in bps/km^2) inside x at the instant t, $r_i(t)$ is the r

expected value over all the area inside i-th beam at the instant t (in bps/km[2]) and Ab_i is the area (in km[2]) of the i-th beam.

The throughput density $r(x,t)$ per square kilometer depends on the population density $D(x,t)$ (in inhabitants/km[2]), the penetration rate $F(x,t)$ (in user/inhabitant), the concurrency ratio $T(x,t)$, bandwidth per user B_u (in Hz/user) and the spectral efficiency η (in bps/Hz), as:

$$r(x,t) = \eta\, B_u\, D(x,t)\, F(x,t)\, T(x,t) \tag{5}$$

On the other hand, the coverage optimization is needed for regular and irregular beam size and grid. When assuming regular beam size allocation, the optimizations problem is shown in the next expression:

$$a^{min}_{\text{lat}_b,\ \text{lon}_b,\ \theta_{3dB},\ \varphi} \rightarrow a = \frac{|a_x(\text{lat}_b,\ \text{lon}_b,\ \theta_{3dB},\ \varphi) - a_y|}{a_y} \tag{6}$$

Subject to

$$a_x(\text{lat}_b,\ \text{lon}_b,\ \theta_{3dB},\ \varphi) \geq a_y \tag{7}$$

$$\emptyset \geq \emptyset_{min} \tag{8}$$

$$\beta \geq \beta_{min} \tag{9}$$

$$\theta_{3dB\,max} \geq \theta_{3dB} \geq \theta_{3dB\,min} \tag{10}$$

Where a_y is the required service area (in square kilometer), a_x is the resulting service area (in square kilometer). The pair of lat_b and lon_b are the geographical position where the satellite antenna is pointing its first beam. The θ_{3dB} parameter is the half-power beamwidth is that depends on the satellite antenna design. The inclination angle φ is beams grid angle with respect to the equator being positive from east to north. This is important to ensure that there will be overlapping between the adjacent beams and thus be able to match the desired coverage. The service area percentage $\emptyset$ is the portion of the service area that lies at least into one beam and it must always be greater or equal to the minimum required. The coverage beam efficiency β is the proportion of each beam is inside the service area and it must be greater or equal to the minimum required for each beam. In this work, we consider a regular coverage where the beams are all inclined in same φ and with the same beam-width.

However, the irregular beam-widths or spot size is needed for non-uniform traffic demand. Assuming irregular beam-width allocation, the optimizations problem can be expressed as follows

$$A^{min}_{\overline{LAT},\overline{LON},\overline{\theta_{3dB}},\overline{\varphi}} \rightarrow A = \left(A_y{}^{-1}\right)^t \left|A_x(\overline{LAT},\overline{LON},\overline{\theta_{3dB}},\overline{\varphi}) - A_y\right| \tag{11}$$

Where

$$A_y = [a_{y_1}\ a_{y_2} \cdots\ a_{y_n}] \tag{12}$$

$$A_x(\overline{LAT},\overline{LON},\overline{\theta_{3dB}},\overline{\varphi}) = [a_{x_1}(\text{lat}_1,\text{lon}_1,\theta_{3dB_1},\varphi_1)\ a_{x_2}(\text{lat}_2,\text{lon}_2,\theta_{3dB_2},\varphi_2) \cdot \cdots a_{x_n}(\text{lat}_n,\text{lon}_n,\theta_{3dB_n},\varphi_n)] \tag{13}$$

$$\overline{LAT} = [lat_1\ lat_2 \cdots\ lat_n] \tag{14}$$

$$\overline{LON} = [lon_1\ lon_2 \cdots\ lon_n] \tag{15}$$

$$\overline{\theta_{3dB}} = [\theta_{3dB_1}\ \theta_{3dB_2} \cdots\ \theta_{3dB_n}] \tag{16}$$

$$\overline{\varphi} = [\varphi_1\ \varphi_2 \cdots\ \varphi_n] \tag{17}$$

Non-uniform traffic demand needs to divide the total service area in n smaller areas and each n-*th* area has different initial conditions and constraints which depend on the required throughput inside this area.

3 Methodology for Adaptive Resource Allocation

We consider two different flexible payload concepts. The first payload can be optimized in terms of bandwidth and power allocation but using regular beams size on the service area. The second payload can be optimized in terms of bandwidth and power allocation but using irregular beams size on the service area. Multiple Feed per Beam (MFB) can be considered for the beam sizes allocation.

For coverage optimization, a_x is one scalar value obtained by a nonlinear relationship between the four variables (*lat*, *lon*, θ_{3dB} and φ). Three different coverage optimization methods are here proposed and compared. The first is a nonlinear least-squares method (*Method 1*) [24], the second is a methodology of nonlinear programming (Method *2*) and the last is a methodology of nonlinear programming with nonlinear constraints (*Method 3*) [25], the non-linear constraint is defined in (7) assuming that a_x is a non-linear function. The optimization methodology for coverage was divided in two parts to obtain better results, θ_{3dB_n} and φ_n are constants in the first optimization part to find the best pair of lat_n and lon_n (the satellite antenna pointing direction is for the first beam over the nth-area), θ_{3dB_n} and φ_n are variable in the second optimization part to find the four variables (lat_n, lon_n, θ_{3dB_n} and φ_n) to coverage optimization. Section 4 shows some results to compare the three different coverage optimizations for regular size beams and it can be observed that it is recommended to use the method 3.

As it is explained in [5], the flexibility in terms of power allocation consists in the possibility to change operating point (input back-off IBO) and power setting of each TWTA independently of the others. As in real systems, the power optimization of each TWTA is done by selecting the IBO which, in turn, gives to certain OBO.

The problem of adaptive resources allocation is very difficult, even assuming full channel state information at the transmitter, finding the optimal resource allocation is not trivial. This is due, on one side, to the non-convexity of the objective function (1) and on the other side to the large number of possible solutions, which makes exhaustive search not viable. Figure 3.1 shows the algorithm for minimizing objective function over changing time in a flexible payload. If T is the instant when objective function (1) must be minimized, the algorithm explains the methodology that payload must apply for adaptive resources allocation.

The algorithm can be applied for the two proposed flexible payloads. However, it is important to notice that n (number of areas over total service area) has a value of 1 and steps 1-3 must be done only when $t = T_0$ for regular beam sizes. Steps 1-3 must be done for any value of t and n is bigger than 1 for irregular beam sizes.

On the other hand, T is the moment in which it is required to adapt again the capacity offered because the required capacity changes.

Algorithm
Over instant t=T
Minimizer cost function

$$F^{min}_{\mathrm{CINR}_i(t)} \rightarrow F = \frac{1}{N_b}\sum_{i=1}^{N_b} |C_i(\mathrm{CINR}_i(t)) - R_i(t)|^{w_i}$$

to define the constraints and initial conditions for coverage
to define n (number of areas over total service area)
coverage **optimization** (Equation 6 for regular beam sizes and equation 7 for irregular beam sizes).

$$a^{min}_{\mathrm{lat}_b,\ \mathrm{lon}_b,\ \theta_{3dB},\ \varphi} \rightarrow a = \frac{|a_x(\mathrm{lat}_b,\ \mathrm{lon}_b,\ \theta_{3dB},\ \varphi) - a_y|}{a_y} \text{ for regular beam size}$$

$$A^{min}_{\overline{LAT},\overline{LON},\overline{\theta_{3dB}},\bar{\varphi}} \rightarrow A = \left(A_y^{-1}\right)^t \left|A_x(\overline{LAT},\overline{LON},\overline{\theta_{3dB}},\bar{\varphi}) - A_y\right| \text{ for irregular beam size}$$

to define the constraints and initial conditions for the system
for i=1:N_b
  if $R_i(t) > C_i(t)$
    if possible to increase the bandwidth allocated to the ith- beam
      to increase bandwidth allocated to the ith- beam
    else
      to increase power assigned to the ith- beam
    end if
  else
    while $C_i(t) \geq z\, R_i(t)$
      to increase bandwidth or/and power allocated to the ith- beam
    end while
  end if
end for

Fig. 3.1 Algorithm for minimizing objective function at instant t=T

4 Study Cases

This section shows the study cases for the methodology proposed to adaptive resources allocation. The first part is the result of beams size assignment (section 4.1) using the three proposed methods over a specific scenario for regular beam sizes and additionally it shows an example of irregular beam sizes coverage. The second part (section 4.2) is a comparison of three different kind of payload, traditional payload, flexible payload but regular beam sizes, and flexible payload with irregular beam sizes.

4.1 Coverage Optimization

Section 3 explained that a_x is a non-linear function and we have proposed three different methods to solve the problem. Figure 4.1.1 shows the finally results for coverage optimization for regular beam size using the three different methods for a specific scenario (table 1). Where *Method 1* is the nonlinear least-squares method, *Method 2* is a methodology nonlinear programming and *Method 3* is a methodology nonlinear programming with nonlinear constraints.

It can be observed that *Method 2* and *Method 3* provide similar results except for parameter φ (the direction of the adjacent beam center for each beam), and this parameter changes the parameter $\emptyset$ (the proportion of the service area that lies within at least one beam) and the orientation satellite antenna. However, it can be observed that *Method 2* and *Method 3* are better than *Method 1* due to better results are obtained, for example, with *Method 2* and *Method 3* beamwidths are smaller (0.04845° less) and therefore more directive which increases the EIRP (0.98 dB less). In addition, the desired coverage using a lower number of beams is achieved which represents a lower cost of the system, however, *Method 1* has a higher service area percentage ($\emptyset$).

Table 1 Scenario 1 to evaluate coverage optimization in regular beam size

Service Area	Portugal, Spain and France
Satellite Position	9° E
Downlink frequency	19.7 GHz
Amplifier power	18 dBW
$\theta_{3dB_{max}} \geq \theta_{3dB} \geq \theta_{3dB_{min}}$	$0.4° \geq \theta_{3dB} \geq 0.66°$
β_{min}	30%
$\emptyset_{min}$	96%
Initial conditions $(\mathrm{lat}_0, \mathrm{lon}_0, \theta_{3dB_0}, \varphi_0)$	(31.5, 7.7, 0.55, 20)

As it is mentioned in section 3, the optimization methodology for coverage has been divided in two parts to obtain better results. The first part of results is shown in figure 4.1.2, equation 6 is the objective function for regular coverage optimization and it is represented on the ordinate axis for each iteration, the three optimizations depend of constraints and initial conditions, that is to say that a local minimum is assured. It is necessary to add an external cycle of iterations that changes the constraints and initial conditions to ensure an absolute minimum. It can be observed that *Method 3* has the best performance because with fewer iterations converge with better results. However, the *Method 1* is the worst in the first part due to after 200 iterations it is still above the other two. The second part is shown in figure 4.1.3, in this part, the three optimizations have a similar performance, *Method 1* and *Method 2* reach values very close to *Method 3*, *Method 1* even gets better results. It is observed that it is obtained better results using nonlinear programming solver with and without nonlinear constraints than nonlinear least-squares solver.

Flexible payload over regular coverage needs to optimize the coverage just once and the flexibility will be given by adaptive power and bandwidth allocation. However, flexible payload over irregular coverage needs to change the coverage over time. Figure 4.1.4 shows an example of a possible scenario of traffic demand over Portugal, Spain, France and Italy when t= T. It is possible that traffic demand will change for the next interval, so the flexible payload must be capable to adaptive the coverage for the new requirements.

To adapt the irregular beam sizes over time is complicate but it has many advantages due to the *EIRP(t)* over each geographical position is different depending of traffic demand and the flexibility will depend less of adaptive power and bandwidth allocation since there will be a previous adaptation changing beamwidth. This is shown in figure 4.1.5, where it can be observed *EIRP(t)* over each geographical position inside serve area for t=T. The coverage optimization was made with nonlinear programming solver without nonlinear constraints.

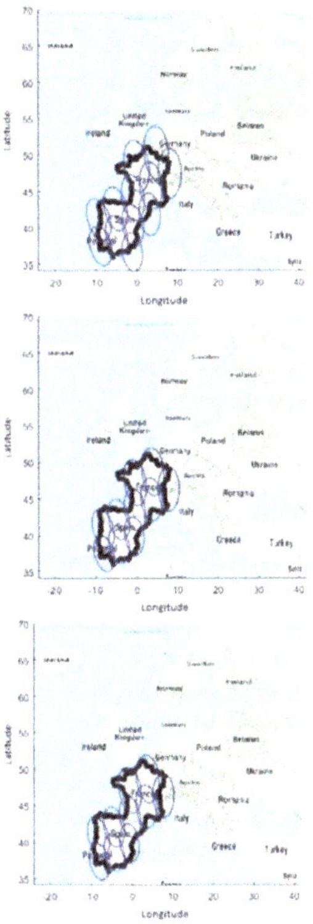

(a)Optimizer 1:
12 beams
$\emptyset$ =96.9036%
θ_{3dB}= 0.65994°
φ =20.0021°
$EIRP_{max}$ = 67.52 dBW
Satellite antenna diameter=
1.616 m

(b) Optimizer 2:
10 beams
$\emptyset$ =96.0267%
θ_{3dB}= 0.61149°
φ =18.6325°
$EIRP_{max}$ = 68.50 dBW
Satellite antenna diameter=
1.74 m

(c) Optimizer 3:
10 beams
$\emptyset$ =96.07%
θ_{3dB}= 0.61149°
φ =19.3856°
$EIRP_{max}$ = 68.50 dBW
Satellite antenna diameter=
1.74 m

Fig. 4.1.1 Results for coverage optimization for regular beam size using the three different optimizations

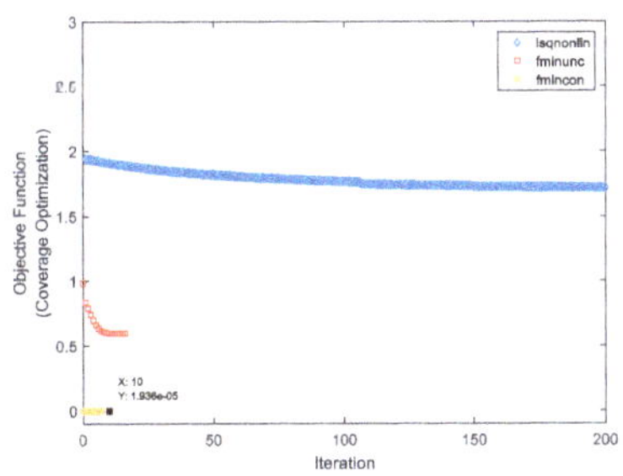

Fig. 4.1.2 First optimization for regular coverage over scenario1. (Method 1: lsqnonfn, Method 2: fminunc, Method 3: fmioncon).

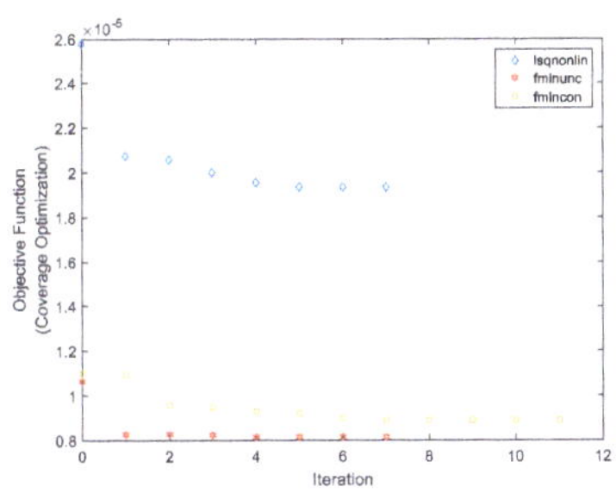

Fig. 4.1.3 Second optimization for regular coverage over scenario 1. (Method 1: lsqnonfn, Method 2: fminunc, Method 3: fmioncon).

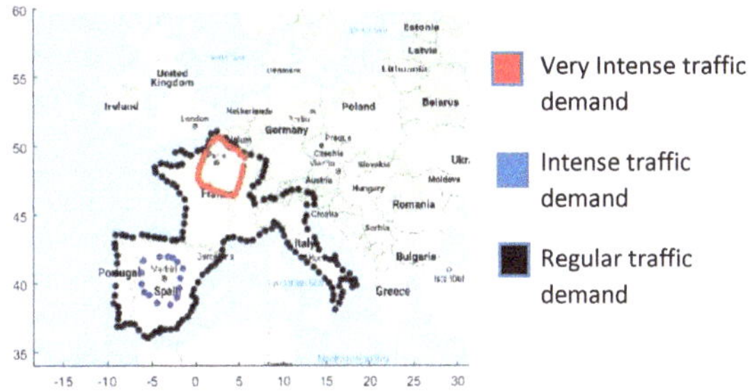

Fig. 4.1.4 Traffic demand when t=T over Portugal, Spain, France and Italy

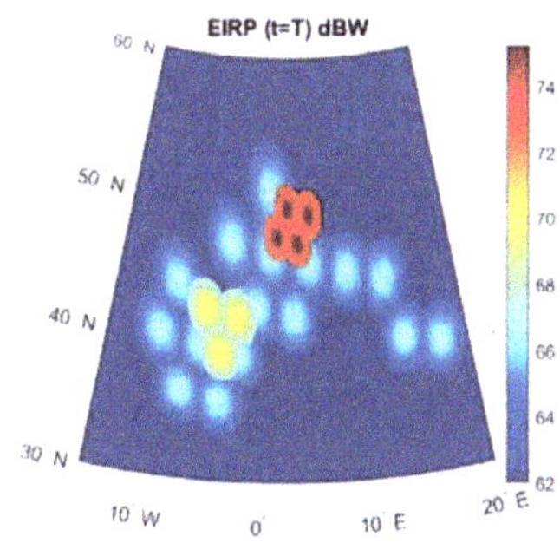

Fig. 4.1.5 EIRP over service area in time t=T for irregular beam size.

4.2 Adaptive Resource Allocation Comparison

Three different payloads have been evaluated for a non-uniform traffic demand changing over time. Traditional multibeam payload, flexible multibeam payload over regular coverage and flexible multibeam payload over flexible irregular coverage. Figure 4.2.1 shows the traffic demand in a specific time moment over a part, it can be observed that is a non-uniform traffic demand. Figures 4.2.2, 4.2.3 and 4.2.4 show the performance of three kind of payload in a specific scenario (table 2) for different ModCods (16APSK 2/3-L, 16APSK 13/18, 32APSK32/45 and 32APSK 11/15).

CINR establishes the possible ModCods that the system can use but ModCods are not very important in the payloads performances, because the dimensioning was in clear sky conditions. The analysis can consider different technical such as VCM or ACM to evaluate the atmospheric conations.

Traditional mutibeam payload has the worst performance for non-uniform traffic demand due to the $C_i(t)$ is constant over time and no matter the $R_i(t)$. There are many beams that offer more capacity than the required wasting resources, but the main problem of this payload appears when the required capacity is much greater than the offered capacity.

The flexible payload over regular beam sizes has better performance than traditional payload. The complexity increases than the anterior payload, because the $Ci(t)$ depends of $R_i(t)$. It can be observed that offered capacity is closer than required, thus the objective function (Equation 1) is minor.

The flexibility payload over adaptive irregular beam sizes has the best performances for non-uniform traffic demand changing over time.

The offered capacity is very similar than required, thus the objective function is even smaller than others. It is needed a minor number of beams for this payload because there are areas inside the coverage that is covering by big beam sizes. There is not excessive traffic demand, due to, where the traffic demand is higher, the beam size is smaller thus the area is minor (Equation 3), and on the other hand the offer capacity is higher so getting a better performance. And where the traffic demand is minor, it happens exactly the opposite.

However, the complexity is higher for the third payload, because this methodology must be applied every time the capacity requirements change, one solution is to use MFB.

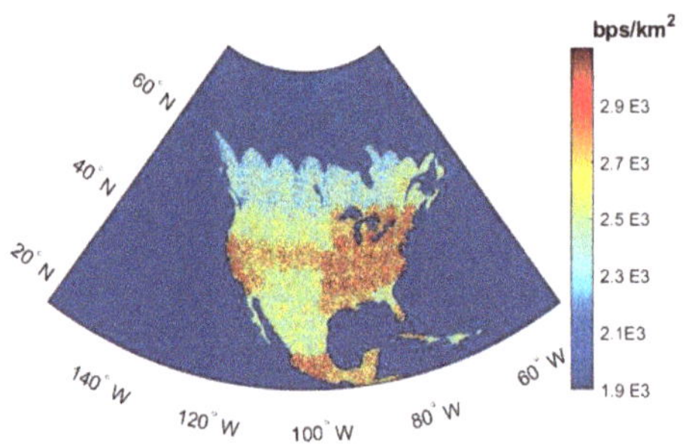

Fig. 4.2.1 Traffic demand in a specific time moment over North America

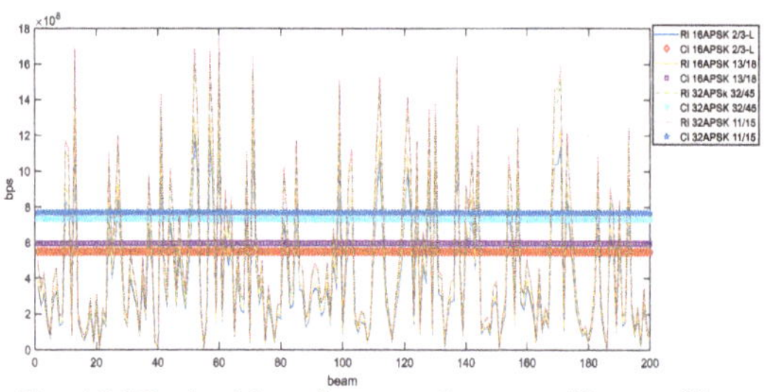

Fig. 4.2.2 Payload 1 performance for non-uniform traffic demand over North America in a specific moment.

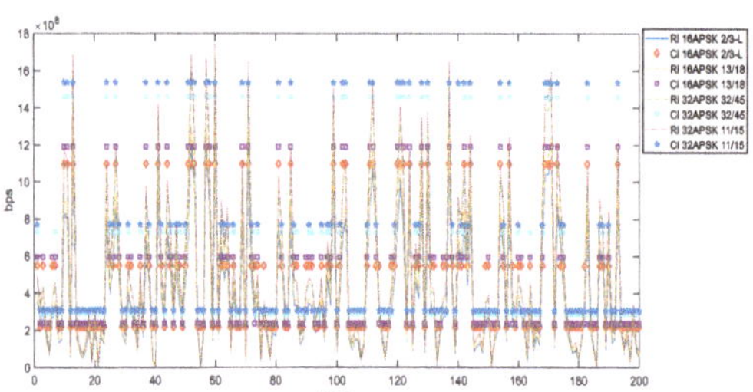

Fig. 4.2.3 Payload 2 performance for non-uniform traffic demand over North America in a specific moment.

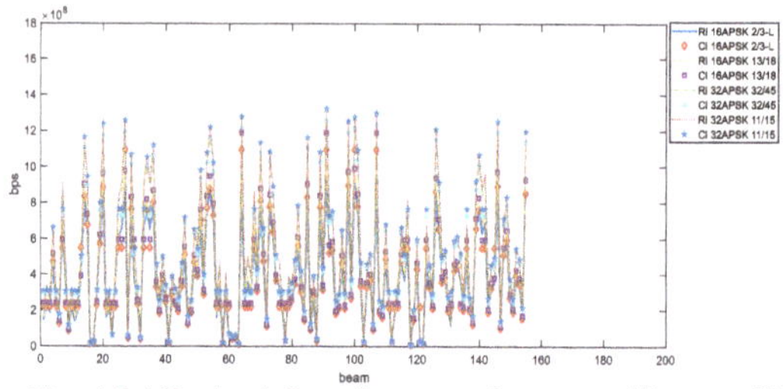

Fig. 4.2.4 Payload 3 performance for non-uniform traffic demand over North America in a specific moment.

Table 2 *Scenario for payloads performance evaluation in a specific interval over North America*

	Payload 1	Payload 2	Payload 3
Characteristics	Traditional Payload	Flexible payload over regular beam sizes	Flexible payload over adaptive irregular beam sizes
Service Area	Mexico and Part of United States		
Frequency	19.7 GHz		
Beamwidth	0.55°	0.55°	0.65°, 0.55° and 0.45°
Bandwidth	250 MHz	100, 250 and 500 MHz	100, 250 and 500 MHz
Power per beam	constant	variable	variable

5 Payload Architecture

Flexible payload architecture is an important limitation that must be studied more carefully. The study of new technologies for flexible payload is a priority. All subsystems of payload satellite must be evaluated to achieve the greatest possible system flexibility.

In [5], authors propose radio resource management for the flexibility payload. In [26], authors defined promising flexible payload solutions. The flexible payload must be capable to separate signals into frequency blocks and then rearrange them, this process requires a channelizer on board the satellite, as the authors mention in [27], the channelizer must have a function of separating and rearranging several thousands to several tens of thousands of blocks depending on the frequency width for the separation and the connection to terrestrial terminals through the connection link. In [26], authors explain that there was no a unique solution, but a set of key units is defined permitting to propose a range of payload solution answering.

We proposed a preliminary architecture for the flexible payload with adaptive set of irregular beam sizes (Figure 5.1 (a)). It is important that bandwidth, power and beamwidth allocation could be adaptive to obtain complete flexibility.

Actually, there are commercial variables TWTA to obtain flexible power allocation, the TWTA IBO must be adaptive. Each TWTA will be an independent unit to obtain flexible bandwidth allocation, each one will be able to feed more than one beam and each one will have a certain number of carriers [5]. If BW_c is the bandwidth per carrier, the bandwidth flexibility allocation will consist to allot a certain number of carriers per beam. Thus, the bandwidth beam will be a multiplication (C_n x BW_c), where C_n is the number of carriers.

MFB (multiple feed per beam) is proposed for adaptive beamwidth allocation (Figure 5.1 (b)). The OMUX of traditional payload must be replaced by a BFN Scenario Switch and BFN for each of the pre-designed and optimized coverage scenarios depending on the projection of traffic demand. BFN Scenario switch is halfway between being able to synthesize any beam and a choice between a set of possible coverages. Although the BFN Scenario switch would not have full flexibility, very satisfactory results can be obtained by optimizing coverages for a previous predictions of traffic demand during the the satellite life, added the flexibility in bandwidth and power will provide an allocation of optimized resources.

a)

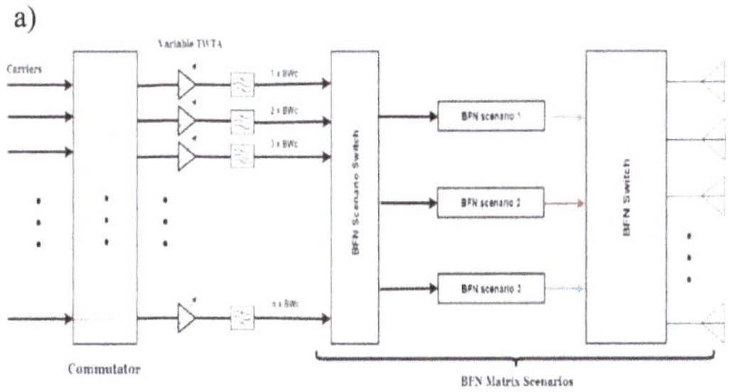

b)

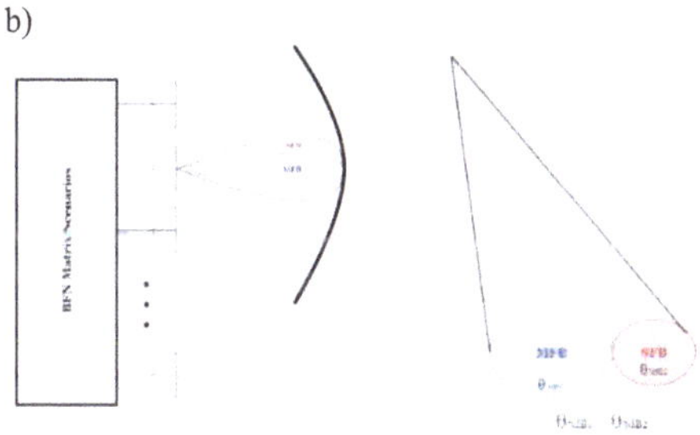

Fig. 5.1 a) *Architecture for Flexible payload (power, bandwidth and beam sizes)*
b) BFN Scenario Switch

MFB designs use small sub-arrays for each beam, and adjacent beam share some of the elements of the matrix. Overlapping power matrices are created, which allow producing superimposed beams using a single reflector aperture [23]. Due to this physical overlap of the feed apertures, overlapping beams can be produced using a single reflector aperture. Thus, mass and cost reduce and simplifies the accommodation of the antennas. But it is important to mention that is obtained at the expense of a complicated beamforming network (BFN) [23]. The MFB system design must be capable to offer flexibles and adaptive coverage for irregular beam sizes. One of the main constraint in flexible payload is the beam size flexibility, for this reason, we propose that MFB system must be capable to modified the multibeam coverage selecting one option of a predesign set.

6 Conclusion

This paper presents a preliminary study about requirements, methodology and architecture for flexible payload implementation, highlighting the importance of coverage optimization and adaptive irregular beam sizes. In this sense, flexible payload is a solution needed for non-uniform traffic demand changing over time an d this paper presented a preliminary proposal methodology and architecture.

Furthermore, the paper results show the variety of possible studies that can be developed on future works, such as, a more exhaustive study about coverage optimization, an analysis about how much time must pass to readjust the payload characteristics, the comparison between different payload technologies to achieve greater flexibility, consideration of traffic demand for 5G networks, and the flexible payload dimensioning for rain conditions. as well as the implication of system costs.

Acknowledgements

Authors would like to acknowledge the Spanish Government, Ministry of Economy, National Program of Research, Development and Innovation for the support of this publication in the projects ENABLING-5G "ENABLING INNOVATIVE RADIO TECHNOLOGIES FOR 5G NETWORKS" (project number TEC2014-55735-C3-1-R) and FUTURE-RADIO "Radio systems and technologies for high capacity terrestrial and satellite communications in an hyperconnected world" (project number TEC2017-85529-C3-1-R).

References

[1] Competitive Digital Single Market,'Towards a European Gigabit Society', COM 587 and Staff Working Document-SWD 300, 2016.

[2] 'New-generation satellite to deliver high-speed broadband across Euroe', [Online]. Available: https://www.eutelsat.com/home/satellites/future-satellites/konnect-vhts.html#. [Acceso: 2 August 2018].

[3] A. Abad Martín y M. A. Salas-Natera, 'The Satellite, a Key Element in Telecommunication Networks'.

[4] M. Bousquet y G. Maral, Satellite Communication Systems. Systems, 'Techniques and Technologies', Wiley, 2010.

[5] G. Cocco, T. de Cola, M. Angelone, Z. Katona y S. Erl, 'Radio Resource Management Optimization of Flexible Satellite Payloads for DVB-S2 Systems', *IEEE Transactions on Broadcasting,* vol. 99, pp. 1-15, 2017.

[6] A. Kyrgiazos, B. Evans y P. Thompson, 'Irregular beam sizes and non-uniform bandwidth allocation in HTS multibeam satellite systems', *AIAA International Communications Satellite Systems Conference*, 2013.

[7] F. Ortiz-Gomez, R. Martinez-Rodriguez y S. Landeros-Ayala, 'Method of Optimizing the Costs of a Satellite Network in Ka and Q/V Bands in the Feeder Link', de

AIAA International Communications Satellite Systems, Trieste, Italia, 2018.

[8] J. Lei y M. A. Vazquez-Castro, 'Joint power and carrier allocation for the multibeam satellite downlink with individual SINR constraints', *IEEE Int. Conf. Commun*, Cape Town, South Africa, 2010.

[9] J. Lei y M. A. Vázquez-Castro, 'Multibeam satellite frequency/time duality study and capacity ptimization', *J. Commun. Netw.*, vol. 13, nº 5, pp. 472-480, 2011.

[10] H. Fenech, A. Tomatis, S. Amos, V. Soumpholphakdy y D. Serrano-Velarde, 'Future High Throughput Satellite systems', de *IEEE First AESS European Conference on Satellite Telecommunications (ESTEL)*, Rome, 2012.

[11] J. Ho Jo, M. Hee You, S. Pal Lee y J. Hoon Kim, 'Manufacturing and Testing of Ka-band Communication Payload of COMS', de *The Third International Conference on Advances in Satellite and Space Communications*, Budapest, 2011.

[12] H.-J. Lee, J. M. Kim, B.-S. Lee, H. Lee y J.-S. RYOO, 'Recent Korean R&D in Satellite Communications', *IEICE Transactions on Communications,* Vols. 1 de 2E92-B, nº 11, pp. 3300 - 3308, November 2009.

[13] S. Yun, M. Uhm, J. Choi y I. Yom, 'Multibeam Reflector Antenna fed by few elements for Ka-band Communication Satellite', *IEEE Antennas and Propagation Society International Symposium (APSURSI)*, Chicago, IL, 2012.

[14] Y. Min Lee, J. Won Eun y S. Pal Lee, 'Conceptual Design of the Ka-band Communication Payload for Communication, Oceanographic, and Meteorological Satellite (COMS)', *Advanced Communication Technology, 2004. The 6th International Conference on*, Phoenix Park, 2004.

[15] S. Pal Lee, J. Ho Jo, M. Hee You y J. Sup Choi, 'Integration, Testing and In Orbit Validation of Ka-band Communication Payload of COMS', *Information and Communication Technology Convergence (ICTC), 2010 International Conference on*, Jeju, 2010.

[16] C. Chandler, L. Hoey, A. Peebles y M. Em, 'Advanced Antenna Technology for a Broadband Ka-Band Communication Satellite', *Technology Review Journal,* pp. 37 - 55, 2002.

[17] C. Chandler, L. Hoey, D. Hixon, R. Chan, T. Smigla, L. Wilson, A. Peebles, M. Em y S. Suleiman, 'Broadband Communications Satellite Antenna Technology for the Emerging Ka-band Market', *TRW Space & Electronics,* pp. 1 - 8, 2001.

[18] E. Pfeiffer, O. Reichmann, A. Ihle, S. Linke, C. Tshepe, N. Nathrath, M. Santos, A. Grillenbeck, J. Santiago-Prowald, P. Rinous y L. Rolo, 'Compact and Stable Earth Deck Multi-Beam Ka-Band Antenna Structure and Dual Gridded Reflector', de *European Conference on Antennas and Propagation, EuCAP*, Roma, 2011.

[19] I. Ohtomo, H. Kumazawa, T. Itanami, K. Ueno, A. Kondo, T. Yasaka, K. Nakajima, Y. Kawakami y M. Misawa, 'On-board Multibeam Deployable Antennas Using Ka, C, and S Frequency Bands', *IEEE Transactions on Aerospace and Electronic Systems,* vol. 28, nº 4, pp. 990 - 1001, October 1991.

[20] C. Rave y A. F. Jacob, 'Architectures for efficient power sharing in active multiple-feed-per-beam satellite antennas', International Conference on Microwave, Radar and Wireless Communications (MIKON)', *International Conference on Microwave, Radar and Wireless Communications (MIKON)*, 2016.

[21] M. Schneider, C. Hartwanger, E. Sommer y H. Wolf, 'The multiple spot beam antenna project Medusa', *European Conference on Antennas and Propagation*, 2009.

[22] G. Caille, Y. Cailloce y J. Torstensson, 'Space multibeam antenna with very high G/T, for ka-band multimedia via satellite missions', de *9th International Symposium on Antenna Technology and Applied Electromagnetics*, Quebec, Canada, 2002.

[23] M. Schneider, C. Hartwanger y H. Wolf, 'Antennas for multiple spot beam satellites', *CEAS Space Journal,* vol. 2, pp. 1-4, 2011.

[24] M. J. Box, 'Non-linear Optimization Techniques', Oliver & Boyd, 1969.

[25] M. S. Bazaraa, H. D. Sherali y C. M. Shetty, 'Nonlinear Programming: Theory and Applications', Wiley-Interscience, 2006.

[26] Thales Alenia Space, 'ARTES ¾ Generic Flexible payload Study Final Report', 2015.

[27] N. Komiyama, A. Miura, T. Orikasa y Y. Fujino, 'Development of Resource Allocation Re-construction Technology (Digital Beam Former and Digital Channelizer', *Journal of the National Institute of Information and Communications Technology,* vol. 62, nº 1, pp. 151-163, 2015.

Adaptive Onboard Compensation of Non-Linear HPAs and Imperfect Butler Matrices in Multiport Amplifiers for High Throughput Satellites

Ovais Bin Usman*, Gerhard Staude† and Andreas Knopp‡

Munich University of the Bundeswehr, Neubiberg, 85577, Germany

Multiport amplifier (MPA) is a key technology for future high throughput satellites (HTS) which aim at delivering extremely high data rates with flexible payloads. MPAs offer more flexible and efficient use of power onboard the satellite. This paper outlines two major issues which degrade the optimal MPA performance, i.e. the non-linear effects of high power amplifiers (HPAs) and the hardware imperfections in implementing input/output-hybrid Butler matrices (INET/ONET). Furthermore, a novel two-step technique to improve MPA performance is presented. The proposed technique compensates the imperfect INET/ONET in the first step and predistorts the non-linear amplifiers in the second step. Simulation results presented indicate that by implementing the proposed adaptive solution, we gain in the bit error ratio (BER) performance and the system robustness to hardware imperfections.

1. Introduction

Payload flexibility will be an important requirement for future high throughput satellites (HTS) to optimize their performance in different operating conditions. Moreover, flexibility in power is of significant importance especially for multi-beam satellite systems, which utilize frequency reuse schemes to deliver high-speed broadband services such as video on demand. Multiport amplifiers (MPAs) provide a practical solution for the payload's power flexibility requirements and have been studied as an effective approach in multi-beam systems[1]. The basic operation of the MPAs is to allow power sharing between different output ports, and can be described by the use of multiple HPAs through signal distribution and re-combination. The incoming input signals are transformed by an input-hybrid-matrix (INET) and inputted to the stack of HPAs. After amplification, the signals are then recombined by an output-hybrid-matrix (ONET)[2].

For MPAs to operate ideally, the INET and ONET should be built perfectly in the payload. Butler matrices are used to implement INET and ONET. Since INET and ONET are analog devices, they suffer from hardware imperfections which lead to ineffective Butler matrix operations. This further causes leakage to adjacent ports, unwanted power loss[3], and port isolation problems[4]. In addition to the ideal INET/ONET, the HPAs are required to be linear with equal gains for successful operation of the MPAs in the conventional MPA models[5]. However, aging of HPAs causes their AM-AM and AM-PM characteristics to vary over time. Different HPAs in the MPA may age at different rates causing unequal gains which does not allow the perfect recombination of the signals at the output ports of the MPA even if perfect INET/ONET Butler matrices exist. To keep the HPAs linear, usually a high input backoff (IBO) is used, but this leads to power inefficient HPAs[6]. In this study, the performance of MPAs is analyzed by removing the aforementioned assumptions made for the optimal MPA performance. Furthermore, the HPAs are considered non-linear and operated closer to saturation. Operating HPA closer to saturation introduces severe intermodulation noise and significant inter-channel/carrier interference (ICI). Moreover, these non-linear effects become more significant when multicarrier signals are used[7]. To counteract these non-linear effects, signal predistortion (SPD) onboard the satellite is considered for MPAs in this study. Conventionally, SPD is implemented using low-cost analog linearizers which over time suffer from degradation in performance due to the aging of the HPAs[8]. However, with the emergence of onboard processors (OBPs), adaptive digital predistortion (DPD) can be implemented. Adaptive DPD not only compensates for the HPA's non-linearities, but also tracks the changes in signal characteristics at the satellite and

*Research assistant, Chair of Signal Processing, ovais.usman@unibw.de

†PD Dr.-Ing, Chair of Signal Processing, gerhard.staude@unibw.de

‡Professor, Chair of Signal Processing, andreas.knopp@unibw.de

adjusts the DPD coefficients accordingly. To keep the payload complexity low, this study considers implementing bandlimited DPD[9] which imposes lower burden on the processing power and bandwidth handling requirements of the analog to digital converters (ADCs) and OBPs[10].

To improve MPA performance, a two-step adaptive method is proposed which presents a novel technique to effectively compensate the effects of imperfect INET/ONET in the first step, and performs bandlimited DPD[6,9] for the non-linear HPAs in the second step. The result section will present the performance gains observed in terms of bit error rates (BERs), when the adaptive two-step scheme is implemented. From now on , this extended abstract is organized as follows: Section II presents the conventional MPA model and highlights the issues related to it. Section II also explains the two-step technique for compensation of the imperfect INET/ONET and the block-based predistortion of the non-linear amplifiers. In Section III and Section IV, simulation results are interpreted and the conclusions are drawn. Notation used throughout the paper: Bold letters indicate vectors and matrices, and non-bold letters express scalars. The operators $(.)^{\mathrm{T}}$ and $(.)^{H}$ represent the transpose and the hermitian operator, respectively. $\boldsymbol{I}_N$ builds the $N \times N$ identity matrix.

2. Multiport Amplifier Architecture

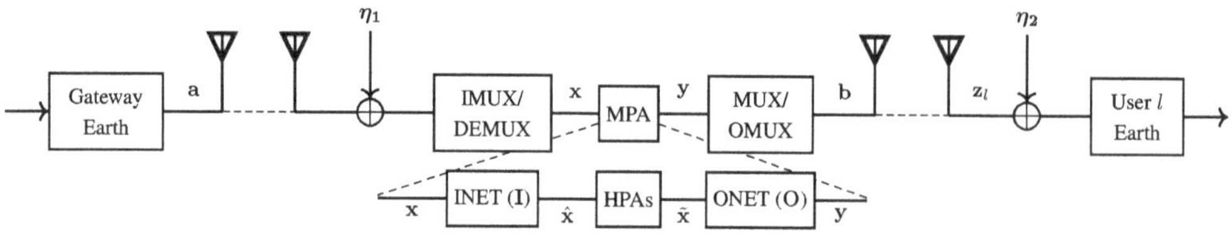

Figure 1: Satellite transmission link with a conventional MPA model.

Fig. 1 represents the considered system model depicting a basic satellite transmission chain. The focus is on the onboard $N \times N$ MPA architecture model zoomed-in on Fig. 1. An IMUX/de-multiplexer is used to capture the bandwidth of interest and to separate the received L-carrier signal into N $\frac{L}{N}$-carrier signals represented by $\mathbf{x} = [\mathrm{x}_1\ \mathrm{x}_2\ \ldots \mathrm{x}_N]^{\mathrm{T}}$. The de-multiplexed signals ($\mathbf{x}$) are then inputed to the INET of the MPA. The amplified signals $\tilde{\mathbf{x}} = [\tilde{\mathrm{x}}_1\ \tilde{\mathrm{x}}_2\ \ldots \tilde{\mathrm{x}}_N]^{\mathrm{T}}$ are then fed to the ONET, which recombines the signals with different phase shifts so that each of its output ports provides an amplified version $\mathbf{y}$ of the INET input signals.

A. Conventional linear MPA model

The $N \times N$ MPA represented by the zoomed-in part of Fig. 1 can be modeled as follows

$$\hat{\mathbf{x}} = \mathbf{I}\mathbf{x}\ (a) \quad \tilde{\mathbf{x}} = \mathbf{G}\hat{\mathbf{x}} = \mathbf{G}\mathbf{I}\mathbf{x}\ (b) \quad \mathbf{y} = \mathbf{O}\tilde{\mathbf{x}} = \mathbf{O}\mathbf{G}\mathbf{I}\mathbf{x}\ (c), \tag{1}$$

where $\mathbf{I} \in \mathcal{C}^{N \times N}$ and $\mathbf{O} \in \mathcal{C}^{N \times N}$ represent the INET and the ONET are Butler matrices. $\mathbf{G} \in \mathcal{C}^{N \times N}$ is the diagonal gain matrix for the N stacked HPAs. If $\mathbf{I}$ and $\mathbf{O}$ are unitary and exact conjugate transpose of each other, i.e., $\mathbf{IO} = \mathbf{OI} = \boldsymbol{I}_N$ and $\mathbf{G}$ is a diagonal matrix with equal linear gains of the form $g_i = \mathrm{A}e^{j\theta}\ \forall i = 1 \cdots N$, then the desired output of the MPA architecture is given as follows

$$\mathbf{y}_d = \mathbf{G}\mathbf{x}. \tag{2}$$

B. Practical issues with conventional MPAs

When the assumptions of accurate unitary INET/ONET and linear HPAs with equal gains are not met the MPA performance is affected. This subsection details the issues which arise when these two assumptions are removed.

1. Non-Linear HPAs

HPAs are generally driven closer to the saturation point to increase their power efficiency. However, this leads to non-linear distortions in HPA's output. In general, the HPA output can be expressed as

$$\mathrm{y}(t) = g[r(t)] \exp\{\theta(t) + \psi[r(t)]\}, \tag{3}$$

where $r(t)$ represents the envelope of the input signal to the HPA. $g[r(t)]$ and $\psi[r(t)]$ are non-linear functions which can be defined using the Saleh model detailed in Ref. 11. Since the Saleh model is non-linear the HPA output (Eq.

3) becomes non-linear as well. This means that Eq. 3 cannot be expressed as $y_i = g_i x_i$, since $g_i \neq Ae^{j\theta}$, therefore making Eq. 1b and Eq. 1c invalid. Furthermore, it can be observed from Eq. 3 that the output of the HPA in the non-linear case depends on the envelope of the input signal. This can lead to not only non-linear but unequal HPA gains. The resulting intermodulation noise (IMD) and spectral regrowth limits the perfect reconstruction of the input signal at the output of the MPA, even if the INET and ONET are implemented accurately in the hardware.

2. *Hardware imperfections in INET or ONET*

The INET and the ONET of the MPA are generally implemented in the hardware using 3 dB couplers (Butler matrices). For the correct working of MPAs, the INET and the ONET must be implemented in the analog domain such that they work as ideal phase shifters. For example, an accurate 2×2 INET and ONET should be able to to perform the following matrix operation on the incoming signal.

$$\mathbf{I} = \frac{1}{\sqrt{2}}\begin{bmatrix} 1 & j \\ j & 1 \end{bmatrix} (a). \quad \mathbf{O} = \frac{1}{\sqrt{2}}\begin{bmatrix} 1 & -j \\ -j & 1 \end{bmatrix} (b). \tag{4}$$

2×2 3 dB couplers can be joined to implement larger 4×4 or 8×8 INETs/ONETs. The 3 dB couplers are analog components which age over time. As a result, they loose their perfect phase shifting capabilities causing the imperfect INET/ONET Butler matrix operations. This leads to inter-branch/port interference and the isolation problem of the input signals at the output of the MPA. In this study, a degraded ONET is considered represented by a matrix $\bar{\mathrm{O}} \in \mathcal{C}^{N\times N}$. Furthermore, the INET is digitally implemented in the OBP.

C. Proposed Two-Step Method

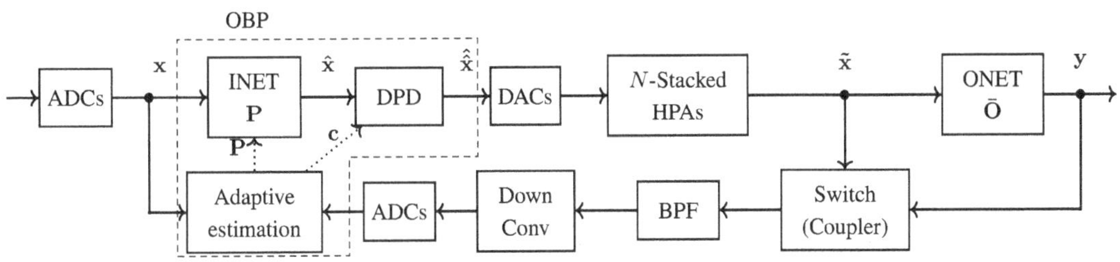

Figure 2: The proposed MPA model for the two-step compensation for pmperfect INET/ONET and DPD for non-linear HPAs.

The previous subsection explained the issues which can arise in MPA's performance in the presence of imperfect INET/ONET and non-linear HPAs. A two-step compensation method is presented here for such problems. Fig. 2 presents the proposed MPA architecture which implements the two-step method. In the first step, a modified INET matrix **P** detailed in Ref. 12 is computed which compensates for the imperfect ONET and interference at the output due to the unequal gains of the HPAs. To compute **P**, first the imperfect ONET ($\bar{\mathrm{O}}$) is computed in the digital domain. Once the new modified INET **P** is computed, it is updated in the MPA model of Fig. 2, and then the HPA output is used to adaptively obtain the DPD coefficients **c** in the second step.

1. *Step 1: Imperfect ONET compensation*

Step 1 implements a modified INET in the digital domain in form of a matrix $\mathbf{P} \in \mathcal{C}^{N\times N}$. The ONET is still built in the analog domain due to the higher power handling requirements. To compute the compensation matrix **P**, it is assumed that the HPAs are linear, therefore the DPD block can be bypassed. Operating HPAs linearly allows the basic MPA model of Eq. 1 to remain applicable, even when ONET imperfections exist. The following set of equations hold for step 1

$$\hat{\mathrm{x}} = \mathbf{P}\mathrm{x} \text{ (a)} \quad \tilde{\mathrm{x}} = \mathbf{G}\hat{\mathrm{x}} = \mathbf{G}\mathbf{P}\mathrm{x} \text{ (b)} \quad \mathrm{y} = \bar{\mathrm{O}}\tilde{\mathrm{x}} = \bar{\mathrm{O}}\mathbf{G}\mathbf{P}\mathrm{x} \text{ (c)}, \tag{5}$$

where $\mathbf{G} \in \mathcal{C}^{N\times N}$ is a diagonal matrix, and the diagonal entries g_i can be unequal. Assuming that the matrix $\bar{\mathbf{O}}$, i.e, the degraded real ONET is known, the digital compensatory matrix $\mathbf{P}$ can be computed as follows. The desired output of the system irrespective of the imperfect ONET is given by Eq. 2, i.e., $\mathbf{y}_d = \mathbf{Gx}$. However, the actual output of the system is $\mathbf{y} = \bar{\mathbf{O}}\mathbf{GPx}$. By equating the actual output of the system to the desired output, the compensatory matrix $\mathbf{P}$ becomes

$$\mathbf{P} = (\bar{\mathbf{O}}\mathbf{G})^{-1}\mathbf{G}. \tag{6}$$

Eq. 6 can be computed accurately if $\bar{\mathbf{O}}$ is known exactly. Since the analog components degrade over time, however, the imperfections in $\bar{\mathbf{O}}$ need to be estimated adaptively. To estimate $\bar{\mathbf{O}}$, a feedback loop is required as depicted in Fig. 2. However, due to the measurement and the quantization noise, the feedback input signal $\hat{\mathbf{y}}$ to the OBP contains measurement errors which is expressed in the digital domain as.

$$\hat{\mathbf{y}} = \bar{\mathbf{O}}\mathbf{GPx} + \eta_Q \quad (a) \qquad \hat{\mathbf{y}} = \bar{\mathbf{O}}\underline{\mathbf{x}} + \eta_Q \quad (b), \tag{7}$$

where η_Q is the added measurement noise and can be modeled as AWGN, and $\underline{\mathbf{x}} = \mathbf{GPx}$. Letting $\mathbf{P} = \mathbf{I}$, i.e., the perfect INET, computing an estimate $\hat{\mathbf{O}}$ for the true degraded ONET ($\bar{\mathbf{O}}$) becomes a system identification problem in the presence of noise, where $\hat{\mathbf{y}}$ and $\mathbf{x}$ are the measured and input signals, respectively. $M >> N$ observations are made to make the system in Eq. 7 an over-determined system expressed as follows

$$\begin{bmatrix} \underline{x}_{1,1} & \underline{x}_{2,1}\cdots & \underline{x}_{i,1} & \underline{x}_{N,1} \\ \underline{x}_{1,2} & \underline{x}_{2,2}\cdots & \underline{x}_{i,2} & \underline{x}_{N,2} \\ \vdots & \vdots & \vdots & \vdots \\ \underline{x}_{1,M} & \underline{x}_{2,M}\cdots & \underline{x}_{i,M} & \underline{x}_{N,M} \end{bmatrix} \begin{bmatrix} \hat{\mathbf{O}}_{i,1} \\ \hat{\mathbf{O}}_{i,2} \\ \vdots \\ \hat{\mathbf{O}}_{i,N} \end{bmatrix} = \begin{bmatrix} \hat{y}_{i,1} \\ \hat{y}_{i,2} \\ \vdots \\ \hat{y}_{i,M} \end{bmatrix},$$

$$\underline{\mathbf{X}}\hat{\mathbf{O}}_i^{\mathrm{T}} = \hat{\mathbf{y}}_i^{\mathrm{T}}, \tag{8}$$

where $\underline{x}_{i,j}$ and $\hat{y}_{i,j}$ represent the jth observation at the ith port of the MPA, respectively. Furthermore, $\hat{\mathbf{O}}_i$ is the ith row vector of the degraded ONET matrix. To ensure sufficient independent linear equations for the over-determined system, an artificial uncorrelated input signal $\mathbf{x}$ is used to excite the MPA model of Fig. 2. A least squares (LS) technique is applied to Eq. 8 to solve for the ith row of the imperfect $\hat{\mathbf{O}}$. This results in

$$\hat{\mathbf{O}}_i^{\mathrm{T}} = (\underline{\mathbf{X}}^H\underline{\mathbf{X}})^{-1}\underline{\mathbf{X}}^H\hat{\mathbf{y}}_i^{\mathrm{T}}, \tag{9}$$

where the solution of Eq. 9 is a row vector estimate of the respective real degraded ONET ($\bar{\mathbf{O}}$) Butler matrix row vector. Note that the ONET ages slowly, therefore, $\hat{\mathbf{O}}$ and $\mathbf{P}$ can simply be computed during the routine IoT measurements.

2. ***Step 2: Digital predistortion for non-linear HPAs***

The proposed architecture depicted by Fig 2 can be used in the second step to perform DPD for MPAs. The DPD linearizes the stacked HPAs leading to better MPA performance[6,9]. After the compensatory INET $\mathbf{P}$ is updated using Eq. 6, the DPD coefficients $\mathbf{c}$ can be updated adaptively during the normal satellite operation. The DPD technique used in this study to show the improvement in the performance of MPAs is detailed in Ref. 6. The implemented DPD is a block-based method which utilizes the LS based approach to compute the DPD coefficients. Note that the DPD coefficients need to be updated adaptively because of the changing signal characteristics of the uplink signal, e.g. higher number of carriers, adaptive coding, increased bandwidth etc. Therefore, the DPD coefficients are updated regularly during the normal operation of the satellite, unlike the compensatory matrix $\mathbf{P}$ which can just be updated during the service or downtime of the satellite. Since each output port of INET contains a portion of the incoming signals, it is sufficient to perform the estimation of DPD coefficients using only one port of the MPA and apply the DPD coefficients to all the ports. As a result only one feedback loop and one bandlimiting filter is required. This significantly reduces the added complexity to perform DPD for MPAs. The bandlimited DPD algorithm implemented in this paper is detailed in Ref. 13. A short summary is provided here.

Using the memory polynomial model[13], the mth observation of the DPD output and the bandlimited HPA output at the ith port is given by

$$\hat{x}_i(m) = \sum_{k=1}^{K}\sum_{q=0}^{Q} c_{kq,BL}\left[\hat{x}_i\,(m-q)\,|\hat{x}_i\,(m-q)|^{k-1}\right] \tag{10a}$$

$$\tilde{x}_i^{BL}\,(m) = \sum_{k=1}^{K}\sum_{q=0}^{Q} w_{kq,BL}\left[\sum_{l=0}^{L}\hat{\hat{x}}_i\,(m-q-l)\left|\hat{\hat{x}}_i\,(m-q-l)\right|^{k-1} g(l)\right] \tag{10b}$$

where K and Q represent the non-linearity order and maximum memory depth, respectively. c_{BL} and w_{BL} are the unknown bandlimited memory polynomial model coefficients for the DPD and the HPA. $\hat{x}_i\,(m)$ is the mth sample at the ith output port of the INET. $g(.)$ is the bandlimiting filter in the feedback loop. Defining,

$$\hat{x}_i(n,k,q) = \hat{x}_i(n-q)\,|\underline{\hat{x}}_i(n-q)|^{k-1} \tag{11}$$

$$\hat{\hat{x}}_i(m,k,q) = \sum_{l=0}^{L}\hat{\hat{x}}_i\,(m-q-l)\left|\hat{\hat{x}}_i\,(m-q-l)\right|^{k-1} g(l), \tag{12}$$

the Eq. 10 can be written compactly as follows

$$\hat{\hat{x}}_i(m) = \underline{\hat{\mathbf{x}}}_i(m)^{\mathrm{T}}\mathbf{c}_{BL}, \tag{13a}$$

$$\tilde{x}_i^{BL}\,(m) = \underline{\hat{\hat{\mathbf{x}}}}_i(m)^{\mathrm{T}}\mathbf{w}_{BL}, \tag{13b}$$

where $\underline{\hat{\mathbf{x}}}_i(m) = [\hat{x}_i(m,1,0)\ \hat{x}_i(m,1,1)\ \cdots\hat{x}_i(m,1,Q)\ \cdots\hat{x}_i(m,K,Q)]^{\mathrm{T}}$. $\underline{\hat{\hat{\mathbf{x}}}}_i(m)$ also has the same definition as $\underline{\hat{\mathbf{x}}}_i(m)$. The DPD vector is defined as $\mathbf{c}_{BL} = [c_{10}\ c_{11}\cdots c_{1Q}\cdots c_{KQ}]^{\mathrm{T}}$, and $\mathbf{w}_{BL}$ is defined in the same way as $\mathbf{c}_{BL}$. By gathering M samples or observations of the input at the ith port, where M is block size, i.e. $\underline{\mathbf{x}}_i = [x_i(0)\ x_i(1)\ \cdots\ x_i(M-1)]^{\mathrm{T}}$, we can write the MP models in matrix form as

$$\underline{\hat{\hat{\mathbf{x}}}}_i = \underline{\hat{\mathbf{X}}}_i\mathbf{c}_{BL}, \tag{14a}$$

$$\underline{\tilde{\mathbf{x}}}_i^{BL} = \underline{\hat{\hat{\mathbf{X}}}}_i\mathbf{w}_{BL}, \tag{14b}$$

where $\underline{\hat{\mathbf{X}}}_i = [\underline{\hat{\mathbf{x}}}_i(0)\ \ \underline{\hat{\mathbf{x}}}_i(1)\ \cdots\ \underline{\hat{\mathbf{x}}}_i(M-1)]^{\mathrm{T}}$. $\underline{\hat{\hat{\mathbf{X}}}}_i$ and $\tilde{\mathbf{X}}_i^{BL}$ can be defined in a similar way as $\underline{\hat{\mathbf{X}}}_i$

Eq. 14 can be solved for $\mathbf{w}_{BL}$ and $\mathbf{c}_{BL}$ by applying the least squares algorithm[14], i.e.

$$\mathbf{w}_{BL} = (\underline{\hat{\hat{\mathbf{X}}}}_i^H\,\underline{\hat{\hat{\mathbf{X}}}}_i)^{-1}\underline{\hat{\hat{\mathbf{X}}}}_i^H\,\underline{\tilde{\mathbf{x}}}_i^{BL}, \tag{15}$$

$$\mathbf{c}_{BL} = (\tilde{\mathbf{X}}_i^{BL^H}\tilde{\mathbf{X}}_i^{BL})^{-1}\tilde{\mathbf{X}}_i^{BL^H}\underline{\hat{\hat{\mathbf{x}}}}_i, \tag{16}$$

The QR decomposition method can be applied to obtain the above inverses as it leads to smaller rounding errors[15].

3. Simulation results

This section compares different simulation configurations in terms of the power spectral densities (PSDs) and BER. For simulation purposes, the ONET is always assumed to be imperfect, and the HPAs are assumed to follow the non-linear Saleh model. A $L = 8$-carrier signal is uplinked from the gateway to the satellite with a total bandwidth of 500 MHz. A 4×4 MPA is considered ($N = 4$) in the presented simulation results, i.e., $\frac{L}{N} = 2$ carriers access each input port of the INET. $N_b = 21600$ QPSK symbols are transmitted per block. For the pulse shaping of the QPSK modulated symbols, a root raised cosine filter with a roll-off factor of 0.35 is used. Unless stated otherwise, a feedback bandwidth of 500 MHz and IBO of 8 dB is used. As a worst case scenario, the imperfections in $\bar{\mathrm{O}}$ are modeled as random deviations from the perfect ONET (O), i.e., $\bar{\mathrm{O}} = \mathrm{O} + \Delta$, where $\Delta_{i,j} \sim \mathcal{N}\left(0, \sigma_d^2\right)$. However, the proposed novel compensation technique in the Step 1 is valid for any other degraded ONET model as well, since, first the degraded ONET is estimated, and then compensated for using Eq. 6. In this abstract, the results are provided for an ONET distortion level of $\sigma_d = 0.2$, which correspond to approximately -14 dB of inter-branch/port interference at the output of the MPA. It should be noted that the DPD algorithm implemented in this paper is valid for any other HPA model too. This is due to the fact that the proposed DPD technique first estimates the HPA model coefficients and then uses them to compute the DPD coefficients.

A. PSD Analysis

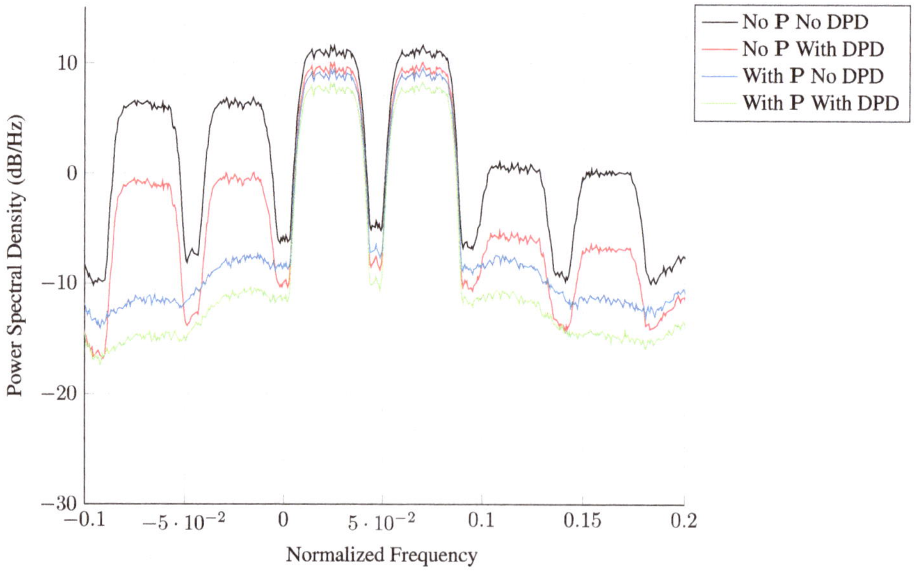

Figure 3: PSDs at the output of ONET, $\sigma_d = 0.2$, i.e inter-branch/port interference=-14 dB.

The PSD results at one of the output ports of the ONET are shown in Fig. 3. The results for only one output port are provided for the clarity of the figure. The other ports exhibit similar results. Note that when only the imperfect ONET is compensated with the new modified INET **P**, inter-branch/port interference is minimized, but the intermodulation noise remains. Furthermore, when only the proposed DPD is implemented and **P** is not used (imperfect ONET not compensated), the intermodulation noise is reduced but significant inter-branch/port interference is experienced. Moreover, the best performance with the lowest sidelobe power is observed, when both the **P** and the DPD are used simultaneously. Note that for the ease of understanding, the four 2-carrier input signals of the MPA were placed at different intermediate frequencies to readily observe the possible spectral regrowth and inter-branch/port interference. Generally, the N inputs to the INET are at the same intermediate frequency. Therefore, if any interference from the neighboring ports exists due to an imperfect ONET or unequal HPA gains, it gets added on top of the desired output signal of a particular port. These summed up interferences cannot be filtered at the receiver leading to significantly high BERs. Table 1 provides the adjacent channel power ratios[16] (ACPR) for the simulated example. The ACPR is lowest when the compensatory INET **P** and the DPD are implemented together.

Case	ACPR-left lobe	ACPR-right lobe
Two-step Scheme	-15.92 dB	-18.56 dB
Without P and With DPD	-4.58 dB	-10.8 dB
With P and Without DPD	-14.91	-15.65 dB
Without P and Without DPD	-4.56 dB	-10.53 dB

Table 1: ACPR Analysis for the 4 × 4 MPA example.

B. BER analysis

Fig. 4 presents the BER vs. $\frac{E_b}{N_o}$ of a single carrier for the simulated MPA example. For the clarity of figure the curves for the other carriers are not drawn on the figure as they exhibit similar results. Note that the worst BER performance is observed for the configuration in which no **P** and no DPD is used. While, the best BER performance is achieved when the two-step scheme is implemented to compensate for both, the imperfect ONET and the non-linear HPAs.

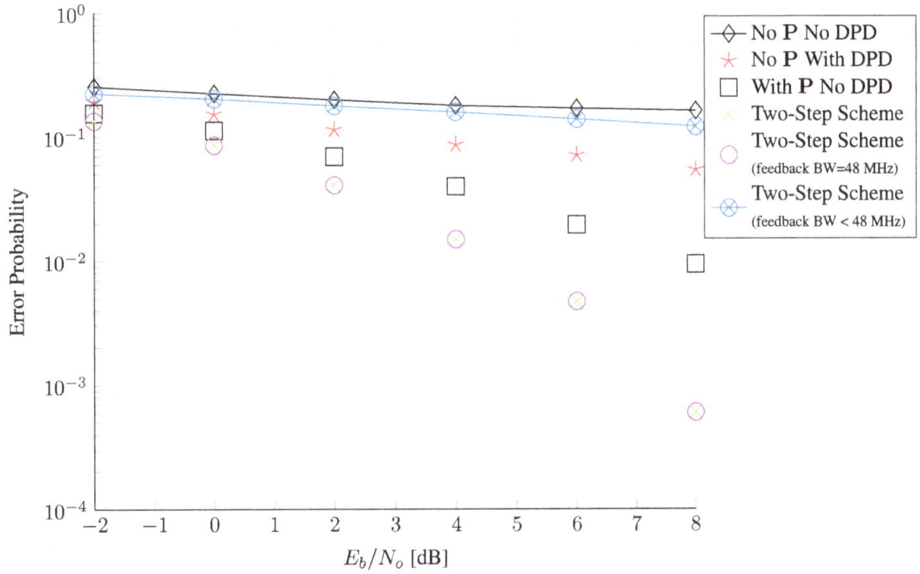

Figure 4: BER performance analysis, $\sigma_d = 0.2$, i.e inter-branch/port interference=-14 dB.

Furthermore, from Fig. 4, it can be seen that that a similar BER performance is achieved for the scenario when only 48 MHz of feedback bandwidth is used. If the feedback bandwidth is reduced further, the BER performance degrades as very little information about the true HPA output signal is available at the OBP to perform accurate estimation of the DPD coefficients. Performing DPD for the ultra-low bandwidths requires additional signal processing and complexity. Several techniques for fullband DPD with bandlimited information exist in the literature,[9,10,17] A complete analysis on the trade-off of reducing the feedback bandwidth and the added computational complexity for the MPAs in HTS scenarios is left as future work.

4. Conclusion

In this paper the practical issues related to the optimal MPA performance were presented. It was observed that the port isolation and BER performance of the MPA suffers in the presence of non-linear HPAs and hardware imperfections in INET/ONET. By simulations it was shown that the proposed adaptive two-step method to perform predistortion and ONET imperfection compensation provides the lowest BERs and ACPRs. Furthermore, bandlimited presdistortion was implemented for the presented HTS scenario to reduce the complexity of the payload and power consumption.

References

[1]Egami, S. and Kawai, M., "An Adaptive Multiple Beam System Concept," *IEEE Journal on Selected Areas in Communications*, Vol. 5, No. 4, May 1987, pp. 630–636.

[2]Mallet, A., Anakabe, A., Sombrin, J., and Rodriguez, R., "Multiport-Amplifier-Based Architecture Versus Classical Architecture for Space Telecommunication Payloads," *IEEE Transactions on Microwave Theory and Techniques*, Vol. 54, No. 12, Dec 2006, pp. 4353–4361.

[3]Lee, H. L., Lee, M. Q., and Yu, J. W., "Analysis of multi-port amplifier calibration for optimal magnitude and phase error detection," *IET Microwaves, Antennas Propagation*, Vol. 10, No. 1, 2016, pp. 102–110.

[4]Tanaka, M. and Egami, S., "Reconfigurable multiport amplifiers for in-orbit use," *IEEE Transactions on Aerospace and Electronic Systems*, Vol. 42, 2006, pp. 228–236.

[5]Lee, M.-Q., Lim Lee, H., and Won Yu, J., "Reconfigurable 4x4 Multi-port Amplifier with Switchable Input and Output Matrices," Vol. 10, 05 2016.

[6]Ding, L., Zhou, G. T., Morgan, D. R., Ma, Z., Kenney, J. S., Kim, J., and Giardina, C. R., "A robust digital baseband predistorter constructed using memory polynomials," *IEEE Transactions on Communications*, Vol. 52, No. 1, Jan 2004, pp. 159–165.

[7]Schetzen, M., "Theory of Pth-order inverses of nonlinear systems," *IEEE Transactions on Circuits and Systems*, Vol. 23, No. 5, May 1976, pp. 285–291.

[8]Zhou, D. and DeBrunner, V. E., "Novel Adaptive Nonlinear Predistorters Based on the Direct Learning Algorithm," *IEEE Transactions on Signal Processing*, Vol. 55, No. 1, Jan 2007, pp. 120–133.

[9]Liu, Y., Pan, W., Shao, S., and Tang, Y., "A General Digital Predistortion Architecture Using Constrained Feedback Bandwidth for Wideband Power Amplifiers," *IEEE Transactions on Microwave Theory and Techniques*, Vol. 63, No. 5, May 2015, pp. 1544–1555.

[10]Su, G., Chen, W., Zhang, S., and Ghannouchi, F. M., "A robust and low sampling rate digital predistortion algorithm for broadband PA modeling and predistortion," *WAMICON 2014*, June 2014, pp. 1–4.

[11]Saleh, A. A. M., "Frequency-Independent and Frequency-Dependent Nonlinear Models of TWT Amplifiers," *IEEE Transactions on Communications*, Vol. 29, No. 11, November 1981, pp. 1715–1720.

[12]Ovais Bin Usman, Thomas Delamotte, A. K., "Low-effort On-board Compensation for Hardware Imperfections in MPAs," *American Institute of Aeronautics and Astronautics*, 2017.

[13]Yu, C., Guan, L., Zhu, E., and Zhu, A., "Band-Limited Volterra Series-Based Digital Predistortion for Wideband RF Power Amplifiers," *IEEE Transactions on Microwave Theory and Techniques*, Vol. 60, No. 12, Dec 2012, pp. 4198–4208.

[14]Ding, L., Ma, Z., Morgan, D. R., Zierdt, M., and Pastalan, J., "A least-squares/Newton method for digital predistortion of wideband signals," *IEEE Transactions on Communications*, Vol. 54, No. 5, May 2006, pp. 833–840.

[15]Golub, G. H. and Van Loan, C. F., *Matrix Computations*, Johns Hopkins, 3rd ed., 1996.

[16]Lin, F.-L., Chen, S.-F., Chen, L.-F., and Chuang, H.-R., "Computer simulation and measurement of error vector magnitude (EVM) and adjacent-channel power ratio (ACPR) for digital wireless communication RF power amplifiers," *Gateway to 21st Century Communications Village. VTC 1999-Fall. IEEE VTS 50th Vehicular Technology Conference (Cat. No.99CH36324)*, Vol. 4, 1999, pp. 2024–2028 vol.4.

[17]Liu, Y., Yan, J. J., Dabag, H. T., and Asbeck, P. M., "Novel Technique for Wideband Digital Predistortion of Power Amplifiers With an Under-Sampling ADC," *IEEE Transactions on Microwave Theory and Techniques*, Vol. 62, No. 11, Nov 2014, pp. 2604–2617.

Distributed precoding for multiple satellite systems with overlapping coverage areas

V. Joroughi[1] B. M. R. Shankar[1] S. Maleki[1] S. Chatzinotas[1] J. Grotz[2] B. Ottersten[1]

[1] Interdisciplinary Centre for Security, Reliability and Trust (SnT), University of Luxembourg, 29 Avenue John F. Kennedy, L-1855, Luxembourg
[2] SES S.A. Betzdorf, Luxembourg
* E-mail: vahid.joroughi@uni.lu

Abstract: This study aims at designing precoding in multiple satellite multibeam systems with overlapping coverage areas where the multiple satellites have the same footprints and overlap in their coverage. Besides, high throughput full frequency reuse pattern among satellites is used. In such an architecture, the key objective lies in the collaboration between satellites and obtains: i) reliable multibeam infrastructure to serve unforeseen changes in the traffic demand through establishing supportive secondary satellites, ii) multiple satellites can provide service to higher spatial diversity by keeping the size of the payload affordable, iii) employing multiple satellites provides hardware redundancy to guarantee uninterrupted service delivery. However, intra-satellite and inter-satellite interference are the bottleneck of the whole network and employing interference mitigation techniques, particularly precoding techniques, is essential. In this paper, we analytically and numerically study designing precoding technique that: a) properly mitigates intra-satellite and inter-satellite interference, b) since the performance of precoding is sensitive to the quality of Channel State Information (CSI), proper low complex CSI exchange mechanism among the satellites is developed.

1 Introduction

Nowadays, the fixed satellite services are used to provide broadband connectivity to a large number of user terminals, distributed over a wide coverage area. The continued growth in demand for this broadband service in the recent years have stimulated the satellite service providers to deploy advanced interactive broadband services [1]. In this contesxt, the success story of Multiuser MIMO applications in terrestrial networks have been mimicked in satellite systems, through the use of multiple spot beams instead of a single global beam in the coverage area so that higher spectrum utilization is obtained by employing fractional frequency reuse among the beams [2].

1.1 Multibeam satellite systems: beyond state-of-the-art

The benefits of multibeam satellite systems can be summarized as:
(i) Obtaining larger antenna gain-to-noise (G/N) ratio than the single beam case.
(ii) Available spectrum in multibeam architecture can be reused among spatially separated beams.
(iii) The multibeam architecture can support different modulations and code rates depending on the user link (i.e. the bidirectional communication link between the satellite and user terminals) quality.
(iv) Possibility to serve simultaneously multiple users in different geographical locations [3].
Nevertheless, as the number of beams increase, the system performance becomes limited by the increased level of interference among the users due to the side lobes of the beams. To cope with this problem, the investigation of aggressive frequency reuse methods comes into play. The term aggressive frequency reuse refers operating a fractional reuse system among the beams so that adjacent beams operate in different frequency bands. N_c, with $N_c > 1$, indicates the number of disjoint frequency bands employed among beams. Clearly, lower the N_c, higher the overall system bandwidth is. On the other hand, higher the N_c, lower the interference level will be.
Another promising solution is to use full frequency reuse method, i.e. $N_c = 1$, within adjacent beams and allocating total available bandwidth to all the beams. However, a high level of inter-beam interference becomes the bottleneck of the whole system and application of interference mitigation techniques such as precoding in the forward link (i.e. the link between gateway and user terminals), and multi-user detection in the return link (i.e. the link between user terminals and gateway) is necessary [4]-[5]. Note that the performance of interference mitigation technique is sensitive to the quality of Channel State Information (CSI) at the gateway. In addition, these schemes are realized at the gateway in order to guarantee a high reduction of the user terminal and payload complexity.

1.2 Multiple satellites systems: benefits and challenges

One of the beneficial application of multiple satellite systems with multibeam coverage can be the case multiple satellites have the same footprints and thus completely overlap in their coverage. This architecture refers to Multiple Satellites with Completely Overlapped coverage (MS-COV). In practice, this MS-COV architecture is equivalent to the case where a new satellite is launched to the orbit before the current one is retired. Indeed, MS-COV closely resembles the collocated satellite systems, e.g. the four Astra1S in 19.2°E orbital location [6]. Concretely, the reasoning behind employing MS-COV architecture is summarized below:

(a) *Orbital slot congestion:* in the evolution of satellite systems, orbital slots are becoming a scarce resource. Therefore, deploying more than one multibeam satellites, in one orbital position becomes relevant [6].
(b) *Traffic demand:* the operational lifetime of a multibeam satellite spans over a period of more than fifteen years. It is probable that unpredictable changes in the traffic demand might dictate the launch of secondary satellites to support existing ones. Further, even if traffic demand is well predicted, the gradual deployment offered by multiple satellite system reduces the upfront investment and the operational cost.
(c) *Payload complexity:* full frequency reuse ($N_c = 1$) increases the communication payload size since a single high power amplifier cannot be shared by multiple beams [7]. Hence, the payload is required to drive a large number of beams that cover large regions (e.g. pan-European coverage) which can be carried by multiple satellites in MS-COV mode.
(d) *Redundancy:* hardware redundancy to guarantee uninterrupted service delivery in case of malfunctions. It implies that the MS-COV

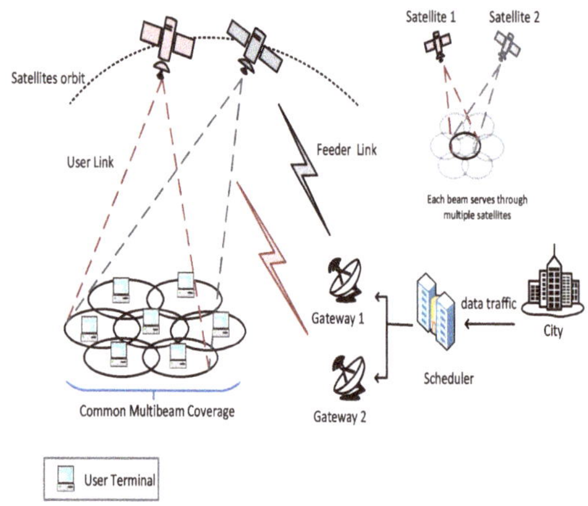

Fig. 1: The MS-COV architecture with a set of 2 gateways.

architecture provides the possibility that in case of failing one satellite, the traffic can be rerouted through other satellites to the coverage area.

(e) *Apriori co-location:* long periods of coexisting satellites appear by default during the satellite replacement phase. For this, a priori co-location can be exploited towards increasing the system capacity.

For the sake of clarity, Figure (1) depicts a MS-COV architecture.
In addition to the above mentioned advantages, one of the current application of the MS-COV configuration is to provide multiplexing and diversity which implies each user terminal receives a part of the desired data from one of the satellites or each user obtains the data through one satellite and copies of the same data through the rest of satellites. Thus, in case of satellite/gateway failure, the traffic can be received through other satellites. For the sake of keeping user terminal complexity low, the satellite orbits shall be close enough together as each user employs a single antenna to manage the multiple data from different satellites. However, MS-COV suffers from high cost level with respect to conventional systems with a single satellite. Moreover, developing signal processing schemes in MS-COV requires considering cooperation among the employed satellites and gateways, leading to a high computational complexity [8]-[11].

1.3 Motivation

This work attempts to design precoding technique in the forward link of MS-COV architecture while full frequency reuse pattern ($N_c = 1$) is employed among the satellites and beams. The objective of implementing precoding scheme in MS-COV architecture is to mitigate intra-satellite interference, i.e. the interference among the beams of each satellite, and inter-satellite interference, i.e. the interference among beams of adjacent satellites. While the performance of precoding has a close relation with the quality of CSI at the transmitting segment, a CSI exchange mechanisms is proposed aiming at properly handling CSI exchange among the gateways that serve the satellites.
Unfortunately, as multiple satellite create different CSIs, establishing CSI feedback mechanism among user and gateways is intuitively cumbersome. In addition, applying perfect CSI exchange mechanism among gateways entails establishing costly inter-gateway connection. To this end, we focus to develop a low complex but effective partial (i.e. non-perfect) CSI feedback and exchange mechanisms.
Concretely, the following MS-COV architecture is explored: each satellite connects to a single gateway whose role is to precode the transmitted signals and send to the satellite. To properly compute precoding matrix at each gateway, we use a partial CSI feedback from users to the gateways. Besides, a partial CSI exchanged mechanism is employed among gateways. For the sake of mathematical convenience, it is conceived that each beam can serve single user per beam.

Remark 1: It should be noted that even with highly directive antennas the feeder link (i.e. the bidirectional communication link between the satellite and gateway) originating at different gateways are partially interfering. However, we assume that gateways are sufficiently separated on the Earth surface and space so that the inter-feeder link interference can be ignored. ∎

In addition to the stated beneftis of MS-COV in Section 1.2, it is expected that the proposed MS-COV configuration can tackle feeder link bandwidth limitation. As a matter of fact, by increasing the demand in the coverage area the feeder link bandwidth resources must be enhanced accordingly. Unfortunately, the feeder link bandwidth resources are scarce due to its costly hardware requirements [4], hence, accommodating enough feeder link resources for increased demand in the multibeam coverage is hard task. For instance, consider that

$$\mathrm{B}_{\text{feeder-link}} = N\mathrm{B}_{\text{beam}}, \tag{1}$$

where N is the number of on-board feed signals. B_{beam} and $\mathrm{B}_{\text{feeder-link}}$ respectively are the per-beam and the required feeder link bandwidths. Denoting K as the total number of beams in multibeam coverage with

$$K < N, \tag{2}$$

it is evident that in case of increasing the bandwidth of each beam B_{beam} the feeder link bandwidth $\mathrm{B}_{\text{feeder-link}}$ should be accordingly increased. Remarkably, considering N number feeds, with $N > K$, reduces the scan losses for large coverage respect to the payload configuration with identical number of feeds and beams $N = K$. The case $N > K$ refers to on-board Multiple Feed per Beam (MFB) configuration. This feeder link limitation bottleneck in MS-COV can be covered by reusing the available feeder link bandwidth across different satellites so that by increasing the demand the new satellite is employed with the same feeder link resources. In this context, the required feeder link bandwidth becomes

$$\mathrm{B}_{\text{feeder-link-MS}} = \frac{N}{G}\mathrm{B}_{\text{beam}}, \tag{3}$$

where G is the number of satellites with $\mathrm{B}_{\text{feeder-link-MS}} < \mathrm{B}_{\text{feeder-link}}$.
The rest of the paper is organized as follows. Sections 2 present the signal model in MS-COV architecture. Precoding scheme for MS-COV architecture is developed in Sections 3. Section 4 provides numerical results. Finally, we derive our conclusion in Section 5.

Notation: Throughout this paper, the following notations will be adopted. Boldface uppercase letters denote matrices and boldface lowercase letters refer to column vectors. $(.)^H$ and $(.)^T$ denote Hermitian transpose and transpose matrices, respectively. $\mathbf{I}_N$ builds $N \times N$ identity matrix. $(\mathbf{A})_{i,i}$ represents the i-th and j-th entry of $\mathbf{A}$. Finally, $\mathrm{E}\{.\}$ and $||.||$ refer to the expected value operator and the Frobenius norm, respectively.

2 Signal model and problem formulation in MS-COV architecture

Consider the forward link of a MS-COV architecture, where G multiple geosynchronous (GEO) satellites, employing one gateway per satellite, provide broadband services to a common coverage (overlapped) area with a large set of users. To this end, each satellite is equipped with an array fed reflector antenna, whose number of feeds is denoted by N. These feeds serve a set of K beams in MFB mode in (2), and the users are assumed to be uniformly distributed within the beams. We assume a total number of $M \triangleq NG$ feeds at G satellites which serve K beams in an overlapped manner.

Following a Time Division Multiplexing (TDM) scheme, in each time instant, a total number of K users is served, i.e. one user per beam. At any time instant, each gateway/satellite transmits a total number of K symbols (traffic streams) to the beams/users, one symbol per beam, such that each user receives G streams, one per satellite. We consider these G streams per user appear considering:
(a) The G satellite orbits are close enough to each other and each user employ a single antenna to receive multiple symbols from different satellites.
(b) The multiple streams at each user can be appeared either (i) each user receives a part of the desired data through each satellite, or (ii) each user receives a data stream through one of the satellites and a set of $(G-1)$ copies of the same data stream through the rest of satellites. Obviously, in both cases a post-processing is required at each user aiming at distinguishing multiple received streams. In this work, we focus on (ii) due to its requirement for no specific post-processing at users.
The notation $B \triangleq KG$ represents the total transmitted signals to K users through G satellites.
Figure 1 shows the overall MS-COV architecture for $G = 2$ satellites.
In such an architecture, assuming Remark 1, the overall received signals can be modeled as

$$\mathbf{y} = \mathbf{Ht} + \mathbf{n}, \tag{4}$$

where $\mathbf{y} = \sum_{f=1}^{G} \mathbf{y}_f \in \mathbb{C}^{K\times 1}$ is a vector containing the received signals at K users through G satellites. The vector $\mathbf{y}_f \in \mathbb{C}^{K\times 1}$ denotes a total of K desired signals received from f-th satellite. $\mathbf{t} = (\mathbf{t}_f, ..., \mathbf{t}_f, ..., \mathbf{t}_G)^T \in \mathbb{C}^{B\times 1}$ is the stacked transmitted signals at K user terminals. Vector $\mathbf{n} \in \mathbb{C}^{K\times 1}$ is the impact of Gaussian distributed zero mean and unit variance noise at K user terminals, i.e. $E\left\{\mathbf{nn}^H\right\} = \mathbf{I}_K$. The matrix $\mathbf{H} \in \mathbb{C}^{K\times M}$ denotes the user link channel matrix and can be decomposed as follows

$$\mathbf{H} = (\mathbf{H}_1, \ldots, \mathbf{H}_f, \ldots, \mathbf{H}_G), \tag{5}$$

where $\mathbf{H}_f \in \mathbb{C}^{K\times N}$ represents the user link channel of f-th satellite with

$$\mathbf{H}_f = \mathbf{D}_f \mathbf{S}_f. \tag{6}$$

The diagonal matrix $\mathbf{D}_f \in \mathbb{C}^{K\times K}$ indicates the atmospheric fading in the user link and can be calculated by

$$\mathbf{D}_f = diag(\frac{1}{\sqrt{A_{f,1}}}, ..., \frac{1}{\sqrt{A_{f,k}}}, ..., \frac{1}{\sqrt{A_{f,K}}}), \tag{7}$$

where $A_{f,k}$ denotes the rain attenuation at user k achieved from the f-th satellite.
$\mathbf{S}_f \in \mathbb{C}^{K\times N}$ models the feed radiation patterns, the path loss and the receive antenna gain

$$\mathbf{S}_f = \left[\mathbf{s}_{f,1}^T, \ldots, \mathbf{s}_{f,k}^T, \ldots, \mathbf{s}_{f,K}^T\right]^T, \tag{8}$$

where row vector $\mathbf{s}_{f,k} \in \mathbb{C}^{1\times N}$ is the gain received from the N feeds of f-th satellite at the k-th user terminal.
The i-th entry of $\mathbf{s}_{f,k}$ is modeled as

$$\mathbf{s}_{f,k} = \left[s_{f,k,1}, \ldots, s_{f,k,i}, \ldots, s_{f,k,N}\right], \tag{9}$$

where $s_{f,k,i}$ refers to the feed transmit gain at the k-th user from the i-th feed of the f-th satellite.
We now proceed with calculating $s_{f,k,i}$.
For an identical number of beams and user terminals, the k-th user position towards the i-th feed of the f-th satellite can be illuminated by an off-axis angle $\phi_{f,k,i}$. It is assumed that the footprint of each beam is circular and the illumination angle at the center of the beam is zero. Then, the illuminating angle $\phi_{f,k,i}$ corresponds to the position of the user in the k-th beam is computed by

$$\phi_{f,k,i} = tan^{-1}(\frac{d_{f,k}}{h_f}), \tag{10}$$

where h_f and $d_{f,k}$ represent the height of the f-th satellite and the distance of user terminal k to the center of beam where it is located. Employing the this illuminating angle establishes the satellite beam gain as follows

$$s_{f,k,i}^2 = \left(\frac{(\rho+1)(1+\omega)}{(\rho+1)(1+\omega)+\omega}\Omega\right)^2 \tag{11}$$

with

$$\Omega \triangleq \left(\frac{2J_1(u)}{u} + 2^{\rho+1}\rho!\frac{\omega}{1-\omega}\frac{J_{\rho+1}(u)}{u^{\rho+1}}\right), \tag{12}$$

where $u = \lambda^{-1}\pi d_a \sin(\phi_{f,k,i})$ [4]. The notation d_a refers to the aperture diameter of the transmitting parabolic antenna, ω is the aperture edge taper and λ is the wavelength. In addition, the operator $J_\rho(u)$ denotes the Bessel function of the first kind with order $\rho = 1, 2, 3, \ldots$.
The reader can refer to [8],[9], for more detail of the user link channel model.
For notational convenience in (5), we also define $\mathbf{h}_k \in \mathbb{C}^{1\times M}$ as the channel vector which contains the channel components between M feeds of G satellites and k-th user. That is

$$\mathbf{H} = \left(\mathbf{h}_1^T, \ldots, \mathbf{h}_k^T, \ldots, \mathbf{h}_K^T\right)^T, \tag{13}$$

and

$$\mathbf{h}_k = \left[\mathbf{h}_{k,1}, \ldots, \mathbf{h}_{k,f}, \ldots, \mathbf{h}_{k,G}\right] \tag{14}$$

where $\mathbf{h}_{k,f} \in \mathbb{C}^{1\times N}$ refers to the N channel components of f-th satellite at the k-th user.
Towards a spectrally efficient communication, all the M feeds at G satellites share the same frequency resources ($N_c = 1$). The user link interference is the communication bottleneck of the whole system. Indeed, this interference appears: (i) among feeds of various satellites which accounts for the inter-satellite interference, (ii) among the footprints of beams which are generated by each satellite. This refers to intra-satellite or inter-beam interference.
To cope with the inter-satellite and intra-satellite interference, each gateway employs a precoding scheme. In this context, the transmitted symbols in (4) can be decomposed as

$$\mathbf{t} = \mathbf{Wx}, \tag{15}$$

where $\mathbf{x} = (\mathbf{x}_1, ..., \mathbf{x}_f, ..., \mathbf{x}_G)^T$ is a vector that contains the transmitted symbols to K users through G satellites/gateways. The vector $\mathbf{x}_f \in \mathbb{C}^{K\times 1}$ is the transmitted signal through f-th gateway with $E\left\{\mathbf{x}_f\mathbf{x}_f^H\right\} = \mathbf{I}_K$. Matrix $\mathbf{W}$ is the block precoding matrix at a total of G gateways. We assume each gateway constructs a part of precoding matrix $\mathbf{W}$, i.e,

$$\mathbf{W} = \text{diag}\left(\sqrt{\alpha_1}\mathbf{W}_1, \ldots, \sqrt{\alpha_f}\mathbf{W}_f, \ldots, \sqrt{\alpha_G}\mathbf{W}_G\right), \tag{16}$$

with $\mathbf{W}_f \in \mathbb{C}^{N\times K}$ is the precoder in f-th gateway. The scalar α_f must comply with

$$\text{trace}(\mathbf{W}_f\mathbf{W}_f^H) \leq P_f, \tag{17}$$

where P_f is the total power at f-th satellite so that $P_T = \sum_{f=1}^{G} P_f$ denotes the available power budget at G satellites. In this context, for f-th gateway we can have that

$$\mathbf{t}_f = \sqrt{\alpha_f}\mathbf{W}_f\mathbf{x}_f. \tag{18}$$

Then, the overall system performance received by K users from f-th satellite can be optimized considering the maximum sum rate, i.e.,

$$\begin{aligned} &\underset{\mathbf{W}_f}{\text{maximize}} \quad \sum_{k=1}^{K} \text{minimum} \quad r_{f,k} \\ &\text{s.t.} \quad \text{trace}\left(\mathbf{W}_f\mathbf{W}_f^H\right) \leq P_f, \end{aligned} \tag{19}$$

where $r_{f,k}$ denotes the achievable rate of k-th user from f-th satellite and can be calculated as

$$r_{f,k} = \log\left(1 + \text{SINR}_{f,k}\right), \tag{20}$$

with $\text{SINR}_{f,k}$ denotes Signal to Interference plus Noise Ratio (SINR) received by k-th user through f-th satellite so as

$$\text{SINR}_{f,k} = \frac{|(\mathbf{H}_f\mathbf{W}_f)_{kk}|^2}{\sum_{j\neq k}|(\mathbf{H}_f\mathbf{W}_f)_{kj}|^2 + \sum_{q\neq k,j}|(\mathbf{H}_q\mathbf{W}_q)_{kq}|^2 + 1}, \tag{21}$$

where index j denotes the interference received by the k-th user from adjacent beam served by the f-th satellite/gateway, i.e. intra-satellite interference in the k-th user. Subscript q represents the interference received by k-th user from adjacent satellites to f-th satellite, i.e. inter-satellite interference at the k-th user. Obviously, to optimize $r_{f,k}$ in (19) one promising solution would be designing the precoding matrix $\mathbf{W}$. This is the objective of the next section.

3 Precoding scheme in MS-COV configuration

As stated above, the performance of precoding is sensitive to the quality of CSI at the transmitting segment. To this end, the user terminals shall estimate the CSI of their user link channel elements in (5) and feedback to the gateways. While each gateway calculates a part of precoding matrix $\mathbf{W}$ in (16), a proper CSI feedback among users and gateways as well as a CSI exchange mechanism among gateways are required.

For a moment, lets assume that there is no CSI exchange mechanism among gateways. Then, conventional Zero Forcing (ZF) design of precoding matrix in f-th gateway, i.e. $\mathbf{W}_f$, can be calculated as

$$\mathbf{W}_f = \mathbf{H}_f(\mathbf{H}_f\mathbf{H}_f^H)^{-1}. \tag{22}$$

However, to design precoding in (22) the CSI of $\mathbf{H}_f$ in (5) shall be calculated by the users and send back to the f-th gateway. Therefore, a total of G CSI feedback mechanisms is required at each user terminal, one CSI feedback mechanism per satellite, leading to high computational complexity at user terminals. To tackle this, using CSI exchange mechanism at the transmitting segment avoids having multiple CSI feedback mechanism at the user.

To exchange CSI among gateways, we propose using a scheduler, Fig 1, that whose role is:

(i) to distribute signals that must be transmitted to each user terminal* through different satellites, e.g. $\mathbf{x}_f$ to f-th gateway. For this, the scheduler selects a set of high quality on-board feeds at each satellite towards each user.

(ii) To handle CSI exchange among the gateways. The scheduler shares the CSI of the selected feeds in (i) with the gateways. Then, precoding at each gateway is done upon the information of selected feeds.

To fulfill (i) and (ii), user terminals feedback the CSI of the channel, e.g. $\mathbf{h}_k$ in (13) from k-th user, to the scheduler through one of the employed satellites so that scheduler has access to the CSI of block channel matrix in (13). In this context, one satellite network is only involved to feedback CSI to the scheduler, leading to low complex CSI feedback mechanism in the considered MS-COV architecture with G satellites.

**Remind that the satellites orbits areclose enough together so that each user employs a single antenna to manage the multiple symbols from the different satellites.*

Two approaches can be conceived in order to embed scheduler at transmitting segment: first, implementing as an individual unit in the transmitting segment and connect it to the gateways (see Figure 1). This is referred to Central-Scheduler scheme. In this way, the Central-Scheduler provides an enough possibility to fulfill the requirements in (i)-(ii) in the cost of higher complexity at the transmitting segment. Second, the scheduler can be employed into the gateways so that each gateway uses an individual scheduler. This is referred to Distributed-Scheduler scheme. Comparing to Central-Scheduler, implementing Distributed-Scheduler provides no additional complexity in the transmitting segment, however, this unit suffers from lower degree of freedom to fulfill (i)-(ii). In this work, we use Central-Scheduler due to its capabilities even if it provides higher complexity.

Concretely, the derivation of precoding matrix $\mathbf{W}$ in (19) goes along the following logic:

First, we present a low complex CSI feedback and consequently a feed selection mechanism based on available CSI at the Central-Scheduler. Then, the information of selected feeds exchanges between corresponding gateways and Central-Scheduler. Eventually, a ZF precoding is deployed where each gateway performs a part of precoding matrix $\mathbf{W}$. We focus on ther ZF precoding due to its better interference rejection capabilities and therefore improve the value of SINR obtained by each users. Moreover, it provides a low computational complexity as depicted in [12].

3.1 *CSI feedback mechanism*

Since each user terminal has access to the CSI of its own channel, it is expected that k-th user terminal, $k = 1, ..., K$, quantize and report the direction of its channel vector [13]. This is referred to Channel Directional Indicator (CDI) and can be calculated as

$$\widetilde{\mathbf{h}}_k = \frac{\mathbf{h}_k}{||\mathbf{h}_k||}, \tag{23}$$

where $\mathbf{h}_k$ is the channel vector for k-th user and already defined in (13). It is considered that each user has an individual quantization codebook formed by unit norm column vectors

$$\mathbf{C}_k = \{\mathcal{C}_{k,1}, ..., \mathcal{C}_{k,i}, ..., \mathcal{C}_{k,D}\}, \tag{24}$$

where $\mathbf{C}_k$ is the codebook in the k-th user. $\mathcal{C}_{k,i}$ refers the i codewords of the codebook $\mathbf{C}_k$. Indeed, the number of codebooks depends on the number of feedback bits which is assumed to be equal to the number of whole feeds at satellites, i.e. $D = M$. The quantization criterion in this context is minimum chordal distance [14], i.e.

$$\breve{\mathbf{h}}_k \triangleq |\widetilde{\mathbf{h}}_k\mathbf{C}_k|. \tag{25}$$

Then, the k-th user shares knowledge of its codebook in (25) with the Central-Scheduler through one of the satellite as discussed above. Each user terminal is assumed to use an independently generated codebook in (24) to avoid that multiple users may return the same quantization vector. Note that this work considers a random vector quantization scheme, whereby the codewords are independently chosen from an isotropic distribution on the M-dimensional unit sphere for existing M feeds. The reader can refer to [14] for further detail on a random vector quantization set.

3.2 *Feed selection method in MS-COV*

The larger number of transmit feeds than the beams provides degree of freedom to employ feed selection diversity to increase channel gain and improving performance of precoding scheme. Indeed, this feed selection method shall guarantee generating maximum achievable rate toward the coverage area through each feed. To this end, while the user terminals feedback the CDI of the user link, the Central-Scheduler can select a set of feeds with favorable quality at each satellite and exchange their information with the gateway that serve the corresponding satellite. Then, precoding scheme at each

gateway can be constructed through the information of these selected feeds.
We consider the following feed selection: the selected feeds should orthogonal to other feeds, i.e. for k-th and j-th users:

$$\breve{\mathbf{h}}_k \breve{\mathbf{h}}_j^H = 0 \qquad \forall k, j \tag{26}$$

with $k \neq j$ in (25). The orthogonaltiy in (26) yeilds minimizing inter- and intra-satellite interferences. To this end, each feed shall keep orthogonality to the set of feeds embedded at the corresponding satellite and also adjacent satellites. For this, the Central-Scheduler employs Gram Schmidt orthogonal subspaces approach [15] which is widly applied in terrestrial networks [13]. For instance, in a set of two feeds, i.e. $M = 2$, it calculates the projection components of second feed orthogonal to the space spanned by the first feed's vector. To use advantages of all N feeds embedded at each satellite, we propose Central-Scheduler selects two groups of feeds at each satellite:
Primary Selected Feeds (PSF), a set of K feeds that are directly served K beams and provides maximum possible gain to corresponding beams, i.e. one feed per beam/user.
Supplementary Selected Feeds (SSF), a set of $(N - K)$ feeds that provides supportive gains to $(N - K)$ out of K beams so that some beams have overlapped coverage generated by PSF and SSF. Each supportive feed is selected to serve the beam that receives the maximum gain from the corresponding feed.
Unfortunately, finding a set of PSF and SSF feeds that are completely orthogonal each other and gurunteeing (26) requires an exhaustive search which is not computationally feasible in a system with M feeds of G satellites. Interestingly, our Central-Scheduler is similar to the case that the authors in [13] have been used in the framework of terrestrial communication networks. In this context, we propose the following low complex sub-optimal orthogonal feed selection procedure:
Step 1, since the $\mathbf{x}_f$ in (15) is the signals transmitted to K user through f-th satellite, the Central-Scheduler is responsible to distribute the signals of vectors $\mathbf{s}$ in (15) among gateways.
Step 2, using CDI of $\breve{\mathbf{h}}_k$ in (25) transferred from k-th user, $k = 1, ..., K$, the Central-Scheduler constructs

$$\breve{\mathbf{H}} = \left[\breve{\mathbf{h}}_1^T, \ldots, \breve{\mathbf{h}}_k^T, \ldots, \breve{\mathbf{h}}_K^T\right]^T, \tag{27}$$

where $\breve{\mathbf{H}}$ is the CDI of $\mathbf{H}$ at the Central-Scheduler. For (27), we can also write

$$\breve{\mathbf{H}} = \left[\breve{\mathbf{h}}_{feed,1}, \ldots, \breve{\mathbf{h}}_{feed,i}, \ldots, \breve{\mathbf{h}}_{feed,M}\right], \tag{28}$$

where $\breve{\mathbf{h}}_{feed,i} \in \mathbb{C}^{K \times 1}$ is the i-th column of $\breve{\mathbf{H}}$ and denotes the CDI of i-th feed at the Central-Scheduler.
Step 3, for the i-th feed, with $i = 1, ..., M$, the Central-Scheduler calculates $\breve{\mathbf{g}}_i$. It refers to the orthogonal components of $\breve{\mathbf{h}}_{feed,i}$ which are orthogonal to the subspace spanned by $\{\breve{\mathbf{g}}_1, ..., \breve{\mathbf{g}}_{(M-1)}\}$. This can be done through using Gram Schmidt theorem for orthogonal subspaces in [15]. Concretely, the Loop 1 computes the orthogonal subspace of M feeds vectors in (28).
Given Loop 1, $\breve{\mathbf{Q}}$ can be decomposed as

$$\breve{\mathbf{Q}} = [\breve{\mathbf{g}}_1, \ldots, \breve{\mathbf{g}}_i, \ldots, \breve{\mathbf{g}}_M] \in \mathbb{C}^{K \times M}, \tag{29}$$

where each vector of $\breve{\mathbf{Q}}$ becomes orthogonal to one another; i.e. $\breve{\mathbf{g}}_1 \perp \ldots \perp \breve{\mathbf{g}}_i \ldots \perp \breve{\mathbf{g}}_M$. For (29), we can also write:

$$\breve{\mathbf{Q}} = [\breve{\mathbf{Q}}_1, ..., \breve{\mathbf{Q}}_f, ..., \breve{\mathbf{Q}}_G]^T \tag{30}$$

where $\breve{\mathbf{Q}}_f \in \mathbb{C}^{K \times N}$ denotes the block matrix of $\breve{\mathbf{Q}}$ concerning f-th satellite in (29).
Step 4, reminding serving single user per beam, at f-th satellite (and respectively at other satellites) the Central-Scheduler selects: (i) PSF, the i-th feed of f-th satellite with largest projected norm towards k-th user. A set of K feeds, one per user, is selected; (ii) SSF, a set of $(N - K)$ feeds that provides maximum supportive gains to $(N - K)$ users is selected.
For the sake of completeness, the feed selection procedure for K users at G satellites is summarized in Loop 2.
After obtaining a set of channel vectors $\mathbf{H}_{\mathcal{S}}\mathbb{C}^{K \times N}$ using Loop 2, the Central-Scheduler transmits the information of the $\mathbf{H}_{\mathcal{S}_f}$ to the f-th gateway. The precoding at f-th gateway, $f = 1, ..., G$, can be then formulated by

$$\mathbf{W}_f = \mathbf{H}_{\mathcal{S}_f}^H \left(\mathbf{H}_{\mathcal{S}_f} \mathbf{H}_{\mathcal{S}_f}^H\right)^{-1}, \tag{31}$$

where $\mathbf{W}_f$ in (31) is a ZF precoding constructed by $\mathbf{H}_{\mathcal{S}_f}$.

Loop 1: Estimation of channel matrix $\breve{\mathbf{H}}$.

Input: Matrix $\breve{\mathbf{H}}$
Result: Matrix $\breve{\mathbf{Q}}$
Initialize;
Empty matrix $\breve{\mathbf{Q}} \in \mathbb{C}^{K \times M}$
Then;
for *i=1, ... ,M* **do**
a- Calculate

$$\breve{\mathbf{g}}_i = \breve{\mathbf{h}}_{feed,i}^H - \sum_{j=1, i \neq j}^{M-1} \frac{\breve{\mathbf{h}}_{feed,i}^H \breve{\mathbf{g}}_j}{||\breve{\mathbf{g}}_j||^2} \breve{\mathbf{g}}_j = \breve{\mathbf{h}}_{feed,i}^H \left(\mathbf{I} - \sum_{j=1}^{i-1} \frac{\breve{\mathbf{g}}_j^H \breve{\mathbf{g}}_j}{||\breve{\mathbf{g}}_j||^2}\right)$$

b- Fill the i-th column of $\breve{\mathbf{Q}}$ with $\breve{\mathbf{g}}_i$

$$\breve{\mathbf{Q}} \leftarrow \breve{\mathbf{g}}_i$$

c- Put

$$i \leftarrow (i + 1)$$

repeat;
End

4 Numerical Results

4.1 Simulation Setup

In order to further compare the performance of the proposed scenarios in this study, Monte Carlo simulations have been carried out. The user link in the downlink is assumed to operate at 20 GHz Ka-band. A baseband block fading as described in [16] models the satellite antenna radiation pattern, the path loss, the receive antenna gain and the noise power.
Each satellite employs a total number of $N = 20$ feeds and $K = 14$ beams. Remarkably, 14-adjacent beams holds circular fashion coverage so that the central beam receives the largest amount of interference. Each user terminal is equipped with a single antenna and a set of two satellites $G = 2$ are presumed.
We evaluate a MS-COV networks where each satellite serves through a single gateway so that the coverage of satellites are overlapped.
Results have been averaged for a total of 500 user link channel realizations and it is also assumed user link of each satellite has a total of 500 MHz (in Ka band) available bandwidth. For the sake of completeness, the rest of parameters are collected in Table 1. We computed the following performance metrics: first the SINR for each user, after interference mitigation, and then its throughput (bit/s) is inferred according to DVB-S2x standard [18]. The relationship between the received SINR and the spectral efficiency achieved by DVB-S2x standard is provided in [18]. Second, the simulation results include also the availability based on the Cumulative Distribution Function (CDF) of the SINR. In this case, the instantaneous

Loop 2: PSF and SSF procedures at a total of G satellites.

Input: Matrix $\check{\mathbf{Q}} = (\check{\mathbf{Q}}_1, ..., \check{\mathbf{Q}}_f, ..., \check{\mathbf{Q}}_G)^T$
Result: Matrix $\mathbf{H}_{\mathcal{S}}$
Initialize;
Empty matrix $\mathbf{H}_{\mathcal{S}} = \text{diag}(\mathbf{H}_{\mathcal{S}_1}, ...\mathbf{H}_{\mathcal{S}_f}, ..., \mathbf{H}_{\mathcal{S}_G})^T$,
$\mathbf{H}_{\mathcal{S}_f} \in \mathbb{C}^{K \times N}$ is the empty block matrix for the f-th satellite,
for f=*1, ... ,G* **do**
 Then;
 A-Find PSF in the f-th satellite;
 for k=*1, ... ,K* **do**
  a- Consider the arbitrary function $g(.)$. The Central-Scheduler selects the i-th feed of the f-th satellite which transmit maximum gain to the k-th user as follows

$$\pi_k(i) \triangleq \arg\max_i g(\check{\mathbf{Q}}_f)$$

  where $\pi_k(i)$ denotes the i-th feed (i.e. i-th column) of $\check{\mathbf{Q}}_f$ which transmits maximum gain towards k-th user;
  b- The Central-Scheduler picks up

$$\mathbf{g}_k(i) \triangleq \check{\mathbf{Q}}_f(\pi_k(i))$$

  where $\mathbf{g}_k(i)$ is defined as the $\pi_k(i)$ column of matrix $\check{\mathbf{Q}}_f$;
  c- Central-Scheduler fills the k-th row of $\mathbf{H}_{\mathcal{S}_f}$ with $\mathbf{g}_k(i)$, i.e.

$$\mathbf{H}_{\mathcal{S}_f} \leftarrow \mathbf{g}_k^T(i)$$

  d- Put

$$k \leftarrow (k+1)$$

  repeat;
  End;
 B- Find SSF for the f-th satellite;
 for $z = (K+1), ..., (N-K)$ **do**
  e) The Central-Scheduler repeats (a)-(c) for remained (N-K) feeds (out of K feeds employed in SSF mode) in the f-th satellite. Indeed, the Central-Scheduler selects z-th feed of f-th satellite to serve user that obtains maximum gain from this z-th feed;

$$z \leftarrow (z+1)$$

  End

$$f \leftarrow (f+1)$$

 repeat;
 End

availability indicator for the k-th user is given by

$$A_k = g(\gamma_k) \tag{32}$$

which is equal to 0 if the user is unavailable, i.e. if its instantaneous SINR is lower than that required by the lowest ModCod of DVB-S2x, $\gamma_k < -2.85$dB [18], and is equal to 1 otherwise.
For a best practice, we also consider two reference scenarios:
Lower bound: a conventional 4-frequency reuse pattern $N_c = 4$ among beams.
Upper bound: we consider the gateways have access to perfect CSI of the user link channel matrix. It implies two satellites ideally behave as one large satellite with perfect CSI available at the transmitting segment which increases the throughput received at each user terminal.

Table 1 Simulation parameters

Parameters	Description
Earth radius	6378.137 Km
Satellite height	35786 km (GEO)
Numer Of Satellites	2
Feed radiation pattern	Bessel Approx. [17]
Number of feeds N per Sat.	20
Number of beams per Sat.	14
Total bandwidth	500 MHz
Roll-off factor	0.25
clear sky gain	17.68 dB/K
user antenna gain	41.7 dBi

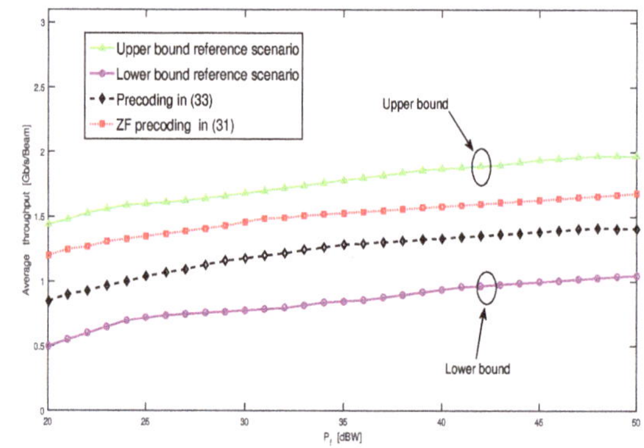

Fig. 2: Average throughput (Gbit/s) of ZF in (31) per beam received based on DVB-S2x MODCOD.

4.2 *Results*

This section presents the simulation results related to the scenarios described in Section 3.
The following schemes are evaluated:
(a) Gateways have no CDI exchange mechanism so that each gateway processes beams independently with available CDI of the users that serves. Given (15), we use a conventional ZF precoding technique at each gateway as

$$\mathbf{W}_f = \check{\mathbf{H}}_f^H(\check{\mathbf{H}}_f\check{\mathbf{H}}_f^H)^{-1} \tag{33}$$

where $\check{\mathbf{H}}_f$ is defined in (27). Obviously, the precoding in (33) can mitigate intra-satellite interference, however, it has no affect on inter-satellite interference.
(b) The gateways collaborate through a scheduler to exchange the CDI among gateways so that the ZF in (31) is obtained.
Fig. 2 depicts the performance of ZF in (31) in terms of average throughput as a function of available power P_f. The lower bound reference scenario has the lowest achievable rate due to its bandwidth limitations. It is also seen that the relative gain for the proposed ZF in (31) goes close to the upper bound reference scenario. In addition, the ZF precoding in (33) obtains the low achievable rate and close to the lower bound due to lack of a CDI exchange mechanism among gateways. As a consequence, the higher the coordination among gateways, the higher the achievable rates is.
Moreover, the corresponding CDF of the SINR for the aforementioned scenarios is provided in Figure 3. It implies the availability of the user link. Obliviously, due to presence of a CDI exchange mechanism, the system performance with the ZF in (31) achieves better availability results and close to the upper reference scenario than the ZF precoding in (33) and the lower bound reference scenario.

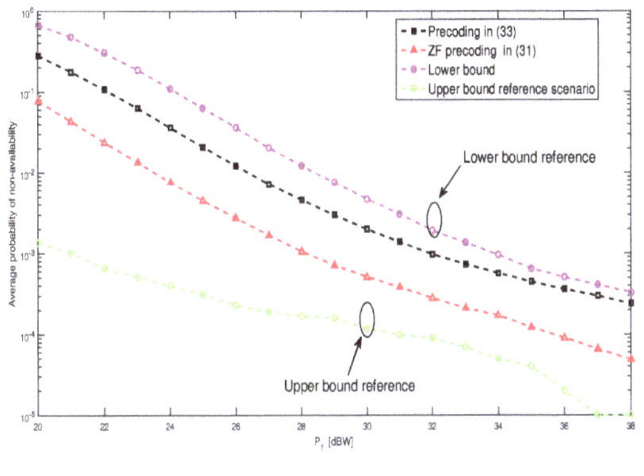

Fig. 3: CDF of SINR comparison when the rain fading is considered.

5 Conclusion

In the present work, the topic of deploying precoding technique in MS-COV architecture discussed since a high throughput full frequency reuse pattern among beams is considered. The objective in the proposed MS-COV architecture lies in the collaboration between multibeam satellites systems and obtains reliable multibeam infrastructure to serve unforeseen changes in the traffic demand through secondary satellites. Moreover, employing multiple satellites in this context can lead higher spatial diversity by keeping the size of the payload affordable. Last but not least, employing multiple satellites in MS-COV configuration provides hardware redundancy to guarantee uninterrupted service delivery. In proposed MS-COV architecture, we developed a precoding technique aiming at mitigating inter- and intra-satellite interference. To establish precoder, we developed low complex CSI exchange and feedback mechanisms respectively among gateways as well as among user terminals and gateways.

Acknowledgments

This work was partially supported by the National Research Fund, Luxembourg under the projects "FNR-DPSAT" and "FNR-PROSAT".

References

1 G. Verelst, O. Vidal, E. Alberty, D. Galinier, N. Metzger, P. Inigo, S. Stirland, and G. Huggins: 'Innovative system architecture to reach the terabit/s satellite', *31st International Communications Satellite Systems Conference (ICSSC)*, 2013, Florence, Italy.
2 D. Minoli: 'Innovations in satellite communications technology', *John Wiley&Sons Inc. Hoboken*, 2015, USA.
3 J. Lizarraga, P. Angeletti, N. Alagha, and M. Aloisio: 'Flexibility performance in advanced Ka-band multibeam satellites', *IEEE International Conference in Vacuum Electronics*, 2014, pp. 45–48
4 M. A . Vazquez, A. I. PeÃt'rez-Neira, D. Christopoulos, S. Chatzinotas, B. E. Ottersten, P. M. Arapoglou, A. Ginesi, and G. Taricco: 'Precoding in multibeam satellite communications: present and future Challenges', *CoRR, vol. abs/1506.08018, 2015. [Online]. Available: http://arxiv.org/abs/1506.08018*, 2016.
5 V. Joroughi, M. A. Vzquez and A. I. Prez-Neira: 'Precoding in multigateway multibeam satellite systems', *IEEE Transactions on Wireless Communications*, 2016, vol. **15**, pp. 1–16
6 C. Forrester: 'High above: the untold story of Astra, Europe's leading satellite company', vol. *plan B communication,*, 2010.
7 D. Christopoulos, S. Chatzinotas, and B. Ottersten: 'Full frequency reuse multibeam satellite systems: Frame based precoding and user scheduling', *IEEE Transaction on Wireless Communications*, 2010, vol. *4*, pp. 1120–1112
8 S. Chatzinotas, B. Ottersten, R. DeGaudenzi, Pappas G.J.: 'Cooperative and cognitive satellite systems', *Elsevier publication*, 2015.
9 D. Christopoulos, S. Chatzinotas,B. Ottersten: 'User scheduling for coordinated dual satellite systems with linear precoding', *IEEE International Conference on Communications Conference (ICC)*, 2013, Hungry.
10 V.Joroughi: 'The design of precoding technique in collocated multibeam satellite systems', *Proceedings of AIAA-16*, 2016, USA.
11 O. Tervo, H. Pennanen, D. Christopoulos, S. Chatzinotas, B. Ottersten: 'Distributed optimization for coordinated beamforming in multicell multigroup multicast systems: Power minimization and SINR balancing', *IEEE Transaction on Wireless Communications*, 2009, vol.**66**, pp. 171–185.
12 SM Kay: 'Fundamentals of statistical signal processing: estimation theory', *Prentice-Hall Inc. Upper Saddle River*, 1993, USA.
13 T. Yoo, N. Jindal and A. Goldsmith: 'Multi-antenna downlink channels with limited feedback and user selection', *IEEE Journal on Selected Areas in Communications*, 2007, vol. **25**.
14 K. Mukkavilli, A. Sabharwal, E. Erkip, and B. Aazhang: 'On beamforming with finite rate feedback in multiple antenna systems', *IEEE Transaction on Information Theory*, 2013, vol. **49**, pp. 2562–2579
15 M. Hazewinkel: 'Orthogonalization', *Encyclopedia of Mathematics, Springer, ISBN 978-1-55608-010-4.*
16 ITU-R Recommendation P.618-10: 'Propagation data and prediction methods required for the design of Earth-space communication systems', 2009, Italy.
17 Call of Order 2-Task 1: 'Fair comparison and combination of advanced interference mitigation techniques ', *Satellite Network of Experts (SatNEx) 3 project of ESA, contract NO:23089/10/NL/CPL.*
18 web site:www.dvb.org: 'DVB-S2x: Digital video broadcasting user guidelines for second generation system for broadcasting interactive services, news gathering and other broadband satellite application', *ETSI TR 102 376.*

Productized Multicarrier Predistortion Total Throughput Gains around 20% over Linearized Channels in True Customer Use Cases

Dieter Duyck, Hadi Gharibdoust, Dirk Breynaert and Apostolos Mitakidis

Abstract—**Efficiency gain is a relevant metric of communication technology directly translating to the same improvement in operational expenditures while sometimes also allowing capital expenditure gains. The latest and best known multicarrier predistortion techniques for satellite communications yielding significant guaranteed throughput gains for all the receivers, regardless of the link budget, even over linearized transponders, have now been productized, tested over the Avanti satellite Hylas 4, and deployed, up to 500 MHz. This paper provides performance results for multiple use cases, including different number of carriers (1, 2, and 4 carriers), balanced and unbalanced power levels. For the relevant use case of two beams with unequal throughput requirements over a completely linearized transponder, total throughput gains around 20% in a single transponder have been achieved. We expect a continued and significant roll-out of this technology in the next 5 years, as long as the carriers in the forward of a high throughput satellite link follows a fixed carrier plan.**

1. INTRODUCTION

Due to the large distances to be bridged in satellite communications, power is a scarce resource. Therefore, the on-board amplifier, typically a travelling wave tube amplifier (TWTA) is driven close to saturation, making the aggregate satellite channel non-linear (NL).

Predistortion improves the error rate performance over nonlinear channels, such as forward links (from hub to terminal) in satellite communications, given high SNR in the uplink (from hub to satellite). An overview of recent predistortion technologies is given in [4].

Due to the presence of the pulse-shaping filter in the modulator and the matched filter in the receiver, the symbol-level channel between the transmitted and received symbols is a non-linear channel with a large memory, especially for low roll-off schemes. With the next generation of satellites, especially high throughput satellites, it is more common to have significantly larger bandwidths per transponder. This leads to a more frequent adoption of multicarrier transmission, also in the forward link from hub to terminal. Typically, these transponders are linearized, meaning that they have a limited small signal gain, e.g. 3 dB, and a limited phase rotation. As efficiency is a key differentiator in the forward link, *multicarrier predistortion for linearized transponders* becomes a very important requirement for advanced modulators.

Technical Labs, Newtec, Laarstraat 5, 9100 Sint-Niklaas.

A. Mitakidis is with Avanti Communications Group plc, Cobham House, 20 Black Friars Lane, London EC4V 6EB.

Last year, Newtec demonstrated its patented multicarrier predistortion for linearized channels with unseen gains [4]. More specifically, before [4], the best HW implementation [7], [8] achieved 1 dB total degradation gain on a very nonlinear channel and 0.2 dB on a typical linearized channel. The best publications [1], [9], [10] achieved 4 dB on a very nonlinear channel and 0.9 dB on a linearized channel. Newtec's multicarrier predistortion achieves 1 dB gain over a linearized channel, outperforming the best publication, but with 10 times less complexity than [1], [9], [10].

This paper builds further upon [4] which is the best known technique in theory and in practice and was recently awarded an ESA ARTES success story [6].

1) Contribution of this paper: This paper presents results for a demonstration that was performed with 2 carriers over a linearized transponder of 230 MHz in Hylas 4 from Avanti, focusing on the impact of predistortion on use cases with unbalanced power per carrier, where total throughput (over the entire transponder) gains up to 25% are not an exception.

2. A SSESSED USE CASES

A. Satellite parameters

The Hylas 4 transponder over which we assessed predistortion was very linear, with a small signal around 2.8 dB, a phase rotation at saturation around -11 degrees and a medium compression beyond saturation (roughly 1 dB compression at an -6 dB input backoff (beyond saturation).

This paper presents results for two transmission scenarios tested over this transponder. The transmission scenarios consider two carriers per transponder, each carrier downlinked to a dedicate beam of 115 MHz:

1) Each carrier has the same power
2) 1 carrier has 10 dB more power than the outer to achieve a target throughput

The rationale behind the presented use case is the following. The considered transponder is designed to serve two separate geographical areas, each illuminated with a dedicate beam. To avoid frequency interference to other services, only the carrier of interest is transmitted to its appropriate beam. To save power and mass, a single travelling wave tube amplifier is used to amplify both carriers. Two output multiplexing (OMUX) filters are used to filter out all frequency components except the useful carrier to avoid frequency interferences. The forward link is schematically illustrated in Fig. 1.

Imagine now the following 2 use cases:

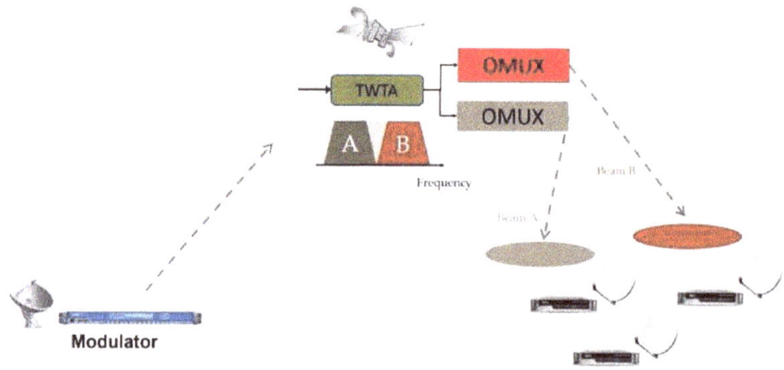

Fig. 1. A forward link with 1 TWTA and 2 OMUX filters is common for 2-beam scenarios.

1) Both geographical regions are equally congested in terms of demand. As a consequence, the power spectral density of both carriers are equal, resulting in the same throughput in both carriers. For example, 500 Mbps is required in each region, which can be delivered by 32APSK34.
2) Beam A covers a major city and the throughput demand exceeds 500 Mbps. In fact, 575 Mbps is needed for which a 64APSK 32/45-L modcod is needed. As such, 2.6 dB more link budget is needed for beam A. As the TWTA total output power is constant, beam B drops 7.4 dB in power spectral density[1], such that beam B is 10 dB below beam A in power spectral density.

An alternative technique than unbalancing power to give more power to beam A, without sacrificing efficiency of beam B, would be to reduce the symbol rate of beam B. Note that it is beyond doubt that unbalancing the power is the better alternative over reducing the symbol rate. This is easy to show using the theoretical (but achievable) bounds on the throughput over a linear channel. More specifically, it is well known that employing a 0% roll-off scheme with a capacity-achieving coded modulation yields a throughput of $\mathrm{SR} * \log_2(1 + \mathrm{SNR})$, where SR is the symbol rate. For beam A, the symbol rate is maximum due to the OMUX filter. As such, a throughput gain can only be increased by an increase of the power spectral density of that beam (yielding an increase in SNR for that beam). That increase of power spectral density directly translates to the same increase in total power spent for that beam, which results in a decrease in power spent for beam B. This decrease either reduces the spectral density when keeping the symbol rate constant (option 1), either reduces the symbol rate when keeping the spectral density constant (option 2). Option 1 results in a logarithmic decrease of the throughput of beam B, while option 2 results in a linear decrease of the throughput of beam B, which is numerically illustrated in table I, using Shannon efficiencies instead of real-life efficiencies.

Hence, the second use case is the most common and appropriate use case in the case of the event that a beam is *full* and requires more power.

[1]Indeed, $10^{2.6/10} + 10^{-7.4/10} = 10^{0/10} + 10^{0/10}$ or the addition of the powers in Watt of both beams in both scenarios equal.

B. *Waveform parameters*

Concerning the modcod, any modcod was applicable, but for the test we chose 32-APSK3/4 as reference modcod. A 32-APSK modcod is often feasible to reach in practice, while it is still a high modcod allowing to see more sensitivity in terms of performance.

The roll-off of the waveform was 5%.

As the transponder allowed 230 MHz, each carrier transmits at 109.5 Mbaud.

3. Measuring methodology

The impact of predistortion on the selected use cases was measured in the lab and over satellite. In both scenarios, a waveform with constant coding and modulation (CCM) was transmitted at 230 MHz. The bandwidth was chosen to match that of an available dual beam transponder. Please note that any bandwidth up till 500 MHz would yield the same conclusions. The data to fill the carrier are generated by an internal PRBS generator inside the FPGA.

A. *Measurement setup in the lab*

Since Newtec's predistortion is productized, Newtec has an FPGA based modulator in which predistortion can be enabled with configurable parameters.

In order to assess predistortion over wideband signals, Newtec implemented an internal wideband channel emulator inside the modulator FPGA, see Fig. 2.

Hence, the lab setup is quite straightforward, including a modulator, with internal channel emulator, outputting an L-band signal, representing the signal that would be present after the receive antenna, LNA and downconverter in a real-life system. Said L-band signal is then received through a wideband FPGA based Newtec demodulator receiving a desired carrier.

TABLE I

IN A REFERENCE SCENARIO, EACH BEAM HAS 643 MBPS. HOWEVER, THE DEMAND IN BEAM A IS 736 MBPS, FOR WHICH 2.6 DB MORE POWER IS NEEDED (AS THE SYMBOL RATE CANNOT BE INCREASED). CLEARLY, REDUCING THE POWER DENSITY OF BEAM B IS THE BEST OPTION IN TERMS OF TOTAL THROUGHPUT AND IN TERMS OF THROUGHPUT FOR BEAM B.

	Reference		Reducing power density		Reducing symbol rate	
Beam	A	B	A	B	A	B
Received Power in carrier [dBm]	-30,0	-30,0	-27,4	-37,4	-27,4	-37,4
Symbol rate [Mbaud]	109,5	109,5	109,5	109,5	109,5	19,9
Power Spectral Density [dBm/Hz]	-110,4	-110,4	-107,8	-117,8	-107,8	-110,4
Noise density [dBm/Hz]	-128,0	-128,0	-128,0	-128,0	-128,0	-128,0
Shannon Efficiency [bpsHz]	5,9	5,9	6,7	3,5	6,7	5,9
Throughput 0% rolloff [Mbps]	643,1	643,1	736,5	385,6	736,5	117,0
Linear power [uW]	1,0	1,0	1,8	0,18	1,8	0,18
Total linear power (constant) [uW]	2,0		2,0		2,0	
Total throughput [Mbps]	1286,3		1122,1		853,5	

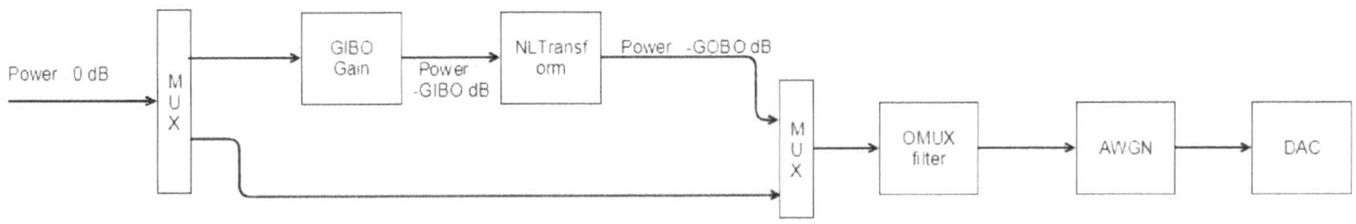

Fig. 2. A block diagram of the very wide internal channel emulator used to assess wideband predistortion. NlTransform is implemented through a look-up table implementing any desired non-linear transformation.

B. Measurement setup over satellite

The satellite test setup is illustrated in Fig. 3. In order to reach the SNR threshold corresponding to a frame error rate of 1e-3, an L-band noise generator is added which can add noise up to 500 MHz wide.

C. Gain measurement

The gain measurement is as follows. Both in the lab as in the satellite test setup, it is possible to modify the input back-off without altering the total satellite power.

For the default link without predistortion, the noise variance is increased until a frame error rate of 1e-3 is achieved. The input-back-off is consecutively optimized, after which the noise variance is again increased, resulting in the reference noise variance corresponding to the SNR threshold for the reference case without predistortion. This noise value is not touched anymore.

Next, when predistortion is enabled, the input back-off is optimized again, after which the modulation and encoding is increased to higher efficient waveforms as long as the frame error rate is smaller than 1e-3 (or essentially zero). The resulting link margin is taken into account to get an average throughput gain over an entire geographical area with variable link budgets (through an integration over the typical efficiency versus SNR step curve in ACM systems, see [5].

4. PREDISTORTION GAIN LEVERAGES IN A LARGE THROUGHPUT GAIN IN AN UNBALANCED CARRIER SCENARIO

In Sec. II-A, we showed that maintaining the symbol rate and reducing the power spectral density of beam B is the best option to give more throughput to beam A.

Here, we will theoretically show that a power gain, that can be for example retrieved through predistortion, results in a large leverage on the total throughput gain in an unbalanced carrier scenario.

We bring back Table I and assume that the total power gain through predistortion is 1 dB, i.e., the TWTA can produce 1 dB more power as the output backoff thanks to predistortion is 1 dB better[2]. As 1 dB translates to a 25% throughput improvement, the total received power in clear sky can be 2.5 uW instead of 2 uW. As beam A requires 1.82 uW to achieve its throughput demand, beam B can now have 0.68 uW instead of 0.18 uW, almost a four-fold (almost 6 dB) improvement in power. This is a typical application of leveraging. This is illustrated in Table II.

Hence, the 1 dB power gain leverages to almost 6 dB power gain for beam B, an almost 100 % throughput gain in beam B, and a total throughput gain around 20%.

The reason for this big leverage is that the power unbalance situation, which is very suboptimal, decreases. Beam A almost needs all TWTA power so beam B has very little power.

[2]Note that this is a simplification of the reality as also the carrier to distortion ratio changes, but we present the simplification as it provides insight.

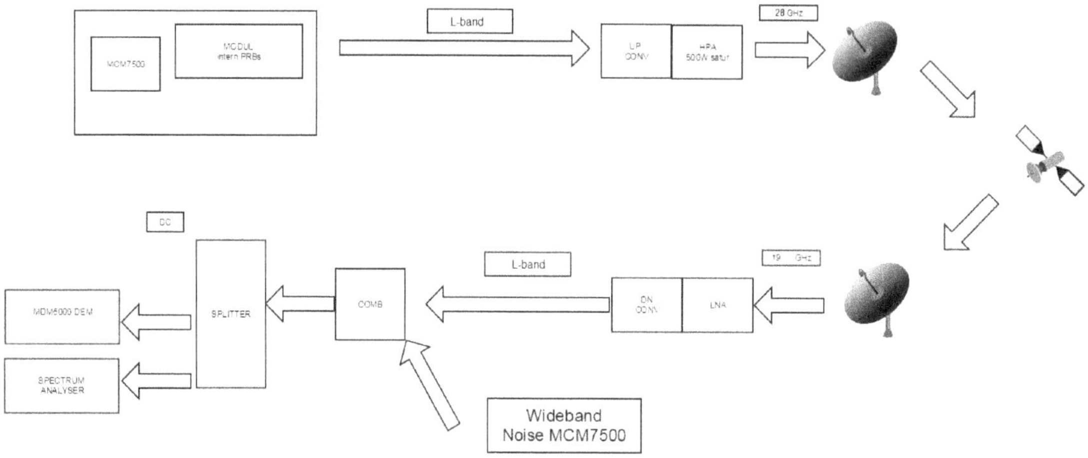

Fig. 3. A block diagram of the satellite test setup to assess the impact of predistortion.

TABLE II
THE TABLE SHOWS THE IMPACT OF A 1 DB POWER GAIN ON THE THROUGHPUT IN AN UNBALANCED POWER SCENARIO. THE UNITS OF THE ROWS ARE THE SAME AS IN TABLE I.

	Reducing power density		1 dB power gain	
Beam	A	B	A	B
Rx Pow	-27,4	-37,4	-27,4	-31,7
SR	109,5	109,5	109,5	109,5
PSD	-107,8	-117,8	-107,8	-112,1
Noise PSD	-128,0	-128,0	-128,0	-128,0
Shannon Eff	6,7	3,5	6,7	5,3
Throughput	736,5	385,6	736,5	582,6
Linear power	1,8	0,18	1,8	6,8
Total power	2,0		2,5	
Total Mbps	1122,1		1319.1	

Thanks to the increase of the TWTA output power (thanks to predistortion) and the fact that beam A does not need more power, beam B suddenly sees its received power increased in a multiplicative way. Denoting the power gain in dB and linear by G_{dB} and G_{lin}, it is easy to prove that this beam B power increase in dB is $G_{\mathrm{dB}} + 10\log_{10}(1 - \frac{U}{G_{\mathrm{lin}}}) - 10\log_{10}(1-U)$, where $U = \frac{\text{power(beam } A)}{\text{total power}}$, hence an indicator for the power unbalance, in the situation without predistortion. The closer U is to 1 (the bigger the initial power unbalance), the larger the maximum gain $G_{\mathrm{dB}} - 3 - 10\log_{10}(1-U)$ which is achieved when $G_{\mathrm{lin}} = 2U$. Note that $G_{\mathrm{lin}} = 2U$ is the power gain corresponding to a restoring of the power unbalanced situation to a power balanced situation. In that case $10\log_{10}(1 - \frac{U}{G_{\mathrm{lin}}}) = -3$. Hence, the maximum gain is $10\log_{10}(U) - 10\log_{10}(1-U) = -10\log_{10}(1/U - 1)$.

Concluding, besides the normal beam power gain G_{dB} that would result from a transponder output power gain of G_{dB}, there is also an additional power gain due to the restoring from a suboptimal unbalanced power situation to an optimal balanced power situation. The maximum of this additional gain is $-3 - 10\log_{10}(1-U)$, where $U > 0.5$ (otherwise, the initial situation would not be unbalanced), which is illustrated in Fig. 4.

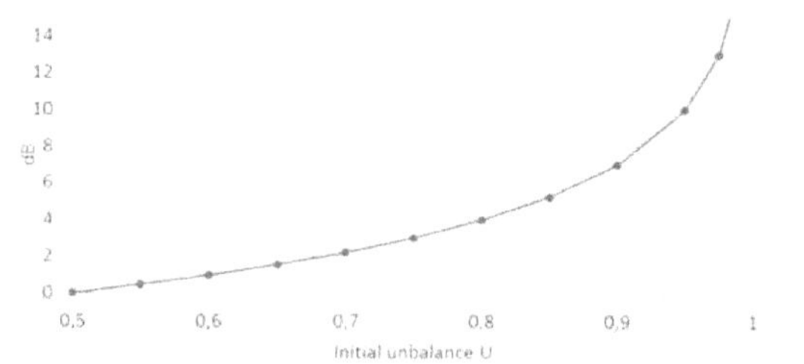

Fig. 4. Maximum additional predistortion gain due to restoration from an initial unbalanced situation.

5. NUMERICAL RESULTS

A. Power balanced scenario over satellite

Using the lab and satellite test setup, we tuned to noise level to have a frame error rate of 1e-3 for a modulation and coding of 32-APSK 3/4 for each of the carriers. When enabling predistortion, we managed to increase the saturation level with 2.6 dB. As a result, we managed to increase the modulation and coding to 32APSK 7/9 with a remaining link margin on each of the carriers. Using the techniques from [5] it was found that the average throughput gain achieved was 7.3 % in the lab 7.4 % over satellite, giving high confidence in the results.

TABLE III
AN OVERVIEW OF THE LAB RESULTS, WITH NORMALIZED THROUGHPUTS IN MBPS. NOTATION: DOWNLINK FADING MARGIN (DLFM), MODCOD (MC), MULTICARRIER PREDISTORTION (PRED).

Pred	Mbps-A	Mbps-B	Mbps-Tot	Mc-A	Mc-B	DLFM-A	DLFM-B	Gain
OFF	500	500	1000	32A34	32A34	15 dB	15 dB	
ON	536	536	1073	32A79	32A79	16 dB	16 dB	7.3%
OFF	575	232	807	64A3245-L	8P59-L	17 dB	8 dB	
ON	575	442	1017	64A3245-L	32A23-L	17 dB	4 dB	26%

B. Power unbalanced scenario over satellite

In the power unbalanced scenario, the modulator output power for each of the carriers was 10 dB different, as set forward in the use case description in Sec. II-A. The downlink noise was configured such that the modcod of the most powerful carrier was 32-APSK 3/4. Due to the 10 dB power difference, the lowest modcod could only achieve QPSK 3/5.

Switching on predistortion, we managed to increase the modcod of the lowest carrier to 16-APSK 8/15-L without any packet loss and remaining link margin, resulting in

- a 6.5 dB increase in availability
- a beam B throughput gain of 77.7 %
- a total throughput gain of 18.9 %

C. Lab results

In the lab, we performed a similar measurement, with the difference that we really tried to get a higher throughput for beam A in the power unbalanced scenario than in the power balanced scenario, to get a global picture. Table III and Fig. 5 present an overview of the measurements. The throughputs in Mbps are normalized to a total throughput of 1000 Mbps in order to have a faster understanding of the results.

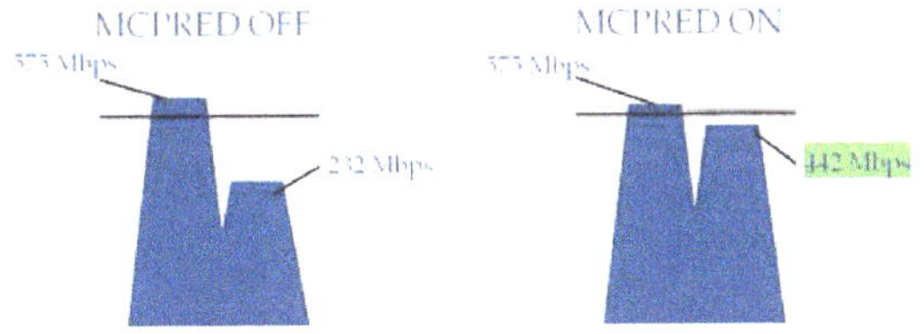

Fig. 5. A visual illustration of the significant throughput improvement of predistortion on a power unbalanced carrier scenario.

VI. CONCLUSION

The forward link throughput of communication networks over satellite is still a bottleneck. Efficiency gains on that link directly translates to the same improvement in operational expenditures while sometimes also allowing capital expenditure gains.

Newtec productized multicarrier non-linear predistortion for linearized satellites and tested its technology over the Avanti satellite Hylas 4. For the relevant use case of two beams with unequal throughput requirements over a completely linearized transponder, total throughput gains around 20% in a single transponder have been achieved. We provide an analytical formula for the maximum additional gain that can be obtained by non-linear predistortion power gains depending the power spectral density unbalance between two carriers.

ACKNOWLEDGMENTS

This research was developed within the framework of ESA-funded project "Every Community Online (ECO)", Ctr. No 4000117721/16/UK/AD.

REFERENCES

[1] B.F. Beidas, R.I. Seshadri and N. Becker, "Multicarrier successive predistortion for nonlinear satellite systems," IEEE Trans. on Comm., vol. 63, no. 4, pp. 1373-1382, April 2015.

[2] G. Colavolpe, G. Montorsi and A. Piemontese, " Spectral Efficiency of Linear and Continuous Phase Modulations over Nonlinear Satellite Channels," in Proc. 2012 IEEE International Conference on Communications, ICC 2012.

[3] Digital Video Broadcasting (DVB), DVB Document A83-2, Second generation framing structure. channel coding and modulation systems for Broadcasting Interactive Services. News Gathering and other Broadband satellite application, Part II: S2-Extensions (DVB-S2X), Mar. 2014.

[4] D. Duyck, A. Mengali et al., "An overview of multicarrier predistortion techniques and associated throughput gain for an actual hardware implementation," AIAA Intern. Comm. Sat. Systems Conf. (ICSSC), Oct. 2017.

[5] D. Duyck, A. Mengali, et al., "The correct gain metric to assess predistortion techniques," AIAA Intern. Comm. Sat. Systems Conf. (ICSSC), Oct. 2017.

[6] https://artes.esa.int/news/new-ground-technology-boosts-link-margin

[7] R. Piazza, B. Shankar, E. Zenteno, D. Rnnow, J. Grotz, F. Zimmer, M. Grasslin, F. Heckmann, and S. Cioni, "Multicarrier Digital Pre-distortion/ Equalization Techniques for Non-linear Satellite Channels," Proc. Of 30th AIAA International Communications Satellite Systems Conference (ICSSC), Ottawa (Canada), 24-27 Sep. 2012.

[8] R. Piazza, B. Shankar, and B. Ottersten, "Non-parametric data predistortion for non-linear channels with memory," in IEEE International Conference on Acoustics, Speech, and Signal Processing (ICASSP), 2013. [Online]. Available: http://orbilu.uni.lu/handle/10993/4862

[9] B.H. Beech and D.G. Edwards, "Method and apparatus for reducing distortion of digital data," patent EP1129556 B1.

[10] B.H. Beech, D.G. Edwards, and R. Perinpanayagam, "Method and apparatus for reducing distortion of digital data," patent EP1371202 B1.

A MITIGATION TECHNIQUE FOR ADJACENT CHANNEL INTERFERENCE IN COMMUNICATION SATELLITES

Leah L. Wang

Lockheed Martin Space, 1111 Lockheed Martin Way, Sunnyvale, CA 94089, USA
email: leah.l.wang@lmco.com

Keywords: COMMUNICATION SATELLITES, ADJACENT CHANNEL INTERFERENCE, INTERFERENCE MITIGATIONS

Abstract

Adjacent channel interference is one of the major impairments that reduces achievable capacity in communication satellites, especially in high capacity throughput satellites (HTS) with densely packed spot beams. Adjacent channel interference (ACI) arises due to leakages in channel filters in any satellite repeaters. Because all filters are finite response filters, adjacent channel interference is always present. The most common technique to suppress ACI is to design multi-pole Butterworth filters with very steep roll-offs and thus very low leakages. However, the steeper the filter is, the higher its insertion loss will be, and the more expensive it will get. A very simple and low cost technique to suppress ACI is presented here. The technique is based on optimizing carrier assignment scheme to users in a ground system. The proposed technique allows maximum bandwidth usage without ACI impairment thus maximizing the achievable capacity throughput in any communication satellite system.

1 Introduction

Mitigating ACI is critical in maximizing communication satellite throughputs [1]. Figure 1 illustrates how ACI arises in HTS systems. Closely packed spot beams provide continuous coverage across the coverage area. Each beam carries a channel shown as channels 1 to 4 for 4 beams shown. For illustration purpose, each beam was given a different color, red, green, blue, and orange, respectively. In a forward link as an example, all channels come from a gateway where those channels are mapped next to each other to minimize spectrum usage. The gateway uplink was then routed through the repeater before each channel is downlinked to each beam. Inside the repeater, all channels are filtered, frequency converted and amplified. There can be multi-stage filtering to separate those channels as well as to filter out spurious contents from passing through various active payload equipment. To maximize the useful bandwidth per channel, the filter passband needs to be as wide as possible. Figure 1 shows the filtering process where filter shape can be approximated as a trapezoid shape with a roll-off slope on each side. Because of the finite slope of the filter roll-off, we can see that each channel carries parasitic spectrum from adjacent channels. Those parasitic spectrum will cause the ACI because any beam not only receive its intended channel but also the parasitic interfering channel from its neighbors at the exact same frequency. For example, blue beam is supposed to receive channel 1 only, but it also receives a second parasitic partial channel 1 carried by the green beam, which leads to multiple path interference or ACI. It is the spectrum near channel edges that is mostly affected by ACI where filter roll-off is smallest. Those edge spectrums impacted by ACI are marked by rectangular boxes for illustration purposes. By making the filters steeper, the parasitic spectrum can become narrower thus help to alleviate the problem, but steeper filters are costly and very lossy as mentioned above.

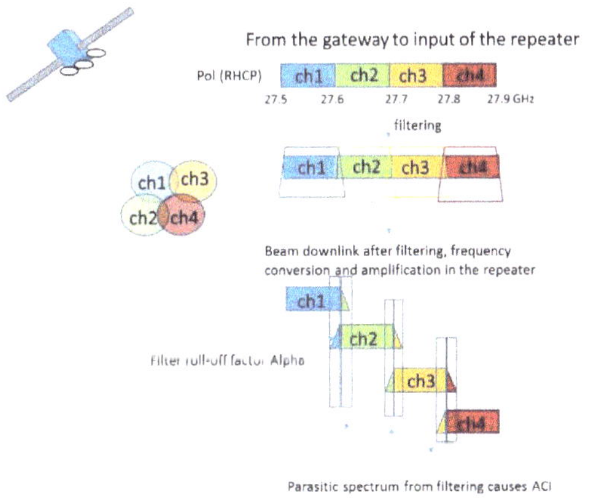

Figure 1 Illustration of the origin of adjacent channel interference (ACI)

2. Methodology

A very simple and low cost solution to suppress ACI without the need for steep channel filters is presented here. The proposed technique allows maximum bandwidth usage without ACI impairment. The technique is based on optimizing carrier assignment scheme to users in a ground system. Conventionally, the ground system randomly assigns carriers from available bandwidth to users requesting services. In this technique, the carrier assignment to users will not be random but will depend on where the user is located at. As shown in Figure 2, the users located near beam centres will be given carriers from edge of the channel where those carriers or spectrum are affected most by ACI. The users located at outer edge of the beam will be given carriers from the centre of the channel where ACI is virtually non-existent. By doing so, the ACI impacted carriers will be physically located farthest from their interferers, which are parasitic channels carried by adjacent neighboring beams. The total ACI suppression at the carrier location is the total carrier power in dBm minus the total interferer power in dBm at the same bandwidth. The total ACI suppression can be expressed as:

$$ACI_{total} = P_{carrier\ _{repeater}} - P_{interferer_{repeater}} + G_{antenna_{carrier}} - G_{antenna_{interferer}} = ACI_{repeater} + \Delta_{antenna\_gain} \quad ;$$

Where $\Delta_{antenna\_gain}$ is the difference in antenna gain at the carrier location between the carrier user beam and its neighboring beams, and $ACI_{repeater}$ is the ACI suppression from the repeater, which is the net filter rejection ratio.

From the above equation, the total ACI suppression is automatically enhanced by the antenna gain difference from the distance separation.

As an example, figure 3 illustrates the antenna elevation pattern plots for two adjacent beams, blue and green. Each beam is 0.9 degrees in diameter and the spacing between the two beam centres is 0.78 degrees. As shown, the carrier gain at blue beam centre is much higher than the interferer gain at the same location from the green beam pattern. The delta in gain is between 12 to 32 dB from ½ radius to beam centre, shown in two back arrows. This delta in gain provides additional ACI suppression beyond that from the repeater. It worths noting that the antenna gain difference gradually diminishes near the beam edge as shown in the plot, therefore, conventional approach of assigning carriers randomly to users will not be able to take advantage of antenna gain based ACI suppression effect. The ACI suppression enhancement achieved by this method can dramatically reduce the ACI suppression requirement at the repeater. A typical less-steep filter with rejection ratio of 5dB at ~4% roll-off from the band edge will translate to total ACI suppression ratio of 17-37dB by adding the antenna gain enhancement, which is more than sufficient to eliminate the ACI impairment. The ACI suppression can be enhanced by at least 12dB if ACI impacted carriers are optimally assigned to users located within the half diameter of the beam from the beam centre.

Those less-steep filters will have lower insertion loss and will be at lower cost, which can lead to simpler, better performance, and lower cost repeaters.

The proposed carrier distribution method can be easily implemented into the current ground systems by very minor software modification. No new equipment development is needed. First a satellite operator or service provider has all the HTS communication system information, including beam coverage and size, frequency/bandwidth allocations per beam, and satellite location information. Upon receiving a service request from a user, the service provider scheduling software will first determine which beam can be used to best service this user based on the geographical location of the user, e. g. longitude and latitude of the user location. The user location coordinates will be compared to the beam location coordinates and the beam that best encompasses the user will be selected. In addition, the distance between the user and the beam centre will be recorded. With the servicing beam (n) selected and the distance (d) known between user and the beam centre, the carrier frequency can be assigned to the user using the following algorithm shown in figure 4. Assuming the total bandwidth of the servicing beam n is B and its radius is r, with starting frequency f1 and ending frequency f2. If the distance d is less than r/2, then carrier frequencies in the ranges of [f1, f1+(f2-f1)*alpha] or [f2-(f2-f1)*alpha, f2] can be assigned to the user, where alpha is the 5-dB roll-off factor of the channel filter inside the repeater. If the distance d is greater than or equal to r/2, then carrier frequencies in the range of [f1+(f2-f1)*alpha, f2-(f2-f1)*alpha] can be assigned to the user.

Once the carrier frequency is determined, the total assigned bandwidth is simply the requested capacity from user divided by the link spectral efficiency. The link spectral efficiency is computed by the link budget at the user location and is a factor of carrier noise ratio, interference level, free space losses, and atmospheric losses. Depending on the actual link budget and capacity requirement at the user, it is possible that roll-off factor at other roll-off points (10dB, 7dB, etc.) and/or distance at other boundaries (r/3, r/4, etc.) can be used to determine the carrier frequency assignment rules described above.

In this paper, we will analyse capacity throughput enhancement by implementing this technique in a typical HTS payload system. We demonstrate that significant capacity improvement, more than 20%, can be achieved in systems where link budget is mostly C/I limited.

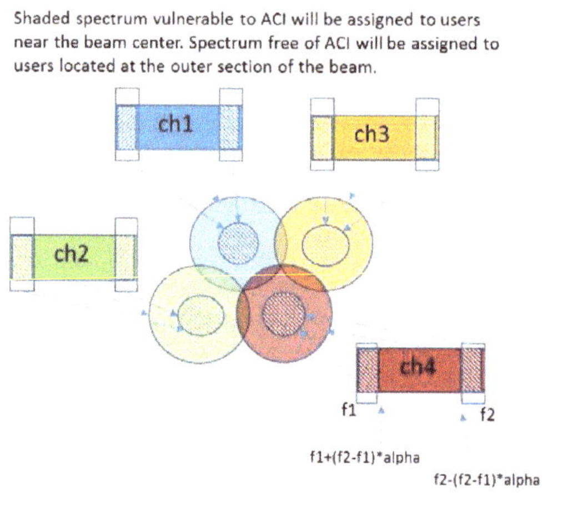

Figure 2 Illustration of the new carrier assignment method to users based on users' geographical location within a beam.

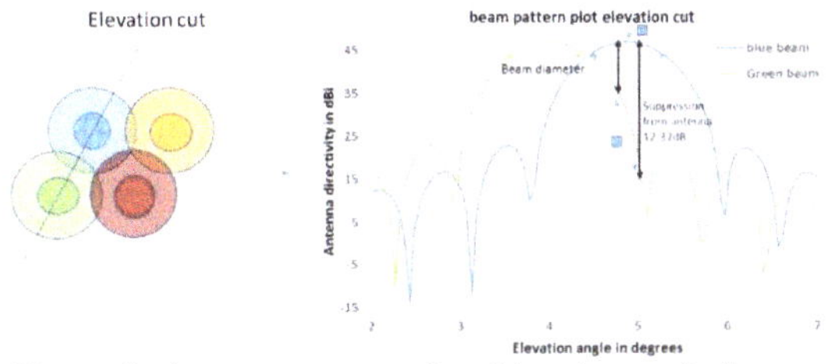

Figure 3 Antenna pattern plot (elevation cut) for two neighboring beams, blue and green, showing increased ACI suppression from antenna gain difference for users near blue beam centre.

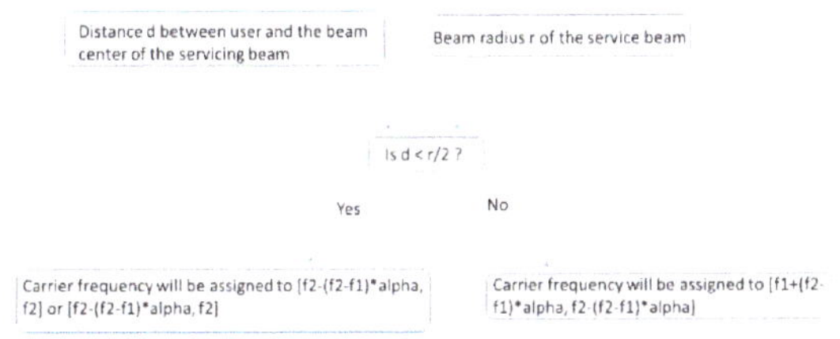

Figure 4 Algorithm for carrier assignment/distribution

3 Results

To assess the capacity enhancement from the proposed technique, an example Ku HTS system was used. Fig. 5 shows the HTS beam layout with 20 beams in a typical 4-color frequency re-use scheme from the orbital location at 25E. Two scenarios were evaluated and compared in assessing the total forward link capacity throughput: one is for C/N limited links using aeronautical terminals and the other one is for C/I limited links using large 1.2 meter user terminals. Under each scenario, the forward link clear-sky capacity was evaluated with and without the proposed ACI mitigation technique. When the proposed ACI suppression technique wasn't implemented, only repeater filter ACI suppression was used and carrier distribution to users was random. For the capacity evaluation, the repeater level ACI suppression from the filters was assumed to be 5dB at 4% roll-off from the band edge. By doing that, the useful bandwidth was at 96% to maximize bandwidth usage and capacity throughput. The gateway uplink was assumed to be non-limiting in the overall forward link budget, and user downlink drives the overall link budget and capacity. Those assumptions are generally true for most HTS systems, therefore, the results presented here are not limited to the example HTS system used, but should apply to any typical HTS systems. Table I summarizes the capacity results for all cases. For the scenario with aeronautical user terminals, the link was mainly C/N limited and the capacity enhancement from ACI mitigation was about 13%. For the scenario with large 1.2 meter user terminals, the link was mainly C/I limited and the capacity enhancement from ACI mitigation was much more significant, about 26%. Fig.6 compares the ACI suppression map with and without implementing the proposed technique. We can see that the ACI suppression level was significantly improved across the entire coverage area by using the proposed technique. The improvement in ACI suppression was achieved by optimally assigning carriers to users with ACI impacted carriers assigned to users near beam centres where ACI suppression was enhanced by the antenna gain of the intended carrier. Fig.7 compares the forward spectral efficiency map with and without ACI mitigation for C/I limited links. We can see that the ACI suppression improvement directly translated to the improved spectral efficiency leading to much higher capacity throughout the coverage area. For simplicity, uniform bandwidth distribution to the coverage area was used in generating the spectral efficiency map, so the spectral efficiency directly scales to the capacity by the same bandwidth at any grid point and the spectral efficiency map is identical to the capacity map. The technique also applies to un-even bandwidth distributions for local hot spot accommodations. In that scenario, the spectral efficiency map will be different from the capacity map because the bandwidth scaling will vary by locations. However, both spectral efficiency and capacity will see similar enhancement from the proposed ACI mitigation technique.

For C/I limited links, ACI mitigation is very critical in optimizing capacity throughput. For C/N limited links, ACI mitigation also helps to optimize capacity throughput, although the enhancement is not as significant as that for C/I limited links. Most links are actually dynamic in nature and can be C/I limited in certain conditions and C/N limited in other conditions, depending on atmospheric conditions, non-geo satellite interferences, and other dynamic link parameters. The proposed ACI mitigation technique is simple to implement and can benefit all types of links.

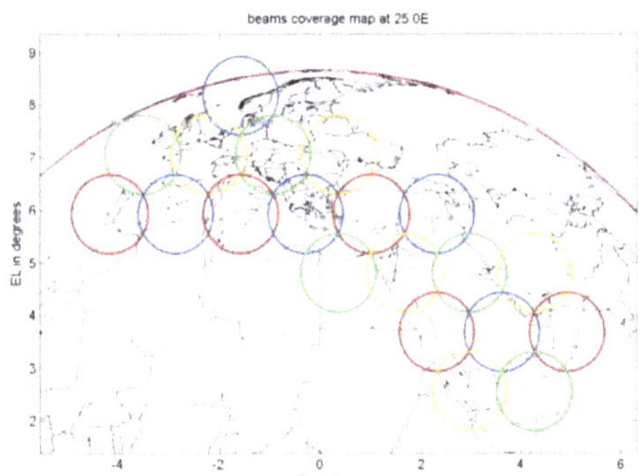

Figure 5 Beam layout for an example HTS system with 20 spot beams at 4-color frequency re-use configuration

Table 1 Throughput capacity comparison with and without ACI mitigation

Capacity comparisons on ACI mitigation		Total Forward bandwidth (GHz)	Average C/I (excluding ACI) (dB)	Average C/N (dB)	Average ACI (dB)	Average spectral efficiency (b/s/Hz)	Total Forward Capacity (Gbps)	Capacity enhancement (%)
Scenario 1 aero terminals, C/N limited link	With ACI mitigation	3.8	13.5	6.2	21	1.22	4.636	**13%**
	No ACI mitigation	3.8	13.5	6.2	12	1.08	4.104	--
Scenario 2: 1.2M terminals, C/I limited link	With ACI mitigation	3.8	13.5	17	21	2.6	9.88	**26%**
	No ACI mitigation	3.8	13.5	17	12	2.07	7.866	--

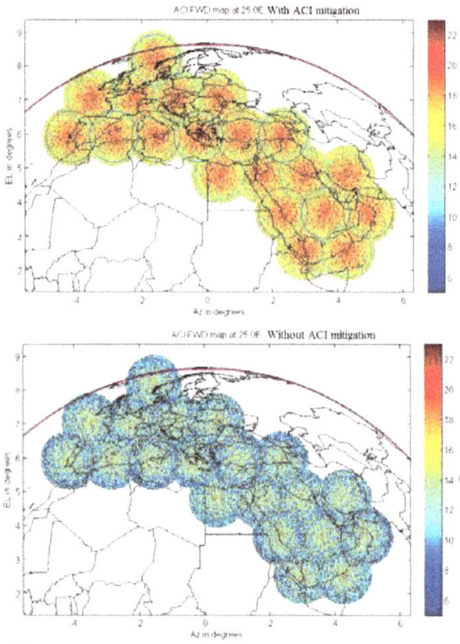

Figure 6 ACI suppression heat map across the coverage area. Top graph: with the proposed ACI mitigation technique; bottom graph: without ACI mitigation

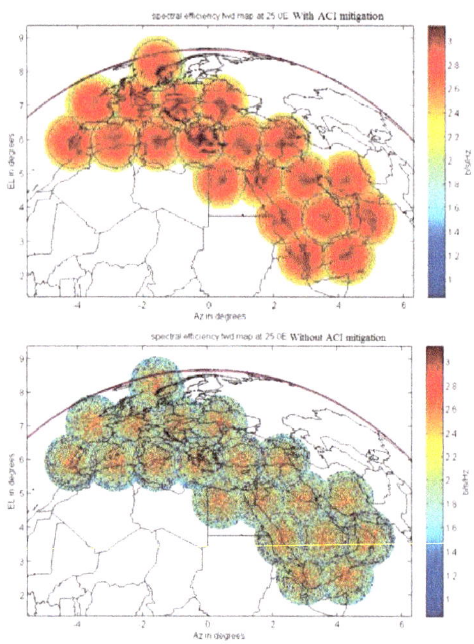

Figure 7 Spectral efficiency heat map across the coverage area. Top graph: with the proposed ACI mitigation technique; bottom graph: without ACI mitigation

4 Conclusion

A simple and low cost technique to mitigate adjacent channel interference has been presented. This technique can maximize useful bandwidth and efficiently translate all useful bandwidth into capacity. Most importantly, this technique requires no hardware development. It can be implemented entirely in the ground system software. Significant capacity enhancement has been demonstrated using a typical HTS system for both C/N limited links and C/I limited links. As the industry is searching for solutions for lower cost per bit satellites, this technique can help to drive down the cost by maximizing the useful bandwidth and capacity throughput.

Acknowledgements

The author would like to thank colleagues at Lockheed Martin Space Systems for their support and discussions.

References

[1] Wang, L., ‘A mitigation method for adjacent channel interference in communication satellites’, patent pending, 2018

Section 4 – New Satellite System Architectures and Components

NOVEL RF ARCHITECTURES AND TECHNOLOGIES FOR VSAT

Fabrizio De Paolis[1*], Enrico Lia[2], Václav Valenta[2], Petar Jankovic[2]

[1] European Space Agency, ECSAT, OX11 0FD Harwell Campus, United Kingdom
[2] European Space Agency, ESTEC, 2200 AG Noordwijk, The Netherlands

**Fabrizio.De.Paolis@esa.int*

Keywords: TRANSCEIVER, SOTM, VSAT, KA-BAND SATCOM, SATELLITES

Abstract

Very Small Aperture Terminal (VSAT) systems provide satellite communication solutions for users where terrestrial technologies are either uneconomical or simply non-existent. VSAT ground terminals conceived to date are largely based on conventional RF architectures, which make use of bulky, high power and/or costly components. The continuous downwards pressure on the price and form factor of VSAT terminals requires a major evolution of the associated RF hardware – in terms of higher packaging density, lower power dissipation and optimized frequency stability. This paper describes the recent developments in the field of ground segment RF architectures and technologies and provides an overview of relevant R&D opportunities for VSAT terminals.

1 Introduction

Advances in satellite communication technology have led to the deployment of Very Small Aperture Terminal (VSAT) networks supporting a variety of applications. Meanwhile, the price of VSAT equipment has reduced considerably over the years. An overview of relevant technology trends is provided in [1], where the new generation of low cost, highly integrated RF ICs and baseband modem System on Chip (SoC) are identified as the key elements for future VSAT evolutions. While RF IC based solutions are certainly attractive for large volumes of Ku-Band terminals, the situation is less clear at higher frequencies such as Ka-Band, especially if mobility is required. For those applications, single-chip solutions do not always offer the best trade-off. Limitations in the RF IC performance, as well as uncertainties in the business case and unit volume, may not always justify the high non-recurrent engineering costs of IC developments. On the other hand, recent developments in the area of passive and active RF technologies pave the way towards innovative VSAT solutions. For small- and medium-volume requirements, discrete or hybrid RF solutions makes the most sense in terms of performance and cost-effectiveness.

As an indication of the scale of the problem, consider a hypothetical Ka-Band VSAT for Satellite Communications on the Move (SOTM), which is broadly representative of the latest generation of terminals currently under development. A basic set of requirements for the terminal is given in Table 1 (scan losses and other parameters are not considered).

Table 1 Basic Requirements for a Hypothetical SOTM VSAT

Parameter	Value	Comments
Rx Frequency	19.7 – 21.2 GHz	
Tx Frequency	29.5 – 31.0 GHz	
IF Frequency	950 – 1950 MHz	
G/T	15 dB/K	At boresight
EIRP	50 dBW	At boresight
Polarization	Dual co-/x-polar	Electronic Setting
Elevation Range	0 to 60°	From boresight
Dissipation	<100 W	

For the sake of generality, this paper describes possible developments for Ka-Band SOTM, where typical applications include: Maritime, Land Mobile, Aeronautical and Machine-to-Machine (M2M) among others. It should be noted that most of the considerations in next sections are also applicable to other use-cases (both fixed and mobile) and terminal grades (consumer and professional).

2 Near and Long Term VSAT Architectures

In order to take full advantage of the aforementioned technology developments, non-conventional RF architectures must be considered for VSAT. Depending on the specific requirements a large number of options are possible.

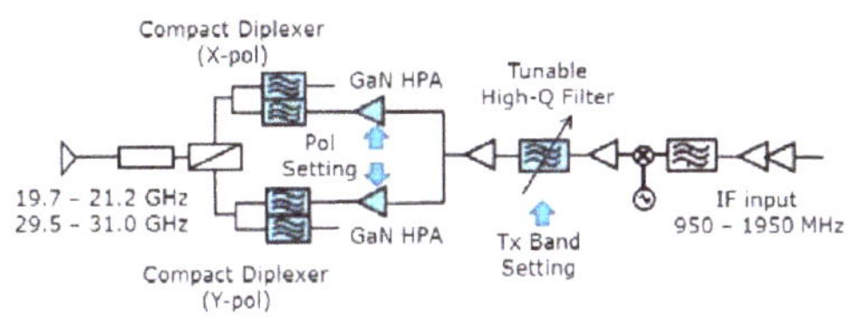

Fig. 1 Near-term approach to VSAT simplification (only the transmitter section is shown).

Figure 1 shows one possible approach based on hardware already existing or under development, ideal for centralized amplification schemes with conventional antenna reflectors (fixed or mechanically steerable).
In this design, compact diplexers are fed by single-device GaN High Power Amplifiers (HPAs) preceded by a tunable high-Q filter, which allows the desired transmit band setting. This filter needs to provide sharp rejection of the spurious signals generated by the heterodyne transmitter. Polarization setting could be implemented by selecting the desired RF chain, or via external RF switching (electromechanical or solid-state).
As a compared to conventional VSAT architectures, this approach has the benefit of improved RF performance and lower power dissipation, while providing a moderate level of architectural simplification due to the reduction in number/size of filters and MMICs.

Another possible architecture more suitable to implement low profile, electronically steerable VSAT terminals is shown in Figure 2. The introductions of low-loss phase shifters allow a passive array architecture, where a single transceiver feeds the antenna array or sub-array. The high-power amplifier is optimized for high-efficiency (e.g., GaN HPA with Doherty techniques). A single integrated Low Noise Amplifier (LNA) is used before down conversion.
This approach may lead to a number of benefits, in particular lower power dissipation and a reduction in the total number of components. In addition, it enables the possibility to introduce input/output filtering, which is normally difficult or impractical for ground terminals based on active arrays.

A third approach, rather long-term, shall be considered. The heterodyne receiver has been the standard option of choice for decades. Designers are now looking for an evolution from the widely adopted heterodyne architecture to a direct RF-sampling approach, which offers a number of advantages when combined with various techniques. In recent years, the rapid advance of analog-to-digital converter (ADC) sampling rates and the integration of matched channels now offers options for the receiver architect that were not practical a few years ago. Similar considerations apply to the transmitter architecture, as the radio transmitters can now be designed with direct digital modulation due to the availability of high-speed DACs, enabling it to operate directly at high frequencies.

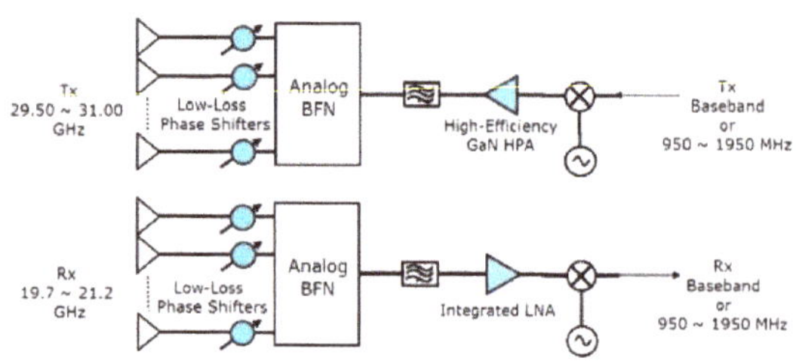

Fig. 2 Example of VSAT architecture based on passive arrays

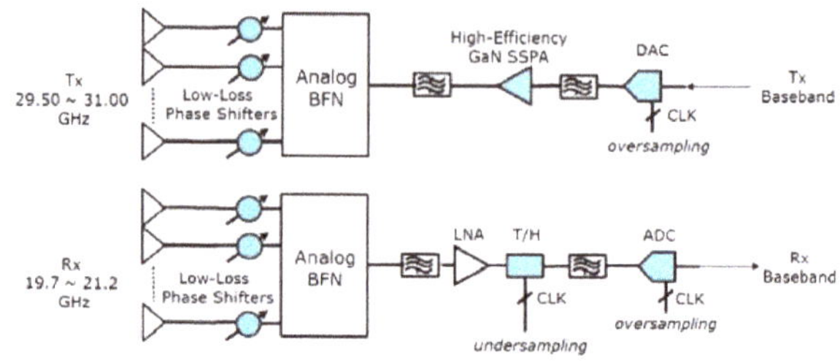

Fig. 3 Possible architecture based on direct RF-sampling

Fig. 3 shows a possible direct-RF sampling approach, for both receiver and transmitter. This is one of the simplest architectures, due to the lower number of components and the absence of image frequency issues. The receiver architecture is based on a direct-RF sub-sampling concept, employing a band-pass filter followed by a track-and-hold amplifier and ADC. This approach carries the advantage of replacing the mixer and local oscillator blocks, usually employed in conventional architectures, by a single chip, enabling a compact and low cost design. The transmitter architecture is based on a high-speed digital-to-analog converter operating directly at high frequency, followed by a solid-state power amplifier. The approach also aims at cost saving and compactness, however there are still a number of aspects to be improved and challenged within this type of architecture, especially when considering semiconductor technology, such as power consumption and faster sampling rates in order to enable operation at higher RF frequency (Ku-/Ka-band and beyond).

Of course there are many other possible solutions, such as those based on digital beamforming. These architectures are not yet feasible at Ka-Band because low power and high speed ADCs and DACs are very challenging to implement. This might change in the future as CMOS technology gets faster and ADCs and DACs can make digital beamforming competitive with other architectures [2]. Hybrid options are possible too, where an RF IC transceiver is used together with external LNA, HPA and diplexer [3], leading to a modular and flexible solution.

Which to use among the above and other possible approaches depends on a number of system level considerations (including the antenna apertures):

- Size and weight constraints
- Required EIRP and G/T
- Scan range and number of beams
- Polarization (linear/circular, single/dual, co-/x-pol)
- Modem compatibility
- Cost targets
- Level of development risk

It is not the objective of this paper to evaluate the relative merits of each of these architectures since it requires a detailed understanding of the individual user requirements, but rather to identify on going technology developments that can potentially play a role in their deployment.

3 Relevant Passive Technologies

3.1 Tunable High-Q Filters

Filter technology has recently experienced a major step forward in the area of on board flexibility. Center frequency and bandwidth tunable filters have reached a good maturity level, such that applications to output circuits are now feasible. Electromechanical waveguide filters with wide tuning range (>500 MHz), low insertion loss (<1 dB) and high-Q factor (>5000) at Ka-Band have been recently reported [4], [5]. Next challenges in this area would be extensions of the tuning range to cover several transmit bands, while keeping complexity and cost to a minimum. As markets for space-qualified components become more competitive, filter designers are under pressure to find ways to provide outstanding performance at lower prices. This may open the door for technology transfer from space to ground segment, with promising technologies becoming more affordable for VSAT manufacturers and integrators.
Other filter types using MEMS technology [6] show promise of a more compact size, wider tuning range and potentially low-cost design, although they have a relatively low Q-factor. Attempts to improve the Q combining more exotic topologies with low-cost manufacturing have been recently published [7], [8].

3.2 Compact Diplexers

Diplexers for Ka-Band VSAT applications are normally realized in waveguide technology. Recent advances in the area of electromagnetic (EM) design and modeling allow the design of increasingly complex diplexers / multiplexers, often integrating multiple functions in a single assembly [9],[10]. Additive manufacturing (AM) is becoming more popular, as it removes many technological barriers that prevent the optimal design of waveguide diplexers [11]. Selective laser melting (SLM) is in particular one of the most suitable AM processes for manufacturing waveguide components operating up to 50 GHz, as it allows manufacturing of metal parts with good accuracy (<30 μm). This may enable compact and lightweight structures for VSAT, with equivalent or even superior RF performance.
Advances in microwave and millimetre wave integrated technologies also facilitate the design of high performance, compact and low cost diplexers. The use of low temperature co-fired ceramic (LTCC) [12], substrate integrated waveguide (SIW) [13] and micromachining [14] has shown promising results in reducing the cost of the feed chain, with the potential of large-scale fabrication compatible with planar monolithic circuits.
The "dream" diplexer is always a single device, integrating filters, orthomode transducer and polarizer in a compact and low cost assembly.

3.3 Low loss Phase Shifters

High-performance integrated phase shifters are key components in an antenna array. They can be implemented using passive or active devices, and using lumped or distributed elements. For VSAT phased arrays, a passive phase shifter design would offer a favorable trade-off in terms of power consumption, noise level, complexity and cost. One of the main challenges of phase shifters at Ka-Band is to obtain a very low loss design. This is currently an important research topic in both academia and industry.
Phase shifters based on co-planar waveguide with a movable dielectric slab are reported in [15], showing average insertion loss slightly above 1 dB with maximum 170° phase shift.
A similar solution could be implemented using MEMS [16], [17]. Although MEMS technology is probably still not mature for commercial VSATs, the performance reported in literature (around 3 dB loss) looks acceptable for passive array Ka-Band applications.
Liquid crystals are also gaining popularity among emerging VSAT players [18], for both active and passive arrays, although more work seems necessary to optimize losses and packaging.

4 Relevant Active Technologies

4.1 GaN HPAs

Historically, Gallium Arsenide (GaAs) technology has been used for high performance and robust implementation of HPAs in VSAT terminals. However, in recent years, Gallium Nitride (GaN) technology has reached a high level of maturity and is becoming a key enabler in modern satellite communication for both space and ground segment systems. The main benefits, namely high power density and improved Power Added Efficiency (PAE), make GaN technology appealing, especially for VSAT terminals with reduced size and improved DC-to-RF conversion. Today, MMIC designers have the opportunity to benefit from well-established GaN processes, featuring a wide variety of transistors in different device configurations with reliable large-signal models. The latter advancements offer RF system engineers a high degree of integration and in some special cases, even system simplification due to inherent robustness of GaN. Unlike the space segment where the MMIC designers must follow the required derating rules (namely in terms of junction temperature and maximum drain voltage levels), VSAT terminals can fully benefit from the potential of the GaN technology. Recently, 10 W class single-chip GaN HPA MMICs with PAE of more than 30% at Ka-band have been reported [19], [20]. However, the commercial offer of rugged, fully passivated, 10 W class HPA MMICs is still very limited [21], [22]. For low-cost design, ready-to-use QFN packaged Ka-band MMICs with saturated power of up to 4 W are commercially available [23], [24]. As of today, commercial GaN MMIC technologies come with two different types of substrate: silicon and silicon carbide. It needs to be underlined that the substrate properties play an essential role for costs, RF losses and thermal management, which is in turn linked to the reliability and overall RF performance. Lowering costs is volume-driven, hence considering the ongoing developments in the arena of 5G where GaN on silicon substrates can become more cost-effective due to high volumes, it is expected that this evolution may positively impact also on developments of modern VSAT terminals. On the other hand, for high performance VSATs and small military terminals, silicon carbide or even diamond substrates will likely be the ultimate choice due to better reliability and overall RF performance. The latter, GaN on Diamond substrate, is still in the development phase under several national DoD contracts and it is expected that once available, this technology will become a radical technology enabler.

The complexity of modulated signals is also increasing, which has an impact on HPA design: advanced modulation schemes and multicarrier operation result in high Peak-to-Average Power Ratio (PAPR), which requires HPAs to operate in larger back-off to guarantee a certain level of signal distortion (linearity). At the same time, the PAE of classical HPA MMICs is strongly degraded under large back-off. Although similar issues have been resolved at lower frequencies by dedicated HPA architectures issues (e.g. Doherty, envelope tracking, etc.), classic HPA architectures still dominate in today's VSAT frontends. As an example, a single-chip HPA MMIC for the space segment at Ka-band demonstrates the Doherty concept [25], which could be applicable also to VSAT HPAs.

For ground SATCOM applications where higher power levels are required (beyond 5-10 W), corporate on-chip power combining becomes impractical due to high losses, which penalizes the overall PAE. As a result, power from several HPA MMICs is combined using typically wideband spatial combiners with proven combining efficiency approaching 90% [26], [27]. Today, spatial power combining technology is well established and is the key enabler in modern power-efficient high power VSAT and high power Block Up-Converters (BUCs).

It should be mentioned that the 30 GHz frequency is a crossover point where GaN and GaAs HPAs are overlapping in many characteristics. For SOTM VSAT applications requiring high PAE and very good thermal performance, GaN is the ideal candidate; however for fixed applications GaAs is still competitive due to its similar RF performance and lower cost.

4.2 High-speed ADCs and DACs

There are a number of topologies for high-speed ADCs and DACs. Commercially available DACs are now featuring 14-16 bit resolution with 2.5 Giga-Sample per second (GSps) enabling a high Spurious-Free Dynamic Range (SFDR), however ground breaking technologies based on Silicon-Germanium Carbon (SiGeC) provides an analog bandwidth extending beyond 7 GHz and direct digital synthesis up to K-band (26.5 GHz), enabling sampling rates up to 6GSps. The introduction of carbon in the intrinsic base of transistors is the key element to achieve high transition frequency and low resistance, two parameters DAC performance strongly relies upon. The ADC receiver architecture is less demanding in terms of sampling rate effort, as sub-sampling techniques can be employed together with a high-performance ADC device. The sub-sampling technique, also known as band-pass sampling, has advantages to reduce the sampling rate when using a sample and hold device (S&H) to produce low-frequency replicas of the RF signal. This is because most of the signal processing is done in the digital domain, which is lower than twice the highest signal frequency, but higher than twice the signal bandwidth. Technology demonstration through the ESA ARTES Programme [28] achieved a frequency of 26 GHz with more than 1 GHz bandwidth, which is suitable to be used with a BFN covering the standard IF in ground segment equipment. Nowadays, commercially available high-linearity Track-and-Hold Amplifiers (THA) offer 25 GHz input bandwidth with THD (Total Harmonic Distortion) better than -40 dBc over the total bandwidth, and 4 GHz sampling rate [29].

In order to cope with future requirements of RF architectures operating at higher frequencies, a solution could be to extend the sampling rate by mixing [30]. This concept is applicable to both ADCs and DACs. The architecture is designed for doubling ADC and DAC sample rate. Single DAC input clock frequency is half its sample rate. Similarly to DACs, it is possible to combine several ADCs to multiply the effective sample rate and addressable bandwidth. Both architectures are suitable to allow double sampling rate, which is currently a limitation within the semiconductor technology, in order to achieve higher frequencies.

4.3 LNAs/LNBs

The first stage of a VSAT receiver has the upmost effect on the overall receiver performance. Noise figure, gain stability over temperature, gain flatness, linearity and low power consumption are the main design drivers in LNAs for VSAT. Working at high frequencies often means additional issues such as low frequency stability, high component losses, and overall limited availability of the MMIC devices with acceptable performance.

Because of performance and cost targets of the available amplifier technology, LNA suppliers prefer a design based on packaged GaAs HEMT transistors, which could provide noise figure near 1 dB over the 17.7 to 22.2GHz range. Gain instability and poor input return loss of such a design may require the use of low loss isolators, leading to a total noise figure of the LNA assembly close to 1.5 dB.

Increasing demand for LNA modules with superior performance, followed by the wider bandwidth requirements for multi-band terminals (such as SOTM and Fly-Away) are leading to new trends in the area of integrated LNA devices and MMICs. The GaAs technology remains the preferred solution of the MMIC designers due to its high gain and low noise figure up to the millimeter wave frequencies. Low noise figure, stable gain and good return loss are achieved by means of adaptive matching or a balanced configuration at the input and the output of the MMIC. Performance of integrated LNA MMICs is comparable with a combination of the discrete HEMT and low loss isolator [31], [32], the advantage of such integrated LNA being robustness, compactness and simplified DC biasing.

On the Low Noise Block Converters (LNB) side, important benefits derive from the development of highly integrated frequency synthesizers. Programmable Phase Locked Loops (PLLs) with integrated VCOs are now operating over wide frequency range (>16 GHz) [33], [34]. The low phase noise and low power consumption of these devices are enabling multi-band performance or simultaneous band conversion without increasing the size of the LNB. Furthermore, these devices provide higher environmental robustness, facilitating the LNB integration on transportable or moving platforms.

5 Conclusion

ESA is actively involved in evaluating novel RF architectures for Ka-Band VSAT applications, while supporting the development of relevant enabling technologies. Results of internal R&D and industrial activities are encouraging and a number of these initiatives should lead to technology validation and product industrialization in the coming years.

References

[1] Gat, Y., Shacham, G., Yoram, B.: 'The dream of sub 100$ VSAT is getting closer to reality', Proc. 21st Ka and Broadband Communications Conference, Bologna, Italy, October 2015

[2] 'ESA ARTES Programme activity: PRIME Beam Former Chip for Phased Array Antenna', https://artes.esa.int/projects/beam-former-chip-phased-array-antenna, accessed 17 August 2018

[3] 'ESA ARTES Programme activity: SITKA Development of Transceiver ASIC for DTH Ka/Ku band terminals', https://artes.esa.int/projects/sitka, accessed 4 September 2018

[4] Arnold, C., Parlebas, J., Zwick, T.: 'Reconfigurable Waveguide Filter with Variable Bandwidth and Center Frequency', IEEE Trans. Microw. Theory Techn., vol. 62, no. 8, pp. 1663–1670, Aug. 2014

[5] Yassini, B., Yu, M., Keats, B.: 'A Ka Band Fully Tunable Cavity Filter', IEEE Trans. Microw. Theory Techn., vol. 60, no. 12, pp. 4002–4012, Dec. 2012

[6] Pelliccia, L., Cacciamani, F., Farinelli, Sorrentino, R.: 'High-Q tunable Waveguide Filters using Ohmic RF MEMS Switches', IEEE Trans. Microw. Theory Techn., vol. 63, no. 10, pp. 3381–3390, Oct. 2015

[7] Bowrish, B., Mansour, R.: 'A Tunable Waveguide Filter Designed with a Constant Absolute Bandwidth Using a Single Tuning Element', Proc. IEEE MTT-S Int. Microw. Symp. (IMS), Philadelphia, PA, USA, June 2018

[8] Nam, S., Lee, B., Kwak, C., Lee, J.: 'A New Class of K-Band High-Q Frequency-Tunable Circular Cavity Filter', IEEE Trans. Microw. Theory Techn., vol. 66, no. 3, pp. 1228–1237, Mar. 2018

[9] 'ESA ARTES Programme activity: Mercury Development of Next Generation Ka-Band VSAT Transceiver', https://artes.esa.int/projects/mercury, accessed 17 August 2018

[10] 'ESA ARTES Programme activity: Triple Band Feed Horn', https://artes.esa.int/projects/triple-band-feed-horn, accessed 17 August 2018

[11] Peverini, O. A. et al.: 'Integration of RF Functionalities in Microwave Waveguide Components Through 3D Metal Printing', Proc. IEEE MTT-S Int. Microw. Symp. (IMS), Honololu, HI, USA, June 2017

[12] Rigaudeau, L. at al.: 'LTCC 3-D Resonators Applied to the Design of Very Compact Filters for Q-Band Applications', IEEE Trans. Microw. Theory Techn., vol. 54, no. 6, pp. 2620-2627, June 2006

[13] Jaschke T., Rohrdantz, B., Mohncke J. P., Jacob, A. F.: 'A Ka-Band Substrate-Integrated Waveguide Diplexer with Wide Frequency Spread', Proc. European Microw. Conference (EuMC), Rome, Italy, Oct. 2009

[14] Nocella, V. et al.: 'E-Band Cavity Diplexer Based on Micromachined Technology', International Journal of Microwave and Wireless Technologies, vol. 8, no. 2, pp. 179-184, Mar. 2016

[15] Abdellatif, A. S. et al.: 'Low Loss Wideband and Compact CPW-Based Phase Shifter for Millimeter-Wave Applications', IEEE Trans. Microw. Theory Techn, vol. 62, no. 12, pp. 3403-3413, Dec. 2014

[16] Farinelli, P. et al.: 'Development of different K-band MEMS Phase Shifter Designs for Satellite COTM Terminals', Proc. European Microw. Conference (EuMC), Rome, Italy, Oct. 2009

[17] 'RF MEMS and Tunable Circuits', http://rfmicrotech.com/products-smart-solutions/rf-mems-and-tunable-circuits/, accessed 29 August 2018

[18] Jost, M. (Invited): 'Electronically Liquid Crystal-based Beamsteering Antennas for SatCom-Applications', Workshop WM07: New Developments for Satellite Communications on the Move, European Microw. Conference (EuMC), London, UK, Oct. 2016

[19] Gasmi, A. et al.: '10W Power amplifier and 3W Transmit/Receive module with 3 dB NF in Ka band using a 100nm GaN/Si process', IEEE Compound Semiconductor Integrated Circuit Symposium (CSICS), Miami, FL, USA, Oct. 2017

[20] Yamaguchi, Y. et al.: 'A CW 20W Ka-band GaN High Power MMIC amplifier with a Gate Pitch Designed by Using One-Finger Large Signal Models', IEEE Compound Semiconductor Integrated Circuit Symposium (CSICS), Miami, FL, USA, Oct. 2017

[21] 'NGC HPA MMIC family APN228/229/248', http://www.northropgrumman.com/BusinessVentures/Microelectronics/Products/Pages/GaNPowerAmplifiers.aspx, accessed 4 September 2018

[22] 'TGA2594', https://www.qorvo.com/products/p/TGA2594, accessed 4 September 2018

[23] 'CGY2650UH', http://www.ommic.com/produits, accessed 6 September 2018
[24] 'TGA4906', https://www.qorvo.com/products/p/TGA4906, accessed 4 September 2018
[25] Valenta, V. et al.: 'High-gain GaN doherty power amplifier for Ka-band satellite communications', Proc. IEEE Topical Conference on RF/Microwave Power Amplifiers for Radio and Wireless Applications (PAWR), Anaheim, CA, USA, Jan. 2018
[26] Kang, Y., Jin Pin X., Zheng Hua, C.: 'A full Ka-band waveguide-based spatial power-combining amplifier using e-plane anti-phase probes ', Proc. IEEE International Wireless Symposium (IWS 2014), X'ian, China, March 2014
[27] Harvey, J., Brown, E. R., Rutledge, D. B., York, R. A.: 'Spatial power combining for high-power transmitters', IEEE Microwave Magazine, vol. 1, no. 4, pp. 48-59, Dec. 2000
[28] 'ESA ARTES Programme, activity: Sample and Hold Amplifier for Pre-Processor Down-Conversion', https://artes.esa.int/projects/shadow, accessed 4 September 2018
[29] 'The world's first K-band capable digital-to-analog converter', https://www.e2v.com/products/semiconductors/dac/ev12ds460, accessed 5 September 2018
[30] Laperle, C, O'Sullivan, M.: 'Advances in High-Speed DACs, ADCs and DSP for Optical Coherent Transceivers', Journal of Light wave Technology, Vol. 32, No. 4, February 2015
[31] 'AD4371', http://www.analog.com/en/products/adf4371.html#product-overview, accessed 29 August 2018
[32] 'STuW81300', https://www.st.com/content/st_com/en/products/wireless-connectivity/rf-solutions/rf-pll-synthesizers/stuw81300.html, accessed 29 August 2018
[33] 'QPA2626', https://www.qorvo.com/products/p/QPA2626, accessed 29 August 2018
[34] 'ESA ARTES Programme activity: K-Band Satcom Ground Terminal LNA with Dual Mode Performance', https://artes.esa.int/projects/k-band-satcom-ground-terminal-lna-dual-mode-performance, accessed 17 August 2018

A Modular Architecture for Low Cost Phased Array Antenna System for Ka-Band Mobile Satellite Communication

Wael M. Abdel-Wahab[1,2], H. Al-Saedi[1], M. Raeiszadeh[1], E. Alian[1], G. Chen[1], A. Ehsandar[1,2], N. Ghafarian[1], H. El-Sawaf[1], A. Palizban[1], M. R. Nezhad-Ahmadi[1], S. Safavi-Naeini[1]*

[1]*University of Waterloo, Centre for Intelligent Antenna and Radio System (CIARS), Electrical and Computer Engineering, 200 University Ave. West, ON., Waterloo, Canada*
[2]*C-COM Satellite Systems Inc., 2547 Sheffield Rd, ON, Ottawa, Canada*
**wmabdelw@uwaterloo.ca*

Keywords: Flat panel, active phased array antenna, Ka/K-band, and modular approach, circular polarization radiation

Abstract

A low-cost and low profile phased array antenna system based on a modular approach is being developed for Ka-band mobile SATCOM applications. A Transmitter (TX) antenna array has been designed to provide the required circular polarized (CP) radiation beam with sufficient gain and a radiation pattern that satisfies the standard (FCC) emission mask. In the proposed modular approach, the intelligent active phased subarray (module) is designed and used as a building block to construct the whole array in any form and any size in any customized platform. In this design, the size of the intelligent subarray is chosen to be 4×4 with half-wavelength separation between the elements for both TX (30 GHz), and RX (20 GHz) array antennas. Larger array antenna with a few thousands of radiating elements can be formed by assembling as many modules as required on one platform. This paper discusses the modular approach aspects and its application in large scale phased array antenna implementation. Super TX / RX array modules of 256 elements are assembled and tested successfully to validate the proposed modular architecture. The measured radiation patterns at 30/ 20 GHz shows that the antenna's main beam can be steered to ± 70° in both azimuth and elevation directions.

1. Introduction

Over recent years, mobile satellite communication (SATCOM) becomes the only means to extend the range of communication and sensing to all corners of the world and far beyond the reach of any terrestrial system SATCOM. New emerging mobile SATCOM, particularly Ka-Band systems are a highly promising solution for mobile networks to establish high data rate communications with reliable and continuous coverage over the globe [1-4]. Ka-band systems have been rapidly growing to accommodate a larger number of users, as compared to existing Ku-band technology, and to cover many emerging applications, such as cellular backhaul services, broadband internet access to remote areas, interactive TV, among numerous other applications. Conventional mechanical scanning parabolic reflector antenna provides high gain and low side-lobe radiation pattern, which satisfies most of the SATCOM requirements. However, it is bulky, heavy, prone to failures and suffers from complex feeding system.

Electronically scanned low-profile phased array (Flat panel) can resolve all these issues particularly for mobile applications [5]. The inherit structure of the phased arrays allows for high gain directive communication while being flexible enough to electronically steer the beam of the array to a desired direction without moving it mechanically; which is referred to as "beamforming". This makes the phased array technology best-suited for high data rate directive communication applications such as satellite communications. The main challenges in developing a commercial phased-array system are its cost and complexity. For commercial satellite terminals, the radiation power density should follow a highly stringent standard mask [6, 7]. In SATCOM applications, antenna with large radiating apertures ($40\lambda_o \times 40\lambda_o$) are usually used to increase the gain to overcome the path loss at Ka-band and sustain a robust communication link with GEO-satellites. Existing large scale phased array solutions are either complicated or expensive [8]. A modular architecture, wherein the entire antenna array is made of identical intelligent subarray modules (blocks) is considered to be the most promising approach to develop cost-effective and flexible systems for mass market applications.

In this paper, a low-cost low profile phased array antenna system based on a modular approach is being developed for Ka/K-band mobile SATCOM applications. As a proof-of-concept, both transmitter (TX) and receiver (RX) array antennas, each consisting of 16 intelligent active modules to form 256 elements, are designed, fabricated, and tested to verify the proposed modular approach. The array antenna shows stable circular polarized (CP) radiation pattern, which can be steered up to an elevation angle of ± 70 degrees off-boresight. The size of array antenna can be scaled to a few thousands of radiating elements to provide the required gain, proper aperture illumination tapering (side-lob level) to satisfy the SATCOM application requirements .

2. System Model and Antenna Aperture Requirements

In active phased array antenna system (APAA), the signal transmitted or received by each element is controlled in phase and magnitude by an intelligent beam-forming network allowing for beam steering to a certain direction (electronic beam steering). Fig. 1 shows the structure of a satellite link where the TX (30 GHz) and RX (20 GHz) active phased array system is introduced. The transmitter (TX) is composed of a modulator, up-converter, splitter, variable gain amplifiers (VGAs), phase shifters, power amplifiers (PAs), and the TX antenna elements. The Receiver (RX) is composed of antenna elements, feeding network, low noise amplifiers (LNAs), phase shifters, amplifiers, the combiner, a down-converter, and a demodulator, as shown in Fig. 1.

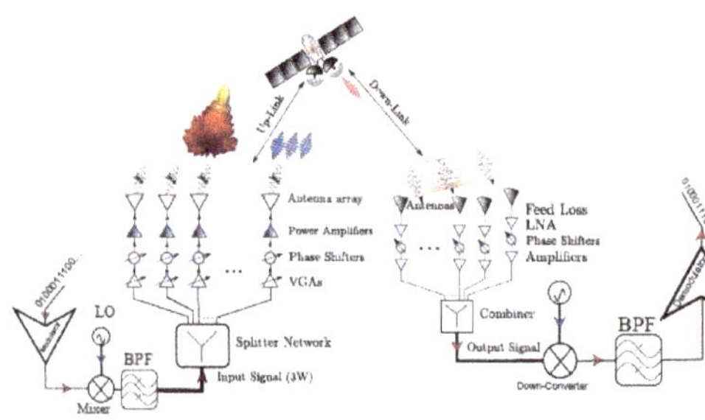

Fig. 1 Active phased array antenna system Architecture.

It is well-known that the signal radiated or combined from the all the antenna elements (in the far-field region) can be written as:

$$\vec{E}(\theta,\emptyset,r) = \vec{E}_{se}(\theta,\emptyset,r)\sum_n I_n e^{j\frac{2\pi f}{c}\hat{r}.\vec{r_n}} \quad (1)$$

Where f is the RF frequency, c is the speed of light, $\vec{E}$ is the total radiated field, $\vec{E}_{se}$ is the single element radiated electric field, $\hat{r}$ is the radial unit vector in spherical coordinate, and $\vec{r}_n$ is the location of antenna number n. The spherical angles $(\theta,\emptyset)$ shows the radiation in space. Coefficients $I_n = |I_n|e^{j<I_n}$ shows the effect of gain and phase control on the constructed beam from all the antenna elements.

Flat panel APAA with large radiating aperture is usually required for TX and RX band SATCOM applications to achieve high gain to overcome the path loss, to enhance the effective isotropic radiated power (EIRP), and to improve signal-to-noise ratio (SNR), and consequently to establish a reliable and robust communication link. Large scale APAA of aperture size of 4000 elements (spacing 0.5λo) with circular truncation are required to achieve ERIP > 48 dBW at large scanning angles up to ± 70 degree off-boresight at Ka band as shown in Fig. 2(a). Furthermore, magnitude variation (tapering) of ~ 12 dB over the aperture is required as shown in Fig. 2(b) to control the side lobes and to satisfy the TX radiation mask as shown in Fig. 3 as the beam is steered to different scanning angles. In practical implementation of large scale array antenna with such huge number of antenna elements in one PCB, the design becomes quite complicated, the fabrication cost becomes extremely high, and the fabrication errors become significant.

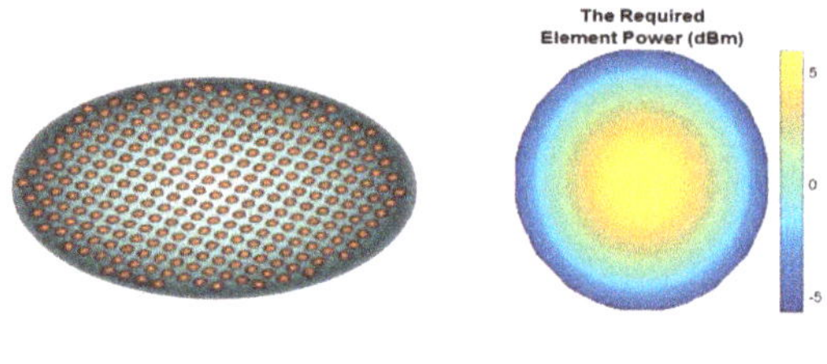

a *b*

Fig. 2 Radiating aperture (a) size, (b) Aperture power distribution.

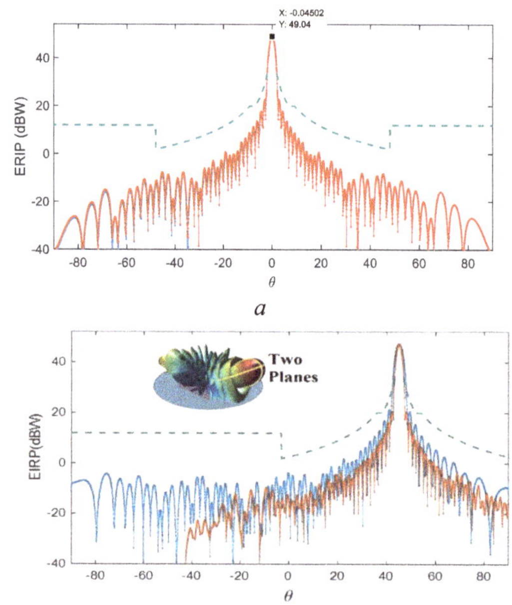

Fig. 3 EIRP Radiation pattern with bandwidth = 6.25 MHz (a) Boresight (12 dB tapering), (b) 45 degree off-boresight.

3. Modular Architecture Approach of large Scale Phased Array Antenna

The design of large scale APAA system becomes very complicated and impractical if traditional methods and architectures are followed. However, applying the modular approach in APAA architecture minimizes the design complexity and reduces the fabrication cost. In this approach, the entire APAA aperture is made of identical intelligent subarray modules (blocks) which can be integrated to any customized platform as shown in Fig. 4. Depending on the SATCOM application, the APAA can be scaled to any size and the module size could be 4×4 or bigger depending on the design complexity and cost. However, sophisticated software system is developed to control the antenna modules. The RF-signals from all the modules are combined by an efficient millimeter-wave (MMW) multi-layer feed circuit. All the control signals of the chips which decide the phased array operational mode are managed by a powerful processing unit(s). For completeness, an embedded built-in calibration system is necessary to monitor the phase and magnitude distribution errors over the entire aperture. In this work, the size of the intelligent subarray is chosen to be 4×4 with half-wavelength separation between the elements for both TX (30 GHz), and RX (20 GHz) array. More details about the sub-systems included in this approach will be discussed during the conference.

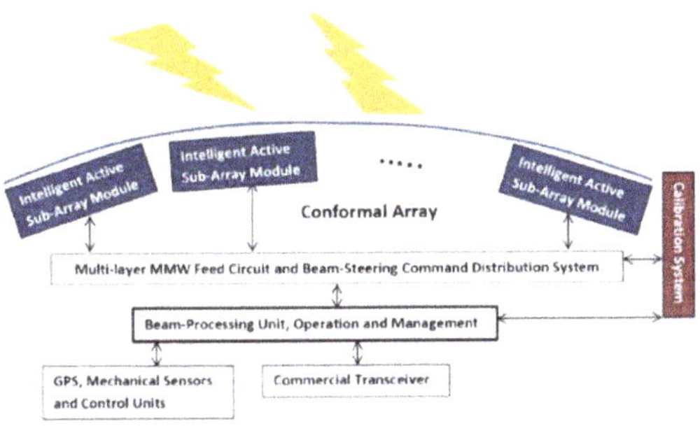

Fig. 4 Intelligent, modular, and scalable architecture of phased array antenna.

Fig. 5 Fabricated modules, (a) RX antenna (20 GHz), (b) TX antenna (30 GHz), (c) MMIC view.

4. Design and Implementation of APAA Module

In 4×4 APAA module designs, patch antennas are used as the radiating aperture elements over TX and RX operating bands as shown in Fig. 5(a) and Fig. 5(b), respectively. The patch antenna provide circular polarized radiation over the operating frequency bands, Ka-band (30 GHz) for TX, and K-band (20.0GHz) for RX, respectively. For beam forming, the aperture's magnitude and phase are controlled by multi-channel MMIC chips which are integrated to the antenna's feeding network as shown in Fig. 4(c). The antenna structures were fabricated, tested, and the MMIC is characterized before the full-integration. Some of these results will be discussed in the following subsections.

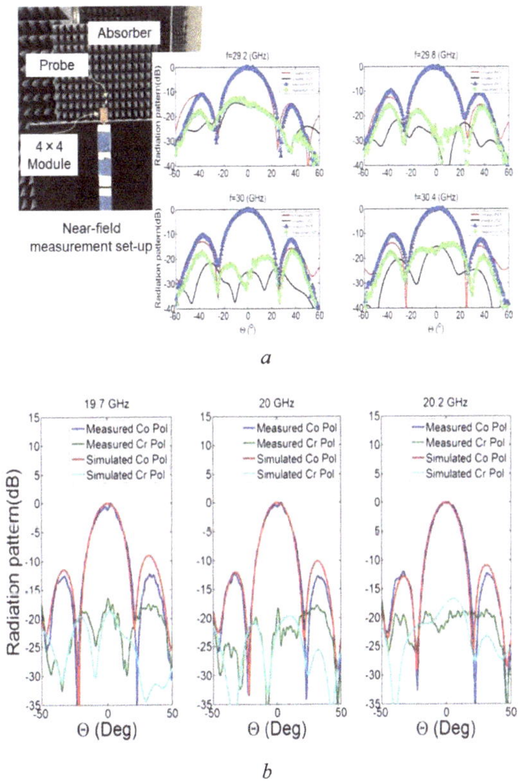

Fig. 6 Radiation pattern of passive 4×4 array antenna at different frequencies, (a) TX, (b) RX.

Sixteen (4×4) passive CP patch antenna arrays (fixed-beam) were designed, fabricated by the PCB low cost technology, and their radiation patterns were tested over the operating frequency TX and RX bands by planar near-field system. Fig. 6(a) and Fig. 6(b) show the simulated and measured radiation patterns of both TX and RX antennas at different frequencies, respectively. Good correlation between the measurement and simulation results are observed. The antennas show an average CP gain ~ 16 dBic and axial-ratio of less than 3 dB at the boresight direction over the operating bands which could be good candidates for some SATCOM applications. However, other techniques are being developed to improve the CP radiation purity not only at the boresight, but also at the desired scanned angles.

5. Active phased array antenna: Super Module

After the 4×4 APAA module (block) in Fig. 5 are calibrated and fully characterized, super-modules can be simply formed by adding a few of these blocks scaled to any size and integrated to any platform depending on the applications. Fig. 7(a) shows an example of a super-module (TX in this case) which includes 16 APAA blocks (256-elements) assembled into a highly efficient feeding network platform. The antenna is low profile and can be integrated to any mobile platform, as shown in Fig. 7(b). Errors in magnitude and phase distribution over the aperture could occur due to the PCB fabrication tolerances, inaccurate integration of MMICs inside the 4×4 APAA module, or the errors caused by the inaccurate positioning of the individual blocks. Therefore, a sophisticated and accurate built-in calibration system is necessary to scan the aperture and correct for any error in both magnitude and phase and point the beam to the correct direction.

Fig. 7 (a) TX super-module APAA (256- elements) based on modular architecture approach, (b) Integration to mobile platform.

6. Results

The super-module APAA was calibrated and its radiation pattern was tested successfully over the operating frequency band by the planar near-field scanner system from NSI at different scanning angles. In this case, uniform amplitude distribution over the radiating aperture is used for excitation with progressive phase distribution. The results show that the antenna's radiation pattern can be steered up to ±70 degree off-boresight at 30 GHz as shown in Fig. 8. Similar results have been observed at 20 GHz (not shown for brevity). Extensive tests are still ongoing and more results will be presented during the conference.

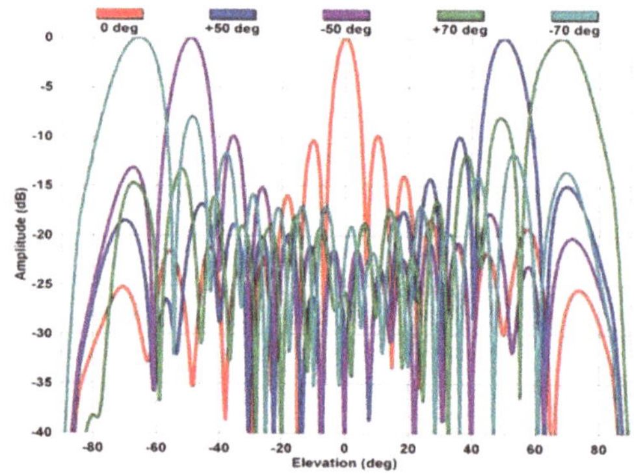

Fig. 8 Measured radiation pattern of TX- super module APAA (256- elements).

7. Conclusion

In this paper, the modular architecture approach used to implement ka/k active phased array antennas for low cost SATCOM applications is presented. Following this approach, phased array antenna can be scaled to any size and integrated to any platform depending on the application. To validate the proposed approach, a 4×4 APAA module was used as a building block to form 256 element APAA super module. The measured results show that the antenna's main beam can be steered to ± 70° off-boresight. The antenna is low profile and low cost compared to other SATCOM solutions.

Acknowledgement

This work was supported by the National Science and Engineering Research Council (NSERC) of Canada and C-COM Satellite Systems Inc.

References

[1] H. Fenech, S. Amos, A. Tomatis, et al.: 'High throughput satellite systems: An analytical approach', IEEE Transactions on Aerospace and Electronic Systems, 2015, 51, (1), pp. 192-202.

[2] Hasan, Mohamed, and Christopher Bianchi.: 'Ka band enabling technologies for high throughput satellite (HTS)

communications' International Journal of Satellite Communications and Networking, 2016, 4, pp. 483-501.

[3] Fenech, H., Alessia T., S. Amos, et al.: "Eutelsat HTS systems," International Journal of Satellite Communications and Networking, 2016, 34, (4), pp. 503-521.

[4] J. Navarro: 'Ultra-small aperture terminals for SATCOM on-the-move applications', 2017 IEEE MTT-S International Microwave Symposium (IMS), Honolulu, HI, 2017, pp. 1152-1154.

[5] Q. Luo and S. Gao: 'Smart antennas for satellite communications on the move', 2017 International Workshop on Antenna Technology: Small Antennas, Innovative Structures, and Applications (iWAT), Athens, 2017, pp. 260-263.

[6] Recommendation, I. T. U. R. S. 524-9 : 'Maximum Permissible Levels of off-Axis EIRP Density from Earth Stations in Geostationary-Satellite Orbit Networks Operation in the Fixed-Satellite Transmitting in 6 GHz, 13 GHz, and 30 GHz Frequency Band'.

[7] L.Gonzale and R. E. Greel,: ' A Regulatory Study and Recommendation for EIRP Spectral Density Requirement/Allowance for STOM terminals at Ka-band on WGS System", 2010-Milcon 2010 Military Communication Conference, San Jose, CA, 2010, pp 1992-1997.

[8] L. Baggen, M. Böttcher, S. Otto, O. Litschke, R. Gieron and S. Holzwarth,"Phased array technology by IMST: A comprehensive overview," 2013 IEEE International Symposium on Phased Array Systems and Technology, Waltham, MA, 2013, pp. 21-28.

[9] L. Baggen, M. Böttcher, S. Otto, et al : 'Phased array technology by IMST: A comprehensive overview', 2013 IEEE International Symposium on Phased Array Systems and Technology, Waltham, MA, pp. 21-28.

A COTS-BASED SOFTWARE-DEFINED COMMUNICATION SYSTEM PLATFORM AND APPLICATIONS IN LEO

Shey Sabripour[1], Jamal Haque[2], Andrew Ciszmar[2], Thomas Magesacher[1]*

[1]CesiumAstro, 13413 Galleria Circle, Suite 225, Austin, Texas, USA
[2]Honeywell Aerospace, Advanced Technology, USA
**thomas@cesiumastro.com*

Keywords: COTS, LEO, active phased array, software defined radio, payload, communication system

Abstract

We present a complete, small form-factor, modular, and flexible software-defined radio communications platform based on carefully selected automotive-grade commercial off-the-shelf (COTS) components. The hardware consists of three credit-card sized modules for software-defined radio signal processing, computing and network processing, and power conditioning, as well as a multibeam active phased array antenna module. Requiring only unregulated spacecraft power supply and a data connection, the platform delivers electronically steerable and programmable radio communications links including encryption, physical-layer and network processing, channelization, and radio-signal conditioning within a low profile, modular, and scalable system. In addition, the platform may host application-specific payload processing tasks using available computing resources and reconfigurable hardware. The combination of programmability and small form factor allows for various use cases ranging from LEO satellites via UAVs and aircraft to missiles.

1 Introduction

Shorter required life spans and constellations with a significantly larger number of satellites in LEO create a demand for lower-cost communication systems compared to GEO. Many "NewSpace" satellite and launch vehicle companies thus choose not to invest resources into full-custom avionics, creating a market for "out-of-the-box" solutions. A similar trend applies to the growing market for airborne communication systems targeting both commercial and military unmanned aerial vehicles (UAVs), where design and manufacturing tend toward lower cost and shorter design cycles.

Cost constraints and development time are the main drivers for adopting the commercial off-the-shelf (COTS) philosophy for flight hardware. Hand in hand with COTS goes the trend towards software defined radio (SDR) platforms based on field programmable gate array (FPGA) hardware [1] combined with single-board computers (SBC) that lend themselves to a modular and model-based system design approach (such as GNU radio [2]).

Leveraging advances made over the past decade in mobile consumer electronics and "intelligent" automotive systems, we present a modular, flexible and scalable system platform for agile communications targeting primarily small- to medium-size LEO spacecraft. The hardware is based on up-screened or hi-reliability automotive-grade parts with wide temperature ranges. A middle path between radiation-hardened design and the buy-and-fly philosophy, known as "careful-COTS" [3], is adopted to address the radiation challenge in LEO, which is a topic that keeps gaining attention and is creating a growing base of reference data (see, e.g., [4,5]). Careful COTS is combined with reliability-enhancing design elements such redundant dual-boot flash or error-correcting code (EEC) memory.

A built-in autonomous self-testing concept stretching from power supplies to radio-signal conditioning (cf. [6]) further supports the careful-COTS approach and enables rapid and cost-efficient system integration. All modules support high shock and vibration ratings suitable for launch vehicles and UAVs.

2 Hardware

Figure 1 shows a block diagram of the platform and a rendering illustrating an exemplary packaging. The main elements are three credit-card sized printed circuit boards (PCBs) for the power conditioning unit (PCU), the single-board computer (SBC), and the software-defined radio (SDR), respectively, as well as an antenna phased-array (APA) module.

2.1 Active phased-array antenna (APA)

The APA is composed of one or several PCB-based tiles each of which contains 64 transmit antenna elements interleaved with 64 receive antenna elements supporting both linear and circular polarization. The backside of a tile is populated with the power amplifiers, low-noise amplifiers, and beamformers, as well as mixers and switches if required. The number of

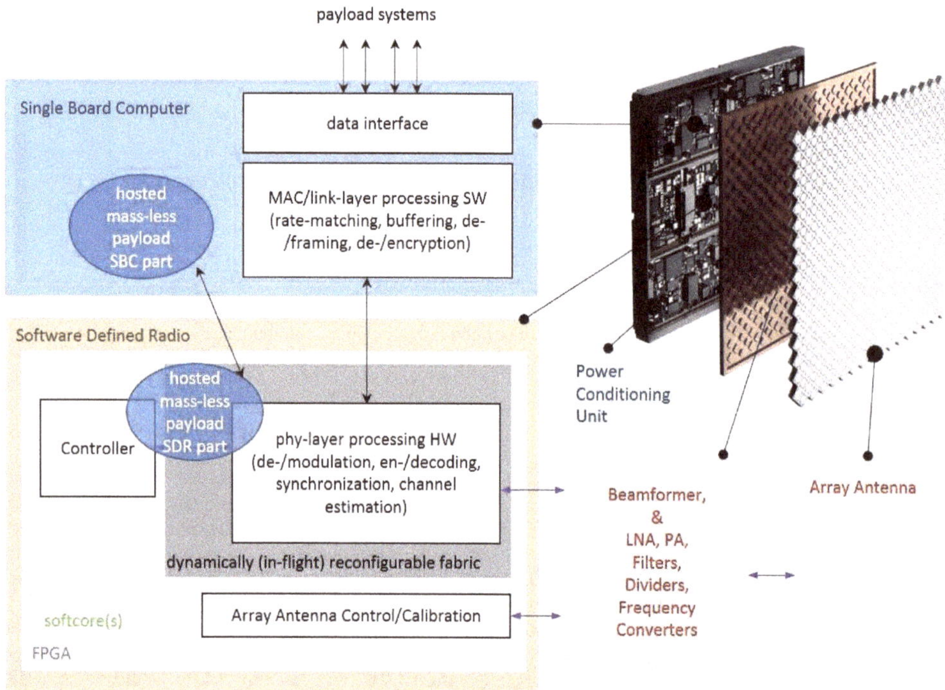

Figure 1: Block diagram and illustration of exemplary packaging of COTS-based software-defined radio communications platform including a steerable 256 element Ku-band active phased array (4 tiles), multi-channel software-defined radio module, quad-core single-board computer module, and isolated power converter ("Nightingale" platform in development at CesiumAstro, Austin, TX).

beams that can be supported is driven by the level of integration of the beamformer circuits, which is steadily growing.

2.2 Software defined radio (SDR)

The SDR module pairs two integrated dual-channel transceivers with an FPGA yielding 4 transmit and 4 receive channels with 100MHz instantaneous bandwidth that can be placed anywhere from 300MHz to 6GHz. In addition, observation ports and sniffer ports can be brought out for digital predistortion and cognitive radio features, respectively. Interface options include 10Gbase-KR and Spacewire

2.3 Single-board computer (SBC)

The SBC module includes point-of-load power converters, DDR4 RAM with ECC, non-volatile storage, dual redundant boot flash, and numerous data interfaces. The processor itself combines ARMv8 cores with dedicated encryption- and datapath-coprocessors to route and encrypt data without loading the processor.

2.4 Power conditioning unit (PCU)

The PCU module is a fully isolated intermediate voltage regulator that converts a 22-36V spacecraft power bus into a 12V system rail. Rated for up to 150W, the PCU operates with 92% efficiency over a wide range of loads. Features include short circuit protection, overcurrent limitation, reverse current protection, and an onboard microcontroller.

The four modules together with the in-flight-programmable hardware running on the SDR's FPGA fabric, the Linux-based operating system on the SBC, and the firmware on various microcontrollers across the boards (including PCU and APA) for monitoring and control constitutes a complete and self-contained spacecraft communication platform. When connected to an unregulated bus providing power and to a data interface of choice (GbE, UART, Spacewire, SPI, IIC, CANbus), the system converts raw (uncoded) information bits into properly "wave-formed" (modulated and encoded) electronically steerable transmit radio beams and vice versa.

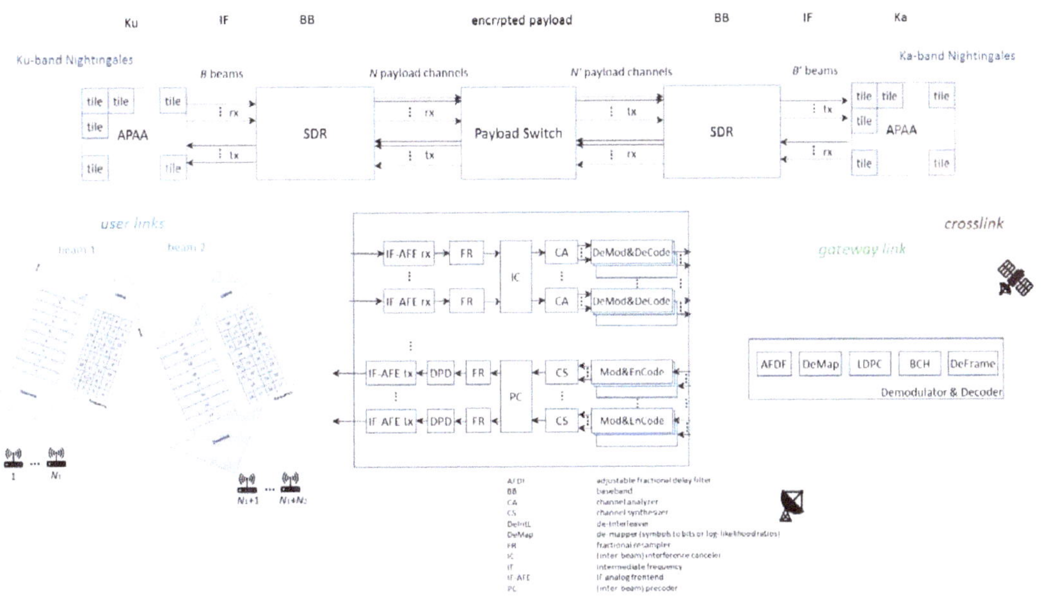

Figure 2: System concept of regenerative payload switch onboard a large-constellation satellite. Received user data allocated in slots across time, frequency, and space (beams) is demodulated, decoded, compactly re-arranged, encoded and properly modulated for transmission over a gateway link or a crosslink.

Partial dynamic reconfiguration of the FPGA [7] enables in-flight modification of mass-less payload applications while keeping vital functionality such as timing-reference maintenance or command-and-control links fully operational.

3 Applications

In the following, we present two application examples that highlight the benefits of an end-to-end software-defined system architecture.

3.1 Example 1: Regenerative active phased array payload

Figure 2 depicts the concept of a regenerative-payload system onboard a satellite of a large constellation. Spatially distributed users on ground have, in general, different and possibly time-varying bandwidth requirements, which can be met by flexible and dynamic allocation of transmission slots across time and across the available frequency band (for example, in Ku band). Spectral re-use across a satellite's footprint is accomplished using multiple beams illuminating areas as required.

The purpose of the payload switch is to extract the users' data received in the reverse uplink, demodulate and decode the individual streams and rearrange them in a more compact from for transmission on the reverse downlink to a gateway on the ground (for example, in Ka band) or on the reverse crosslink to a neighbouring satellite of the constellation. In forward direction, the switch demodulates and decodes the forward-uplink or forward-crosslink feed, extracts packets addressed to individual users, directs them to the appropriate forward-downlink beam, places them on the time-frequency grid according to the desired allocation, encodes and modulates individual users' streams, and transmits the resulting multiplex via the phased array forming a beam pointing at the desired location.

Figure 3 illustrates the benefit of regenerative payload

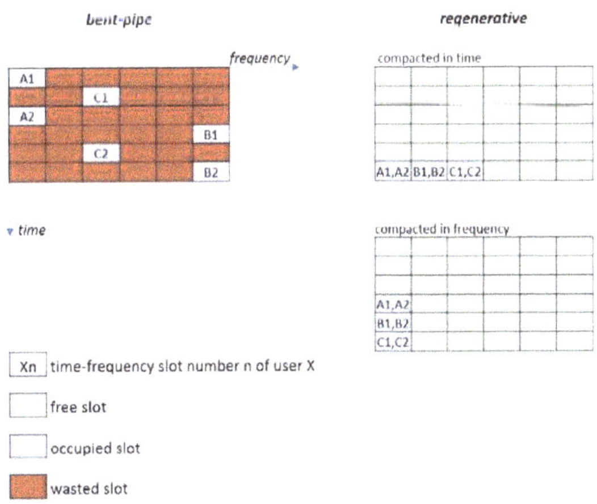

Figure 3: Advantage of regenerative payload processing (right) over bent-pipe processing (left): user data packets scattered over time and frequency slots can be rearranged more compactly and exploit slots with higher capacity available on the gateway link or on the crosslink.

processing. The user data may be scattered across the available time/frequency resource grid. Channelization, i.e., extraction of the users' packets from the received reverse uplink and synthesis of the transmit multiplex in forward downlink according to the allocation map allows for a more efficient use of the gateway link or crosslink since spectral gaps can be eliminated by re-arranging packets more compactly in time and/or frequency. Furthermore, data from more than one user can be packed into reverse downlink/crosslink slots that offer higher capacity.
In this application, the SDR performs frequency-domain channelization and physical-layer processing (following, for example, modulation and coding formats used in DVB-S2 [8]). The SBC is responsible for time-domain channelization, buffering, and re-allocation of packets.

3.2 Example 2: Physical-layer assisted ranging

Besides the prime applications, the tight integration of RF-frontend and transceiver hardware in combination with the possibility to design custom digital signal processing blocks running on the FPGA fabric allows for tailored solutions tackling specific challenges. The available resources can be used to host other payloads (sometimes referred to as "mass-less payloads"), which opens opportunities for applications that make use of or work closely together with the communications system. An example is high-accuracy range estimation between two communicating spacecraft using physical-layer-based time-of-flight estimation.
The principle of time-of-flight estimation is straightforward. Assume two transceivers share a common time basis. Transceiver 1 sends a suitable pulse of bandwidth B at time t_1 to transceiver 2, which detects the pulse and associates it with the time of reception t_2. Sharing t_1 and t_2 across transceivers allows to compute the time of flight $t_2 - t_1$, which is proportional to distance.
Various error sources contribute to the aggregate error of the distance estimate [9]. First, there is the challenge of transceiver synchronization. In case a timing synchronization error exists, t_1 at transceiver 1 does not coincide with t_1 at transceiver 2. Such timing offsets directly affect the time-of-flight estimate and must be avoided, for example, by employing a two-way time transfer method [10]. Second, the detection of the pulse is corrupted by channel noise, which results in a receive signal-energy-to-noise-power-density ratio E_S/N_0. Third, there is quantization noise introduced by sampling with frequency f_S, which fundamentally limits the achievable temporal resolution unless oversampling or any form of interpolation between samples is employed. Finally, a drift between the clocks of the two transceivers adds a timing offset: even if the two transceivers have the exact same notion of t_1, the same number of counted clock pulses corresponds to two slightly different time periods calculated for $t_2 - t_1$ at the two transceivers if their clocks drift while counting. Except when estimating very long distances, the clock drift induced by off-the-shelf oscillators is usually not the dominant source of error. Pulse bandwidth B and E_S/N_0 at the receiver are elementary link-design parameters and SDR-based receiver design allows some flexibility in the choice of sampling frequency f_S.

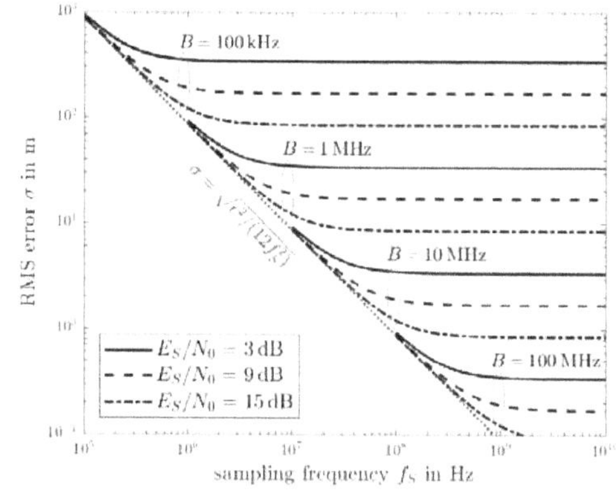

Figure 4: Fundamental limits of achievable ranging error induced by pulse bandwidth B, sampling frequency f_S, and receive signal-energy-to-noise-power-density ratio E_S/N_0. Since different error sources take on the dominating role for different operating points, control over the physical layer is of vital importance for accurate range estimation.

Figure 4 shows the elementary error bounds induced by bandwidth and sampling frequency [9,11]. Since different error sources take on the dominating role for different operating points, control over the physical layer is of vital importance for the design of a ranging system that can achieve the best possible accuracy over a wide range of distances. Furthermore, oversampling, interpolation, and averaging can be employed to improve ranging accuracy.

4 Summary

A "careful COTS"-based reconfigurable hardware platform for communications and hosted payload processing has been presented. Applications range from classic communications processing via features that are closely linked to and benefit from the ability to integrate them into physical-layer processing to custom payload-processing tasks that can be split between reconfigurable hardware (running on the SDR's FPGA fabric) and software running on the SBC's processor.

References

[1] C. Moy and J. Palicot, "Software radio: a catalyst for wireless innovation," *IEEE Communications Magazine*, vol. 53, no. 9, pp. 24-30, September 2015.

[2] Eric Blossom, "GNU radio: tools for exploring the radio frequency spectrum," *Linux J.*, no. 122, June 2004.

[3] D. Sinclair, J. Dyer, "Radiation effects and COTS parts in Small Sats," *27th Annual AIAA/USU Conference on Small Satellites*, Logan, Utah, 2013.

[4] J. Budroweit and A. Koelpin, "Design challenges of a highly integrated SDR platform for multiband spacecraft applications in radiation enviroments," *2018 IEEE Topical Workshop on Internet of Space (TWIOS)*, Anaheim, CA, 2018, pp. 9-12.

[5] D. S. Lee *et al.*, "Single-Event Characterization of the 20 nm Xilinx Kintex UltraScale Field-Programmable Gate Array under Heavy Ion Irradiation," *2015 IEEE Radiation Effects Data Workshop (REDW)*, Boston, MA, 2015, pp. 1-6.

[6] E. Dogaru, F. Vinci dos Santos and W. Rebernak, "A flexible BIST strategy for SDR transmitters," *2014 Design, Automation & Test in Europe Conference & Exhibition (DATE)*, Dresden, 2014, pp. 1-6.

[7] J. Becker, M. Hubner, G. Hettich, R. Constapel, J. Eisenmann and J. Luka, "Dynamic and Partial FPGA Exploitation," *Proceedings of the IEEE*, vol. 95, no. 2, pp. 438-452, Feb. 2007.

[8] A. Morello and V. Mignone, "DVB-S2: The Second Generation Standard for Satellite Broad-Band Services," *Proceedings of the IEEE*, vol. 94, no. 1, pp. 210-227, Jan. 2006.

[9] S. Lanzisera, D. Zats and K. S. J. Pister, "Radio Frequency Time-of-Flight Distance Measurement for Low-Cost Wireless Sensor Localization," *IEEE Sensors Journal*, vol. 11, no. 3, pp. 837-845, March 2011.

[10] D. Kirchner, "Two-way time transfer via communication satellites," *Proceedings of the IEEE*, vol. 79, no. 7, pp. 983–990, Jul. 1991.

[11] H. L. Van Trees, Detection, Estimation, and Modulation Theory. New York: Wiley, 2001.

V-BAND LOW-NOISE AMPLIFIER MODULE FOR HIGH THROUGHPUT SATELLITE APPLICATIONS

Leonardo Pantoli[1], Alessandro Barigelli[2], Giorgio Leuzzi[1], Francesco Vitulli[2], Andrea Suriani[2]*

[1] *Dept. of Industrial and Information Engineering and Economics, University of L'Aquila, L'Aquila, Italy*
[2] *Thales Alenia Space, Via Saccomuro, Rome, Italy*
**leonardo.pantoli@univaq.it*

Keywords: low-noise amplifier, noise figure, V-band, Qband, GaAs pHEMT technology.

Abstract

This paper presents the final test results of the Low-Noise Amplifier (LNA) module developed in Thales under a contract with the European Space Agency (ESA) operating in the frequency range 47.2 to 51.6 GHz. The LNA, featuring an unsurpassed noise figure of less than 2.8dB all over the band, is a complete Engineering Model (EM), ready to be employed as a front-end in advanced Q/V band payload on board of modern High Throughput Satellites (HTS). It provides 45 dB of gain automatically compensated in temperature (range -30 to +65 °C) and high linearity, performing a third order intercept point (IP3) of 24dBm with a power consumption of 870 mW. All the MMICs composing the line-up have been fabricated in Europe by using the recently space qualified process 0.1µm GaAs PHEMT provided by UMS

1 Introduction

The LNA module presented is the finalization of the work initiated in 2015 with the development of several MMICs as basic building blocks of the LNA RF line-up [1, 2]. Since then a number of improvements have been implemented, namely the redesign of the LNA and MLA MMIC with the latest PDK provided by UMS foundry, the design of the Voltage Variable Attenuator (VVA) and the design of the waveguide to microstrip transition and intermediate filter.

The achieved noise figure is better than 2.8 dB from 47.2 to 50 GHz, including the input waveguide isolator losses, which makes this product the state-of-art, particularly for application on board of HTS.

As a matter of fact, the demand for payload components in Q and V band is increasing more and more, so demonstrating that the space market is strongly geared towards the extended use of such frequencies [3-5]. The evolution of satellite-based telecommunications services in the past decade has been such that the main consulting firms agree that Broadband Access services will overtake TV Broadcast as the primary application of communications satellites by 2022. Such services are urging the need for new Satellite systems where "High Throughput (>100 Gbps)" as well as "Very High Throughput (targeting 1 Tbps)" is a keyword going together with the use of higher frequencies, which allow to maximize the overall communication bandwidth. Given such a scenario, the development and space qualification of LNA units and other components for Q-V band payloads is considered a priority for the major companies operating in the space communication market [6].

2. LNA unit design

A picture of the EM of the LNA unit is presented in Fig. 1. The size is 145x26x36 (LxWxH) mm and the weight 151 gr, including the WR19 waveguide input and output isolators. The module is powered by a single positive voltage 3.0-4.0 V for a maximum power consumption of 870 mW at ambient temperature and 900 mW at -30 °C. A negative voltage between 0 and -4 V is required as well to drive the internal VVA and compensate for gain variation of the LNA over temperature. The LNA module is prepared to be assembled in group of 4 or 6 on a baseplate, which houses a thermal compensation network driving the LNA modules' VVAs'.

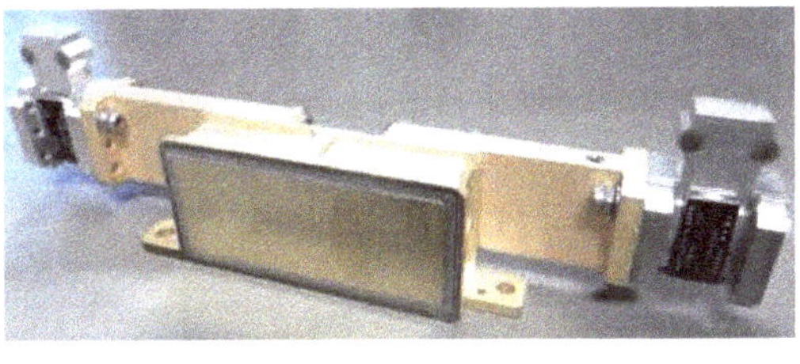

Fig. 1 V-band LNA unit Engineering Model

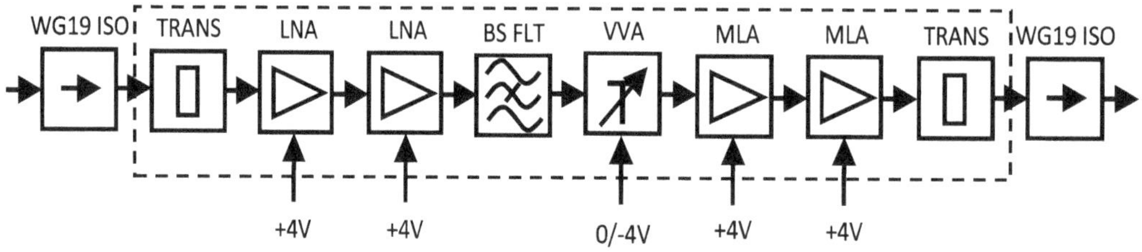

Fig. 2 LNA module RF line-up.

The RF line-up of the LNA module is presented in Fig. 2 showing all MMICs and subassemblies. Input and output WR19 isolators are used to guarantee matching and suitable interfacing with the feed antenna and downconverter, respectively. They have been developed by Trak Inc. UK for this application upon Thales specification. The insertion loss (IL), which is the key performance, is less than 0.25 dB in the operating band from 47.2 to 51.6 GHz, while isolation and input/output matching are better than 20 dB.

According to the space rules, the MMIC's shall be packaged for reliable operation over the lifetime of the mission. That is why a hermetic ceramic waveguide to microstrip transition is employed to connect the isolator to the RF chain and all the MMIC's are assembled in a hermetic hybrid module. Low-losses and adequate bandwidth are the key requirements for the transition. An insertion loss of 0.25 dB has been achieved in the range 47.2-51.6 GHz. The first two amplifier stages are MMIC's featuring 2.0 dB of noise figure and a gain of 16 dB. They establish the noise figure of the whole LNA module together with the low insertion loss of transition and isolator. A ceramic planar band-stop filter follows, which provides 30 dB of attenuation in the transmitting frequency range 37-42.5 GHz. In case a payload in Q-band is embarked on satellite, this filter helps in cleaning up the received frequency spectrum. The following two MMIC stages are Medium Level Amplifiers (MLA), as they feature a high third order intercept point (IP3) of +26 dBm, associated with gain of 15 dB. The overall linearity of the LNA module is mainly determined by such amplifiers and this is a crucial requirement for multi-carrier operations. At the output, the same transition is used to deliver the RF signal to the output waveguide isolator.

The fabrication technology of the LNA module is well established and fully space qualified by Thales Alenia Space. The hermetic housing is made of Kovar and all MMICs and ceramics are glued into the RF cavity, with exception of WGMS transitions and DC feedthroughs that are brazed. Gold 18 μm wires of controlled length are used to bond the items of the line-up. An aluminium lid is seam welded to hermetically seal the module.

In the following section a more detailed description of the key elements of the LNA module is given.

3 Key elements of the LNA module

3.1 WG-MS Transition

The input/output transitions act as a mechanical interface between waveguide and hybrid module and provides the necessary matching of the waveguide impedance to 50 Ohm. The design adopted is quite insensitive to the presence of the lid, as the top of the ceramic substrate is almost fully gold plated.

The type of transition selected is based on the proximity coupling effect, which exploits the coupling of the waveguide mode TE10 with a proper mode of the resonant cavity that is in turns coupled to the coplanar line and following microstrip on the upper surface of the ceramic.

Such a design is also insensitive to manufacturing tolerances, which makes this choice well suited for large scale production. Simulated performances with CST Studio™ are presented in Fig. 3, showing an insertion loss of 0.25 dB and a return loss better than 20 dB covering the whole frequency range with reasonable implementation margin.

3.2 LNA MMIC

The low-noise amplifier MMIC design exploits at the best the features of the novel GaAs 0.1 μm PHEMT UMS process (PH10), recently space qualified and updated in the foundry models of the transistors. The MMIC comprises three stages in cascade having a single bias pad for the drain-source voltage and gate voltages set to zero. The first and the second FETs are chosen for the minimum noise figure. As in the PH10 technology the optimum loads for noise and gain are close together, the associated gain is quite large (8.2 dB), and therefore the third FET in the chain can be designed for high linearity. The design includes a compensation network for the RF bonding wires at input and output. Special attention has been paid to the design for stability [7-8]. Small resistors have been used for each bias loop to avoid positive feedback and higher-value resistors have been included in the RF stubs near each transistor to ensure stable operation at high frequencies.

The test results of the LNA MMIC measured on a test jig including the RF bonding wires are presented in Fig. 4. The noise figure is around 2.0 dB, associated with a gain of 15-16 dB in a larger band from 39 to 53 GHz that makes this die suitable for low-noise amplification in Q-band as well (42.5-

45 GHz). The measured 1dB compression point is +10 dBm, which allows to manage higher signal levels with adequate linearity.

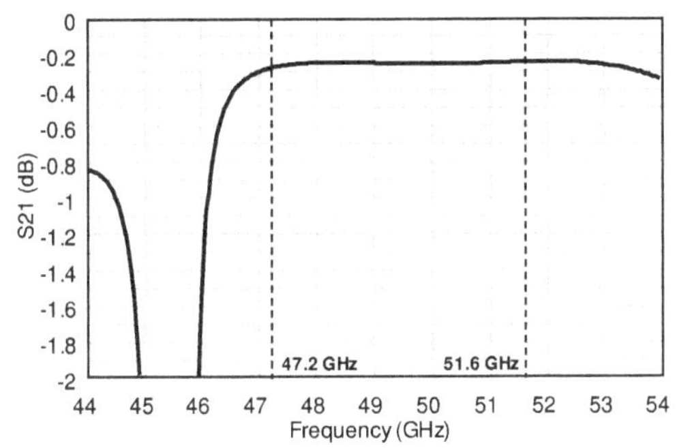

a

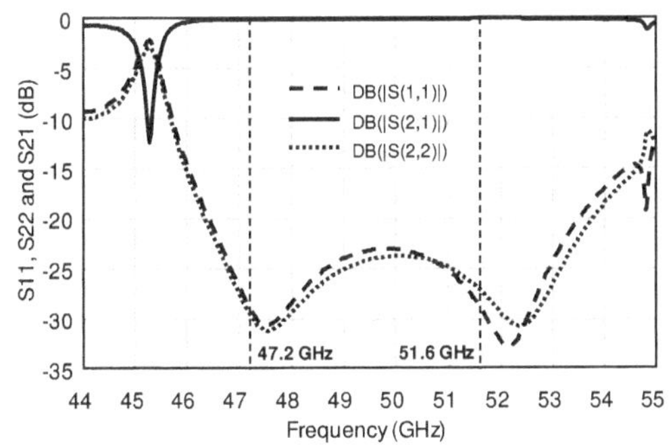

b

Fig. 3 Waveguide to microstrip transition response:
a) magnification of the insertion loss;
b) reflection and transmission coefficients.

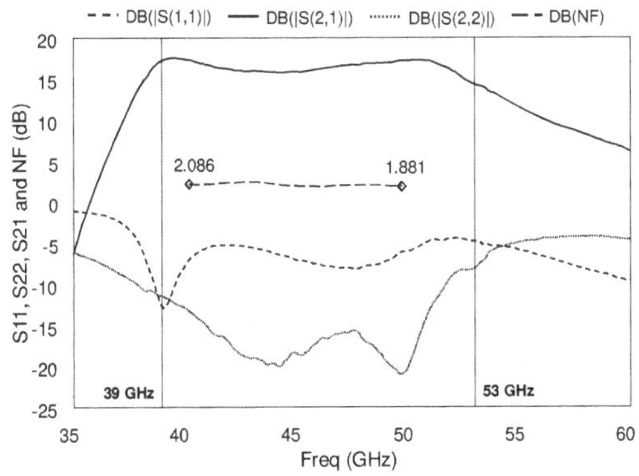

Fig. 4 LNA MMIC Performance measured on test jig.

3.3 Voltage Variable Attenuator (VVA)

The VVA design is based on GaAs MESFETs implemented on PH10 UMS process. A special design effort has been employed to achieve the best linearity. For this purpose the chosen topology features two reflective-type attenuators arranged in a balanced configuration using a Lange coupler. Each attenuator employs two "equivalent" transistors, and each equivalent transistor is made up of three individual transistors of the largest periphery stacked together to enhance the overall linearity. As for the amplifier MMIC, matching networks are included in the die to compensate the RF bonding wire. The measured attenuation and return loss as a function of the control voltage is given in Fig. 5.

It should be noted that the attenuator response is quite flat over a very large band ranging from 35 to 56 GHz, still presenting a more than acceptable matching. The minimum and maximum attenuations of -4 dB and -20 dB are achieved with a control voltage of -4 V and 0 V, respectively (with negligible current draw). The measured input 1 dB compression point, which is a key parameter for such applications, is +5 dBm.

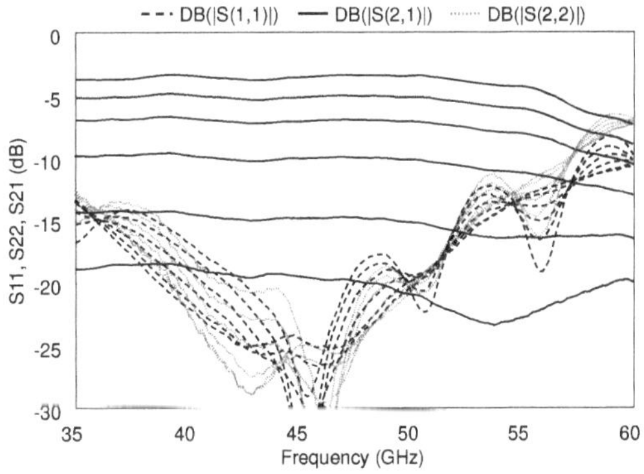

Fig. 5 VVA MMIC Performance measured on test jig.

3.4 MLA MMIC

The MLA, as well the LNA, has been re-designed with the new PDK, enhancing the good performance already achieved with the initial design kit and the preliminary designs, as shown in [1]. A three-stages approach has been chosen also for this MMIC. Taking advantage of the technology characteristics, the same FET has been chosen for each stage, slightly changing only the bias point of the last stage in order to obtain a larger dynamic range. In detail, the FET has 6 fingers with a total gate periphery of 120 um. The first two transistors have been biased with a Drain to Source voltage (Vds) equal to 2.18 V and a drain current of 18 mA; while the third one has Vds=2.35 V and a drain current of 19 mA. This choice allows to achieve at the same time a limited noise figure (NF=2.2 dB) and an output power 1 dB compression point of about 14 dBm, with an output third order intercept point of 26 dBm, considerably greater with respect to the preliminary design [1]. The linear gain is between 15 and 16 dB over a very large bandwidth (see Fig. 6) with a measured gain flatness of only 0.5 dBpp in the full range. Also the matching conditions are satisfactory, making the MMIC suitable to be used both in Q and V bands. As well as the LNA, this design embeds the compensation networks for the RF bonding wires at input and output ports. The design has been optimised taking into account a couple of bond wires in "V" connection at each RF port and also the corresponding external pads on Alumina. The circuit has a single DC pad for the Drains biasing, while the Gates of the transistors are grounded, so simplifying the mechanical structure of the module.

Particular care has been devoted to the stability requirements. Due to the specific structure of the circuit for biasing requirements, an inner stability analysis has been carried out considering both the linear and nonlinear behaviour of voltages and currents in each loop of the circuit involving an active device and considering a frequency range spanning from DC to the cut-off frequency of the transistors. Beyond of using the approaches described in [7-9] for the analysis of the nonlinear regime, the linear regime has been analysed following the method described in [2] by the same Authors. Small resistors have been used to avoid positive feedback and also an inductive feedback has been realized on the Source terminals to obtain stable operations.

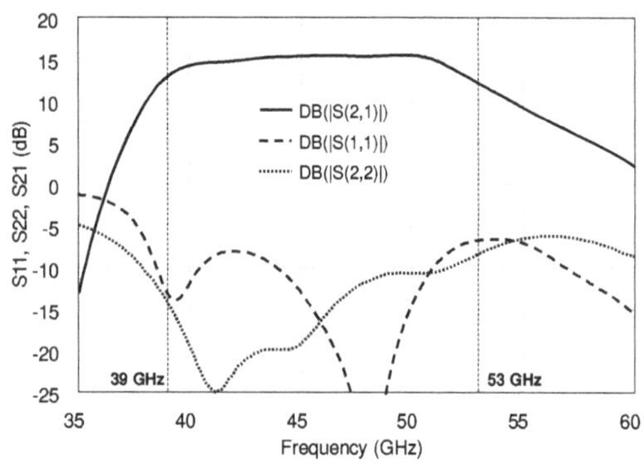

Fig. 6 MLA MMIC Performance measured on test jig.

4 LNA module test results

The whole LNA module has been extensively tested over temperature range. Gain response (Fig. 7) has resulted flat within 1.5 dBpp and stable even without lid, demonstrating that transition is truly insensitive to the presence of the cover.

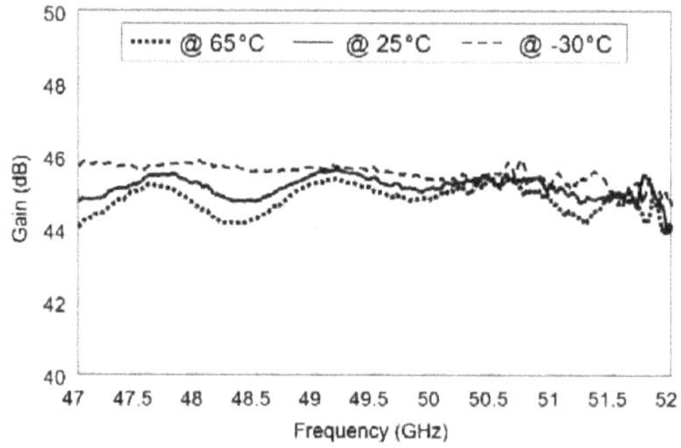

Fig. 7. LNA Module Gain response in temperature

The automatic gain compensation makes the gain stable in the temperature range -30 to +65 °C within 1.5 dBpp over the useful band 47.2 to 51.6 GHz. Nominal gain is 45 dB. Noise figure is presented in Fig. 8 as tested at temperature extremes. The performance is 2.7 dB in average with a maximum value of 2.85 dB in band. This result is congruent with the test and simulation of the individual items in the daisy chain, such as the insertion loss of the isolator (0.25 dB), the losses of the WG-MS transition (0.25 dB) and the noise figure of the LNA MMIC (2.0 dB).

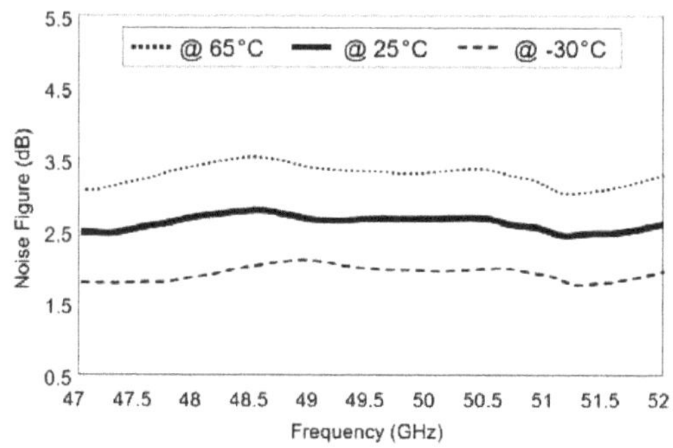

Fig. 8. LNA Module Noise Figure in temperature.

The IP3 of the LNA module has been evaluated by means of the two-tone test shown in Fig. 9. For this application the max input level is -45 dBm that results in two carriers having a level of -3 dBm each at the output. The third order intermodulation product are at -47.67 dBc, resulting in a IP3 of +24 dBm.

The module was powered at +4.0V with a DC consumption of 870 mW at 25°C and 900 mW at -30 °C. The test instrumentation used was Keysight PNA-X 5247A for Sparameters and two-tone test and Spectrum analyzer RohdeSchwartz FSW67 for noise figure measurement.

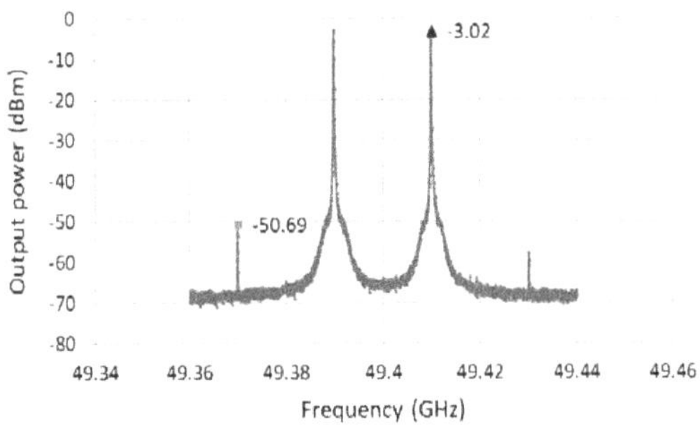

Fig. 9. LNA Module Two-tone test.

5 Conclusions

Test results of a 45 dB gain LNA Module to be used on modern Q/V payload embarked on board of HT satellites have been presented. The LNA features a state-of-art noise figure of 2.7 dB in the band 47.2 to 51.6 GHz and a high IP3 of +24 dBm for linear operation in multicarrier condition. The LNA module is a full Engineering Model, ready to be space qualified and placed on the market.

Acknowledgement

The authors wish to acknowledge the Italian Space Agency (ASI) for funding this work in the context of project KALOS. Particular thanks are due to François Deborgies from ESA/ESTEC for his expertise and technical support.

References

[1] L. Pantoli, G. Leuzzi, A. Barigelli, F. Vitulli and A. Suriani, "An ultrawideband LNA module for space applications," 2015 10th European Microwave Integrated Circuits Conference (EuMIC), Paris, 2015, pp. 160-163.

[2] L. Pantoli, A. Barigelli, G. Leuzzi and F. Vitulli, "Analysis and design of a Q/V-band low-noise amplifier in GaAs-based 0.1 μm pHEMT technology," in IET Microwaves, Antennas & Propagation, vol. 10, no. 14, pp. 1500-1506, 11 19 2016.

[3] Sung Rae Park, Rick-Nghia Nguyen, Steve-Trung Nguyen, Norman H. Chiang and James J. Sowers, V-band receiver for commercial space applications, Microwave Symposium (IMS), 2017 IEEE MTT-S International, 4-9 June 2017.

[4] Aloisio, M., Angeletti, P., Coromina, F., et al.: 'Exploitation of Q/Vband for future broadband telecommunication satellites'. IEEE Thirteenth Int. Vacuum Electronics Conf. (IVEC), Monterey, CA, 24–26 April 2012, pp. 351–352.

[5] Apollonio, D., Arena, S., Biondi, A., et al.: 'QV band receiver converter for satellite communications'. IEEE 11th European Radar Conf. (EuRAD), Rome, IT, 8–10 October 2014, pp. 360–363.

[6] P. Ranieri et al., "V Band Receiver For Future High Throughput Satellites", 21st Ka and Broadband Satellites Conference, Bologna Italy, October 12-14, 2015.

[7] Pantoli, L., Leuzzi, G., Santarelli, A., et al.: 'Stability analysis and design criteria of paralleled-device power amplifiers under large-signal regime', IEEE Trans. Microw. Theory Tech., 2016, 64, (5), pp. 1442–1455.

[8] L. Pantoli, G. Leuzzi, A. Santarelli, F. Filicori and R. Giofrè, "Stabilisation approach for multi-device parallel power amplifiers under large-signal regime," 2011 6th European Microwave Integrated Circuit Conference, Manchester, 2011, pp. 144-147.

[9] Pantoli, L.: 'Transient-based conversion matrix approach for nonlinear stability analysis', Electron. Lett., 2014, 50, (13), pp. 923–925.

SATELLITE PAYLOAD DESIGN FOR CISLUNAR COMMUNICATIONS

Vladimir Lemos, Francisco Javier De Pablos Martin, David Gómez Otero, Tomás Navarro, Octavio Camino, Xavier Geneste
European Space Research and Technology Centre,
Keplerlaan 1, 2201 AZ Noordwijk, The Netherlands
European Centre for Space Applications and Telecommunications,
Harwell Campus, Didcot, OX11 0FD, United Kingdom
E-mail: Vladimir.Lemos@esa.int, Francisco.Javier.De.Pablos.Martin@esa.int, David.Gomez.Otero@esa.int, Tomas.Navarro@esa.int, Octavio.Camino@esa.int, Xavier.Geneste@esa.int

Abstract

This paper presents the design and performance of a communications payload module equipped with dedicated antennas for K-band, UHF and S-band links, which is envisaged as part of a possible future satellite constellation. The proposed design enables high availability and high capacity digital communication links between the Moon and Earth for a wide range of lunar missions; manned and un-manned, orbiters (including cubesats), landers, sample return vehicles, and includes compatibility to missions such as Lunar Orbital Platform-Gateway and Orion.

Design requirements

The main objective of the payload design is to enable high capacity digital communication links between the far side of the Moon and Earth for manned and un-manned lunar missions. These missions are likely to employ radar and high-resolution hyperspectral imaging instruments, requiring high date rates in excess of 20Mbps to return science data and enable other communication services (e.g. voice, video etc.).

The requirements include the provisioning of dedicated digital communication links between orbiting platforms such as the planned NASA Lunar Orbital Platform-Gateway, Orion and assets on the Moon and the provisioning of low rate data links operating at UHF and S-band.

Coverage area and constellation design

The coverage area focuses on the South Pole Aitken Basin region on the far side of the Moon. The region has an estimated surface area in excess of 5 million squared kilometers. Full coverage is then possible for orbits above 875Km. In order to deliver a viable space based network infrastructure around the Moon, the design focuses on the so-called Frozen Lunar Orbits [RD2]. The trajectory considered for the satellites in the constellation is an inclined quasi-circular orbit with radius of 9,750Km with a field of 8.7° from nadir, which provides full visibility of the far side of the Moon and a total coverage area in excess of 14.5 million square kilometers. The South Pole Aitken Basin region is within a reduced +/- 6.7° field of view as seen from the satellite for links with assets above 5.0° elevation angle.

A suggested constellation design places three satellites symmetrically distributed in equally inclined circular orbit planes. The orbital planes are 120° apart as in a Walker Constellation to increase system availability and reduce coverage gaps. An example of three satellites in an inclined orbit about the South Pole Aitken Basin region on the far side of the Moon is shown in Figure 1: the image from NASA Clementine Mission shows the South Pole at the centre and extends to 70° South latitude at the edge. 2.0° K-band spot beams are projected over the coverage, where the yellow spot beam is pointed towards the Schrodinger crater.

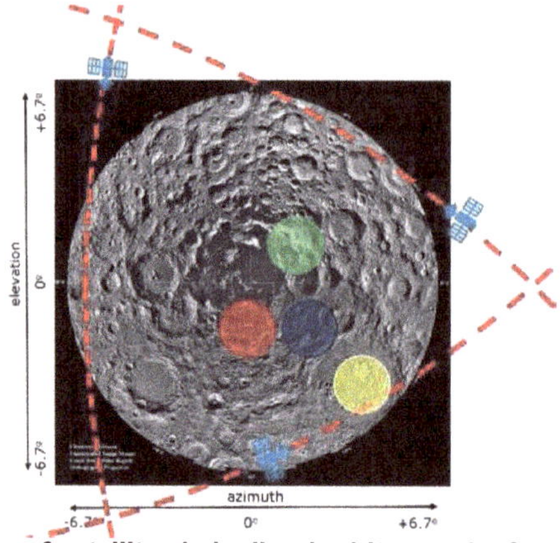

Figure 1 – Top view of satellites in inclined orbit over the far side of the Moon

Radio propagation

Since there is no significant atmosphere on the Moon, essentially a classical exosphere [RD1], there is negligible attenuation due to gaseous absorption of radio frequency waves. The main propagation impairments foreseen are at low UHF and S-band frequencies due to scattering off the Moon surface. Satellite antennas operating at low frequencies are also susceptible to unwanted interference or jamming from transmissions originating on the Earth or other spaceborne platforms.

Spectrum resources

The links between Earth and the satellite are within 22.65-23.15GHz and 25.5-26.1GHz following the recommendation from the Space Agency Frequency Coordination Group [RD7]. The links between the satellites and the assets on the Moon are within 22.55-23.05GHz and 26.5-27GHz bands. Provisioning of satellite capacity at UHF band is limited [RD3] because of (HI) and (OH) spectral lines for radio astronomy.

Frequency plan

The allocated 500MHz of bandwidth at K-band is divided into 125MHz sub-bands as depicted in Figure 2. UHF and S-band traffic to Earth on the forward link is routed to the orthogonal polarization antenna port. The channelization plan takes into account an initial mass and power envelope for the communications payload of 200Kg and 1.0kW respectively.

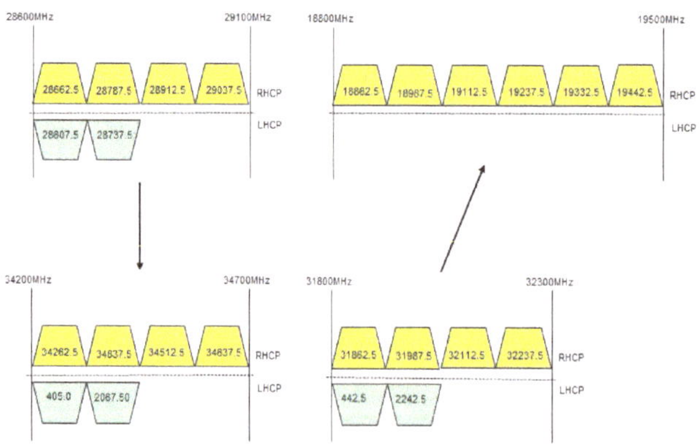

Figure 2 – Channelization plan

Reference scenario

The reference scenario depicted on Figure 3 summarises the design drivers for a satellite constellation in a Frozen Lunar Orbit; the return link capacity is mostly constrained by the limited power resources allocated for the uplink of the assets on the Moon's environment. The satellite employs small 0.5m K-band steerable reflector antennas shaped to optimize the coverage using 2.0° spot beams.

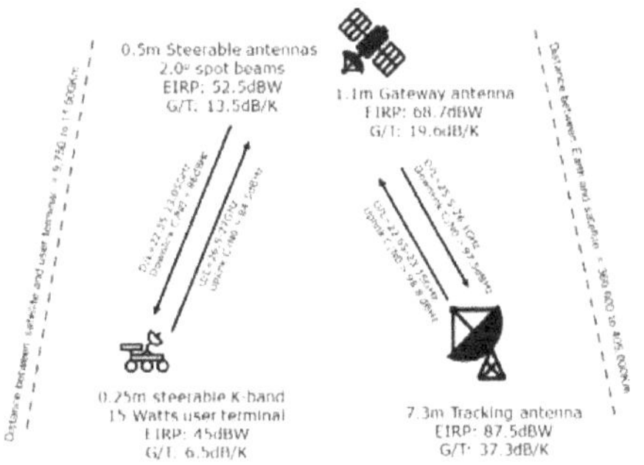

Figure 3 – Operational scenario for K-band satellite

On the first hop of the return link, the operational scenario translates into a C/N0 of 84dB/Hz. A rover with a small steerable antenna could sustain uplink rates of at least 20Mbps, with its transponder employing digital modulation using QPSK with ¾ FEC. On the forward link, the satellite links maintains C/N0 of 86dB/Hz that allows downlink rates of at least 35Mbps. For purposes of link budget analysis, the target Eb/No is specified at 11dB to maintain 1.0E-06 bit error rates.

The use of higher frequencies in the return link allows for higher data rates. Communication between assets on the Moon's environment and a satellite in the lunar region lay within 31.8-32.3GHz and 34.2-34.7GHz bands, allocated by the ITU for Space Research Services (SRS). This would enable a rover to reach uplink at rates up to 60Mbps (C/N0 of 88.7dBHz).

The system provides continuous communications to a multiplicity of assets geographically dispersed across the far side of the Moon. Alternatively, very high communications capacity over a narrow portion of the coverage can be achieved by overlapping the spot beams.

Capacity planning

Increase capacity is achieved by coverage planning using radio cells for ¼ frequency re-use, where a minimum of 61 spot beams are required for full coverage of the South Pole Aitken basin. The spectrum resources can be re-used such that no neighboring cell's area operates at the same frequency. Coverage planning using radio cells allows for system capacity estimations based on multiple access schemes such as TDMA, FDMA and CMDA. A number of scenarios can be exploited by either operating the system at a lower portion of the beam, employing antennas of different sizes or a combination of both. An exhaustive capacity planning is outside the scope of this paper.

The overall system capacity depends on the number of simultaneous assets accessing the network, the relative signal level between the uplink carriers, data rates required and the multiple access scheme adopted. The transponder loading characteristics are a design driver because they impose operational constraints on the amplifiers. A mix of narrow and broadband carriers, sparsely distributed across the channel, is assumed for definition of requirements and performance analysis.

A 1.1m K-band reflector antenna is proposed on the satellites for the return link to Earth. Further increase in aggregate data rate on the return link is achievable by using high RF power on the K-band transponder. For the reference scenario a 200W linearized TWTA is used, which maintains C/N0 levels of at least 97dB/Hz on the downlink to Earth under multicarrier operations. Each satellite can provide at least 450Mbps of aggregate capacity on the return link, with carriers modulated using QPSK with ¾ FEC. Using the L-TWTAs in parallel provides a means for increased aggregate system capacity on the return link.

UHF and S-band service area consists of the entire +/-8.7° field-of-view of the satellite as seen from a Frozen Lunar Orbit because at low frequencies directional antennas provide wide beams.

Baseline design

The satellite consists of a single deck communications module for the repeater equipment, with dedicated antennas for UHF, S-band and K-band links. All antennas provide high gain beams to decrease DC power requirements and enable high data rate digital communication. A Master Reference Oscillator (MRO) provides a 10MHz input signal for all frequency converter units, maintaining coherency in both forward and return links. On the forward link, a 4-channel linear phase multiplexer routes the K-band signals to the steerable antennas. The forward link is switchable by tele-command for the provisioning of dedicated links to orbiting platforms such as the planned NASA Lunar Gateway. On the return link, the K-band uplinks from the assets are combined by a common low power output multiplexer. The system is designed to minimize signal distortion by means of linearized amplifier and filter technology. A dedicated digital signal processor provides sub-channelization and gain control at UHF band to minimize signal degradation due to propagation impairments and unwanted interference over the Moon. The satellite can simultaneously operate up to six coherent transponders throughout its operational lifetime and implements the channelization plan shown in Figure 2. The design features a switchable TT&C sub-system designed for LEOP and cruise operations at S-band and at X-band.

K-band communications payload

All repeater equipment and thermal hardware is installed on the inner -(Z) panel of the satellite. The K-band sub-system includes five 0.5m mechanically steerable antennas: four are installed on the +(Z) panel and one on the –(Y) panel that is dedicated for links with higher orbiting platforms such as the NASA Lunar Gateway. The steerable antennas incorporate self-diplexed transmit and receive feed chains that support simultaneous operation in both left- and right-hand circular polarisations. A coaxial diplexer provides key functionality for the combined Tx/Rx operation.

On the forward link the uplink signal is down-converted by a K-band receiver. A SPDT coaxial switch provides routing via tele-command to the forward links using the orbiting platforms. The input signal is split by a multiplexer and the individual carriers are routed to each of the four steerable antennas. On the return link, the K-band uplinks from the assets are up-converted by dedicated receivers on each beam and combined into a single carrier by a multiplexer that is common to all transponders in the payload. On the output section, a high power multiplexer is not required and the same filter technology used at the input section can be used because of the low signal levels after the frequency up-conversion. The output signal is routed to the high power amplifier, which consists of a compact TWT-based microwave power module (MPM) with a linearized channel amplifier (LCAMP) in a 2:1 redundancy scheme. A high power K-band waveguide isolator is installed at the output section to dissipate any return signals between the transmit antenna and the MPM. A 1.1m X/K-band antenna is employed for transmission of a wideband carrier (supporting aggregated data communication links) to the Mission Operations Network. This antenna is placed on the –(Y) panel and incorporates a positioning mechanism to maintain constant pointing to the Earth along the orbit. X-band TT&C communications capability is implemented by means of an X/K-band coaxial triplexer in the design. All reflector antennas are in an axial displaced Gregorian optical configuration to minimize the mechanical accommodation envelope. An antenna control electronics unit provides interface to the positioning mechanisms.

UHF-band communications payload

All repeater equipment and thermal hardware is installed on the inner -(Z) panel of the satellite. A self-deployed helix, based on the storable tubular extendible member (STEM) antenna, is installed on the +(Z) panel for communications between the satellite and assets. This concept is extremely low mass and can extend up to 3.0m axial length providing high gain. On the forward link, the uplink signal is down-converted to a lower intermediate frequency (IF) at S-band that is input to a diplexer. A low pass filter is integrated with the diplexer to avoid undesired signals in the receive path. The IF signal is input to a triple-balanced band mixer for conversion to UHF. A phase-locked dielectric resonator source provides a reference for the frequency translation. The output signal is routed to the UHF amplifier, which consists of a GaN-based solid State Power Amplifier (SSPA) with a lineariser in a 2:1 redundancy scheme. A coaxial isolator is installed at the output section to dissipate any return signals between the transmit antenna and the SSPA, due to mismatches in the design. On the return link, the UHF sub-system incorporates a small digital signal processor that provides gain control and basic anti-jamming capabilities. The RF front-end down-converts the asset uplinks at UHF to a low intermediate frequency centered at 30MHz and provides analogue I/Q demodulation to the conversion stage. An additional gain stage and bandpass filtering is included in the RF front-end to condition the asset uplink for A/D conversion. The output of the baseband processor is up-converted to K-band by a dedicated frequency converter and combined by the multiplexer that is common to all transponders in the payload. The output signal is routed together with the aggregated K-band data traffic.

S-band communications payload

All repeater equipment and thermal hardware is installed on the inner -(Z) panel of the satellite. A small unfurlable S-band high gain antenna in a short back-fire configuration is installed on the +(Z) panel for communications between the satellite and assets. This antenna is extremely low mass and deploys a 0.5m mesh reflector. On the forward link, the uplink signal is down-converted to S-band and as such is input to a UHF/S-band diplexer. The output signal is routed to the high power S-band amplifier, which consists of a GaN-based solid State Power Amplifier (SSPA) with a lineariser in a 2:1 redundancy scheme. A coaxial isolator is installed at the output section to dissipate any return signals between the transmit antenna and SSPA, due to mismatches in the design. On the return link, the uplinks from the assets are up-converted to K-band by a dedicated unit and combined by the multiplexer that is common to all transponders in the payload. The output signal is routed together with the aggregated K-band traffic.

Payload Schematic diagram

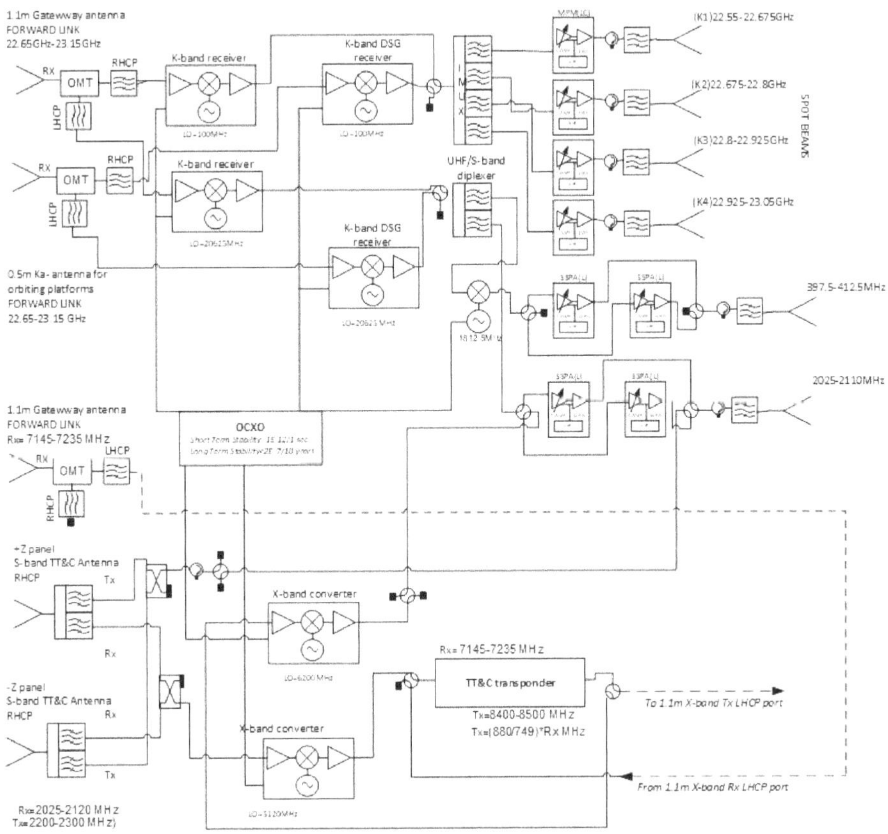

Figure 4 – Forward Link schematic diagram

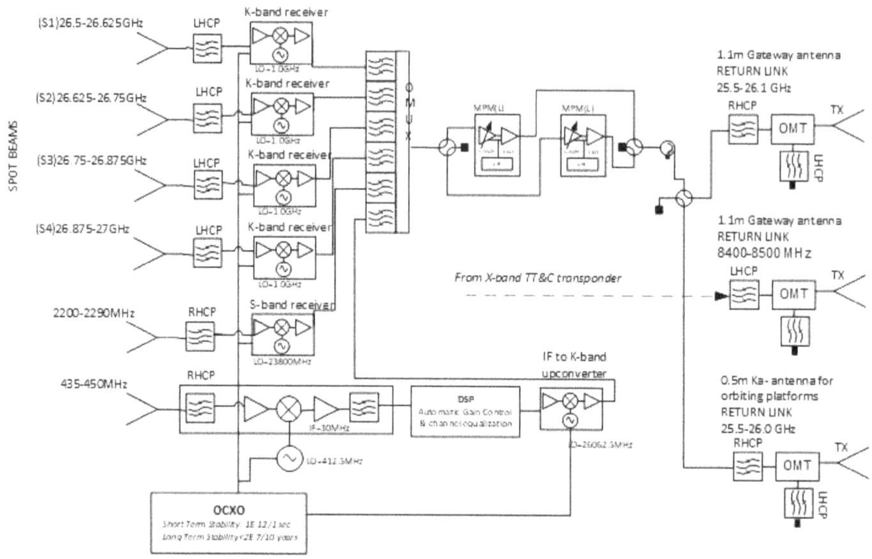

Figure 5 – Return link schematic diagram

K-band spot beam antennas

The K-band spot beam steerable antennas are implemented based on a displaced axis reflector design [RD4], which allows for a reduced mechanical accommodation envelope when compared to single offset or other dual reflector configurations. The reflector surfaces are shaped to optimize the coverage, with a reduction on gain. The feed chain of the 0.5m steerable antennas consists of a horn, an ortho-mode transducer (OMT) and polariser network together with a coaxial diplexer for simultaneous Tx/Rx application in both LHCP/RHCP polarisations. The spot beams can be steered anywhere within the field of view of the satellite by means of 2-axis gimbal positioning mechanism (APM). In this configuration, the feed chain is fixed and the only moveable part is the main reflector, which can be steered about 360° in azimuth and over the desired elevation. The total antenna assembly mass is estimated at 3.5kg.

K-band Gateway antenna

The 1.1m X/K-band antenna is also implemented as a displaced axis reflector design, but requires a different feed to cover the 7145-8500MHz, 22.65-23.15GHz and 25.5-26.1GHz frequency bands. Amongst different techniques, the preferred implementation is a modification of the coaxial feed design [RD5]. In the original design, the feed is a coaxial microwave structure where the low-frequency signals propagate in TE11 mode and the higher frequency signals propagate through the hollow centre conductor in a TE11 circular waveguide mode. A modification of this design by using a corrugated horn and a dielectric rod provides better performance at the higher frequencies.

The 1.1m X/K-band antenna antenna is placed on the –(Y) panel together with a deployment boom to ensure continuous pointing towards the Earth is possible without mechanical interference within the field of view. An antenna positioning mechanism is installed on the deployment boom. The total antenna mass is estimated at 8.5kg.

K-band antenna control electronics unit (ACU)

The ACU provides an electrical interface to the positioning mechanisms and performs digital signal processing for steering and tracking. It provides manual discrete positioning and tracking capabilities via tele-command or automatic positioning at a fixed rate. Different tracking modes are implemented in digital signal processing. The unit can interface with platform avionics, such as sun and star trackers, for orbit determination. The ACU requires less than 10Watts DC Power and 2.0Kg mass.

UHF-band antenna

The UHF high gain antenna is based on a self-deployed storable tubular extendible member (STEM) helix antenna [RD6]. A set of 11.m and 22.5m UHF STEM antennas was deployed in the first Canadian ionospheric sounding Alouette satellite in 1962. The technology has since been used in a number of defence and scientific missions.

S-band antenna

The S-band miniature unfurlable antenna provides a deployed aperture of 0.5m. The technology was originally conceived by a consortium between the US Central Intelligence Agency and defence contractor TRW for the Rhyolite satellite program in the 1970s. The antenna design is a short back fire configuration, with a reflector structure at the focal point excited by short dipoles.

Frequency conversion

A 10MHz Master Reference Oscillator (MRO) is a key equipment on the communications payload to maintain coherence amongst the signals on both forward and return links. A high precision oven controlled oscillator (OCXO) is mounted on an anti-vibration baseplate. A quartz crystal dielectric resonator circuit is employed in the OCXO to maintain low frequency variation over temperature. The MRO achieves Allan deviation of 1E-12/second (short term stability) and less < 2E-7 over 10 years (long term stability). It provides a reference signal to all frequency converters on the payload with 3:1 redundancy. An internal passive network distribution provides 10MHz signal distribution. The equipment requires 25W DC power and about 3.5Kg mass. All frequency converters employed in the payload synthesize the translation frequency using a common 10 MHz reference.

Channel multiplexing and filtering

The multiplexers are implemented with self-equalized high order filters (10 to 12 pole) and provide linear phase frequency response characteristics. These type of multiplexers have reduced mechanical envelope (50%) and mass (30%), when compared to traditional filter design. The concept of manifold-coupled multiplexer works also in reverse configuration where the channels are combined to a common output port. Isolators between hybrid and each manifold decouple the blocks of non-contiguous channels and suppress any interactions between adjacent channels.

High power amplification

The K-band transponders are fitted with MPMs that combine TWTA (EPC and TWT) and a Linearized Channel Amplifier (LCAMP) in one housing. A solid state driver amplifier provides input to a miniaturized TWTA. The linearizer is mounted external to the unit, together with the EPC that provides DC-DC conversion. K-band MPMs typically achieve better than 17dB NPR (noise power ratio), which is a figure of merit for channel linearity, at 2dB gain compression point. Each 35Watts unit require about 80Watts at 2dB compression point with PAE (efficiency) of 43%, dissipating an estimated 40Watts through heat pipes. A higher power 200Watts MPM requires about 420Watts DC power and dissipates 210Watts. Each unit is about 3.5Kg, including the high voltage cables.

A set of S-band and UHF SSPAs are employed in the amplification section of the payload. The SSPAs incorporate GaN solid state devices on the driver and power modules. SSPAs achieve about 51% PAE, including the EPC. SSPA units are compact, without a lineariser are less than 1Kg mass.

UHF Digital Signal Processor

The UHF digital signal processor is implemented in Field Programmable Gate Array Technology (FPGA). A high performance I/Q demodulator provides the analogue inputs for A/D conversion. The FPGA consists of channelization, digital gain, power measurement and quadrature mixing blocks. Channelization is done using half band linear phase IIR filter design. Basic I/Q correction is implemented to improve the EVM (error vector magnitude) to less than 0.5% rms. The unit provides adjustable gain between 70 to 120dB, with an estimate 7.0dB noise figure.

Signal level analysis

Table 1 and Table 2 shows the end-end signal of the payload communications sub-system, with the input Power Flux Density (PFD) dynamic range defined at the antenna input.

	PFD dynamic range (dBW/m^2)	Transponder gain (dB)	Noise Figure (dB)	System Noise Temperature (Kelvin)	G/T (dB/K)	EIRP (dBW)
K-band U/L (22.65-23.15GHz) D/L (22.5-23.05GHz)	-130 to -100	> 165				52.50
UHF U/L (22.835-22.850GHz) D/L (397.5-412.5MHz)	-125 to -95	>130	2.7	450	19.6	26.50
S-band U/L (22.65-22.735GHz) D/L (2025-2110MHz)	-130 to -100	> 137				27.25

Table 1 – Forward link signal budget

	PFD dynamic range (dBW/m^2)	Transponder gain (dB)	Noise Figure (dB)	System Noise Temperature (Kelvin)	G/T (dB/K)	EIRP (dBW)
K-band U/L (26.5-27GHz) D/L (25.5-26.0GHz)	-142 to -110	> 175	2.8	462	13.5	
UHF U/L (435-450MHz) D/L (26.0-26.085GHz)	-165 to -125	> 187	2.0	372.6	-9.0	68.7
S-band U/L (2200-2285MHz) D/L (26.085-26.1GHz)	-140 to -110	> 173	2.8	456	-9.0	

Table 2 - Return link signal budget

Conclusions

A payload design for providing UHF, S-band and K-band communications to assets in the Cis-lunar environment, optimised for the South Pole Aitken Basic region, has been presented in detail in this paper.

It includes a comprehensive breakdown of the payload components, reference use case scenarios, payload diagram, signal budgets, coverage area, capacity sizing, frequency plan, and envisages compatibility with a constellation of three identical payloads on Frozen Lunar Orbits.

As further steps, access scheme techniques for the different communication services shall be analyzed in detail to cope with the flexibility driven by the market demand evolution.

References

RD[1] "Formation of the lunar atmosphere", R. R. HODGES, Conference Report, 1975

RD[2] "Constellations of Elliptical Inclined Lunar Orbits Providing Polar and Global Coverage", Todd A., Proceedings AAS/AIAA Astrodynamics Specialists Conference, August 7-11, 2005

RD[3] Protection criteria for deep-space research, ITU-R SA.1157-1

RD[4] "Analysis of two-Mirror antennas of a general type", Y.A. YERUKHIMOVICH. Begell House, Telecommunications Radio Engineering 27, 97, November 1972

RD[5] "Coaxially fed dual-frequency horn for offset parabolic reflector", Per-Simon KILDAL, Antennas and Propagation Society International Symposium, 1995. AP-S. Digest

RD[6] Coilable extensible apparatus, George J KLEIN, US Patent US3144215A granted 27 April 1961

RD[7] Communication frequency allocations and sharing in the lunar region, Recommendation CCSDS SFCG REC 32-2R1, July 2013

Section 5 – High Speed Optical Communications and Feeder Links

Alphasat, Sentinel-1A/B, Sentinel-2A/B, and EDRS Paving the Way for Systematic Optical Data Transfer for Earth Observation Missions

Edoardo Benzi
European Space Agency, ESTEC, Noordwijk, The Netherlands

F. Cipollini, B. Rosich, O. Colin
European Space Agency, ESRIN, Frascati, Italy

I. Shurmer, F. Marchese
European Space Agency, ESOC, Darmstadt, Germany

Thomas Marynowski
TESAT Spacecom GmbH & Co.KG, Backnang, Germany

Mark James
Inmarsat plc, London, United Kingdom

Since late 2013, Inmarsat operated Alphasat GEO communication satellite has been providing a reliable platform for the experimental activities of its 4 hosted Technical Demonstration Payloads (TDPs), procured and operated by ESA. Amongst them, TDP1 is capable to receive observation data from a lower orbit (LEO) spacecraft thanks to its Laser Communication Terminal (LCT), and route the data to ground via a Ka band RF link.

With the launch of Sentinel-1A in April 2014, of Sentinel-2A in June 2015, of Sentinel-1B in April 2016, and of Sentinel-2B in March 2017 in their sun-synchronous orbits, an increasing number of in-flight companions for the ASA terminal were made available. Part of the Copernicus program, the Sentinel satellites perform their earth observation missions thanks to a Synthetic Aperture Radar (S1A and S1B) and a Multi-Spectral Imager (S2A and S2B) respectively, and carry on board an Optical Communication Payload, based on the same TESAT LCT embarked on Alphasat.

An Optical Inter-Satellite Link demonstration campaign in late 2014 with ASA and S1A using lead to the first data transmission from S1A in November 2014. The campaign was completed with results well beyond expectation in terms of link performances.

Following the completion of the in-orbit commissioning of its Optical Communication Payload, S2A joined S1A in the following "Experimentation Phase", aiming at the characterisation of the paired terminal performances over its operational envelope, that has been going on since the beginning of 2015.

S1B and recently S2B have also been used as companion to Alphasat TDP1 terminal for their in orbit testing and calibration activities in the light of their commissioning into the now operational European Data Relay Satellite system.

The Alphasat and Sentinels experience has provided the ultimate demonstration of the technology underlying the EDRS, and also made available an important opportunity to test the system operations in a realistic scenario and gain valuable experience to be put to fruition for the EDRS GS and operations, as well as for any future optical intersatellite communication system.

1. Introduction

In order to fulfil their missions, LEO Earth Observation satellites are increasingly asked to respect very demanding standards of performances in terms of image resolution, coverage, data latency, and timeliness.

Although the usual data downlink architecture based on a network of ground stations operating at RF frequencies can provide an answer, this increases ground-segment complexity & cost. It is also limited geographically to land-based locations and further restrictions such as stringent ITU interference regulations and geo-political considerations.

An obvious complement to the GS use can be provided by GEO relay stations. The advantage of this architecture provided by the increased visibility is further boosted in terms of data return if very high data rate can be achieved using optical inter-satellite communication. Furthermore, optical GEO-relay networks enable near-real-time data access.

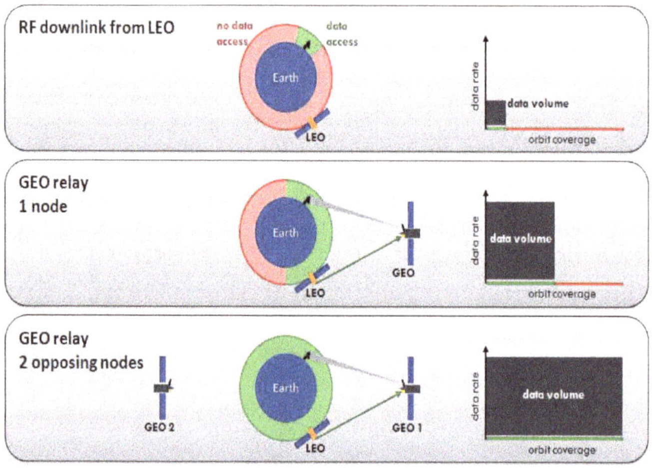

Figure 1: RF downlink from LEO compared with GEO relay: orbit coverage and data volume (schematic)

In the GEO-relay architecture presented here, the LEO satellite transmits data to a GEO satellite which receives the data and immediately forwards them to ground. For the LEO-GEO inter-satellite segment, a laser link has been chosen as the ideal solution offering secure point-to-point communication and superior data rates not subjected to ITU frequency or bandwidth regulations.

For the downlink leg of the "Intersatellite Communication" the system here presented relies on a traditional high-performance RF spot beam link architecture. This avoids the main constraint of optical communication (need of a clear line of sight), with the possible use of multiple RF channels to ensure the end to end system high data throughput. The downlink from GEO to ground ideally uses a high-performance RF spot beam. It shall be noted that the very same GEO optical communication terminal has been successfully used to prove GEO to ground optical communication via a series of ad-hoc links to an optical ground station.

Considering the typical orbital geometry for an Earth Observation mission (eg sun-synchronous LEO), a GEO relay with one single GEO node already provides nearly hemispherical coverage. A second GEO node located at the opposite GEO position will enable nearly global coverage (Figure 1, center and bottom). Waiting for the next satellite ground pass is no longer required, and data can be made available in quasi-real-time with only latency times of GEO transmission applying. The ground stations can be at any convenient location reachable by the spot beam of the GEO nodes.

The Key technology for the implementation of the Inter-Satellite Link data transfer system presented here is the Laser Communication Terminal used for the Optical Inter-Satellite Link (OISL), briefly presented in the next section.

2. The Laser Communication Terminal (LCT)

TESAT Laser Communication Terminals provide full-duplex inter-satellite communication at 1.8 Gbps. For comparison, this data rate equals to USB-3.0 consumer electronics standard. The existing system design covers both LEO-GEO as well as GEO-GEO links with up to 72000 km link distance at full data rate. The optical links are using infra-red laser light at 1064 nm wavelength. Coherent BPSK homodyne reception ensures robustness against sun blinding and jamming. A 7.2 μrad beam divergence angle, comparing to a beam diameter of 320 m at LEO-GEO distance, enables "stealth" communication and prevents interception of the link by other parties.

The LCT contains all functions for data handling (lasers, modulator, de-modulator, fiber amplifier and receiver), all optical elements for beam shaping and steering, and all electronics necessary for mechanism command and control, trajectory calculation and link quality measurements. After loading a link scenario into the LCT internal controller (~4 Kbytes / link), the LCT is able to autonomously manage the pointing, acquisition and data transfer.

The LCT uses a two-axis gimbal for coarse beam pointing and two small two axis mirrors for fine pointing and point-ahead angle control. A clever algorithm allows to use the beam steering mechanisms for spatial acquisition, saving size, weight, and power that would otherwise be needed for an additional beacon laser.

Figure 2: Laser Communication Terminal

The LCT is a single unit with a four-footed quasi-isostatic mechanical mounting interface allowing for fast and easy integration onboard spacecraft. Thermal isolation from spacecraft by a dedicated radiator facilitates thermal control and stabilization of the laser-optical components. The standardized modular design and interfaces make the LCT suitable for embarkation on basically all types of LEO and GEO spacecraft and allows for adaptation to various scenarios and customer applications. Design lifetime is more than 15 years in GEO orbit. The system budgets and margins validated in orbit and the flexible design allow for both simplified LEO-terminals ('LEO-light') and for increased data rates up to 7.2 Gbps in future designs.

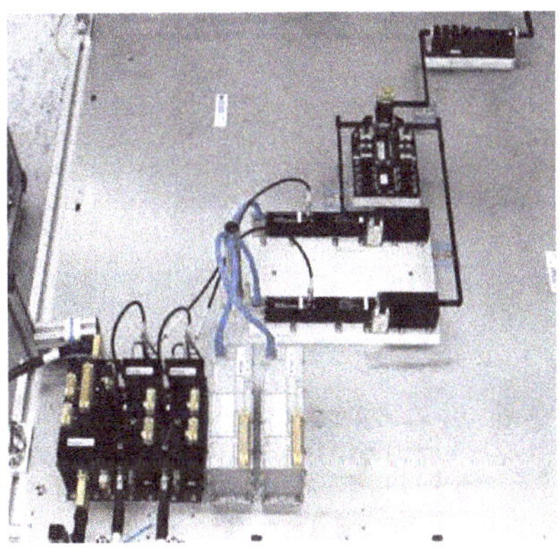

Figure 3: TDP1 Ka-band transmitter mounted on test panel

The Ka-Band Transmitter

In the case of Alphasat, the data received by optical inter-satellite links is relayed to ground by means of a Ka-band transmitter operating at 26.5 GHz carrier frequency. The modular system consists of the following: a Digital Switch Matrix (DSM) unitthat serves as the data interface to the LCT and allows data channel switching/routing; two RF chains in cold redundancy with each other; a modulator; a TWTA (EPC+TWT); and an RF isolator. A waveguide switch routes the RF signal through a bandpass and highpass filter towards a Ka-band antenna.

As a special feature unique to TDP1, the LCT can use the Ka-band downlink to continuously transmit its complete set of internal telemetries with sample rates up to 25kHz, allowing a large variety of experiments and detailed analysis of internal processes.

3. Optical Inter-Satellite Link Execution Concept

In order to perform an Optical Inter-Satellite Link, the two laser terminals involved need to share exact position information, as well as being precisely synchronised. The very high directionality of the laser beam also implies that the two terminals' optical axes need to be accurately aligned, translating in the need for a real-time precise knowledge of the two hosting spacecraft's relative position and attitude throughout the link.

The time synchronisation necessary is achieved using OTS system (notably GPS), while accurate attitude is obtained via interfacing directly with the satellite ACS sensors (notably star-tracker data).

The link itself is carried out in five main phases: (1) Preparation, (2) Spatial Acquisition, (3) Tracking and Frequency acquisition, (4) Communication, and (5) Link termination.

1. Link Preparation

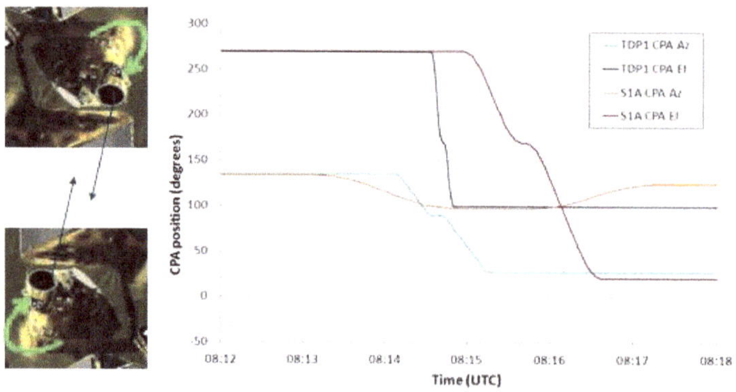

Figure 4: LCT preparation for a link – CPA movement to link start position

At a suitable time before link start time, the two terminals independently prepare for the operations by un-parking (in case they have been previously parked, see (5) below) and reaching a predetermined "link start" position, close to the calculated initial point of mutual acquisition (fig. 4), and of the "link trajectory".

During this phase, the two terminals' Coarse Pointing Assembly (CPA) move to pre-align the telescopes' optical axes. Figure 6 shows the actual telemetry of the Azimuth and Elevation angles change for the Alphasat TDP1 and Sentinel-1A OCP CPA's in this phase.

2. Spatial Acquisition

Shortly prior to the start of the link, CPAs begin to follow the pre-calculated trajectory, and at the start time of the link (T0) the beaconless spatial "Coarse Acquisition" phase starts (see fig. 5). In the first phase of the coarse acquisition (phase 1), the "Master" terminal initiates spiraling its TX laser beam around the target trajectory line of sight with a total uncertainty cone large enough to score hits at the counter terminal ("Slave"). The "Slave" LCT detects the laser signal with a four-quadrant sensor system and coarsely aligns its optical path in accordance to the detected hits density distribution. In the subsequent "phase 2", the roles are inverted, and the "Slave" terminal performs the same spiraling in the direction of highest hits density. The process is

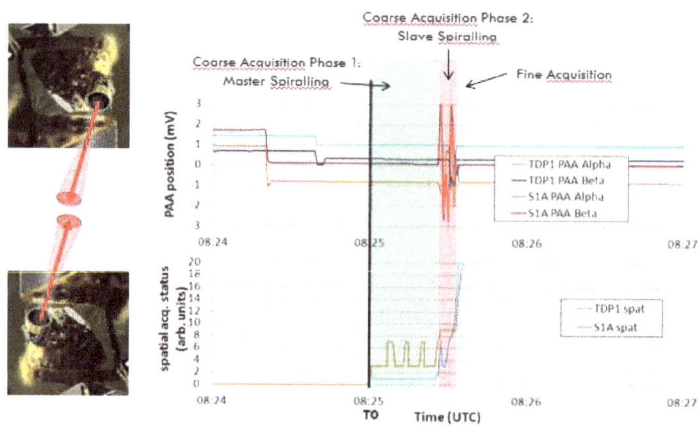

Figure 5: Spatial Acquisition: coarse and fine acquisition sub-phases

iterated in the "Fine Acquisition" phase, where the alignment is refined until both LCTs are aligned better than 10 μrad about the mutual line of sight, to bring the signal onto the "tracking sensor", and achieve and maintain active tracking.

3. Tracking and Frequency Acquisition

Once the alignment of the optical axes is completed, the two terminals start actively tracking each other and perform frequency acquisition, locking their local oscillator laser coherently onto the received signal by an optical phase locked loop.

Fig. 6 shows the Tracking Sensor signal for the two terminals, as well as the "frequency status" (from unlocked to optical phase lock).

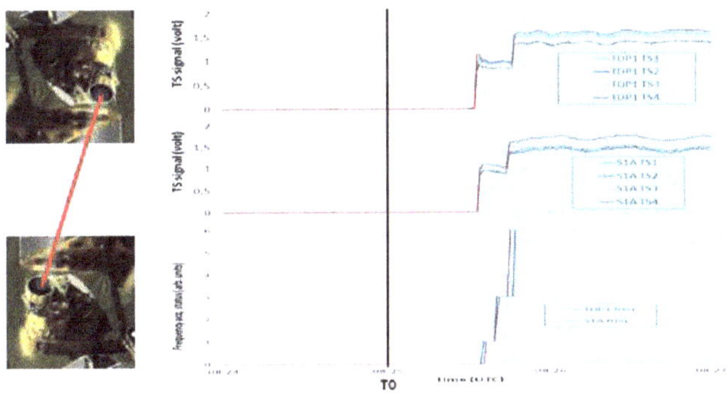

Figure 6: Tracking and Frequency acquisition phase

4. Communication

Once the frequency acquisition is complete, the two terminals start exchanging data modulated (BPSK) on the optical signal. During this phase, active tracking of the counter-terminal signal is performed, with the "Point Ahead Angle" assembly providing the necessary compensation for the along-track movement of the satellites during the signal travel time.

Accurate optimisation of all the parameters involved permits error free optical communication, as shown in fig. 7, reporting the terminal mode and "User data error count" (ie bit errors before correction on the user data channel of the optical transmission).

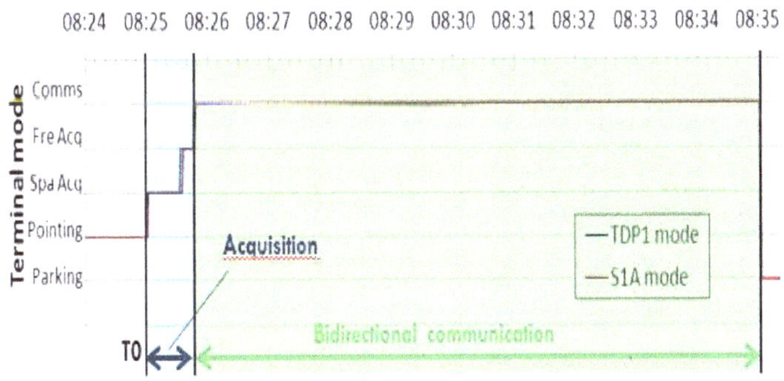

Figure 7: Terminal modes transition during a link

5. Link Termination

At the end of the (pre-loaded) link duration, both terminals independently terminate communication. They return to "Terminal Ready" mode, ready to start a new link by directly going to the next "link start position", or to return to the "safe" parking position should this be required.

4. Alphasat TDP1, Sentinel-1A … and -2A, -1B, and -2B: optical partners in orbit

Launched in July 2013, owned and operated by Inmarsat plc. of London, Alphasat is the first satellite to be launched using the new European high-power telecommunications platform known as Alphabus, jointly developed by Airbus defense and space and Thales Alenia Space and initiated by a partnership between ESA and CNES.

In additions to the primary mission goal of providing mobile communications as part of the Inmarsat geo-mobile fleet, Alphasat is hosting 4 payloads aiming at in flight demonstration of new technologies and/or constituting the flight segment of specific experiments.

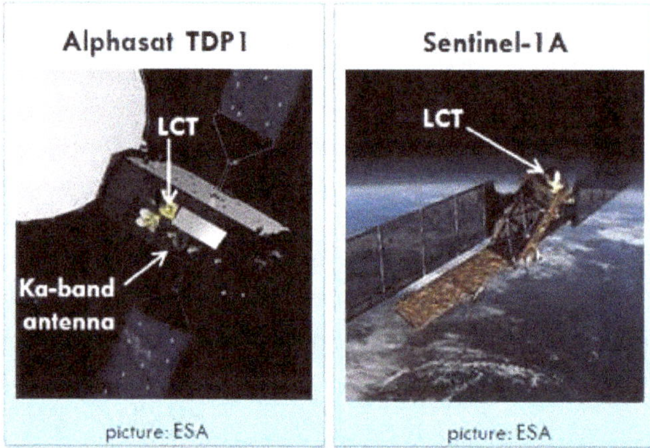

Figure 8: LCTs on Alphasat and Sentinel-1A

Amongst these "technology demonstration payloads", TDP1, developed by TESAT for DLR, represented the world's first available high-speed optical GEO-relay system, consisting of a LCT and the Ka-Band RF downlink transmitter with link capability at data rates up to 1.8 Gbps .

Figure 9: (first) Sentinel-1A SAR image (Berlin, Germany), transmitted via laser

Since completion of its in orbit commissioning, TDP1 has used the ESA Optical Ground Station (OGS) of Tenerife (Canary Islands, Spain) to characterize the optical link performances and optimise its configuration parameters, while awaiting the launch of a proper companion in orbit for performing its primary inter-satellite optical link demonstration mission.

Launched in April 2014, Sentinel-1A is the first European satellite contributing to the Copernicus program. Procured and operated by ESA on behalf of the European Commission, it performs its earth observation mission from its heliosynchronous LEO orbit thanks to a Synthetic Aperture Radar. As an additional payload and part of an effort to complement and improve the standard data downlink strategy with an innovative inter-satellite solution, an "Optical Communication Payload" (OCP), based on a LCT identical to the one on Alphasat has been embarked onboard Sentinel-1A.

In November 2014, the first optical link between Alphasat and Sentinel-1A was established. Since then, a large number of links have been performed, demonstrating performance and reliability, and validating system budgets in orbit, meeting the requirements of a commercial relay service.

Sentinel-2A, the second satellite of the Copernicus Sentinel constellation, joined

Alphasat and Sentinel-1A in space on June 23rd, 2015. Sentinel-2A main payload is composed by a multi-spectral instrument (MSI) with 13 spectral channels in the visible/near infrared (VNIR) and short wave infrared spectral range (SWIR). In addition, similarly to Sentinel-1A, it embarks an OCP based on the same LCT embarked on Alphasat and Sentinel-1A, with the aim again to complement and possibly enhance the mission return by using a high speed optical intersatellite link to downlink its images. Alphasat's TDP1 was used for the its Optical Communication Terminal in-orbit commissioning, completed in March 2016.

Sentinel-1A and -2A were joined in orbit by their twin satellites Sentinel-1B and -2B respectively on April 25th, 2016 and March 7th, 2017. Completing the Copernicus first constellation, they are also equipped by an OCP permitting to enhance their nominal performances by providing the opportunity to increase the overall data return and reduce latency.

Sentinel-2A

picture - ESA

Figure 10: Sentinel-2A

5. Alphasat (ASA) and Sentinels OISL Operations

1. General OISL Operations

The execution of an Optical Inter-Satellite Link implies coordination amongst several actors and systems involved in the exchange of the relevant data and information. Fig. 10 illustrates the generic operations scenario. Although apparently complex, a clear upfront definition of the interfaces between the different actors renders the system operations relatively simple and prone to be automated.

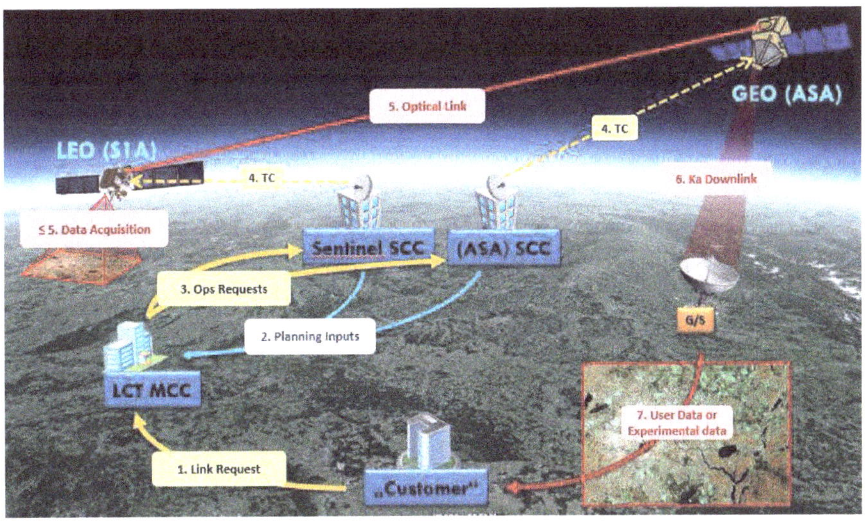

Figure 10: Generic link operations workflow visualisation

Referring to the visualization provided in figure 10, the following steps can be identified at high level:

Step 1 (7): The "Customer" provides the initial Link request to the "LCT Mission Control Centre", and receives the "User Data"

Step 2: The LCT-MCC, in parallel, receives from the Satellites Control Centres (SCC) the planning constraints and inputs for the planning period of interest

Step 3: The LCT-MCC, based on the planning inputs, compiles the "Ops Requests" and delivers them to the corresponding SCC.

Step 4: The two satellites' SCC, independently command the flight segment (and prepare the ground stations) for the execution of the link

Step 5: At the defined link execution time, the OISL takes place, relying the user data acquired at or before the link execution time (step ≤ 5).

Step 6: During the link itself, the data transmitted on the optical inter-satellite segment is downlinked to the involved GS.

Step 7: Finally, the data are delivered to the "customer" from the GS.

For the specific case of ASA and the Sentinels experimental links, the "Customer" is replaced by the LCT experiment operations team at TES, and the user data are composed of actual scientific data as well as by performance and housekeeping parameters acquired by the LCT during the exercise.

2. TDP1 Operations on board Alphasat

As stated above, TDP1 is one of four hosted payloads on the Inmarsat operated Alphasat geostationary communication satellite. As such, the concept of operations has to satisfy the basic requirement of non-interference with the main mission (commercial) objective. To achieve this, the integration of TDP1 in the Inmarsat fleet operations workflow has been pursued and achieved thanks to an automated operations scenario coordinated by the "TDPs ESA Coordination Office" (TECO).

The weekly planning cycle begins with the availability of (preliminary and final) operations planning input by Inmarsat (platform operational "exclusion periods" and orbital information) that are used by the TDPs operations centres to produce their operations requests. These are de-conflicted and harmonised by the TECO planning system to produce a unique TDPs Activities Request File that is delivered to Inmarsat. This request file is imported directly into the activity schedule from where, at the time indicated in the request, it is initiated and the corresponding automatic execution procedure(s) started. Although not nominally foreseen, INM SCC is able to intervene at run time in case of need.

For TDP1, the experiment operations team at TES compiles the ops request and delivers them to a LCT Mission Control Centre (MCC) located in DLR-GSOC (German Space Operations Centre) to prepare the actual "Input Operations request" and corresponding "Procedure Parameters Files" (PPF).

It shall be noted that, although the Activities Request is nominally delivered weekly, the "Procedure Parameters File" associated with each activity scheduled can be updated up to 2 hours before the task execution, thus providing the possibility to specify "fresh" procedures' executions parameters.

For monitoring purposes, a dedicated server of the INM MCS has been customized to provide remote access to Real-Time Telemetry to the TDP operations teams, both via an UDP streaming of the raw TM, as well as directly using the TM de-commutation and visualization capabilities of INM MCS. Daily delivery of both TM and operations ancillary data, as well as a TM archive access on demand are in place to complement the RT monitoring tools and fulfill the needs in terms of data provision to carry out the (experimental) mission of each TDP.

3. OCP Operations on board Sentinels

The (heliosynchronous) Sentinels mission calls for a high degree of autonomy with reduced ground centre intervention, so that commanding by ground with immediate execution is minimized. For OCP operations, the spacecraft supports an on-board Mission Timeline (MTL), where the activities are scheduled with respect to time. An Orbit Position Schedule (OPS) is also provided for scheduling activities with respect to orbit position (useful for data acquisition purposes).

The GMES Sentinels Ground Segment is in charge of the overall commanding and monitoring of the various spacecraft constellations as well as the acquisition, processing and dissemination of their observational data. The two primary components of the Ground Segment are the Flight Operations Segment (FOS) and the Payload Data Ground Segment (PDGS).

Each working day, the PDGS delivers an input plan (Nominal Payload Planning File) covering by default a period corresponding to one repeat cycle (12 days for Sentinel-1A and B, and 10 days for Sentinel-2A and B). If necessary PDGS will modify the plan before distribution to the FOS, e.g. incorporating data "take" requests resulting from user orders submitted since the last update.

The FOS processes the received inputs, generating a schedule covering the complete planned interval, but only uploads commands covering the next 96 hours to comply with the spacecraft on-board queue size limits. As soon as the on-board resources are available, the FOS "tops-up" the command queue by releasing an additional 24 hours' worth of commands, thus ensuring that a plan covering between approx. 72 and 96 hrs is always available on board. This strategy avoids the need for real-time commanding of the payload or subsystems other than unforeseen special operations or recovery activities.

Taking the example for the products specific names of Sentinel-1A and 1B (minor differences apply for the case of Sentinel-2A and 2B), for the OCP operations, the predicted orbit file and the "Skeleton Plan File" (SPF) are provided by the Flight Dynamic System (FDS). The orbit file, together with orbital information of the counter-terminal hosting satellite (provided by the operator of this latter) are used by PDGS to compile the LCT plan. The LCT Plan, SPF, and the Operation Planning File (OPF) provided by the Flight Control Team are harmonized in the Mission Planning System that creates the Plan Increment File and On-Board and Ground Telecommand (TC) schedules.

During the first phases of operations with ASA-TDP1, the LCT Plans provided by PDGS were substituted by "Special Operation Requests" delivered directly by the OCP experiment operations team of Tesat (this is the same as for TDP1). Also similar to the TDP1 case, in order to prepare these SORs, the SPF and orbit file were distributed to the LCT Mission Control Centre and to DLR-GSOC for the preparation of all relevant operational products (notably, link parameters).

Monitoring of S1A OCP takes place at the dedicated control centre located at ESA-ESOC. Off line access to the relevant TM and ancillary data is provided to TES for performance analysis.

4. Joint Sentinels & ASA OISL Operations

Given the experimental nature of the activities conducted, during the first period of operations between TDP1 and Sentinel-1A, as well as for the subsequent Sentinels OCP commissioning activities (see below) a joint concept was agreed for the ASA-S1A Optical Inter-Satellite Links operations, based on the two missions' planning cycle and operations execution scenario. Figure 12 illustrates the concept and workflow.

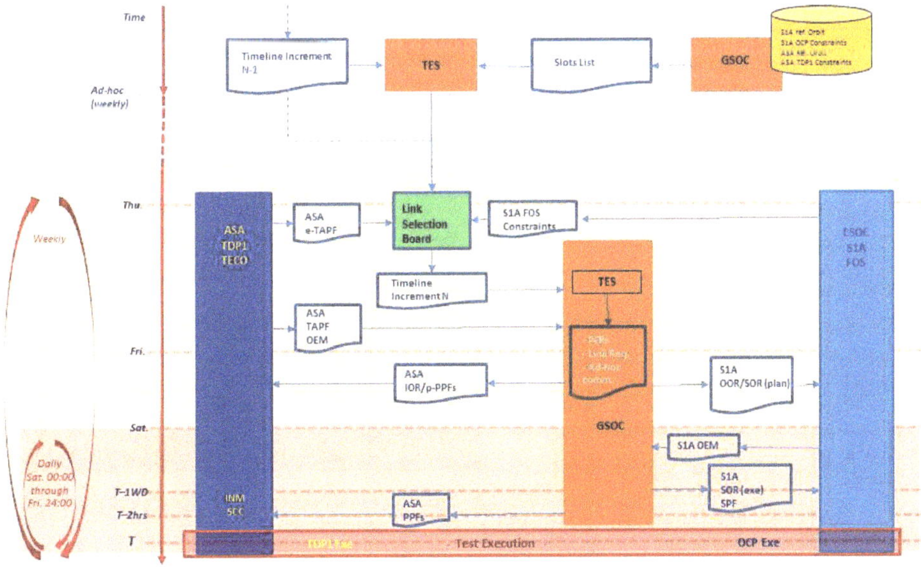

Figure 12: ASA and S1A Joint planning and operations workflow for OISL execution

On a weekly basis (planning period for ASA), a "Link Selection Board" (LSB) picks the interesting links to execute in the next planning increment amongst a list of opportunity provided by the LCT Mission Control Centre (at DLR/GSOC). The output of the weekly LSB is the Timeline for increment N, that the LCT Operations team and LCT MCC (TES and GSOC -see above-) translate into the relevant IOR entries and PPFs for ASA, and Special Operations Requests for the Sentinels.

This simple concept has been permitting an effective and easy coordination of the several actors involved in the links planning and execution, while maintaining the needed independence of each mission's activities. Moreover, thanks to the unique coordination point in the process represented by the LSB, consistent visibility of the on-going activities is provided also to all stake-holders not directly involved with an active role in the OISL operations.

After the completion of the commissioning activities for Sentinel OCP's, an adaptation of the operations planning systems and interfaces has been performed in order to include the planning and execution of TDP1 links with the Sentinels in the nominal Sentinels' OCP operations workflow. Due to the use of the OCP for their primary earth observation mission, it is the Sentinels Payload Data Ground Segment (PDGS) that provides directly the operations requests for the links to the TDP1 MCC and the Sentinels FOS based on the available slots for experimental links. The links requested by the PDGS are henceforth parametrized for TDP1 in order to carry out the preliminary agreed experimentation plan.

It shall be noted that, although the actual link times are not known until the PDGS provide their scheduling, a rough timeline for the links slots is known in advance. In fact the intensive use of the Sentinels OCP for mission data return links with the European Data Relay Satellite system (EDRS, see below) implies that the experimental links are placed in windows corresponding to predictable gaps of availability for links with EDRS. Together with the flexibility built in the operations concept and planning system for TDP1, this permits the necessary level of experimental links preparation by the ASA TDP1 team.

6. The European Data Relay Satellite System

While the mission of Alphasat TDP1 was intended and served as the precursor for the European Data Relay Satellite System (EDRS), the Sentinels OCPs were embarked with the objective to provide an innovative, complementary data downlink possibility by using optical intersatellite data relay offered by the EDRS based "SpaceDataHighway". The "SpaceDataHighway" is a public–private partnership between ESA (European Space Agency) and Airbus Defence and Space, who operates the system and commercializes the data relay service. This latter enable users to transfer their data (imagery, video, voice…) from their Earth Observation satellites, UAVs, multi-mission aircraft by means of optical communication via the EDRS-A (and soon EDRS-C) geostationary nodes to receiving ground stations located in Europe.

Thanks to the availability of Sentinel-1A and Sentinel-2A, the first "dual link" test was performed on April 7th, 2016, demonstrating the operational and technical feasibility of consecutive links to different targets (without returning the LCT CPA to parking), in a first real-life EDRS service scenario.

The first node of the EDRS constellation, EDRS-A, was launched on January 29th 2016 as a hosted payload onboard Eutelsat EB9B. It is now providing data relay service to all the 4 Sentinels satellites equipped with an OCP. An average of 1000 links per month are currently executed between the Sentinels and EDRS-A.

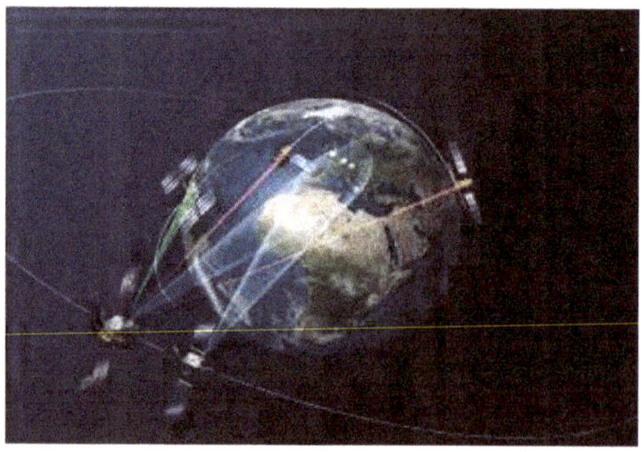

Figure 13: the European Data Relay Satellite System concept (EDRS)

The second node of the constellation, EDRS-C, is planned to be launched in mid 2019, permitting to double the system data relaying capacity.

EDRS will eventually consist of a network of dedicated geostationary terminals providing data relay services. It will provide commercial GEO data relay service for up to 26 LEO satellites, with the European Commission Copernicus Sentinel-1 and -2 being the first (anchor) clients. EDRS LEO-GEO optical links provide user data rates up to 1.8 Gbps for optical LEO-GEO inter-satellite link and Ka-band RF downlink. Each EDRS GEO-relay satellite can downlink data volumes of up to 16.2 TByte per day.

Thanks to the use of Optical Inter-Satellite Links, EDRS gives the user on the ground high-speed access to real-time Earth Observation data, on-demand at the right place and time. Applications for EDRS service include emergency response, open ocean surveillance, UAS communication, weather forecasting and wide-area monitoring.

7. Links and System Performance

Alphasat TDP1 GEO-relay has executed up to the present date (end of July 2018) more than 1800 successful links of both quasi-operational and experimental scope. More than 600 of these were conducted with one of the 4 sentinels OCPs, while the rest were links to the Optical Ground Station or test-links (fig.14). These links permitted initially to verify the expected performances of the LCT, and subsequently, thanks to the increasing amount of data acquired with subsequent links, fine-tune the system parameters to achieve improved end-to-end OISL performances.

During optical link commissioning between Alphasat TDP1 and the Sentinels, the total uncertainty cone was initially set to 2250µrad and reduced to 1000µrad. During all links, pointing accuracy was found better than 500µrad for both link partners. Key for this performance is the hosting platform attitude knowledge, provided to the Sentinels OCPs by the satellites AOCS and, for TDP1, by the Jena-Optronik ASTRO APS star sensor embarked on Alphasat another technology demonstration payload (TDP6).

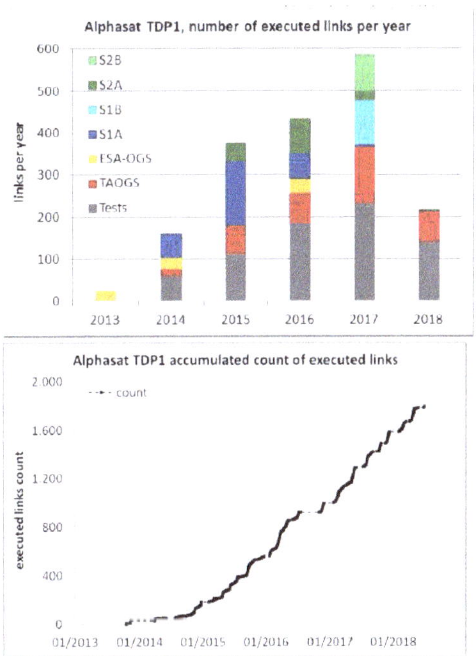

Figure 14: ASA TDP1 links statistics

In parallel to the uncertainty cone, also spatial acquisition duration was reduced from 240s at the very first tests to 34s for default links. During dedicated experimental links, spatial acquisition was reliably achieved within 7s . Even during links with intentionally adverse conditions like ultra-low TX laser power of 100mW, spatial acquisition was achieved in no more than 80s. Figure 15 (next page) shows the 100th link as a typical example of spatial acquisition within 34 seconds for a link with total uncertainty cone of 1000µrad, and 1.1W optical TX power.

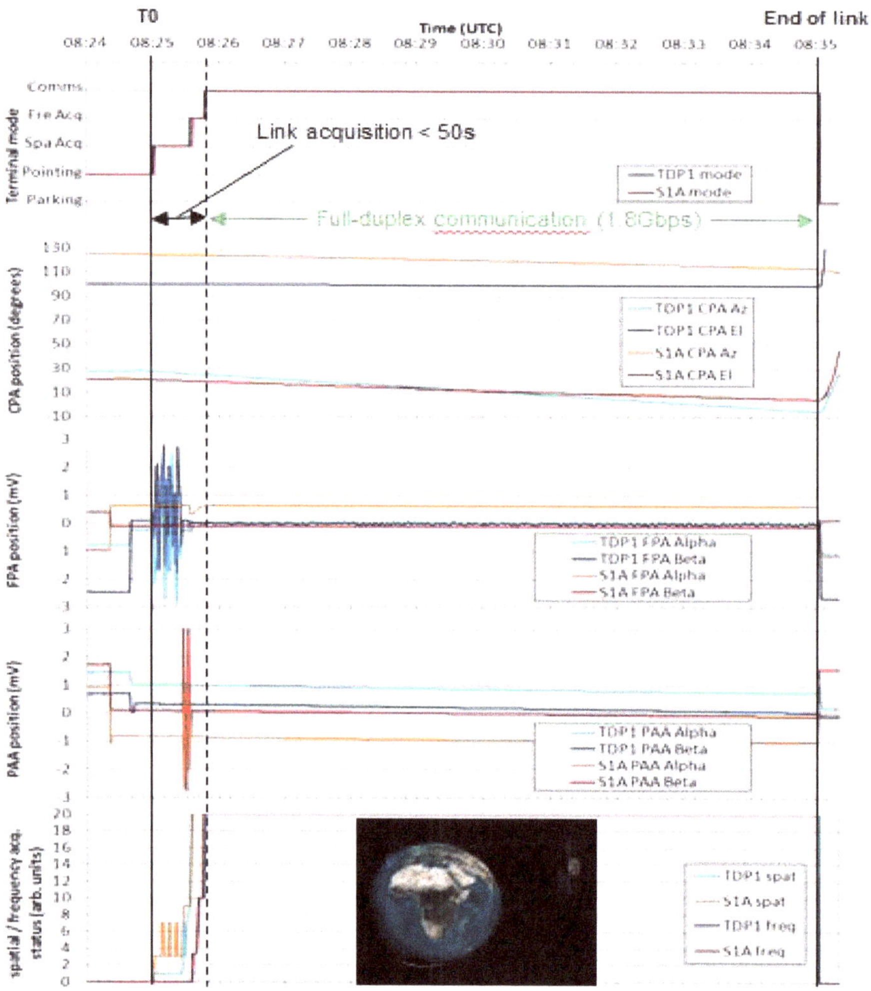

Figure 15: ASA-S1A 100th link (10 mins duration)
TM shows modes transitions, Coarse and Fine Pointing Asembly positions, Point Ahead Angle position, and Spatial and Frequency Acquisition status.

Link acquisition between Alphasat TDP1 and the Sentinels' OCP is reliably and repeatability achieved within less than 50 seconds after T0. During some specific link sessions, acquisition durations of only 20s have even been demonstrated (Figure16, next page). No impact on acquisition duration could be found due to satellite-induced micro-vibrations environment.

During communication, a built-in LPC error detection and correction algorithm ensures integrity of transmitted data even at low signal levels. Using 1.1W optical TX power, during a large number of links not even a single user data bit error has been recorded. This correlates to a user data bit error rate of much lower than 10^{-12}. Additional Reed-Solomon encoding of the user data provide further margins.

Taking into account the maximum optical TX power of 5W, links with GEO-GEO distance are easily possible with full performance.

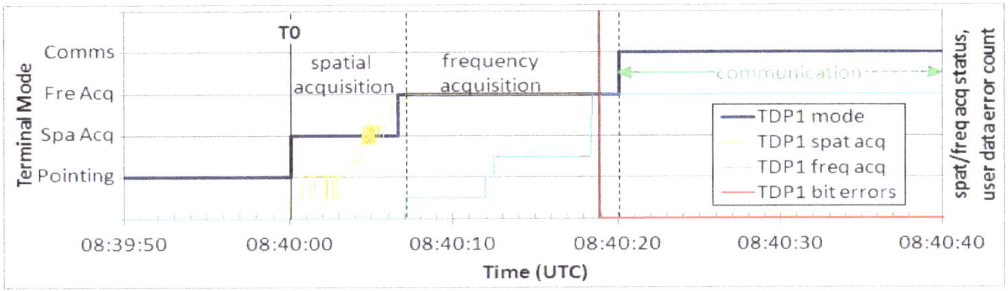

Figure 16: Acquisition within 20s, spatial acquisition 7s. TM recorded at 100Hz sampling rate. Only TDP1 is shown (link performed on Aug. 18th, 2015)

8. Conclusions

With their joint operations, Alphasat TDP1 and the Sentinels OCPs have successfully demonstrated and validated the possibility to include Optical Inter-Satellite Links as a complementary data return strategy for LEO high data volume missions, providing reliable high-speed, near-real-time access to the mission data of LEO satellites outside the range of direct RF ground stations.

The presence in orbit of TDP1 and the Sentinels has permitted to perform in-orbit validation and commissioning of the 4 Sentinels OCPs, and has thus played a key role in allowing the 4 Copernicus Sentinel missions to benefit from the use of the EDRS system and SpaceDataHighway service in operation since 2016.

Beyond providing the most representative technology demonstration test-bench possible, the operational experience being gained with ASA TDP1 and the Sentinels has been proving very beneficial for the full understanding of the various specific issues characterizing Optical Inter-Satellite Links operations. This has permitted to optimise system parameters, and to fine tune links execution scenarios, in particular for the utilisation of the specific Laser Communication Terminal embarked on both Alphasat and the Sentinels spacecraft's.

The links performed in 4 years of in-orbit common operations have also permitted acquisition of an invaluable experience in the Ground Segment and operations specific features implied by the planning and execution of Optical Inter-Satellite Links.

In the light of the benefit and the results obtained, it has been agreed to extend the "experimentation phase" of the first Optical intersatellite link partners (Alp and S1A), beyond the originally planned end of the activities, and including also the other Sentinel satellites.

References

[1] E. Benzi, I. Shurmer, N. Policella, D. Troendle, M. Lutzer, S. Kuhlmann, M. James, "Optical Inter-Satellite Communication: the Alphasat and Sentinel-1A in-orbit experience", AIAA SpaceOps 2016 Conference, May 2016

[2] ESA Telecommunication and Integrated Application, "ARTES8",URL: http://telecom.esa.int/telecom/www/area/index.cfm

[3] D. Troendle, C. Rochow, P Martin Pimentel, H. Zech, F. Heine, H. Kaempfner, M. Gregory, M. Motzigemba, U. Sterr, R. Meyer, M. Lutzer, S. Philipp-May, "Optical LEO-GEO Data Relays: From Demonstrator to Commercial Application", AIAA Space and ICSSC 2014

[4] G. Muehlnikel, H. Kaempfner, F. Heine, H. Zech, D. Troendle, R. Meyer, S. Philipp-May, „The Alphasat GEO Laser Communication Terminal Flight Acceptance Tests", ICSOS 2012

[5] F. Heine, G. Muehlnikel, H. Zech, D. Troendle, S. Seel, M. Motzigemba, R. Meyer, S. Philipp-May, E. Benzi, „LCT for the European data relay system: in orbit commissioning of the Alphasat and Sentinel 1A LCTs ", Proc. SPIE 9354, 2015

[6] M. Witting, H. Hauschildt, A. Murrell, J-P. Lejault, J. Perdigues, J.M. Lautier, C. Salenc, K. Kably, H. Greus, F. Garat, H.L. Moeller, S. Mezzasoma, R. Meyer, B. Guetlich, S. Philipp-May, A. Pagels-Kerp, B. Theelen, M. Wiegand, M. Leadstone, G. Eckert, G. Wuetschner, L. Laux, O. Gerard, D. Poncet, R. Mager, K. Schoenherr, F. Heine, S. Seel, K. Panzlaff, H. Zech, H. Kaempfner, A. Schneider, I. Gutierrez Canaz, C. Arias Perez, H. Schuff, "Status for the European Data Relay Satellite System", ICSOS 2012

[7] E. Benzi, A. Cacioni, "Private Public Cooperation for Hosted Payload Operations: the Alphasat Concept", AIAA SpaceOps 2014 Conference, May 2014

[8] E. Benzi, N. Policella, M. James, A. Cacioni, "Alphasat Mission: TDPs Ground Segment and Operations Concept & Implementation", 20th Ka-band Conference, Oct. 2014

[9] Marina Bernard, Edoardo Benzi, Juan Rivera Castro, Philippe Sivac, Eloy Torres, Simon Weinberg, "Alphasat: Innovation in Orbit", ESA Bulletin 159, Aug. 2014

[10] P. Bargellini, P.P. Emanuelli, I. Shurmer, F. Marchese, C. Steiger, H.L. Moeller, "The GMES-Sentinels Flight Operations Concept", AIAA SpaceOps-2012 Conference, June 2012

[11] http://www.edrs-spacedatahighway.com/

[12] https://www.airbus.com/space/telecommunications-satellites/space-data-highway.html

[13] https://www.esa.int/Our_Activities/Telecommunications_Integrated_Applications/EDRS/EDRS_Global

Diversity Architectures for High Data Rate Ground-to-Satellite Optical and EHF Links

Rajeev Gopal[1]
Hughes Network Systems, LLC, Germantown, Maryland 20876

Architectural framework for next-generation satellite and ground terminal nodes using optical and EHF links will require diversity, redundancy, and network layer switchover designs to efficiently mitigate for signal losses due to atmospheric attenuation and scintillation. Our architecture leverages proven technologies, including Software-Defined Networking (SDN), Wavelength Division Multiplexing (WDM), linear programming optimization, and cost-effective IP/Ethernet packet processing within a unified network model. It is compatible with an agile and efficient control plane that is based on SDN OpenFlow (OF)-based link state measurements, centralized traffic route determination to address optical link fading, and standardized OF-based configuration of commodity hardware. Baseline and contingency traffic routing plans are used for rapid configuration of satellite payload, gateway and ground network nodes to leverage a make-before-break approach for dealing with high data rate link failures. This architectural framework is based on a linear algebraic traffic transport model with the following features: diversity gateway sites, proactive switchovers, optimal traffic routing, traffic engineering per traffic class, and optimized capital and operational costs. The next-generation systems with optical links can thus leverage cost-effective software and networking technologies, scalable cloud computing for optimal routing, and mature SDN protocols that can reduce the overall network implementation risks and costs.

1. Introduction

Optical and EHF links can further enhance High Throughput Satellite (HTS) communications [1], next-generation high-data-rate-sensing satellite capability, and individual user and sensor data rates compared to what has been possible so far with spectrally efficient RF waveforms and innovative space and ground component RF designs. The HTS systems have revolutionized communications for underserved and unserved areas in many regions and now even higher capacity, exceeding 1 Tb/s per satellite, is needed to address the growing needs of consumers, enterprises, and mobile users as well as potential convergence with the 4G/5G systems [2]. Earth-sensing satellite payloads with multispectral sensors and expanding demand for near-real-time data collection also require high capacity relay links. Spectral reuse, enabled by spot beams, has been a key facet of the HTS satellite design, and almost all existing HTS satellites have typically used Ka- or Ku-bands for both user and gateway terminals that are typically placed in their respective user and gateway beams. While the user beams within a system are designed to maximize coverage for all potential users, the gateway RF beams are placed in locations with sparse population, access to terrestrial fiber links, and drier climate.

Use of EHF bands (>30 GHz including Q-, V- and W-bands) and optical frequencies for gateway (feeder) links can release additional spectrum for the user terminals and improve coverage by not requiring disjoint gateway and user beams. The EHF bands (with shorter RF wavelength and beam size) are additionally attractive because of potentially smaller space- and ground-equipment Size Weight and Power (SWaP), but are constrained by significantly higher attenuation associated with weather-dependent factors, such as moisture and precipitation. In addition to higher data rates, also possible with EHF, optical links have additional advantages because they are not regulated by ITU, are immune to interference, consume less power, and have Low Probability of Intercept/Low Probability of Detection (LPI/LPD) properties due to narrow optical beams. With 100s of GHz (visible and infrared) available in optical band, a single optical link can easily exceed 100 Gbps. EHF bands are already in limited use for GSL, and optical wavelengths are in orbit for Inter-Satellite Links (ISLs), immune to atmospheric effects. The ISLs can thus provide heritage space-qualified optical amplifiers, connectors, frequency convertors, and tracking antenna/telescope technologies that can be used for designing communications payloads supporting the next-generation optical GSLs.

Optical GSLs are susceptible to atmospheric disturbances, including scattering and scintillation that affect photonic propagation. Unlike Ku- and Ka-band GSLs that have successfully utilized dynamic power, coding, and modulation

[1] Sr. Technical Director, DISD, rajeev.gopal@hughes.com, 11717 Exploration Lane, Germantown, Maryland 1 (301) 548 1977.

adjustments to mitigate Line of Sight (LOS) attenuation, EHF and optical links will require adaptive diversity architecture with gateways located in distant regions to exploit uncorrelated atmospheric impairments. A key part of system diversity design will be to dynamically assess environmental conditions, rapidly reroute traffic, and implement systemwide configuration to minimize overall unavailability. Unlike transponded satellite designs that have infrequent payload configuration changes (e.g., to implement component failure mitigation), now both ground (gateway, network) and satellite payload configurations must be dynamically adjusted several times a day (or even hour) to deal with the varying atmospheric conditions. This requires network-level load balancing, subnormality detection, and responsive and proactive recovery steps.

Both EHF and optical links can get severely attenuated and incur increased ambient noise because of various atmospheric phenomena. Atmospheric turbulence significantly affects optical signals because of the scintillation effect. Due to narrow beams, especially for optical signals, the transmitting and receiving stations must also be pointed accurately. Precise alignment and Pointing, Acquisition, and Tracking (PAT) require careful design of optical antennas (telescopes). Optical GSLs are affected because of scintillation (and various other weather phenomena involving varying air pressure, temperature and humidity), even during clear weather, as temperature affects the air refractive index structure parameter.

Optical wavelengths in the infrared region can use a relatively high-power threshold for health (retinal protection) reasons and thus provide better performance against attenuation and scintillation. Optical beams are highly directional because the diffraction angle is directly proportional to the wavelength. This can provide more than a billionfold advantage in reducing the footprint of a laser beam operating at 1.54 μm compared to S-band RF that reduces energy requirements for the optical transmission and minimizes inadvertent interference with other communications links. Cooperative systems, under the control of a Software-Defined Diversity Network (SDDN) controller, can leverage a common set of optical terminals that can be used for multiple satellites for cost sharing, as shown in Figure 1. Optimal traffic engineering, even for dynamically changing traffic loads, can be performed to efficiently route traffic over various nodes and links supporting differentiated services. A finer-grained optimization scheme can also include individual Quality-of-Service (QoS) specifications (e.g., smaller delay, or assured delivery) for various traffic flows.

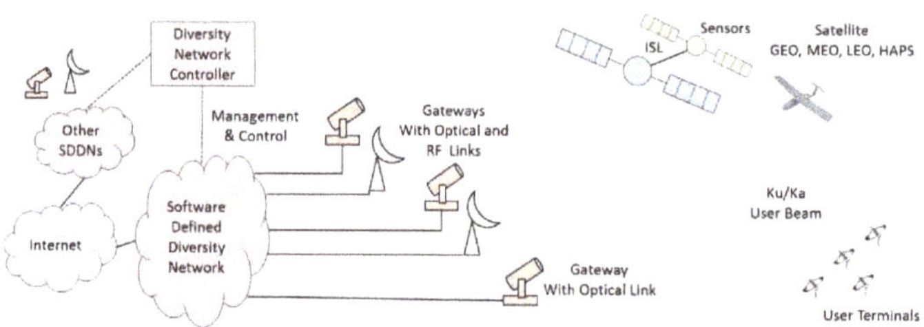

Figure 1. SDDN for RF and optical ground to space links

The RF and optical links sometimes have complementary behavior with respect to commonly encountered weather adversities. For example, fog and haze can significantly attenuate optical links, compared to the RF links. On the other hand, rain affects the EHF band more intensively. Since fog and rain are not temporally correlated, the use of hybrid RF and optical links is likely to have better overall availability. Such diversity is useful in ensuring that a hybrid GSL has a better chance of providing some capacity for basic system management, such as measuring channel state, sending highest priority traffic, or configuring satellite payload. Optical link designs, similar to their RF counterparts, are now exploring a variety of Layer 1 and Layer 2 techniques to increase link capacity and availability, including the following: noncoherent (e.g., on-off key, direct detection) and coherent (e.g., QPSK) modulation schemes; dynamic power management to address attenuation; use of Forward Error Coding (FEC) schemes (e.g., Low Density Parity Codes- LDPC) for Bit Error Rate (BER) management; antenna spatial diversity (e.g., multiple input and multiple output); adaptive optics to counteract scintillation effects; and link layer error correction (e.g., automatic repeat request—ARQ) for error correction. In addition to spatial diversity, temporal diversity can also be leveraged since the scintillation effects typically last .1 s to 10 s of ms. For delay-insensitive traffic, retransmission and use of rateless code (e.g., fountain codes) can also be performed at the link layer.

Any network layer design will require flexibility to accommodate a large collection of rapidly evolving lower layer techniques for optical link performance and availability enhancements. Network layer designs and associated optimizations with respect to QoS, cost, and resource utilization for terrestrial networks have been actively researched and practiced for the last 3 decades. Traditional routing and switchover designs have catered to terrestrial link and node failures, misconfigurations, and congestion. The associated timelines for network topology changes and traffic rerouting typically range from 10 s of seconds to minutes. A direct application of these distributed routing techniques at packet or circuit levels will create large connectivity gaps for high data rate GSLs. On the other hand, a centralized routing and traffic engineering approach, driven by GSL-specific cost and performance optimization, will be able to react in milliseconds and is a key theme of this paper.

The rest of this paper is organized as follows. We first provide an architectural description of satellite systems incorporating optical and EHF GSLs with SDN-based centralized traffic routing that can support representative use cases for diversity switchovers. This is followed by a unified (across packet and circuit processing) model that can be used in optimal switchover decision making for an integrated system comprising optical, RF, and terrestrial links. We then discuss an implementation approach utilizing currently existing SDN, packet processing, and optical circuit-switching building blocks. In conclusion, we provide some guidelines for future work.

2. Satellite Systems with EHF and Optical Ground Space Links

The key components of the next-generation optical and RF high data rate communications payloads along with some representative application payloads are shown in Figure 2. Both space and ground nodes (with similar packet and circuit-switching capabilities) are under the control of the same SDN controller. Typically, the entire communication band is received by an antenna and amplified with low noise amplifiers. A channelizer enables the individual channels within the band to be further processed. In a transponded scheme, only analog signal amplification is performed (along with frequency conversion) while a regenerative payload also includes full (de)modulation and de(coding) to enable packet-level processing, (classification, queuing, policing, switching, and routing) across various channels. While high speed (de)modulation, (de)coding, and packet processing are mature technologies on ground, their space deployment is still not a common practice (prominent exception being the Ka-band SPACEWAY™ satellites that were launched around 2007) and most of the current communications satellites still use a transponded design. This approach has well served the commercial GEO markets that do not benefit from the ISLs. The next-generation NGSO constellations, especially in LEO orbits, have limited field of view and will benefit from ISLs to facilitate flexible gateway locations. Recent NGSO investments and innovations are providing impetus to regenerative architectures for next-generation satellite payloads, including optical technologies. Even with transponded optical payloads, the SDN-based control of optical channel wavelength switching and multiplexing could provide sufficient payload configurability to better deal with GSL fading due to adverse conditions.

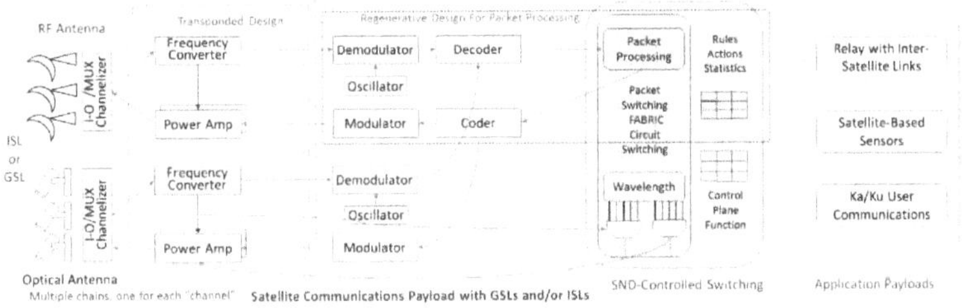

Figure 2. Canonical architecture for next-generation satellites with optical links

Optical communications payloads are likely to be smaller in size and mass since fiber-based signal distribution can replace bulky RF coaxial cable and/or waveguides. With sustained contemporary innovations, especially with frequent launches of smaller test satellites, a wider deployment of transponded optical and hybrid RF/optical payload is becoming more feasible. Diversity-related optical switching and configurability (e.g., add, drop, multiplex wavelengths) typically involve the whole optical or fiber link (or constituent wavelengths) carrying an aggregation of packets as opposed to finer-grained packet processing associated with a regenerative design. The original terrestrial SDN scheme originated at the packet level and since then has been expanded to include circuit-switching for optical wavelengths over local and wide-area fiber links. Such SDN extensions can involve wavelength switching under the control of SDN-based control of nodes over OF protocols with centralized decision-making.

A. Use Cases for Ground Space Link Diversity

The design objectives for a unified network model can be analyzed by considering some representative use cases, as summarized in Figure 3. A single optical link cannot guarantee high availability. This is the norm for Ka-band and Ku-band feeder links (>99% availability) that have sufficient power margins to mitigate most rain-based attenuations. To counteract deep optical link fades, system design must include multiple backup links on the same (with hybrid RF backup links) or geographically distant sites to ensure that at least one of the optical or RF GSLs will be available most of the time to sustain connectivity with a satellite communication payload.

Use Case (a) summarizes the scenarios when a weather phenomenon such as light fog compromises an optical link O_1, but an RF link from the same site is still able to sustain the current level of user and/or management data to the space node, benefiting from a joint use of compromised optical link and better performing RF link.

Use Case (b) represents the scenarios where weather phenomenon or atmospheric turbulence renders optical link O_2 at site 2 totally unusable, and another optical link O_3 from site 3 must be switched in to carry the traffic from ground node n_b to the space node.

Use Case (c) shows the distribution of traffic, based on individual traffic classes, from node n_x to the space node across multiple sites. Here optical links O_a and O_b, at two different sites, are partially compromised because of the environmental conditions, and the resulting congestion has resulted in higher packet loss and longer delay. Thus, both links O_a and O_b are not suitable for high-priority traffic that can be routed via the third site with optical link O_c as shown in Use Case (d).

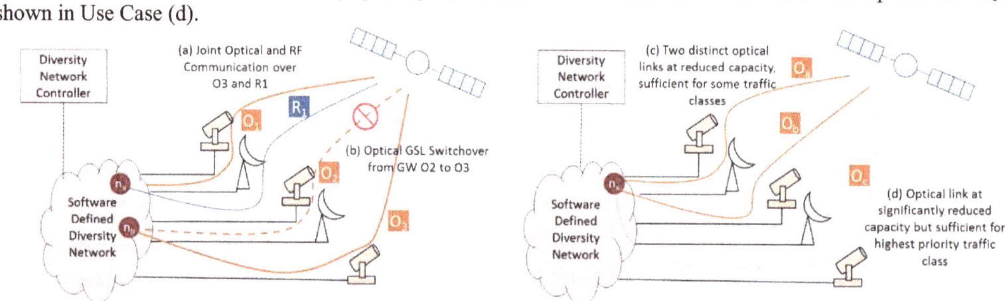

Figure 3. Representative use cases for diversity switchovers of optical and RF links

3. Unified Network Layer Model for Diversity Switching

The proposed network layer model comprises optical, RF, and terrestrial (fiber) links interconnecting packet and/or circuit processing nodes. Link-specific transceivers are used for traffic input and output at each node. The optical and RF GSL capacity is dependent on the current environmental conditions that create variability in the instantaneous network capacity for transporting traffic from the ingress to the egress nodes. By using a single unit (e.g., Mbps) to characterize time-dependent link capacity, it is possible to create a unified network model, described in this section, for a hybrid network. The capacity of a GSL depends on the power received by a transceiver over free space. Aperture size of the transmitting and receiving antennas increase the channel gain, while path loss and atmospheric attenuation decrease the received power because of absorption, scattering, and scintillation effects. A general expression [4] can be used across the RF and optical links to characterize the amount of received power P_R, as function of transmitted power P_T, transmitter (G_T) and receiver (G_R) gains, transmit (D_T) and receive (D_R) antenna sizes, propagation distance (L), total atmospheric (L_a), beam divergence (θ), and other losses (L_m), as follows.

Received Power
$$P_R = P_T G_T G_R \left(\frac{\lambda}{4\pi L}\right)^2 \left(\frac{D_R}{D_T + \theta L}\right)^2 10^{-(L_a+L_m)/10}$$

The total atmospheric loss L_a includes attenuation due to fog, rain, snow, and scintillation effects, each of which depends on the RF or optical signal wavelength λ. Analytical expressions can be determined to quantify spectral efficiency of a selected modulation and coding scheme and a target BER. Spectral efficiency multiplied by available bandwidth can then provide the capacity of the link in units of Mbps. For both RF and optical links, the waveform can be designed to be adaptive, allowing a terminal controller to select a specific combination of coding, modulation, and transmit power P_T to deal with atmospheric attenuation, pointing losses, and/or ambient noise. Thus, the instantaneous capacity of a GSL is a function of time and ranges from a maximum (under the ideal environmental condition, highest transmit power, highest modulation scheme, and least robust FEC) to zero (when dense fog or strong scintillation does not allow an optical link to close). Both RF and optical link capacities are governed by Shannon's information rate limit and implementation losses associated with electromagnetic propagation over free space. Similar to the RF links, analytical and numerical models can be developed to quantify average optical link capacities and outage probabilities [4] based on various communication schemes and environmental conditions.

A network layer Mbps-based unified model accommodates the various performance attributes and constraints related to the space and ground nodes and the links interconnecting them. This model is used for selecting specific links and nodes for transporting traffic for the baseline plan and rerouting traffic when link capacities diminish. The model description in this section is generic, and more specific time-based instantiations are subsequently used for optimizations associated with baseline traffic engineering and failure-related recovery plans.

A. Linear Programming-Based Optimization Model

The network layer representation of the system comprises two types of nodes: space nodes, represented by set $\boldsymbol{S}$, and ground nodes ($\boldsymbol{G}$), which are connected by three types of links: RF ($\boldsymbol{R}$), optical ($\boldsymbol{O}$), and fiber ($\boldsymbol{F}$). The set of all links is thus represented as $\boldsymbol{L} = \boldsymbol{R} \cup \boldsymbol{O} \cup \boldsymbol{F}$. A node belongs to $\boldsymbol{I}$, the set of all ingress nodes, if it has at least one port where external traffic from an external node enters the network $\langle \boldsymbol{N}, \boldsymbol{L} \rangle$. Similarly, $\boldsymbol{E}$ represents the set of all egress nodes such that $\boldsymbol{N} = \boldsymbol{G} \cup \boldsymbol{S}$ and $\boldsymbol{I} \cup \boldsymbol{E} \subseteq \boldsymbol{N}$. The type of a link is indicated by its superscript. For example, l_{ij}^{o} represents an optical link connecting the source node i and the destination node j. Link l_{ij}^{τ} has several attributes to characterize its capacity c_{ij}^{τ}, propagation delay δ_{ij}^{τ}, and usage $u_{ij}^{\tau}(q)$, for traffic class $q\epsilon\boldsymbol{Q}$, the set of all traffic classes. The instantaneous capacity of a GSL depends on the environmental conditions, while the current usage depends on the node configurations based on decisions made at the network layer to optimize a specific objective. The propagation delay on a link is a constant, chiefly dependent on the physical length of the link.

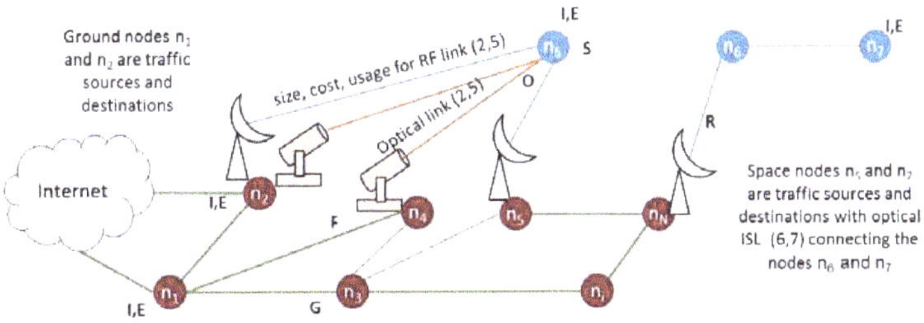

Figure 4. Network layer graph with space and ground nodes and optical, RF, and fiber links

The purpose of a communication network comprising nodes and links is to carry traffic, described by a traffic matrix, from the source nodes in $\boldsymbol{I}$ to the destination nodes in $\boldsymbol{E}$. Traffic of a specific class $q\epsilon\boldsymbol{Q}$ that needs to be transported from node s to node d is represented by $T(s,d,q)$, such that $\sum_{q\epsilon Q} T(s,d,q) = \boldsymbol{T}(s,d)$ and the total traffic $\boldsymbol{T}^{Total} = \sum_{s\epsilon I, d\epsilon E} \boldsymbol{T}(s,d)$. A specific traffic class between a source-destination pair can be divided into multiple subflows $t_{ij}^{\tau}(s,d,q)$ for each link of a specific type to best utilize the network resources and optimize a specific performance objective under various constraints as defined below.

Flow-Constraint

$$\sum_{j\epsilon N} t_{ij}^{\tau}(s,d,q) - \sum_{j\epsilon N} t_{ji}^{t}(s,d,q) = \begin{cases} T(s,d,q), & i = s \\ -T(s,d,q), & i = d \\ 0 & \end{cases}$$

$$\forall\ i\epsilon\boldsymbol{N},\ s\epsilon\boldsymbol{I},\ d\epsilon\boldsymbol{E},\ \tau\epsilon\{f,o,r\}$$

The aggregate traffic $U_{ij}^{t} = \sum_{q\epsilon Q} u_{ij}^{t}(q)$ over a link of type τ from node i to node j is constrained because of the link (c_{ij}^{τ}) and node (C_i) capacities as follows:

Link Capacity Constraint

$$U_{ij}^{\tau} = \sum_{q\epsilon Q} \sum_{s\epsilon I, d\epsilon E} t_{ij}^{\tau}(s,d,q) \leq c_{ij}^{\tau} \quad \forall\, i,j\epsilon\boldsymbol{N}, \tau\epsilon\{f,o,r\}$$

Node Capacity Constraint

$$\sum_{j\epsilon N} \sum_{q\epsilon Q} u_{ij}^{\tau}(q) + \sum_{j\epsilon N} \sum_{q\epsilon Q} u_{ji}^{\tau}(q) \leq C_i \quad \forall\ i\epsilon\boldsymbol{N}, \tau\epsilon\{f,o,r\}$$

Non-Negative Constraint

$$t_{ij}^{\tau}(s,d,q) \geq 0, u_{ij}^{\tau}(q) \geq 0, c_{ij}^{\tau} \geq 0, C_i \geq 0 \ \forall\ i,j\epsilon\boldsymbol{N}, s\epsilon\boldsymbol{I}, d\epsilon\boldsymbol{E}, \tau\epsilon\{f,o,r\}$$

A minimum bound for both capacity and/or usage of a link can also be provided to force a certain amount of traffic to flow on that link. The network layer decision-making problem can be summarized as the determination of the individual source-destination subflows of a specific QoS type $t_{ij}^{\tau}(s,d,q)$ for each link $l_{ij}^{\tau}\epsilon\ \boldsymbol{L}$, subject to multiple linear algebraic constraints and with a specific objective; for example, reducing the overall cost that is a function of current usage $u_{ij}^{\tau}(q)$ of a traffic class q carried over respective links and nodes. The total operational cost depends on link and node unit costs, multiplied with respective usage and aggregated over all QoS types, link types, and nodes as described by the following linear algebraic formulation.

Minimize Operational Cost

$$\sum_{i,j\epsilon N} \sum_{\tau\epsilon\{f,o,r\}} \sum_{q\epsilon Q} \gamma^{\tau}(q).u_{ij}^{\tau}(q) + \sum_{i\epsilon N} \left(\sum_{j\epsilon N} \sum_{\tau\epsilon\{f,o,r\}} \sum_{q\epsilon Q} \Gamma^{\tau} u_{ij}^{\tau}(q) + \sum_{j\epsilon N} \sum_{\tau\epsilon\{f,o,r\}} \sum_{q\epsilon Q} \Gamma^{\tau} u_{ji}^{\tau}(q) \right)$$

Here, $\gamma^{\tau}(q)$ is the unit cost of processing traffic class q link over a link of type τ, and $\Gamma^{\tau}(q)$ is the unit cost at a node for processing outgoing or incoming traffic over a link of type τ and QoS q. This optimization problem, with a linear algebraic objective function, available link, node capacity bounds, and other linear algebraic constraints, can be transformed into an efficient linear programming representation where several techniques (e.g., SIMPLEX) and software algorithms are widely available for efficient implementation.

During the system design (or enhancement) phase, fixed system deployment costs, which are typically proportional to capacities of links and nodes, can also be analyzed. Similar to operational cost optimization, deployment costs can be minimized with the following objective function where $\pi^{\tau}(q)$ and $\Pi^{\tau}(q)$ are the unit prices for acquiring and deploying link and node processing capabilities, respectively. In the following objective function representation, the first term aggregates all costs related to links of various types, and the second term aggregates all processing costs at the various nodes required to accommodate both incoming and outgoing links of various types.

$$\text{Minimize Deployment Cost} \quad \sum_{i,j\epsilon N}\sum_{\tau\epsilon\{f,o,r\}}\sum_{q\epsilon Q}\pi^{\tau}(q).c_{ij}^{\tau}(q) + \sum_{i\epsilon N}\left(\sum_{j\epsilon N}\sum_{\tau\epsilon\{f,o,r\}}\sum_{q\epsilon Q}\Pi^{\tau}(q)c_{ij}^{\tau}(q) + \sum_{j\epsilon N}\sum_{\tau\epsilon\{f,o,r\}}\sum_{q\epsilon Q}\Pi^{\tau}(q)c_{ji}^{\tau}(q)\right)$$

System design is likely to be an iterative process where cost minimization and performance maximization are traded repetitively to arrive at the final values of individual link and node capacities that are then used during operations. In general, system performance can be optimized during original development, operation, and capacity enhancement phases. New or enhanced equipment (implementing software and hardware functions) are carefully selected to support a range of operational performance. As the traffic grows, extra equipment may need to be added to enhance this operational range. During operations, system performance can be tuned to meet instantaneous traffic demands, while minimizing operational costs that depend on usage levels of the individual links and nodes within the capacity bounds determined during system design. For example, operational expense related to third-party fiber links can be reduced by carefully minimizing the number of wavelengths in use, or reduced RF power for a GSL could help with the reduction of on-the-spot purchase of RF spectrum costs in a competitive environment.

B. System Design Phase

During system design, the RF and optical link, and ground network node capacities, locations, and terrestrial connectivity over fiber links are determined. This identifies all nodes and links needed to create a baseline topology associated with the network model. Note that not every node in the network will have links of all types, or to every other node in the network. Typically, a handful of ground nodes (gateways) will be connected to the space nodes (satellites) using optical and/or RF links. All other ground nodes will have sufficient terrestrial fiber connectivity with the neighbors for normal operations and for handling terrestrial equipment and link failures. In general, the network will correspond to a sparse graph (with nodes as vertices) with enough links so that full single or multi-hop connectivity across ingress and egress nodes is preserved even with link and/or node failures. For optical links to a space node, overall availability is acceptable with suitably selected multiple (typically three) diversity gateway locations with no or minimal weather-related correlation. The expected maximum traffic of various classes with respect to the defined ingress and egress nodes of the network can be used to populate the traffic matrix $T(s,d,q)$. Thus, system design can establish the baseline values for various parameters for the network $\langle \boldsymbol{N}, \boldsymbol{L}\rangle$, as introduced in pervious section, that can then be used for optimal decision-making for diversity path determination. The *minimize deployment cost* objective function can be used to determine capacities of all links and nodes in support of (maximum) traffic matrix $T(s,d,q)$. Each link can be analyzed with respect to its location, traffic requirements, and satellite payload design/capacity to determine the selection of antenna and power amplifiers to support link capacity $c(l_{ij}^{t})$ as a function of modulation, coding, and power. These link capacity functions can then be used to determine the total capacity of the system at different operating points, corresponding to the minimum and maximum capacities $c^{Max}(l_{ij}^{\tau})$, and $c^{min}(l_{ij}^{\tau})$ aggregated across all traffic QoS classes. These link capacity bounds are then used during operation to determine optimal link usage levels to carry the time-dependent traffic.

For RF (especially Ka-band and below), this system design approach has worked well since well-designed feeder links have enough link power margins to ride through most of the weather (rain)-related attenuation. For optical links, however, the network must have more diversity so that traffic can be routed quickly to secondary and, if needed, tertiary backup links to address single and double failures. Accordingly, the network needs to identify and provision for additional link and node capacity margins to deal with frequent optical link fades. These capacity margins can be determined after the traditional system design where link usage $u_{ij}^{\tau}(q)$ has been identified for the baseline traffic matrix $T(s,d,q)$. This corresponds to the objective of reducing total operational cost for the anticipated traffic under normal conditions. Only optical link failures, susceptible to scintillation, typically need to be considered for these margins required for double failures (otherwise only single node and link failures require contingency plans). The individual capacities c_{ij}^{o} each optical link l_{ij}^{o} are sorted, and two links from the top are selected for the following two cases: (1) space node, say m, is the destination, and (2) space node w is the source. Each link in our model is unidirectional, and both uplink and downlink using optical transport are likely to fade in a correlated fashion. Since one or both highest bidirectional, high-capacity links can essentially become nonfunctional, the combined capacity $c_{km}^{o} + c_{lm}^{o}$ of these two links to a space node (for the first case) needs to be distributed across all other links (both RF and optical), that terminate in the selected space node proportional to their original values.

Add Uplink Capacity Margin $$c_{im}^{\tau^U} \leftarrow c_{im}^{\tau} + c_{km}^{o}\left(\frac{c_{im}^{\tau}}{\sum_{p\epsilon N, p\neq k, p\neq l} c_{pm}^{\tau}}\right) + c_{lm}^{o}\left(\frac{c_{im}^{\tau}}{\sum_{p\epsilon N, p\neq k, p\neq l} c_{pm}^{\tau}}\right)\ \tau\epsilon\{o,r\}, \forall i\epsilon \boldsymbol{N}$$ *Step 1*

Similarly, the downlink failure case for the two highest capacity optical links from a specific source space node, say w, are handled by selecting the two highest capacity links and distributing their cumulative capacity as a margin for all other RF and optical links originating from the same source node. The updated link capacities for both uplinks are then used to update the node capacities that provide lower bounds for the links and nodes with margins.

Add Downlink Capacity Margin $$c_{wj}^{\tau^D} \leftarrow c_{wj}^{\tau} + c_{wx}^{o}\left(\frac{c_{wj}^{\tau}}{\sum_{p\epsilon N, p\neq x, p\neq z} c_{wp}^{\tau}}\right) + c_{wz}^{o}\left(\frac{c_{wj}^{\tau}}{\sum_{p\epsilon N, p\neq z, p\neq x} c_{wp}^{\tau}}\right)\ \tau\epsilon\{o,r\}, \forall j\epsilon \boldsymbol{N}$$ *Step 2*

Add Node Margin $$C_i^{UD} \leftarrow C_i + \left(\sum_{p\epsilon N}\left(c_{pm}^{\tau^U} - c_{pm}^{\tau}\right) + \sum_{p\epsilon N}\left(c_{wp}^{\tau^D} - c_{wp}^{\tau}\right)\right), \tau\epsilon\{o,r\}, \forall\, i\epsilon \boldsymbol{N}$$ *Step 3*

These minimum capacity constraints are then used to revise nodes and link capacities on traffic paths that are needed to support these extra margins during failures (obtained by running the linear programming problem with these additional minimum capacity constraints).

C. Operational Phase

The operational phase uses the node and link capacity boundaries provided by the preceding system design phase, where deployment costs are minimized and nominal values of link and node usage, normal case capacities, and extra margins to deal with optical link failures are determined. Resultant link and node capacities are then used during operations to estimate link and node usage that minimizes the overall operational costs for a specific traffic matrix bounded by the maximum values used during system design. Depending on the temporal variability of the traffic matrix, the *minimize-operational-cost* can be solved during appropriate management timelines, typically running from minutes to hours for a fast-changing traffic matrix. For a more static network, even a daily optimization is likely to provide a reasonable operational adaptation in link usage that minimizes run-time costs associated with third-party terrestrial link payments, RF spectrum payments (for dynamic allocation), and power consumption in the various network nodes. Each management cycle begins with the traffic matrix for the next cycle, based on historical network performance data, service-level agreements with the customers, and business decisions regarding additional margins for better QoS. The linear programming optimization problem is then solved within the management timeline to ensure that all expected usage values for each link and node have been determined and can be used to update the configuration of each node in advance before the next management cycle begins. The operation continues with the baseline configuration for the duration of the next management cycle. After the determination of the baseline network configuration, contingency plans are created to address random failure of optical links. With precomputation, baseline plans are ready for network configuration prior to the cycle start. These proactive contingency plans, one each for a potential link failure, are also available for rapid reconfiguration of the network when an optical link fails.

4. Implementation Considerations

Today, the bulk of the Internet is implemented as a collection of packet processing nodes, which handle addressing, routing, and switchover using distributed control plane protocols. Most of the links are fiber-based and typically with static configuration, provisioned for the worst-case normal and failure scenarios. Open Shortest Path First (OSPF) is the dominant intraorganization protocol that uses Dijkstra's algorithm to determine the shortest (but not necessarily optimal) path between every node-pair using the link state information provided by each node that is flooded in the network. Using incremental versions to quickly deal with minor network edge and node changes, distributed OSPF route determination can converge in a few seconds for small networks. However, OSPF is typically configured for slower convergence taking several minutes to avoid route flapping and to increase network stability. Border Gateway Protocol (BGP) is the dominant interorganization (typically connecting Autonomous Systems [ASs]) protocol for sharing routing information and changes resulting due to link or node failures. Though rich in describing various kinds of policies for traffic ingress, egress, and associated priorities, BGP is even slower to converge and can take several minutes using widely practiced configurations. Distributed implementation of OSPF- and BGP-based packet routing is well aligned with the original goal of keeping the Internet architecture less dependent on centralized functions. Either IP protocols is not suitable for next-generation satellite GSLs because of the slow convergence timelines inherent with distributed decision making and protocol design aimed at infrequent failures. Even on the terrestrial network side, many organizations are moving away from a distributed IP control plane to the centralized SDN approach. SDN decouples control and data planes, and a centralized SDN orchestration function directly controls the switching fabric of the network nodes. The SDN controller optimizes network performance (e.g., routing for better resource utilization) based on link and node status information sent by each node to the controller. OF groups in each switch are used to perform more complex operations on packets not possible in a flow alone. Each group receives packets as input and performs any OF actions on these packets. A group contains lists of actions, called OF buckets, that can be applied to entering packets. Multiple group types can be specified including SELECT for load balancing and FAST-FAILOVER (with watch Port and/or Group parameters) for rapid recovery. Starting with Layer 2 Ethernet switching devices and OF and OF-Config protocols between the switches and the controller (and related management functions), now the SDN framework has been expanded to include optical links and nodes [5] as well. Cloud-based centralized decision making can be more responsive and can deterministically achieve strict optimization objectives. Optimality was earlier not even considered in distributed approaches since routing function implemented on every node had to run in a computationally starved environment.

The network model can be implemented in a (logically) centralized SDDN controller with execution timelines aligned with the various design and operational management phases. Recent SDN approaches have focused on using hybrid WDM-based circuit and packet switching for core parts of networks and keeping IP packet processing only for access networks. The core network is like the high-capacity terrestrial diversity network for next-generation satellite systems. Most of the packets flowing through the core network nodes are in transit from one access node to another access node and do not require full packet processing (and associated costs) at each core node. This approach reduces cost and enables deterministic and optimal rerouting to mitigate optical link fading. A centralized SDN-type approach with dynamically configurable hybrid WDM circuit and packet switching is needed to efficiently utilize network resources while managing costs. The SDDN controller can maintain E-BGP sessions with its peer ASs and configure the networking nodes in the diversity network so that they can get the advertised IP prefixes from the controller. Multi-Protocol Label Switching (MPLS) or L2 VLAN tags can be used to direct a packet flow to take a specific path within the network where every node is consistently configured with appropriate tagging to manage individual traffic subflows. Besides simpler handling of fewer optical circuits from rules and actions perspectives, the optical circuit switches (fiber, WDM, or OTN) are typically an order of magnitude lower in both acquisition cost and power consumption compared to their packet switch counterparts [6] and easily scale beyond 1 Tbps capacity per node.

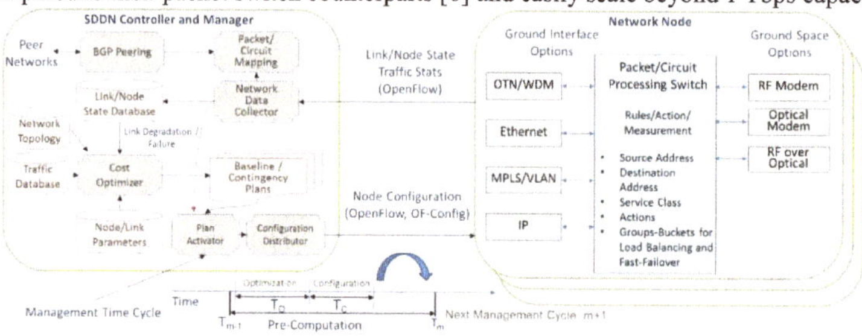

Figure 5. Centralized diversity-switchover implementation with centralized packet/circuit switch controller

A contingency plan for dealing with the failure of a link l_{ij}^{o} is determined by running *minimize-operational-cost* linear programming optimization problem for the baseline link and node capacities and $c_{ij}^{o} \leftarrow 0$. This ensures that all other nodes and backup links are suitably adjusted when the link l_{ij}^{o} under consideration has completed faded. For a dynamically varying traffic matrix (e.g., hourly), the management timeline will likely be in minutes and contingency plans for both first and second failures can be predetermined. This requires a family of contingency plans for each link and associated second link failures to be determined in advance. This is achieved by resetting the respective capacity of each secondary link to 0 while creating secondary all contingency plans for the specific primary link l_{ij}^{o}. This results in $O(|\boldsymbol{O}|^2)$ number of contingency plans where $|\boldsymbol{O}|$ is the size of the set comprising all optical links. All contingency plans for optical link failures can be computed in parallel, so the overall clock-time complexity for contingency plan determination scales with the number of optical link failures that are mitigated instead of the total number of optical links. So SDN-based diversity scheme can easily scale beyond the double optical link failures that is illustrated here. As mentioned earlier, three optical (bidirectional) links from geographically uncorrelated locations are typically needed to achieve high availability of optical GSLs to a space node. So, storage space requirements and computational complexity of contingency plan generation are expected to be manageable with the availability of scalable cloud-based computation facilities where an SDDN controller can be implemented.

It is possible to start moving traffic from a link that has started to fade to the backup link(s), by changing network configurations based on the contingency plan (already predetermined) for the affected link. The usage values for the various nodes and links from the baseline and the contingency plans are used to identify the nodes and links that would be reconfigured when a link l_{ij}^o completely fades. Instead of waiting for its complete failure, traffic from a partially failed link can be diverted by changing the configuration of the source node i of the link so that less traffic is sent to node j over the optical link l_{ij}^o which has started to fade. At the same time, all links and upstream nodes that are affected in the contingency plans for l_{ij}^o are reconfigured so that partial traffic can be diverted per the contingency plan for l_{ij}^o. To minimize frequent network reconfigurations, the changes can be made only at a small number of threshold values. Link capacity reduction can be determined by assessing channel condition based on actual packet loss rate or using link capacity models based on power, modulation, and coding scheme that are currently in use.

The chief differentiator of our SDN-based optical satellite network control is three fold: (1) rapid collection of link and node states (single message from a node goes to the controller vs. link state flooding required for OSPF), (2) fast determination of optimized network configuration (cloud-based implementation of controller vs. CPU limited distributed environment), and (3) rapid reconfiguration of all nodes and links by a single message from the controller to each node vs. depending on incremental information updates by sending information only to neighbors, such as in BGP. Recent tests have confirmed that optical device switchovers (e.g., WDM-based) can be conducted in 25 ms, which is faster than hand-crafted MPLS preprogrammed reroutes [7]. Without predetermination and precomputed contingency plans for rapid reconfiguration, on-demand path determination for rerouting disrupted flows and recovery of the affected path will take an order of magnitude more time. With the increasing role of optical technology in satellite design, WDM-type circuit switching will start becoming available on the space nodes especially when NGSO constellations launch optical ISLs and/or optical space ground links. Combined with optimal network route predetermination based on our scheme, fast packet rerouting over dynamically configured optical switches can minimize data loss over fast optical links operating at 10s or 100s of Gbps.

5. Conclusion

We have defined a unified network model for optimizing traffic route determination required for next-generation systems with hybrid optical and RF links in support of user communication, sensing, and relay application. This model complements the ongoing industrywide work in physical layer in devising modulation, coding, adaptive optics, and PAT techniques for narrow optical beams that are prone to significant fading because of scintillation. Unlike distributed routing and switchover schemes used for the terrestrial packet network, a centralized SDN-based decision making is more responsive and optimal in dealing with frequent weather-related configuration adjustments and switchovers. The model is represented as a linear programming optimization problem involving link and node capacities that need to be determined for minimum deployment (during system design) and operational costs. Procedural descriptions on the use of the model are provided to help with the future implementation within an SDDN controller and packet/circuit-switching nodes, as satellite networks start incorporating optical links. We are considering exploratory analysis of laboratory-based data related to the characterization of high-capacity optical links. This can be combined with the assessment of the ongoing studies of weather and atmospheric impact on optical propagation [8]. This will help confirm our basic assumptions related to the number of backup optical links needed to ensure very high overall availability of a GSL. We are identifying the next steps for refining the model and converting it into algorithmic representation to better analyze the requirements for computation and storage resources needed by the network controller for the next-generation systems with optical links.

References

1. Yash Vasavada, Rajeev Gopal, C. Ravishankar, Nassir BenAmmar, and Gaguk Zakaria, "Architectures for next generation high throughput satellite systems," International Journal of Satellite Communications and Networking, January 2016.
2. Rajeev Gopal and Nassir BenAmmar, "Framework for Unifying 5G and Next Generation Satellite Communications," IEEE Network, September/October 2018.
3. In Keun Sonb, Shiwen Maoa, "A survey of free space optical networks," Digital Communications and Networks, 2017.
4. Theodore D. Katsilieris, George P. Latsas, Hector E. Nistazakis, and George S. Tombras. "An Accurate Computational Tool for Performance Estimation of FSO Communication Links over Weak to Strong Atmospheric Turbulent Channels," Journal of Computation, 2017.
5. Akhilesh Thyagaturu et al., "Software Defined Optical Networks (SDONs): A Comprehensive Survey," IEEE Communications Surveys & Tutorials, Vol 18, No. 4, 2016.
6. Saurav Das, Guru Parulkar, and Nick McKeown, "Rethinking IP Core Networks," Vol. 5, No. 12, Journal of Optical Communication Networking, 2013.
7. Behzad Mirkhanzadeh, et al., "Demonstration of an SDN Orchestrator for Both Flow Provisioning and Fault Handling in an Ethernet-over-WDM Network," 19th International Conference on Transparent Optical Networks (ICTON), 2017.
8. Kenji Suzuki, et al., "Environmental Data Gathering System for Satellite-to-Ground Station Optical Communications," Trans. JSASS Aerospace Tech. Japan Vol. 16, No. 1, 2018.

Research and development approach to realize flexible optical ground network operations for effective data downlink from space to ground

Tatsuya Mukai[1], Yoshihisa Takayama[2], Tomohiro Araki[3]

[1]*Research Unit 1, Research and Development Directorate, Japan Aerospace Exploration Agency, 2-1-1, Sengen, Tsukuba, Ibaraki, 305-8505*

[2]*Department of Communication and Network Engineering, School of Information and Telecommunication Engineering, Tokai University, 2-3-23, Takanawa, Minato-ku, Tokyo, 108-8619*

[3] *Research Unit 1, Research and Development Directorate, Japan Aerospace Exploration Agency, 2-1-1, Sengen, Tsukuba, Ibaraki, 305-8505*

**email mukai.tatsuya@jaxa.jp*

Keywords: Space explorations, Optical direct communications, Cloud blocking and avoidance

Abstract

The application of optical communications technology for intersatellite and direct links is increasing in the fields of remote sensing and space explorations. Optical communication experiments have been conducted internationally and there are plans to use related technologies for space operations. However, there are not many experiments or plans for the network of optical ground stations focused on the particular issue of how to avoid cloud blocking in optical links between space and the ground for effective data downlink.

In 2015, JAXA reported on a future experimental plan for optical direct communication links from space to ground in order to acquire the fundamental technologies of spatial pointing and tracking with the International Space Station and the experimental concept for the avoidance of cloud blockings by optical ground stations, cloud sensors and network planning systems [3]. To forward the study on flexible optical ground network operations, JAXA began joint research with Tokai University in 2017 and has studied more specific experiment method with ground fiber networks, cloud sensors, network planning systems and meteorological satellite data.

This paper presents the research and development approach to realize flexible optical ground network operations for effective data downlink from space to ground.

1. Introduction

In the development history of space optical communications technologies, JAXA launched a LEO satellite called OICETS (**O**ptical **I**nter-orbit **C**ommunications **E**ngineering **T**est **S**atellite) in 2006 aboard a Dnepr rocket from the Baikonur Cosmodrome, in Kazakhstan, and optical direct communication experiments from space to ground were globally conducted in 2009 by JAXA, ESA, DLR, JPL and NICT after the success of the inter-satellite communications experiment between JAXA's OICETS satellite and ESA's data relay satellite called ARTEMIS (Advanced Relay and Technology Mission Satellite) [4]. Through these laser experiments, the operational possibilities of space optical communications in addition to existing RF communications technologies were recognized as communication services by space agencies, and the concept of operations for expected mission scenarios was summarized and published [14] by Optical Link Study Group (OLSG) of Interagency Operations Advisory Group (IOAG) from 2011 to 2012, following the Ground-to-OICETS laser Communications Experiments (GOLCE) workshop held on Tenerife island, Spain [15]. In addition, the CCSDS Green Book of real time weather and atmospheric characterization data was published [16].

In the optical direct communication experiments conducted from the OICETS satellite to optical ground stations (OGS), experimental schedules were coordinated, and sufficient experimental daily slots were secured in advance by a non-automatic method [5] in consideration of future weather conditions. When it came to unexpected cloudy weather compared with that for the predetermined experimental daily slots, rescheduling was basically necessary every time such weather conditions were encountered. In other words, the need for a solution to for cloud blocking between space and the ground was globally recognized for real operations.

Global site diversity has been studied as one of the basic solutions for cloud blocking in the OLSG, and a ground support tool for OGS selection under better weather conditions, called Lasercom Network Optimization Tool (LNOT), was introduced based on meteorological geostationary satellites (GOES) [6], and subsequently used for space optical direct communications from the LADEE Moon explorer and OGSs (in the USA and Europe) in 2014 [10].

It is generally known that at least three sites separated globally by 120 degrees on the earth are required for deep space RF communications to ensure effective contact with probes. Therefore, the following combinations of three geographically separated sites selected according to weather and atmospheric conditions for the RF band, are set for each space agency's deep space RF communications. Fig 1. shows a concept of DSN for the RF network covering three different regions with 120 degree separation.

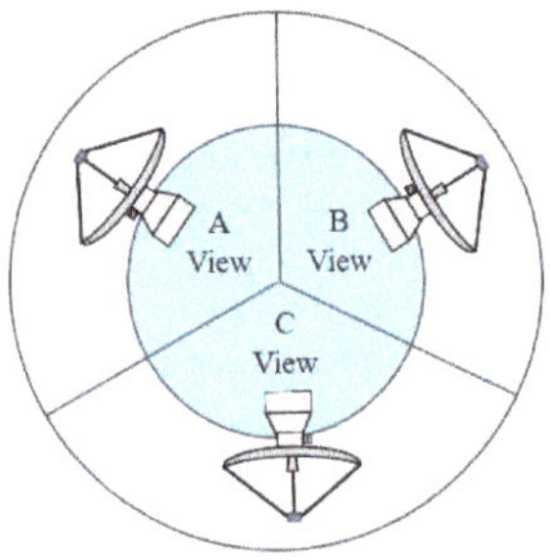

Fig 1. DSN for RF communications

1)NASA Deep Space Network
Goldstone (North America), Canberra (Eastern Australia) and Madrid (Spain)
2)ESA Deep Space Network
Argentina (South America), New Norcia (Western Australia) and Cebreros (Spain)
3)JAXA Deep Space Network
Usuda (Japan), Uchinoura (Japan) and other 2 ground stations of NASA or ESA Deep space network under operational cooperation

For space optical direct communications studied in the OLSG, several sites selected based on weather (cloud coverage) and atmospheric conditions were introduced [14]. However, there are only two regions for deep space optical communications (e.g., USA and Europe). This means there is no optical communication coverage between these two regions. Therefore, a 3rd optical ground station in Asia-pacific region is needed to form a global optical communication network. Fig 2. shows the concept of DSN for the optical communications network to be formed by three different regions, based on the trends in cloud coverage and atmospheric conditions impinging on laser links between the probe and optical ground station.

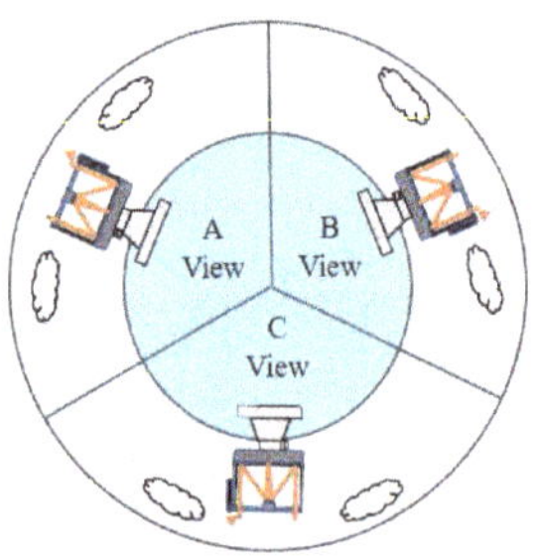

Fig 2. DSN for optical communications

Assuming that it would be possible to form the network globally just as for deep space RF communications, optical global coverage would be established. However, only the site diversity technique used to set optical ground stations at geographically separated locations with uncorrelated cloud coverages is insufficient for a complete solution to cloud blocking. And as mentioned in reference to the Lasercom Network Optimization Tool (LNOT) using the meteorological data of geostationary Satellites (GOES) [6], a method of selecting OGSs based on cloud coverage conditions in a given network is also required. [7], [8], [9], [10]. This kind of method requires not only the meteorological data taken from space, but also the ground-based cloud monitoring data specifically dedicated to given OGS locations, and entails handling the automatic network switching control system with all cloud coverage data, and making decisions on OGS selection based on a calculation of the expected volume of downlink data.

Section 2 introduces the method of network control with functions for cloud blocking avoidance over the OGS network. First, an overview of the network is given. Then the functions of each node are described. And finally, the operation sequence for the avoidance of cloud blocking is presented. This method is under study and the system is partially developed for experimental evaluation and validation purposes.

2. Methodology

2.1 Optical ground network system

Unlike the existing RF ground station network, the optical ground stations network will need a function for cloud monitoring and determining the percentage of cloud coverage overlaid on a given probe orbit, as viewed from the given OGS locations. This function entails two separate roles: monitoring both lower clouds and higher clouds that could possibly block a given link from a probe to the optical ground stations. A local cloud monitoring system could serve the role of monitoring lower clouds, while the full disc images from a geostationary meteorological satellite, in addition to this cloud monitoring and determination system, could be used to monitor higher clouds. Moreover, a network control function will be needed to switch the ground stations connections based on a given situation of cloud coverage during operation. Fig 3. shows a system overview of the optical ground network system including these functional nodes.

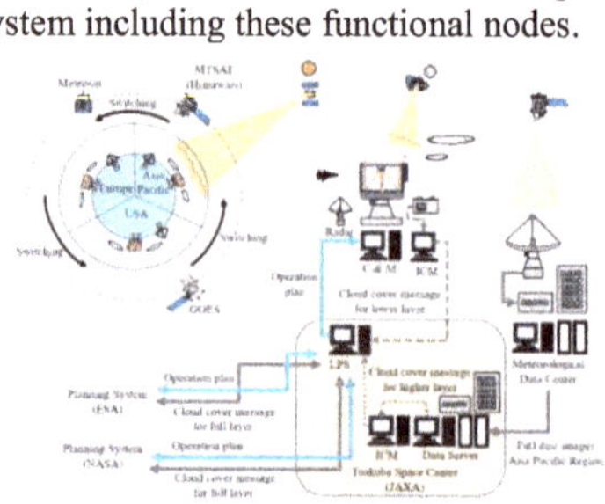

Fig 3. System overview of optical ground network system

2.2 Cloud monitoring and determination method

In order to avoid the cloud blockings in a given link between a probe and an optical ground station, it is necessary to consider the following vertical atmospheric structure shown in Fig 4. The cloud types vary according to height and are categorized according to three different heights, where lower clouds, middle clouds and high clouds exist. When any such clouds cover a given link, the optical communication link will be blocked, possibly leading to an interruption of operations. Therefore, it is necessary to monitor clouds and determine the cloud cover that may block the link between the probe and the OGS, and switch the ground communication link among other optical ground stations with less expected cloud coverage for the avoidance of cloud blocking.

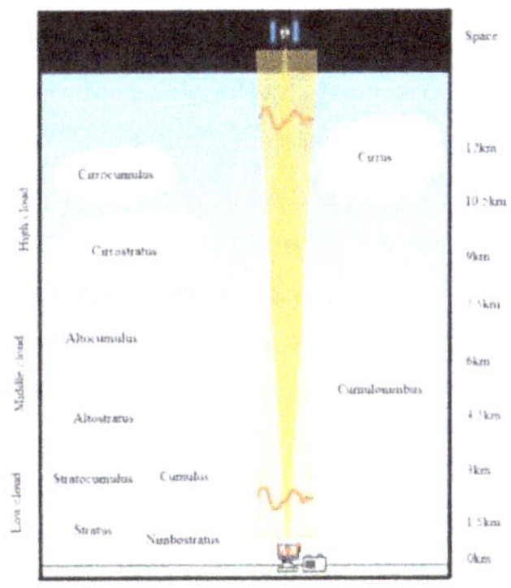

Fig 4. Cloud types and heights in a given path

Fig 5. shows the infrared cloud monitoring (ICM) system for local cloud monitoring, which is composed of visible and infrared sensors connected to laser planning systems (LPSs) and transfers cloud coverage information only as an operation message in a defined transfer cycle, thereby minimizing the volume of data. Other cloud monitoring systems have been introduced [2], [12], [13], [17], but none are equipped with this kind of analysis function. One prototype model was developed by JAXA through field observation testing during night at different locations with different brightness and will be more improved with these functions.

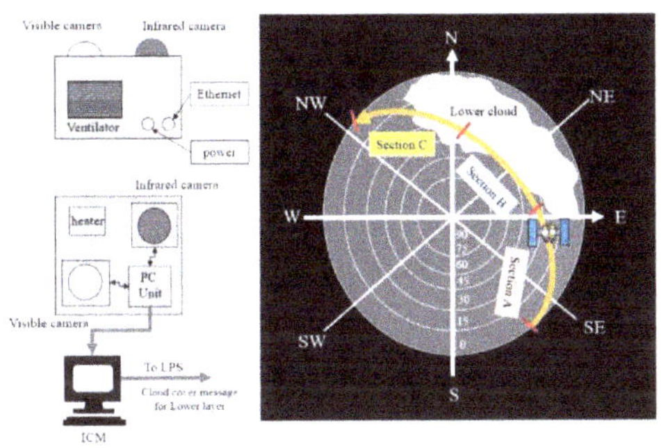

Fig 5. ICM system for lower clouds

This system takes images of lower clouds mainly at night using an infrared sensor with a fish eye lens and overlays a given orbital arc of the targeted probe on them, and the analyzes the cloud coverage of three separated sections on the orbital arc to create a cloud cover message. As shown in Fig 5., the percentage of lower cloud coverage on the given orbital arc is analyzed on three different sections. The reason for setting three different sections on the orbital arc, is for the distinction of the order of event occurrence on a given time axis to the probe flight direction. If this distinction would not be set on the orbital arc, the order of occurrence of cloud blocking would be unknown and recommended OGS selection would be failed. It is necessary to know the relation between the probe's position on the arc and the cloud coverage position in these three different sections, because it will be possible to know the order of event occurrence. The visible sensor is attached for future operational extension.

Fig 6. shows the infrared cloud monitoring (ICM) system for higher clouds. This system monitors and determines the cloud coverage at higher altitudes, where the infrared cloud monitoring system for lower clouds can not monitor, due to sensors sensitivity. The full disk images are captured by a geostationary meteorological satellite called MTSAT (Himawari), mainly using the images of clouds on the IR bands at night and the probe's orbital arc will be overlaid for an analysis of cloud coverage on the arc. To fit the cloud coverage in a small spatial area analyzed by the cloud monitoring system for lower clouds and that in a larger spatial area as done by this system for higher clouds, the cloud monitoring area from the ground will be depicted there and the same concept of three different sections will also be set for the prediction of direction, speed, distance, and arrival time of incoming clouds into the arc in a larger area. This can be analyzed by defined distance according to each pixel pitch size and the update cycle of the satellite images.

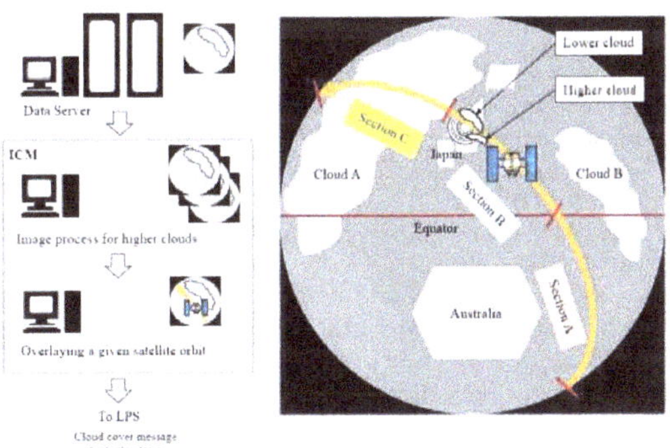

Fig 6. ICM system for higher clouds

For combining 2 types of images for lower and higher clouds, it will be possible to know the whole cloud coverage in a given link and the network switching timing will be determined for a targeted OGS with less cloud coverage. The method to know whole cloud coverage in a given link is based on the combination of 2 type of images and overlaying process. There are research papers on analysis of cloud coverage images taken on the IR bands by meteorological satellites [1], [6], [7], [8], [9], [10], [11], and it is also under study by JAXA. Fig 7. shows the relationship between two different types of images taken by separate monitoring systems on the ground and in space.

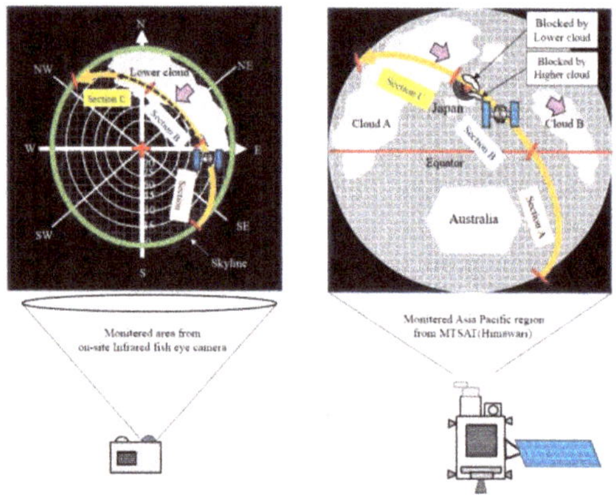

Fig 7. Relationship between two types of images

The infrared cloud sensor has IR sensitivity up to -40 degrees Celcius and it was confirmed that it can sense the lower clouds based on the field testing at night. From the field of view (FOV) and the altitude where it can sense clouds, the diameter of monitoring area can be known and also the distance corresponding to the pixel pitch size and number of pixels on vertical and horizontal directions. For the meteorological satellite called MTSAT (Himawari), the pixel pitch size and the number of pixels on vertical and horizontal directions on the full disk images are determined. One pixel corresponds to 2 km and transfer cycle is once per 10 minutes.

For both information, direction, speed, distance of cloud and arrival time to the monitoring area and the probe orbital arc can be calculated, and it is possible to predict the impact on communication link between space and ground. This type of prediction of cloud coverage is also under study by JAXA.

2.3 Optical network planning method

Only cloud monitoring and its determination can not avoid the cloud blocking over optical ground stations. Fig 8. shows the optical ground network systems operation as divided into three different optical network planning systems, which are laser planning systems (LPS) A, B and C that gather cloud cover messages from each cloud monitoring system for lower and higher clouds. The three coloured bar graphs denote accumulated services at each optical ground station. This assumes that one probe's operation needs three optical ground stations in a first operation pass, where one LPS will serve as the master node and the other two will serve as supporting nodes. These roles are switched among the three different LPS nodes according to the probe operated by space agencies. This is based on the mutual cross support as is the case in the existing DSN RF operations.

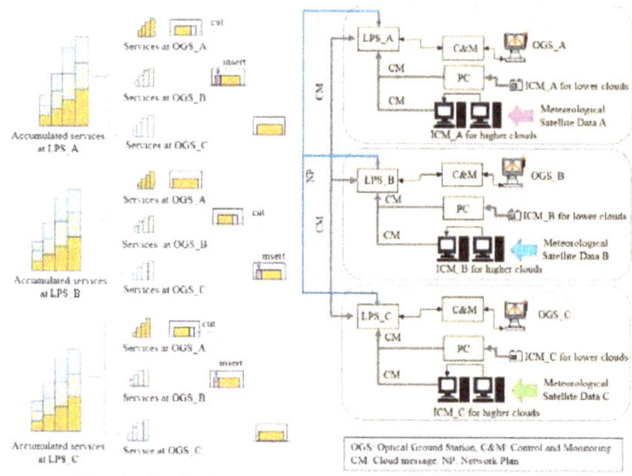

Fig 8. Optical ground network systems operation

Fig 9. shows the operations management of the laser planning system (LPS), with scheduling management at the top and station management at the bottom, and entailing the management of cloud coverage and performance results.

This prototype system was developed and tested by us. For the next step, the design of graphics on the screen and each function will be improved.

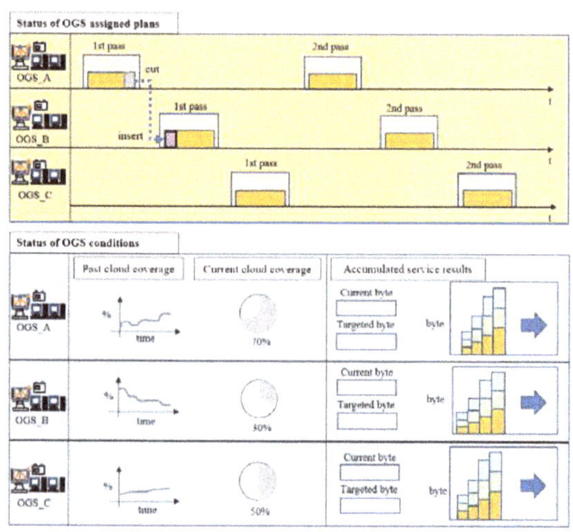

Fig 9. Operations management of the LPS

Especially, the graphs of past and current cloud coverage status will be based on the cloud cover message from each infrared cloud monitoring system connected to each LPS and accumulated service results at each OGS will be shown as graphs and indicate current data byte and targeted byte in downlink. This system is designed for future implementation with the existing RF ground system for TT & C.

2.4 Optical network control method

The LPS serves the roles of not only the managing schedules, cloud coverage and performance results, but also the control of ground network connections among three optical ground stations. Fig 10. shows the control process, where the expected time slot by cloud blocking is cut and inserted by an automatic process through network communications among the LPS nodes. The blocked pass period needs to be inserted into the first time slot of the second ground station, due to sequential data numbering of the data recorder on the probe.

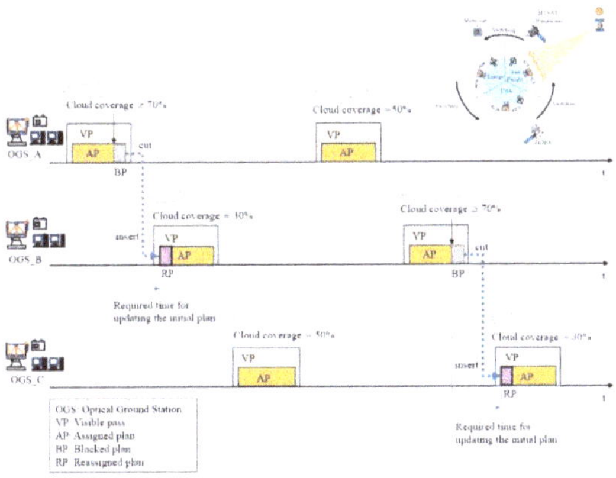

Fig 10. Rescheduling for avoidance of cloud blocking

Fig 11. shows the nominal sequence of operations, where each operation pass is set as the initial plan. On left side, cloud cover message is transferred by the defined cycle to each LPS node and then each LPS node will create the initial operation plan and finally they will be set at each OGS node on right side.

For deep space network operation, the master LPS node and the other supporting LPS nodes will be determined according to the probe belonging to each space agency. In Fig 11., LPS_A node is a master node and the other two nodes are supporting nodes. In this way, they exchange cloud cover messages and network operations plans.

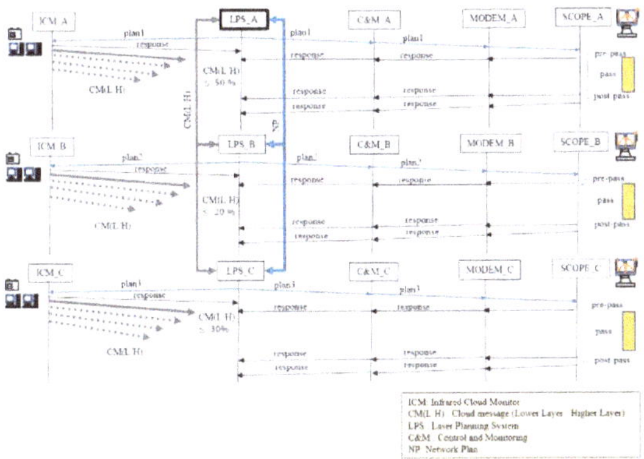

Fig 11. Nominal sequence of operations

Fig 12. shows the sequence when a network switch occurs, due to cloud coverage reaching to the defined percentage of cloud coverage. It is important to secure sufficient the process time for updating plans and changing network connections before the pre-pass time of the targeted optical ground station. This time slot will be larger as each OGS distance becomes larger.

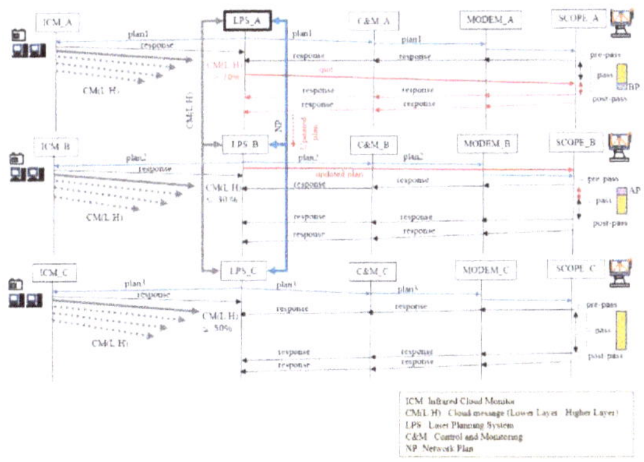

Fig 12. Sequence of operations for cloud avoidance

As the method of cloud monitoring, scheduling and control of OGSs, it is basically necessary to consider each agency's existing operation environment and conditions, because each agency knows its own ground network system for implementation of new ground network system. The idea presented here in this paper is only to exchange network planning files and cloud cover messages on mutually connected ground fiber networks to realize a global optical ground network for DSN optical communications focused on high speed and effective data downlink from each probe.

3. Results

The research and development approach as a methodology for network control with avoidance functions for cloud coverage over the OGS network was explained along with the network overview, functions of each node, and sequence of operations for the avoidance of cloud coverage.

In particular, it was explained that the relationship, message and control sequence between the LPS and each node of the infrared cloud monitoring (ICM) system located at each OGSs site and the infrared cloud monitoring (ICM) system connected to the data server of the meteorological satellite (Himawari), are important for decision making regarding flexible network control in selecting one of the OGSs having a better Cloud Free Line of Sight (CFLOS) to downlink the data from the relevant probe.

This means, that whatever good cloud monitoring systems are implemented, final decision-making is quickly needed and network control must be executed based on the volume of cloud coverage over relevant probe's orbital arc as viewed from the both types of infrared cloud monitoring (ICM) system, and the basic prediction of increased cloud coverage over each OGS must be made with larger image taken from a meteorological geostationary satellite.

The local infrared cloud monitoring system for each OGS is constrained regarding the prediction of the impact on incoming clouds entering the probe's orbital track, due to the smaller field of view than that taken by the meteorological geostationary satellite, and the monitoring limitation of cloud height for lower layer cloud relating to the thermal sensitivity of the infrared cloud monitoring sensor, and the meteorological satellite data to be used for the compensation of the higher regions with medium and higher layer clouds that the infrared cloud monitoring (ICM) systems can not monitor well, given the sensor's sensing capability.

In terms of network control, it was also explained that it is important to secure as much time as possible to complete the network switching before the starting time of the pre-pass process given to each OGS, which differs from such missions as LEO direct link, inter-satellite link and deep space direct links. To secure as much time as possible before the pre-pass starting time, the separation distance among each OGS must be set larger than the general cloud spread length to create an uncorrelated relationship of cloud coverage with a given ground network.

In order to test, evaluate and validate the network control function with cloud monitoring and network switching, it is considered that at least three infrared cloud monitoring (ICM) systems will be set at three different locations in Japan (e.g., Kumamoto, Tokyo and Sapporo) separated by around 1000 km apart from each other and they will be selected from campuses of Tokai University working in cooperation with JAXA for this research theme. Moreover, the laser planning system (LPS) will be set in a given research room on the Takanawa (Tokyo) campus of Tokai University, where a preliminary operation pass will be created and set to the LPS node and simulated OGSs. Then, real-time cloud monitoring will begin, and those programmed network control sequences will run based on messages coming from each infrared cloud monitoring (ICM) system, including the infrared cloud monitoring ICM) system related to the meteorological geostationary satellite.

This is our proposed methodology for the operation model having ground network control with an avoidance function for cloud blocking, with which we can first approach this kind of laser ground system from our operations standpoint to a possible framework of implementing this ground network system in the existing RF operation system.

4. Conclusion

This method is under study and the system is partially developed. The entire system for this research and development of a ground network system with an avoidance function for cloud blocking will be shortly established, and then this method for effective data downlink with an avoidance function for cloud blocking will be transferred to JAXA's optical ground stations network in the future. It is also expected to be applied to operations scenarios where optical direct communications, satellite laser ranging, and passive optical observations of space debris will be interrupted by cloud coverage over each telescope's field of views.

This advanced kind of ground system has been used and tested for optical direct communications from the LADEE moon orbiter to OGSs (e.g., Table Mountain, California, White sands, New Mexico and Tenerife Spain) by NASA/JPL and ESA in 2014, and a supporting ground system called the Lasercom Network Optimization Tool (LNOT) was used for the best selection of OGSs based on the US and European meteorological satellite data [10], However, there is still a loss of communication coverage in the Asia-Pacific region that inhibits achieving full communication coverage for forming an global optical deep space network. Therefore, this approach for research and development is also expected to make contributions toward working together with other space agencies (e.g., NASA/JPL and ESA), in order to finally form an internationally connected optical deep space ground network that is the same as the existing RF deep space ground network.

Each agency has its own existing network and many technical points of views on network constraints and operation requirements toward avoiding a drastic network and operation structure and complex sequences. These are not directly related to the research, but such conditions must be fully considered so as to smoothly transfer the developed and validated system for actual operations. This kind of paper regarding the method of effective data downlink from space to ground by avoiding cloud blocking was first presented by JAXA (Japan), and it is understood that this is just our first step in seeking a solution to this issue of cloud blocking on a given network, However, this work will lead to future collaborative work with NASA/JPL and ESA toward achieving the same objectives.

Acknowledgements

The authors are very grateful to domestic companies for their cooperation through positive discussions about this ground system study and on-going development, the scientists of the National Astronomical Observatory of Japan (NAOJ) for their valuable advices on ground-based infrared cloud monitoring based on experiences with the MAGUNUM robotic telescope at Haleakala, Hawaii, USA and also the scientists of JAXA's Earth Observation Research Centre (EORC) for their advices on satellite-based cloud data retrieved from Japanese meteorological satellites (Himawari).

References

Journal articles

[1] Hironobu IWABUCHI,et al.,:'Cloud Property Retrieval from Multiband Infrared Measurements by Himawari-8', Center for Atmospheric and Oceanic Studies, Graduate School of Science, Tohoku University, Sendai, Japan, Journal of the Meteorological Society of Japan, Vol. 96B, pp. 27–42, DOI:10.2151/jmsj.2018-001, 2018 Special Issue on Meteorology and Climate Change Studies by Using the Geostationary Meteorological Satellite Himawari-8.

[2] MASAHIRO SUGANUMA, YUKIYASU KOBAYASHI, et al.,:'The Infrared Cloud Monitor for the MAGNUM Robotic Telescope at Haleakala', Publications of the Astronomical Society of the Pacific, 119: 567–582, 2007 May, The Astronomical Society of the Pacific, Volume 119, Number 855, DOI: 10.1086/518697.

Conference papers

[3] Tatsuya Mukai, Shinichi Inagawa, Kiyohisa Suzuki,: 'A study of free space laser communication experiment on the ISS Japanese experiment module for space explorations', Research Unit 1, Research and Development Directorate, Space Tracking and Communications Center, Japan Aerospace Exploration Agency, 2-1-1 Sengen, Tsukuba-shi, Ibaraki 305-8505, Japan, 26-28 October 2015, 2015 IEEE International Conference on Space Optical Systems and Applications (ICSOS), New Orleans, LA, USA, DOI:10.1109 /ICSOS.2015.7425083.

[4] Yoshihisa Takayama, Morio Toyoshima, et al.,:'Expanded laser communications demonstrations with OICETS and ground stations', National Institute of Information and Communications Technology (NICT), 4-2-1 Nukui-Kitamachi, Koganei, Tokyo 184-8795, Japan, SPIE LASE, 2010, San Francisco, California, United States, Proc. SPIE 7587, Free-Space laser communication Technologies XXII, 75870D (26 February 2010); doi:10.1117112.845991

[5] Nobuhiro Kura, Takashi Jono (JAXA), et al.,:'OICETS Mission Planning and Scheduling System 2D16', Proceedings of the 50th Space Sciences and Technology Conference, Kokura, Japan.

[6] Robert Link, Mary Ellen Craddock, Randall J. Alliss,: 'Mitigating the Impact of Clouds on Optical Communications', Northrop Grumman Information Technology TASC, 4801 Stonecroft Blvd. Chantilly, VA 20151, (703) 633-8300, 2005 IEEE Aerospace Conference, DOI:10.1109/AERO.2005. 1559416.

[7] Gary S. Wojcik, Heather L. Szymczak, Randall J. Alliss, et al.,:'Deep-space to Ground Laser Communications in a Cloudy World', Atmospheric Effects Group, Information Technology Sector TASC/Northrop Grumman Corporation, 4801 Stonecroft Blvd. Chantilly, VA 20151, Optics and Photonics 2005, 2005 San Diego, California, United States, Proc. SPIE 5892, Free-Space Laser Communications V, 589203 (31 August 2005) doi:10.1117/12.615435.

[8] Randall Alliss, Robert Link, Duane Apling, et al.,: 'Introducing the Renewable Energy Network Optimization Tool (ReNOT): Part I.',Northrop Grumman Information Systems, 3975 Virginia Mallory Drive, Chantilly, VA, 91st American Meteorological Society Annual Meeting.

[9] Randall J. Alliss and Billy Felton,:'The mitigation of Atmospheric Impacts on Free Space Optical Communications', Northrop Grumman Corporation, 15010 Conference Center Dr, Chantilly, VA 20151, Proc. International Conference on Space Optical Systems and Applications (ICSOS) 2012, 3-4, Ajaccio, Corsica, France, October 9-12 (2012).

[10] Randall J. Alliss and Billy Felton,:'Realtime Atmospheric Decision Aids in Support of the Lunar Laser Communications Demonstration', Northrop Grumman Corporation, 7555 Colshire Drive, McLean, Virginia 22102, Proc. International Conference on Space Optical Systems and Applications (ICSOS) 2014, S2-3, Kobe, Japan, May 7-9 (2014).

[11] Hirotsugu Ninomiya, Yoshihisa Takayama, Hajime Fukuchi,:'Diversity Effects in Satellite-Ground Laser Communications using Satellite Images', Tokyo Metropolitan University, Hino, Tokyo, 191-0065, Japan, National Institute of Information and Communications Technology, Koganei, Tokyo, 184-0015, Japan, 29th AIAA International Communications Satellite Systems Conference (ICSSC-2011) 28 November-01 December 2011, Nara, Japan, AIAA 2011-8033.

[12] Kenji Suzuki, Hiroo Kunimori, Morio Toyoshima,: 'Report of the Environmental-data Statistical Processing for Satellite-to-Ground Stations Optical Communications', Wireless Network Laboratory, National Institute of Information and Communications Technology (NICT), IEICE (The Institute of Electronics, Information and Communication Engineers) Communications Society Conference 2015, B-3-9.

[13] Clément Bertin, Sylvain Cros, Laurent Saint-Antonin, Nicolas Schmutz,:'Prediction of optical communication link availability:real-time observation of cloud patterns using a ground-based thermal infrared camera', IRT Saint Exupéry, 118, route de Narbonne, 31423 Toulouse, France, Reuniwatt SAS, 14 rue de la Guadeloupe, 97490 St-Clotilde, Reunion Island, France, Optics in Atmospheric Propagation and Adaptive Systems XVIII, Proc. of SPIE Vol.9641, 96410A, doi:10.1117/12.2194920.

Reports

[14] Interagency Operations Advisory Group, Optical Link Study Group, 'Optical Link Study Group Final Report', 5 June 2012, IOAG.T.OLSG.2012.V1.

[15] NICT supported by ESA,'International Workshop on Ground-to-OICETS Laser Communications Experiments 2010 (GOLCE 2010)', 13th-15th May, 2010, Tenerife, Spain, ISSN2185-1484.

[16] CCSDS,'REAL-TIME WEATHER AND ATMOSPHERIC CHARACTERIZATION DATA', INFORMATIONAL REPORT, CCSDS 140.1-G-1, GREEN BOOK, May 2017.

[17] Paul W. Nugent, Joseph A. Shaw, and Sabino Piazzolla,: 'Infrared Cloud Imager Development for Atmospheric Optical Communication Characterization, and Measurements at the JPL Table Mountain Facility', IPN Progress Report 42-192, February 15, 2013.

Section 6 – VHF Data Exchange Systems

ON THE VHF RADIO CHANNEL FOR THE DATA EXCHANGE SYSTEM VIA SATELLITE (VDE-SAT); EXPERIMENTAL RESULTS FROM THE NORSAT-2 SATELLITE EXPERIMENT

Lars E. Bråten[1], Torkild Eriksen[1], Andreas Nordmo Skauen[1], Anders Bjørnevik[2], Hans Christian Haugli[3], Lars Løge[3]

[1] *Defence Systems Division, Norwegian Defence Research Establishment (FFI), Kjeller, Norway*
[2] *Kongsberg Seatex, Trondheim, Norway*
[3] *Space Norway, Oslo, Norway*

Keywords: MARITIME, SATCOM, VHF, MEASUREMENTS, DOWNLINK

Abstract

Measurements of the Very High Frequency (VHF) downlink from NorSat-2 has been analysed in the context of development of the waveforms for VHF Data Exchange via Satellite. Received carrier, noise and interference power were analysed and compared with theory and simulations. The received power levels are as expected for most of the time. Slowly varying fading is observed with peak-to-peak variations in the order of 20 dB and a standard deviation of 4.2 dB. The fading has a close to normal distribution (in dB) and is mainly caused by sea reflections and to a lesser degree shadowing by the ship structure. Although time periods with significant interference were observed, most of the time the noise floor can be classified as residential or city according to ITU-R Rec. 372 on Radio noise. The noise plus interference level measured did not represent white Gaussian noise, and it is beneficial if the waveform(s) is robust with respect to both narrowband and wideband noise. The received C/N_0 showed a dynamic range of about 25 dB, exceeding 34.7 dBHz for 95 per cent of time.

1. Introduction

This paper presents the results of the first test campaign for the proposed VHF Data Exchange via Satellite (VDE-SAT), carried out in the ESA project VDE-SAT Downlink Verification. VDE-SAT is the satellite component of the VHF Data Exchange System (VDES) [1] in the VHF maritime mobile band (156-162 MHz). Details and plans for the development of VDE-SAT can be found in [2] and [3]. The campaign involved conducting tests to experimentally study the performance of the proposed VDE-SAT waveforms, as well as contributing to the verification and design optimisation for VDE-SAT [4].

The international community, led by the International Telecommunication Union (ITU), the International Maritime Organization (IMO), and the International Association of Lighthouse Authorities (IALA) has decided to implement VDES as a future maritime communications system in accordance with the Radio Regulations for the maritime services. ITU Resolution 360 [5] from WRC-15 stated that a VDES satellite component is necessary to expand the system from coastal areas to a global coverage, and that studies to show that VDE-SAT will not cause harmful interference to incumbent services in the same and adjacent frequency bands need to be conducted. Likewise, it must be studied whether VDE-SAT is resilient to harmful interference from such systems. All members of the ITU Radiocommunication Sector (ITU-R) and other named organisations were invited to provide such results in time for WRC-19.

The Norwegian microsatellite NorSat-2 was launched 14th July 2017 into a sun-synchronous orbit with a height of 600 km. The main payload is an AIS receiver for tracking maritime traffic from space. NorSat-2 is also equipped with a VDE-SAT test transponder. NorSat-2 transmits a signal for validation of the VDE-SAT downlink waveform performance. The satellite is primarily financed by the Norwegian Coastal Administration and the Norwegian Space Centre. Other partners include Space Norway, Kongsberg Seatex and the Norwegian Defence Research Establishment (FFI). The test transmitter has been developed by Kongsberg Seatex and is owned by Space Norway. FFI is responsible for carrying out the VDE-SAT downlink experiment and analysis.

The VDE-SAT development is supported by international and national agencies, such as the European Space Agency (ESA), the Norwegian Maritime Authority and the Norwegian Space Centre.

A brief overview of the downlink experiment design is given in Section 2. A performance analysis is carried out in Section 3 based on about 100 satellite passes with data recorded on board the Coast Guard vessel KV Harstad. Conclusions are given in Section 4 based on the observations and analyses made.

2. Experiment Overview

The Norwegian VDE-SAT downlink demonstration concept is depicted in Fig. 1. The VDE-SAT payload provides the signals for the test campaign as well as the first demonstrations of the services. The transmitted right hand circular polarised signal contains four carriers, three modulated (spread BPSK, Π/4-QPSK and 8-PSK, see [6]) and one unmodulated beacon. The signal is received as baseband sample data and logged by a receiver employing a vertically polarised antenna.

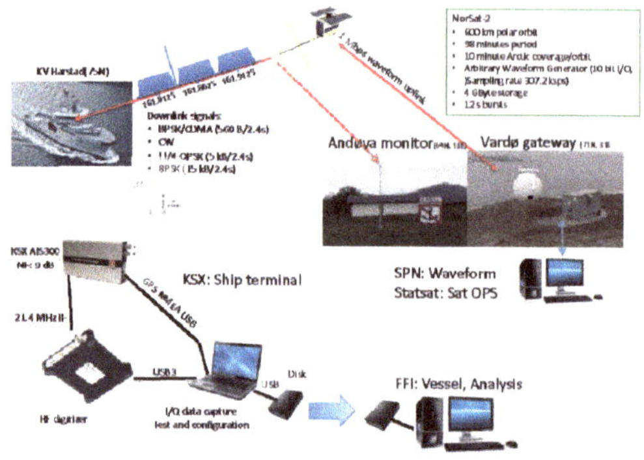

Fig. 1 Experiment setup.

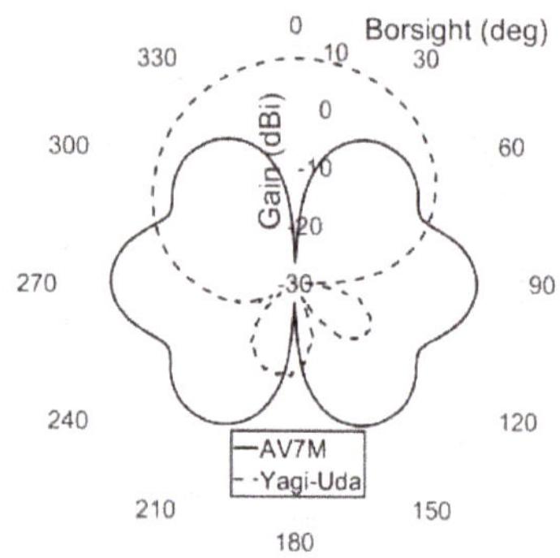

Fig. 2 Comrod AV7M and Yagi-Uda antenna gains.

2.1. NorSat-2 transmitter

The VDE-SAT transmitter is on the same satellite as the SAT-AIS receiver used for operational maritime surveillance. The SAT-AIS payload cannot receive messages when the VDE-SAT payload transmits, and a time sharing compromise was made for the downlink verification campaign: The transmitter alternates on and off for 12 seconds. The four carriers are each transmitted with a power of 20.9 dBm through the satellite antenna. This includes a loss of 1.1 dB between the antenna and the amplifier, delivering 22 dBm per carrier.

NorSat-2 is equipped with a deployable 3-element Yagi-Uda antenna with circular polarisation having a peak gain of about 8 dBi and an axial ratio less than 2 dB. We have utilised a rotation symmetric Yagi-Uda satellite antenna gain pattern in the link budget calculations, although the measurements show minor asymmetries. The satellite attitude control mode is limb pointing towards a specified location, depending on the position of KV Harstad.

2.2. Ship receiver

The VDE-SAT receiver on board KV Harstad is mounted on top of the wheelhouse. The VHF antenna is mounted on the starboard railing. The height above sea level is about 15 meters. The battery room and the top deck with more antennas constitute a blocking structure in part of the field of view of the VDE-SAT antenna. The receiver is installed in a rack in the battery room on the same level. The antenna coaxial cable length is 8 m, corresponding to a loss of 0.3 dB. The vertically polarised ship antenna employed is a Comrod AV7M [7]. The polarisation loss is about 3 dB. The antenna gains as function of boresight angle are displayed in Fig. 2 for both satellite transmit and ship receive antennas.

In our System Tool Kit simulation model we utilise the ship antenna gain pattern measured by the vendor. It should be noted that the Comrod dipole antenna pattern is representative for free space only, but in practice the ship structure and sea surface may modify the antenna diagram.

The receive antenna is connected to an amplifier having a noise figure of 9 dB through a coaxial cable. The receiver was tested in the laboratory utilising a channel model as described in [8] with satisfactory results. The model included Doppler frequency shift and fading of the signal as well as calibrated Additive White Gaussian Noise (AWGN). The ship latitude, longitude and elevation are logged using a GPS receiver and a separate antenna.

3. Results

In this section we present results from the analysis for the time period 14th – 30th Nov. 2017 utilising a dataset obtained on board KV Harstad. The reported sailing route is given in Fig. 3. The vessel left Tromsø 15th Nov., travelling north, presumably for fishing inspections.

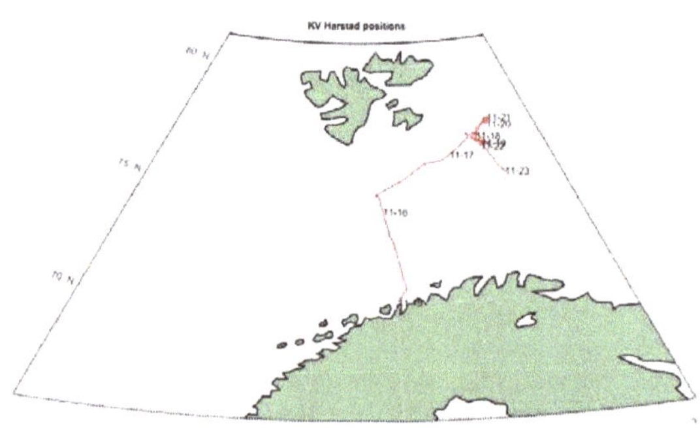

Fig. 3 Reported track, KV Harstad.

3.1. Received beacon carrier power

The beacon data set from KV Harstad contains 178107 samples. The sampling rate is 5 samples/s, and the data correspond to 103 satellite passes with a corresponding duration of 9.9 hours.

The Probability Density Function (PDF) and Complementary Cumulative Distribution Function (CCDF) of the received beacon power for the investigated time period are given in Figs. 4 and 5.

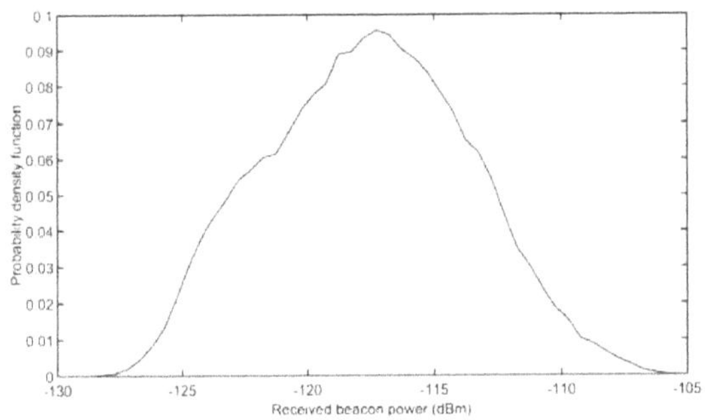

Fig. 4 Beacon power probability density distribution.

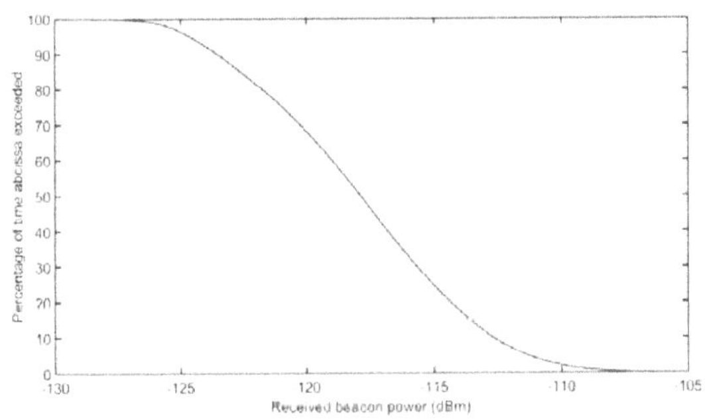

Fig. 5 Beacon power complementary cumulative distribution.

The median of the received beacon power is -117.6 dBm, the mean is -115.8 dBm and the standard deviation (calculated from the power in dBm) is 4.0 dBm.

The measured received power was sorted into elevation bins with boundaries 0, 5, 10,... 90 degrees, and the mean of the linear beacon power was calculated. The standard deviation was calculated based on the values in dB. The average received power is highest within the elevation range 40 to 50 degrees, see Fig. 6.

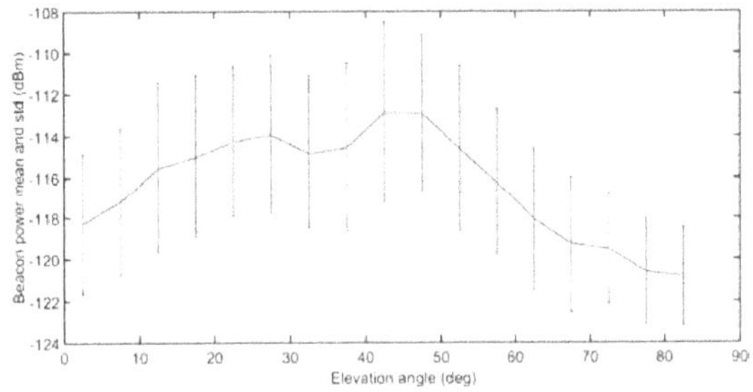

Fig. 6 Mean and standard deviation of beacon power.

The PDFs of received beacon power for the Line-of-Sight (LOS) and Non-Line-of-Sight (NLOS) cases based on the ship 3-dimentional structure are given in Fig. 7.

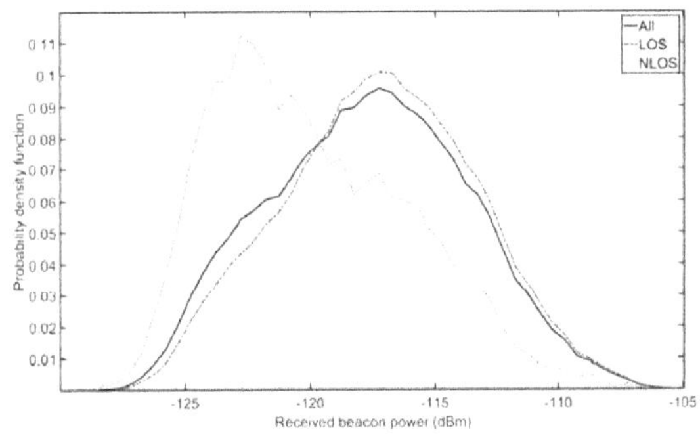

Fig. 7 PDFs of received LOS and NLOS beacon power.

The PDFs show that there are lower signal power levels in the direction of the obstruction mask. The CDFs are shown in Fig. 8.

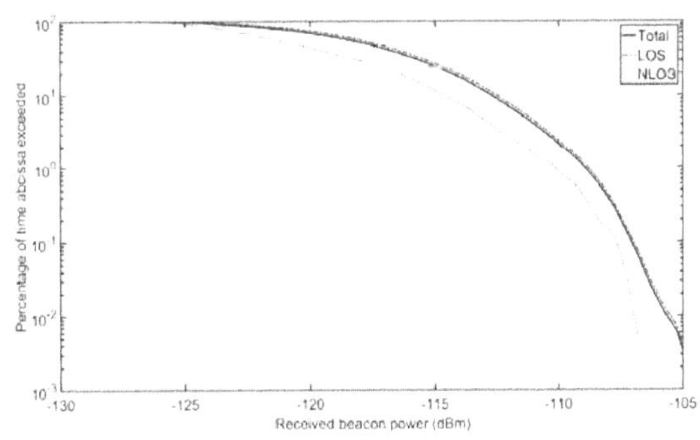

Fig. 8 CCDFs received LOS and NLOS beacon power.

The difference between LOS and NLOS is relatively low, indicating that partly the blocking ship structure is small compared to the Fresnel zone and thereby is not blocking the signal and partly that significant diffracted components exist.

3.2. *Signal fading*

In the following analysis the satellite attitude data is utilised to calculate link budgets and measurements obtained when no attitude data was recorded is removed. This results in valid beacon data in 100 passes. The time duration with measurements equals 537 minutes (~9 hours) and the number of valid samples is 161085.

The received beacon signal power is shown together with the calculated direct line-of-sight power and the resulting coherent sea surface reflection model [8] (see Fig. 9 for a single pass example). The antenna height for KV Harstad is estimated from the GPS measurements as the average height reported when values are between 10 and 30 m, resulting in an antenna height of 15.6 m. The fading model incorporates a number of additional parameters, including the wave height, elevation angle and ionospheric scintillation index S_4.

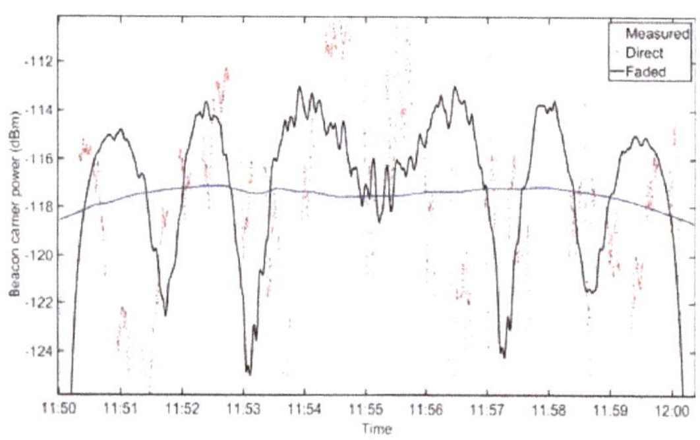

Fig. 9 Received beacon power and sea reflection model, 20th Nov. 2017. Red: measured, blue: direct calculated line-of-sight, black: calculated fading with one coherent specular reflection.

The resulting direct plus coherently reflected fading model represents to some degree the received signal fade depths, although not able to reproduce the variations exactly. The model does not take into account the effect of the ship structure nor the ship movements, and the ionospheric scintillation component is not included here.

We select to calculate the fading as the difference between the calculated LOS component (taking into account satellite position and attitude, antenna gains and free space loss) and the actually recorded beacon power, see Fig. 10. Positive fading implies reduced signal level while negative implies enhanced signal power levels.

The density distribution and cumulative distribution of fading are shown in Fig. 11 a) and b) respectively. The median (and mean) fading value found is -0.1 dB, having a standard deviation of 4.2 dB. The distribution of fading (in dB) follows relatively closely a normal distribution in the range -15 (enhancement) to 5 dB (reduction). A fit between the normal distribution and the fading (in dB) resulted in a mean value of 0.05 and a standard deviation of 4.16 dB.

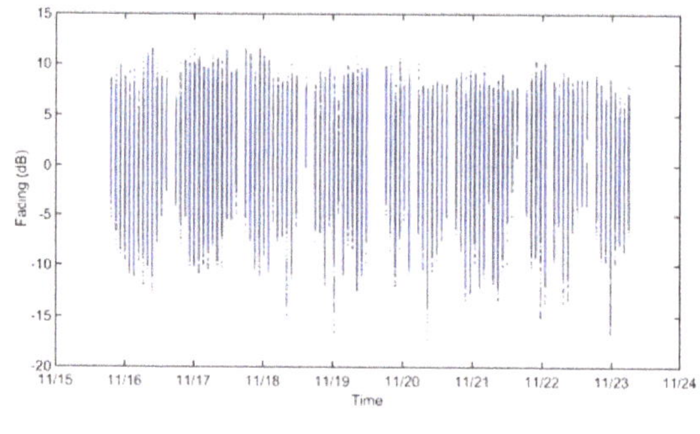

Fig. 10 Estimated fading, 15th – 23rd Nov. 2017.

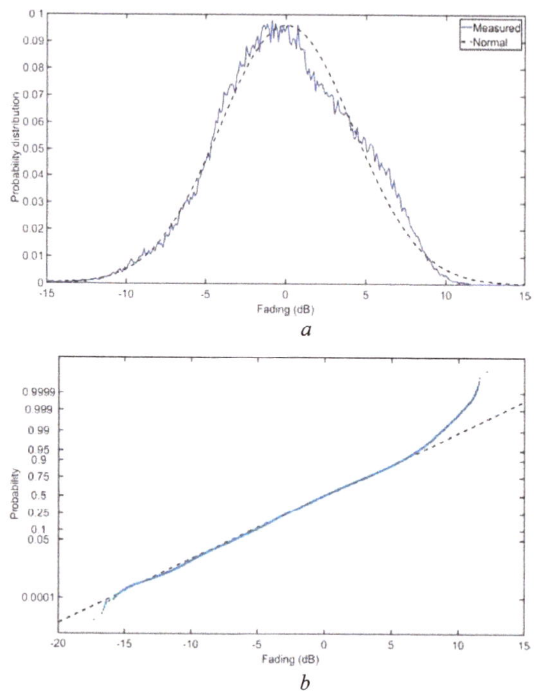

Fig. 11 Estimated fading distributions (a) PDF (b) CDF.

The 95% confidence interval for the mean was [0.028, 0.069] dB and for the standard deviation [4.15, 4.17] dB.

An attempt was also carried out to test whether the Nakagami-Rice distribution could represent the envelope of the fading. Note that in the previous figures negative fading values (in dB) represents enhancements, while for the fading envelope as utilised in a multiplicative channel simulator, small envelopes corresponds to low received signal power. Shown in Fig. 12 is the measured envelope probability density function together with fitted Nakagami-Rice and lognormal distributions.

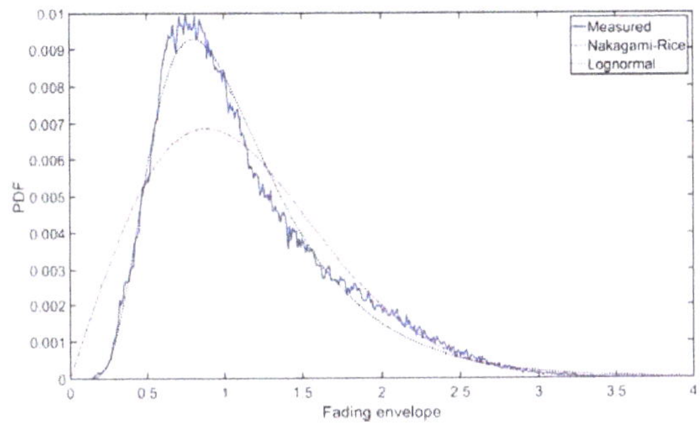

Fig. 12 Fading envelope distributions.

Although no hypothesis test for distribution fits has been carried out, the lognormal model seems to better represent the measured fading envelope compared to the Nakagami-Rice distribution. A lognormal distributed fading envelope corresponds to a normal distributed fading envelope in dB. Note that the definition of fading utilised relies on accurate modelling of the LOS component, and deviations between the model of antenna gains and losses and actual gains and losses influence the results.

The measured fading was sorted into elevation bins and the mean was calculated in the same way as for the beacon power. The standard deviation was calculated based on the values in dB. The average fading varies with elevation angle, with maximum occurring at low elevations angles with a second peak around 55 - 65 degrees, see Fig. 13.

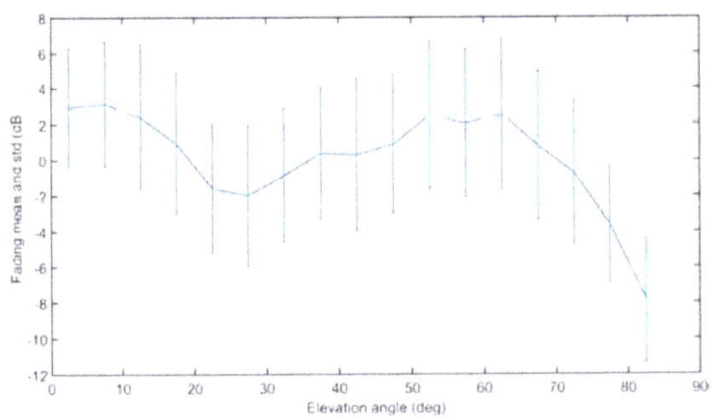

Fig. 13 Mean and standard deviation of beacon fading.

The first decrease above 5 - 10 degrees elevation angle corresponds well with the assumption that the power in the coherent reflection decreases with increasing elevation angle.

3.3. *Noise and interference*

The noise plus interference spectral density is available in two versions from estimators implemented in the receiver. One is extracted as the median from eight spectral bins with 1 kHz spacing within ±5 kHz of the CW beacon ('Beacon'). The other is computed from a 500 kHz spectrum estimate as the median of four bins located between and adjacent to the modulated carriers ('Matlab spectrum'). For simplicity we denote this summed power density as N_0. The measured probability density of N_0 is given in Fig. 14 and the cumulative distribution in Fig. 15.

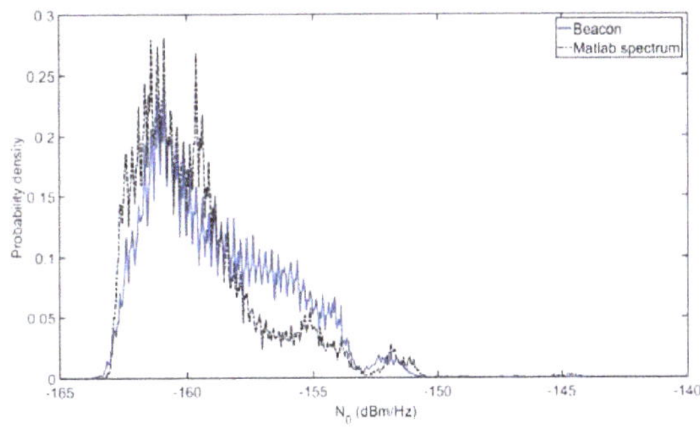

Fig. 14 N_0 Probability density distributions.

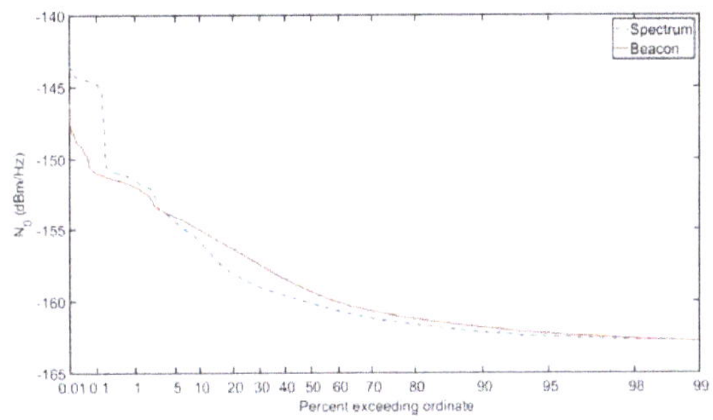

Fig. 15 N_0 cumulative distributions, Rayleigh plot.

On a Rayleigh plot, AWGN is expected to become a straight line [10]. For the larger values N_0, corresponding to small time percentages, it is clear that the interfering narrowband components deviate significantly from AWGN. Also for time percentages exceeding 5, the measured noise floor deviates from AWGN, although to a lesser extent. The median (and mean) value of N_0 extracted from the spectrum estimates is -160 dBm/Hz, while the standard deviation is 2.5 dBm/Hz. For reference, the calculated N_0 utilising ITU-R Rec. P.372 [9] is -155 dBm/Hz for a city environment, -159 dBm/Hz for residential environment, -162 dBm/Hz for rural and -165 dBm/Hz for galactic noise only.

In the draft recommendation [1] the monopole antenna noise temperature for galactic noise is given as an example, equal to 245 K. If a dipole is utilised, as is the case here, the galactic noise temperature is 3.4 dB higher [11] (526 K). It may be useful to characterise the noise as referred to after a lossless dipole antenna to enable comparison with theoretical calculations, and also produce statistics that does not depend on the losses and noise figure in a specific implementation. The cumulative distribution of the measured antenna temperature is given in Fig. 16. The median value is 5541 K, the mean equals 9240 K and the standard deviation is 25304 K. Note that this antenna temperature includes both

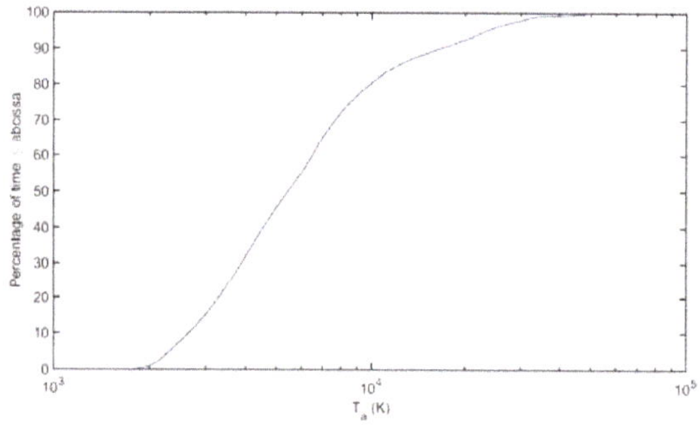

Fig. 16 CDF for estimated antenna noise temperature.

interference and man-made noise, resulting in short time periods with high antenna noise temperatures. The measured antenna noise temperatures are summarised in Table 1.

Table 1 Measured dipole antenna noise temperatures.

Percentile	50	80	90	95	99
T_a (K)	5539	10408	18985	26800	55633

For comparison, the estimated antenna noise temperatures from ITU-R Rec. 372 [9] are 8581 K for residential and 23096 K for city, given a dipole antenna.

3.4. *Signal-to-noise ratio*

The measured ratio of beacon carrier power and noise spectral density (C/N_0) time series utilising the noise floor estimated from the spectrum is presented in Fig. 17. The term N_0 includes both thermal and man-made noise as well as interference.

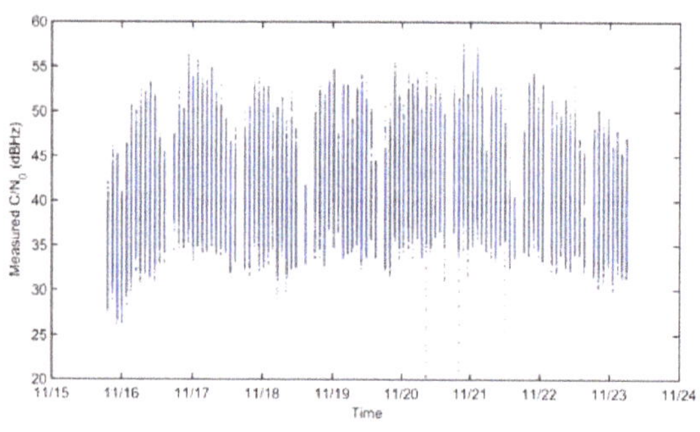

Fig. 17 C/N_0 time series, 15th – 23rd Nov. 2017.

The complementary distribution of C/N_0 is given in Fig. 18. This result may be utilised as one of the main design criteria for the system when specifying the forward error correction and link margin for achieving the link availability and quality objectives. Note, however, that the result is specific for the current experimental set-up. The result in a tabular form is given in Table 2.

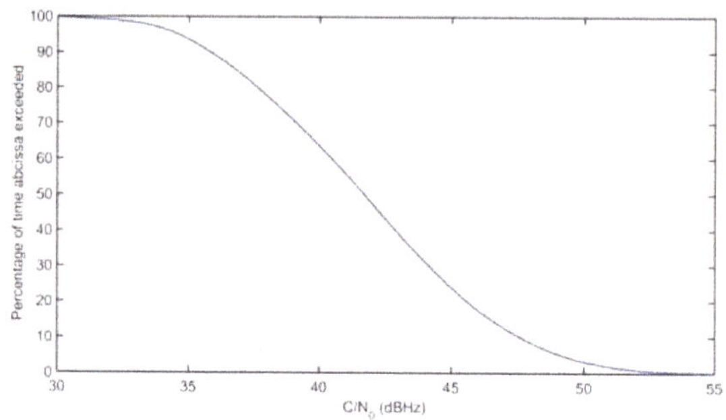

Fig. 18 C/N_0 complementary cumulative distribution.

Table 2 C/N_0 exceeded for a given time percentage.

Availability (%)	**C/N_0 (dBHz)**
100	10.3
99.9	28.5
99.5	30.8
99	32.2
98	33.3
95	34.7
90	35.9
80	37.7

If a link availability of 95 % of the time is required, the waveform should function satisfactory at C/N_0 of 34.7 dBHz, while a link availability of 99.9 % requires the waveform to handle a C/N_0 of 28.5 dBHz. Change of equipment, such as improved antennas, passband filters, and amplifiers, or higher satellite EIRP lead to less stringent C/N_0 requirements.

3.5. *Time variability*

Initially it was planned to utilise a vector of C/N_0 values to extract second order statistics in the form of duration statistics. The corresponding non-connection duration would represent the connection time. The effect of the transmission on/off schedule hinders the development of this statistic, as it is somewhat questionable to interpolate over 12 second periods with no transmissions.

The channel model developed gives some insight into the expected time durations, where peaks are separated in time with about one minute between them. The occurrence of deep fades corresponds to 180 degrees phase shifts between the direct and the coherently reflected component. This is mainly an effect of the elevation angle towards the satellite, but also depends on the antenna height. Interleavers with time duration less than the fade durations are unable to distribute the demodulation errors potentially occurring during low C/N_0 periods. Interleavers operating on a minute scale introduce time significant delays, but might be considered, if necessary, to combat the fading process.

To increase the probability of successful reception of information, one might select to employ time diversity, which means transmitting the same information after a selected time delay. The time delay is selected to ensure a reasonable de-correlated value of channel degradation, in our case the channel fading. Similar considerations may be carried out for an interleaver intended to combat fading. The length in terms of interleaver time duration should be long enough to include reasonably de-correlated channel fade events, increasing the probability that not all symbols within the interleaved symbol block are faded simultaneously.

An attempt to describe the time variations has been carried out by calculating the normalised autocorrelation function (ACF) for the fading (in dBm) for every pass. The measurements were first resampled to obtain equal time spacing between samples, and 12 s periods without logged results were interpolated by using a moving median of 10 samples and linear interpolation at 1 second intervals. The resulting surface plot and average and median ACF are shown in Fig. 19 a) and b), respectively.

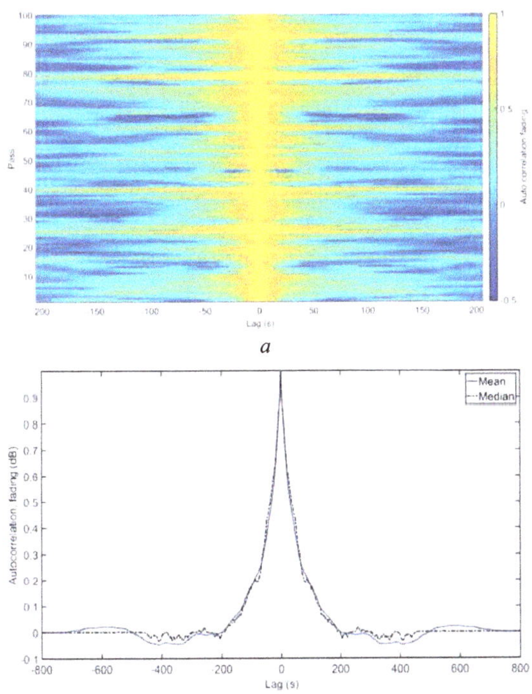

Fig. 19 (a) Fading ACFs as function of time lag and pass number (b) Mean and median fading ACF, all passes.

The mean and median fading ACFs decrease with increasing lag as expected. The mean correlation reaches 0.5 for a lag of 31 seconds and a value of 0.1 at 129 seconds. A one minute lag corresponds to a fading AFC of 0.3. If calculating the ACF on the linear version of fading envelope, an ACF of 0.5 is reached after a time lag of 85 seconds. This implies that a retransmission in a time diversity scheme should occur no earlier than for example 31 seconds if a decrease in autocorrelation of 0.5 is acceptable, or similarly that an interleaver should have a time duration of at least 31 seconds for the same de-correlation threshold.

4. Conclusion

The purpose of the reported work was to characterise the VHF downlink channel expected for the satellite downlink of VDE-SAT. The satellite NorSat-2 transmitted a VHF signal for validation of the downlink (satellite-ship) waveform performance. FFI has analysed the measurements recorded during November 2017.

Received carrier power, noise environment including interference and performance of three modulated waveforms were analysed and compared with theory and simulations. The received power levels are as expected for most of the time. Slowly varying fading is observed with peak-to-peak variations in order of 20 dB. The fading has a close to normal distribution (in dB) with a standard deviation of 4.2 dB. The dominating fading mechanism seems to be a strong coherent specular sea reflection received by the vertically polarised ship antenna. No major degradation was identified due to ionospheric amplitude scintillation.

Measured noise and interference levels revealed significant man-made time varying interference. Although time periods with significant interference were observed, most of the time the noise floor can be classified as residential or city according to ITU-R Rec. 372 on Radio noise. The noise plus interference level measured did not represent white Gaussian noise, and it is beneficial if the waveform(s) is robust with respect to both narrowband and wideband noise.

The received C/N_0 showed a dynamic range larger than 20 dB, exceeding 34.7 dBHz for 95 per cent of time. The measured signal-to-noise levels can be used to estimate expected performance in terms of error rates and availability for the modulated waveforms.

Recommendations for later similar studies include logging of ship movements with waves and also measuring the actual antenna gain diagram when installed on the ship. It would also be beneficial to transmit a channel sounding waveform continuously throughout the passes to enable a more detailed study of the channel time dynamics. Ionospheric phase variations at high latitudes may be another topic of interest for future studies.

Acknowledgements

This work is partly funded by the European Space Agency, ESA-ESTEC, Noordwijk, The Netherlands, under contract no. 4000113274. Opinions, interpretations, recommendations and conclusions expressed herein are those of the authors and are not necessarily endorsed by the European Space Agency.

We acknowledge the effort by Nader Alagha, ESA. He has contributed with useful comments and suggestions during campaign and has also reviewed the current paper.

References

[1] Recommendation ITU-R M.2092-0, "Technical characteristics for a VHF data exchange system in the VHF maritime mobile band," International Telecommunication Union, Geneva, October 2015

[2] Nader Alagha, Juan Lizarraga, Frank Zeppenfeldt, Wouter Jan Ubbels, Gaetan Fabritius, Jean-jacques Valette, Joost Elstak, "THE ROLE OF SATELLITE IN THE EMERGING MARITIME VHF DATA EXCHANGE SYSTEM," 65th International Astronautical Congress, Toronto, Canada, 29. Sep. – 03. Oct., 2014

[3] Torkild Eriksen, Lars Erling Bråten, Hans Christian Haugli, Frode A. S. Storesund, "VDE-SAT – A new maritime communications system, "The 4S SYMPOSIUM, Valletta, Malta, 30. May - 3. June 2016

[4] Hans C. Haugli, Lars Løge, Torkild Eriksen, Stig E. Christiansen, Frode Storesund, Lars E. Bråten, Anders Bjørnevik, "The VHF Data Exchange System (VDES) and Norsat-2 Satellite Testing," In Proc. 35th AIAA International Communications Satellite Systems Conference, International Communications Satellite Systems Conferences (ICSSC), (AIAA 2017-5419)

[5] RESOLUTION 360 (REV.WRC-15), Consideration of regulatory provisions and spectrum allocations to the maritime mobile-satellite service to enable the satellite component of the VHF Data Exchange System and enhanced maritime radiocommunication, CMR15/2015-E, 244 – 246, The World Radiocommunication Conference, Geneva, 2015

[6] F. Xiong, "Digital Phase Modulation and Demodulation, Encyclopedia of Telecommunications" Wiley, 2003

[7] Comrod AV7M Maritime VHF dipole antenna, [Online:] http://www.comrod.com/getfile.php/13179/Datasheets/T%20 Antennas%20-%20Marine/AV7M.pdf

[8] L. E. Bråten, V. Arneson, K. Svenes, T. Eriksen and Ø. Olsen, "Channel Modelling for VHF Data Exchange System via Satellite,“ In Proc.12th European Conference on Antennas and Propagation, 9-13 April, 2018

[9] Recommendation ITU-R P.372-13,"Radio noise," Geneva, 2016

[10] A. Wagstaff and N. Merricks, "Man-made noise measurement programme," IEEE Proc. Commun., Vol. 152, No. 3, June 2005

[11] B. Skeie and B. Solberg, "External man-made radio noise measurements," FFI report, 16/00869, [Online:] http://www.ffi.no/no/Rapporter/16-00869.pdf

Field trials of the VHF data exchange system (VDES) satellite downlink component

Iago Gómez[1], Francisco Valdés[2], Brais Ares[1], Javier Taibo[2], Jorge M. El Malek[1], Nader Alagha[3]

[1]*Gradiant, Vigo, Spain, email: {igomez, bares, jmunir}@gradiant.org*
[2]*Egatel, Ourense, Spain, email:{fvaldes, jtaibo}@egatel.es*
[3]*European Space Agency, Noordwijk, Netherlands, email: nader.alagha@esa.int*

Keywords: VDES, VDE-SAT, SATCOM, FIELD TRIALS, MARITIME.

Abstract

This paper focuses on analysing and characterising the communication channel of the VDES satellite downlink component by carrying out a measurement campaign of the VDE-SAT signal. The VHF downlink signal was transmitted from the Norwegian satellite NORSAT-2 operated by Space Norway. Three simultaneous carriers, specified in ITU-R M.2092-0, were transmitted in the test waveform: carrier 1 with BPSK/CDMA-8 modulation, carrier 2 with π/4 QPSK modulation and carrier 3 with 8PSK modulation. In addition, the satellite also transmits a CW carrier for test purposes.

During this work a COTS-based receiver station has been designed, and installed on the roof of a house next to the sea. An embedded software application was implemented to automatically capture the test waveform every time the satellite was over the horizon. All the captured data was processed offline and the received waveform was thoroughly analysed, in order to assess the quality of the received signal and model the communications channel. The most relevant results derived from this work are presented in the paper.

1 Introduction

The VHF Data Exchange System (VDES) is a maritime communication system (ITU-R M.2092-0 [1]) which has two main goals, on the one hand, protect the AIS (Automatic Identification System) basic function of ship to ship collision avoidance by off-loading short messaging and data traffic from AIS to other channels, and on the other hand, enhance maritime communication applications, based on robust and efficient digital transmission at a much higher rate than the current AIS. In addition, the VDE Satellite Component (VDE-SAT) will allow long range communications between ships and shore.

As part of agenda items for the World Radio Conference 2015 (WRC-15) [2], a re-allocation of VHF channels within the maritime frequency bands was granted. This re-allocation has enabled a VHF Data Exchange (VDE) for ship-to-ship, ship-to-shore, and shore-to-ship communications. The exchange between ship and shore in both directions via satellite at high seas and areas without local shore infrastructure is the subject of a new agenda item for WRC 2019. The VHF Data Exchange System (VDES) aims to provide an integrated terrestrial and satellite communications system to exchange maritime information. The proposed VDES relies on digital data links over a number of maritime VHF channels. Bundling two or more VHF channels (25 kHz each), and using new modulation, coding and access schemes would yield an increased information throughput and enhanced service availability compared to the existing analogue communication links used in such VHF channels.

On 14 July 2017, the first Low Earth Orbit small satellite with VDE payload was launched successfully [3]. ESA in collaboration with Space Norway who is operating NORSAT-2, currently performs preliminary measurements on the VDE-SAT and VHF maritime channel using the NORSAT-2 VDE payload [4].

The main goal of this paper is to describe the measurement campaign of VDE-SAT downlink that was carried out in the period of March to April 2018. The VHF downlink signal was transmitted from the Norwegian satellite NORSAT-2 operated by Space Norway. Three simultaneous carriers, specified in ITU-R M.2092-0 [1], were transmitted in the test waveform: carrier 1 with BPSK/CDMA-8 modulation, carrier 2 with π/4 QPSK modulation and carrier 3 with 8PSK modulation. In addition, the satellite also transmits a CW carrier for test purposes.

The remainder of the paper is organized in the following way. Section 2 gathers different aspects about the measurement campaign such as the receiver location, receiver station hardware and the process of data collection during the field trials. Then, Section 3 explains the data analyses done with the captured signal. After that, the results are presented in Section 4 and finally Section 5 collects the conclusions.

2 Measurement campaign

2.1 Location

The measurement campaign was carried out in Cesantes, a coastal village near Vigo (Pontevedra), in Spain. The receiver station was installed on a beach house's roof, barely 20 meters away from the sea surface, where radio propagation conditions would be similar to those found on a ship.

2.2 Receiver station hardware

The receiver station was designed trying to replicate the same operation conditions a shipborne VDES (or AIS) receiver would have. The approach during the component selection was to use COTS components with an affordable price and easy to find in general electronics distributors, keeping the building process as simple as possible. Therefore, the receiver can be easily replicated by any other research group for similar measurement campaigns. Fig. 1 shows a simplified diagram of the receiver station architecture.

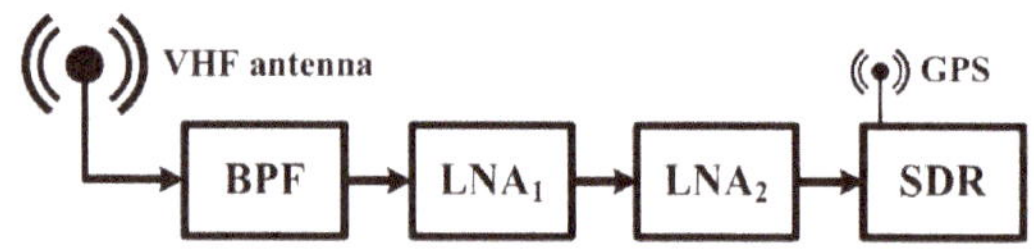

Fig. 1 Hardware architecture of the receiver station

The receiver station uses a commercial dipole VHF antenna[1] specifically designed to be used in maritime communications. This antenna feeds the reception chain with both the VDE components from satellite and the AIS signals from ships and coastal stations (as any shipboard VDES receiver would do). Its vertical radiation pattern is shown in Fig. 2.

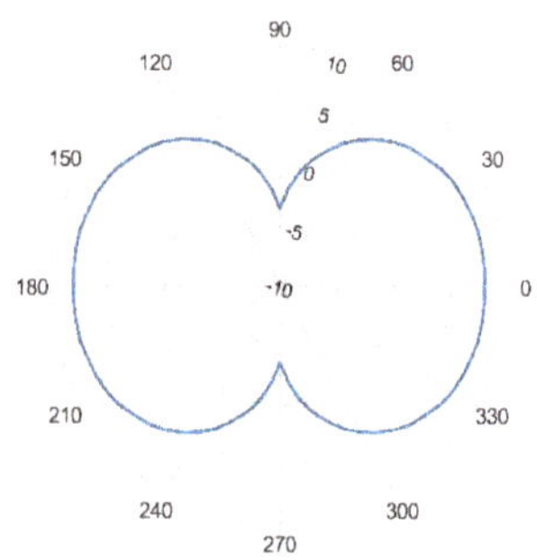

Fig. 2 Vertical radiation pattern (f = 162 MHz)

The receiver chain consists of a band-pass filter for the band of interest, and two consecutive low noise amplifiers for an aggregated amplification of 42.9 dB. Then a SDR platform down-converts, digitizes and stores the signal for further analysis. Furthermore, a GPS is used not only to acquire the UTC time and exact location of the receiver, but also to drive the SDR internal oscillator in order to precisely tune the central frequency.

The estimated noise figure of the proposed receiver station is 3.7 dB. The external appearance of the finished receiver station is shown in Fig. 3.

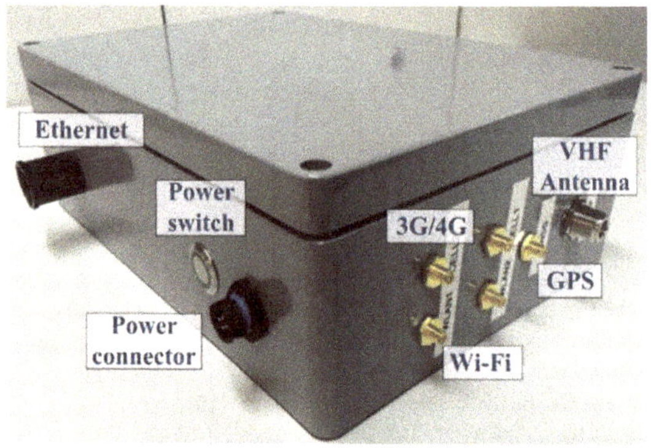

Fig. 3 External appearance of the receiver station

2.3 Captures configuration

The receiver station has been configured so that all VDES upper-leg is captured, including the four carriers transmitted by NORSAT-2 (three VDE components plus CW for testing) along with ASM and AIS channels, emulating the behaviour of a typical shipboard receiver. To this end, the SDR platform is configured to use a filter bandwidth of 350 KHz and a sampling rate of 500 KHz. The frequency spectrum of a baseband captured data is illustrated in Fig. 4.

The capturing process has been automated in such a way that the receiver station starts capturing whenever NORSAT-2 is visible from the location, given that its ephemerides are known. The captured signals are saved to disk for later analysis offline.

2.4 Data collection summary

The measurement campaign started on 22nd of March and finished on 13th of April (2018). From all the satellite passes with visibility over the receiver station location, only the ones that have maximum elevations above 5° were considered for

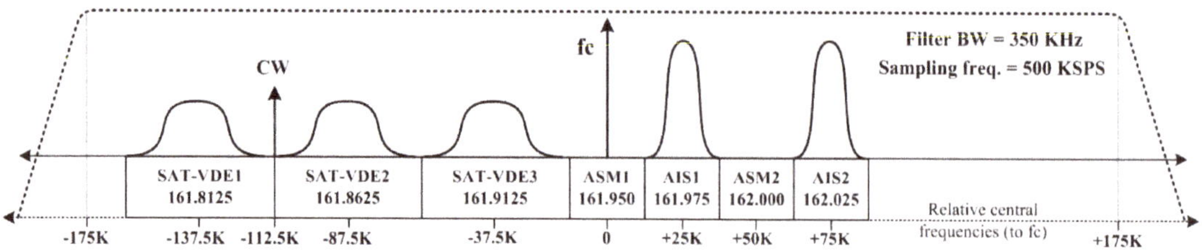

Fig. 4 Frequency spectrum scheme of a baseband captured data

[1] https://www.comrod.com/wp-content/uploads/AV7M.pdf

post-analysis. Thus, the accumulated captured time during the whole period adds up to 790 minutes (over 13 hours).

Given the trajectory of the satellite most of the captures show maximum elevations lower than 30º, whilst there are barely a few captures for elevation angles between 60º and 90º. This can be noticed in Fig. 5, which shows elevation angle versus azimuth angle for all the trajectories collected during the measurement campaign.

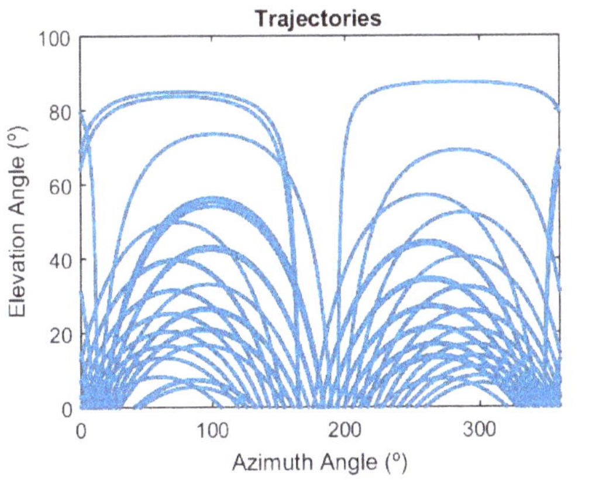

Fig. 5 Satellite elevation angle versus azimuth angle (all passes)

Consequently, more data was collected at lower elevations than at higher elevations. Fig. 6 shows the percentage of captured time where satellite was seen above certain elevation angle. It can be noted 90% of the samples recorded from satellite have an elevation lower than 30 degrees and only 1% of the samples have been recorded with an elevation above 60º. An average capture is around 12 minutes long with a maximum satellite elevation of 29.73 degrees.

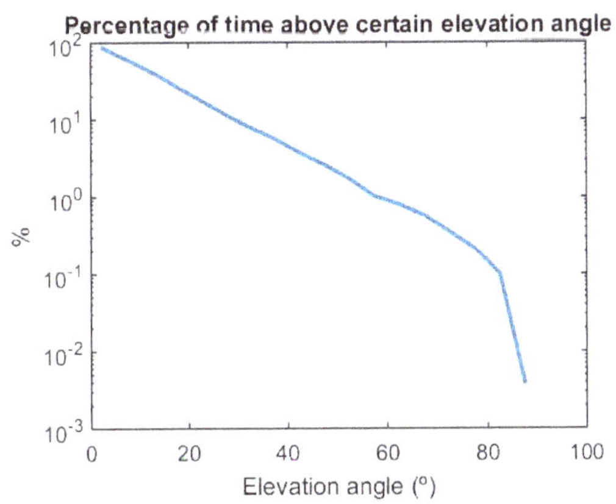

Fig. 6 Percentage of captured time where satellite was seen above certain elevation angle

3 Data processing

This section introduces the processing steps applied over the captured baseband raw samples to be able to derive the results and draw the conclusions that will be presented in the following section. Only two of the four VDES satellite components have been analysed in this work: the CW testing signal and the BPSK/CDMA modulated signal. The following subsections focus on describing how each capture is processed.

3.1 VDES components conditioning

The captured signals require some conditioning before being able to analyse each component independently. Since all VDES upper-leg was down-converted to baseband, each component needs to be isolated prior to the demodulation. This pre-processing mainly involves applying a frequency shift to the desired component in order to center it at baseband, filtering out the unwanted signals and then decimating the resulting signal by an appropriate factor so the targeted component is sampled at a convenient rate for the following stages. A simplified version of this scheme is shown in Fig. 7.

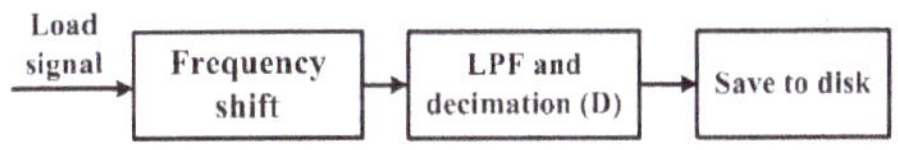

Fig. 7 VDES components pre-processing

The frequency shift for the CW carrier is -112.5 KHz and the decimation factor is 32, resulting a sampling rate of 15.625 KHz. In the case of the BPSK/CDMA carrier the frequency shift is -137.5 KHz, whereas the decimation factor is 4, resulting a sampling rate of 125 KHz.

3.2 CW carrier

The main goal of the CW processing stage is to be able to analyse the constant wave to obtain a noise plus interference density and power level estimation of the capture, and eventually compute the carrier to noise plus interference density ratio ($C/(N_0+I_0)$). Fig. 8 shows the processing steps required in order to estimate such metric.

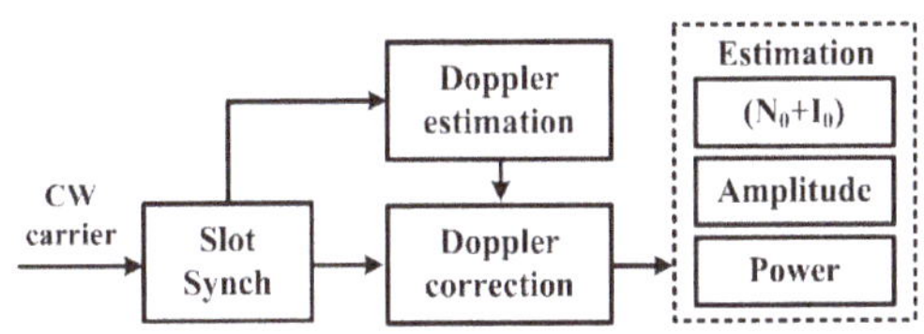

Fig. 8 CW carrier processing stage

NORSAT-2 transmits during bursts of 12 seconds (slots), with a known pattern. Therefore the first step is to detect where this pattern starts within a capture (slot synch). Once the beginning of a pattern is identified, it is straightforward to determine the position of any other ON or OFF period throughout the rest of the capture.

NORSAT-2 moves at a speed of about 8 km/s and this can cause a maximum Doppler of ±4 kHz at VHF [1]. This Doppler shift is estimated and then compensated. From the resulting signal the following metrics are estimated: noise plus interference density (N_0+I_0), amplitude and power.

3.3 BPSK/CDMA modulated carrier

This processing stage aims to measure the experimented quality signal of the captures in terms of signal to interference plus noise ratio (SINR) for the BPSK/CDMA modulated carrier. Although BPSK/CDMA uses a spread spectrum modulation, SINR was computed at chip rate (i.e. before despreading), so spreading gain of BPSK/CDMA is not taken into consideration. Fig. 9 shows the block diagram of the BPSK/CDMA component processing stage, which is applied to every subslot.

First of all, the Doppler shift is compensated by using the frequency offset estimated during the CW processing stage. Next, an SRRC filter is applied to the signal, using a roll-off factor of 0.25.

Another side issue related to Doppler Effect is that the actual symbol rate changes over time, so there is need not only for a timing recovery algorithm but also for an oversampling rate estimator. For this purpose the timing recovery algorithm specified in [5] is used, which outputs a precise oversampling rate estimation of the received signal. Once the oversampling rate has been estimated, final timing error is estimated again using the same algorithm to finally obtain symbols from the samples by applying linear interpolation.

The burst detection block is in charge of the alignment of the transmitted chips (known pattern) with the received ones, using correlation.

The equalizer consists on an adaptive filter that uses as error the difference between the received symbols minus the transmitted BPSK pattern (data aided). The adaptation of the filter employs least mean square (LMS) algorithm. It is important to mention that the receiver will use the whole transmitted sequence for different purposes (time synchronization, equalization, etc.) which would not be possible in a practical receiver. This means that obtained SINR should be considered as an upper bound for any practical solution.

Lastly the SNR of the resulting signal is estimated using a Maximum-Likelihood (ML) data aided estimator [6] [7].

4 Results

Throughout this section, the results obtained during the measurement campaign are presented, both for the CW signal and for the BPSK/CDMA modulated signal. In this section, the statistics of the received signal, the communication channel and the SINR are also described.

4.1 Signal level measurements and channel characterization

Fig. 10 presents the mean signal level (in log scale) and confident levels (10% and 90%) versus elevation angle. As expected, due to the antenna pattern and the VDE-SAT Power Flux Density (PFD) mask, the highest power is received when the satellite is seen with an elevation angle between 30 and 40 degrees.

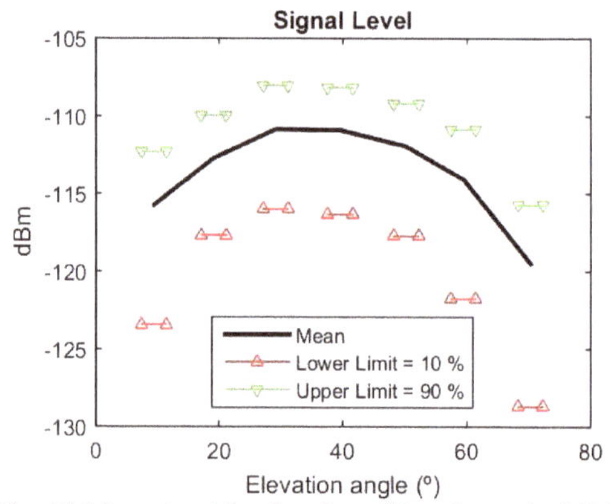

Fig. 10 Mean signal level and confident intervals (10% and 90%) versus elevation angle

In the subsequent figures, only elevation angles between 10 and 50 degrees are considered.

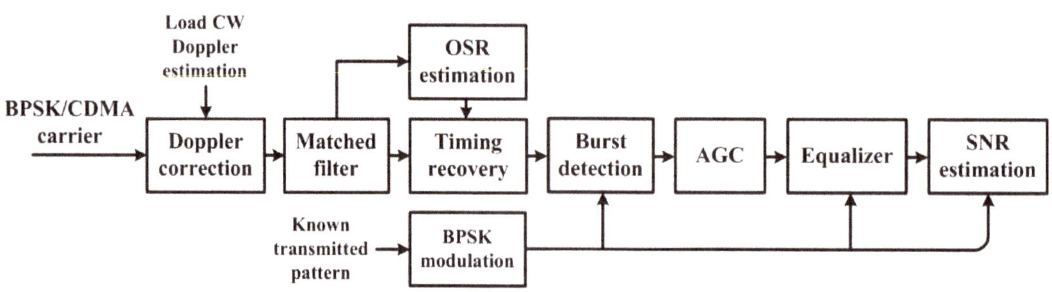

Fig. 9 Block diagram of the processing stage of BPSK/CDMA component

Fig. 11 shows the probability density function (PDF) of the received signal level in mW. This figure also shows three typical mathematical distributions for modelling attenuation of wireless signals traversing shadowing and multipath channels (Gamma [8], Lognormal [8] and Nakagami [9]), the parameters of these distributions are selected so that best fit the PDF of the received signal level. The fitting has been made using the Matlab function `fitdist`[2], provided in the Statistics and Machine Learning Toolbox.

Although both *Lognormal*($\mu = -113.79$, $\sigma = 3.57$)[3] and *Nakagami*($m = 0.62$, $\Omega = 4.72 \times 10^{-23}$)[4] distributions fit quite well with the received signal level, *Gamma*($\alpha = 1.92$, $\theta = 2.88 \times 10^{-12}$)[5] fits perfectly to the signal level PDF.

Fig. 12 shows the cumulative density function (CDF) for the signal level in mW and the CDF for Gamma, Lognormal and Nakagami distributions with the same parameters as in the previous figure.

Finally, Fig. 13 shows the probability density function (PDF) of the received signal amplitude in V. This figure also shows three distributions (Beta [8], Generalized extreme value (GEV) [8] and Nakagami [9]) that fit well with the received signal amplitude. It can be seen that general purpose distributions as *Beta*($\sigma = 6.73$, $\beta = 9.65$) and *GEV*($k = -0.13$, $\sigma = 2.32 \times 10^{-8}$, $\mu = 5.89$) are able to model quite well the amplitude of the received signal. However, *Nakagami*($\mu = 1.92$, $\Omega = 5.53 \times 10^{-15}$), which is a typical distribution for modelling multipath channels fits perfectly the signal amplitude.

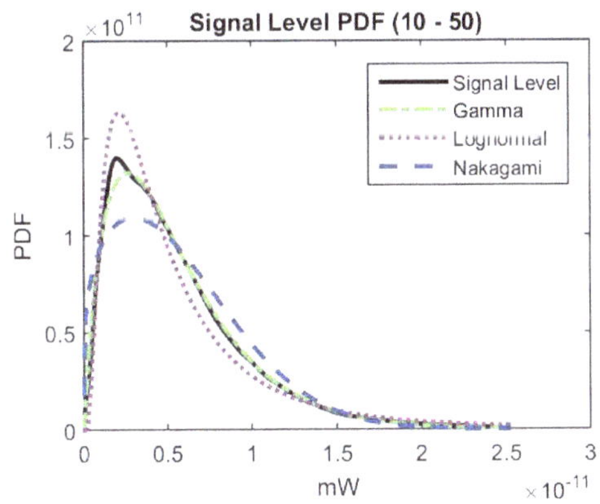

Fig. 11 Normal distribution PDF with parameters $\mu = -113.79$, $\sigma = 3.57$ (blue/dashed) that best fits the PDF of the signal level in dBm (black/solid)

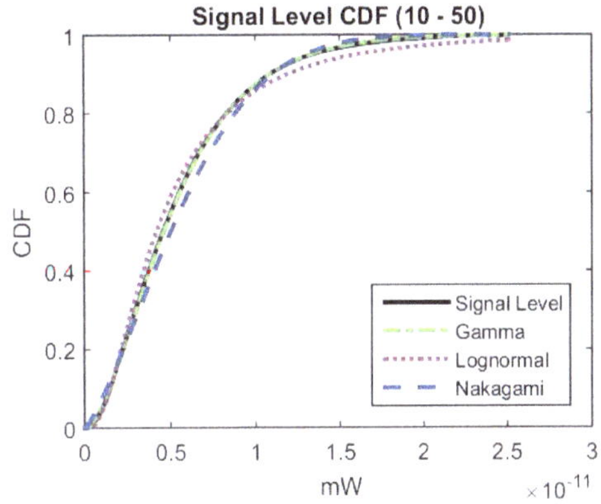

Fig. 12 Normal distribution CDF with parameters $\mu = -113.79$, $\sigma = 3.57$ (blue/dashed) that best fits the CDF of the signal level in dBm (black/solid)

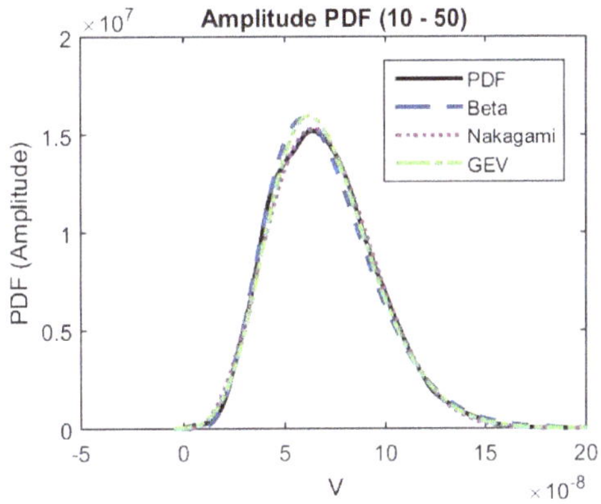

Fig. 13 Lognormal distribution PDF with parameters $\mu = -143.79$, $\sigma = 3.57$ (blue/dashed) and Nakagami distribution PDF with parameters $m = 1.92$ and $\Omega = 5.53e{-15}$ (pink/dotted) that best fits the PDF of the signal amplitude in Volts (black/solid)

4.2 Carrier to noise plus interference density ratio estimation

Taking advantage of the transmission pattern of the satellite[6], the estimation of the (N_0+I_0) level has been computed as the median signal level during all the quiet segments of the satellite along the measurement campaign. The estimated (N_0+I_0) is -162 dBm/Hz, this value will be used in the rest of the report. Fig. 14 and Fig. 15 show the probability density function (PDF) and cumulative density function (CDF) of the (N_0+I_0) respectively. PDF shows two Gaussian shape distributions, bigger one is due to the noise while the smaller is due to a narrowband interference present during the whole measurement campaign. In the CDF trace it can be observed

[2] https://www.mathworks.com/help/stats/fitdist.html

[3] μ and σ are, respectively, the mean in dBm and standard deviation (dB).

[4] *m* and *Ω* are, respectively, the shape parameter (aka fading figure) and scale. These parameters can be found in other references as μ and ω respectively.

[5] α and θ are, respectively, the shape parameter and scale.

[6] The satellite has a transmission pattern which is repeated each minute, the transmission is on for 12s, off for 12s, on for 12s, off for 12s and on for 12s. According to this duty cycle, only the 60% of the time of the passes is used for signal level analysis.

that noise density is above -169 dBm/Hz and below -154 dBm/Hz the 80% of the time.

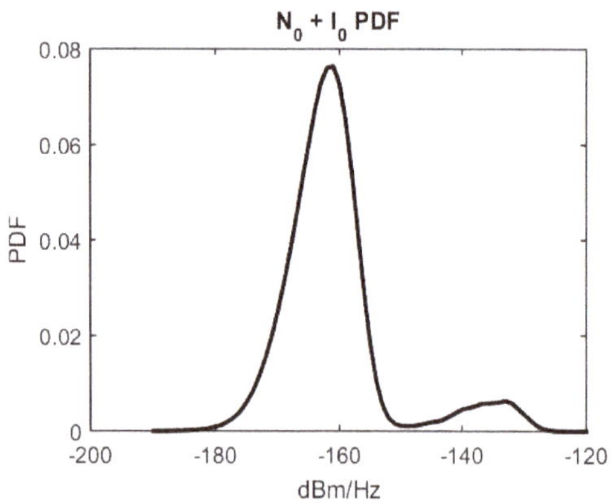

Fig. 14 (N_0+I_0) PDF

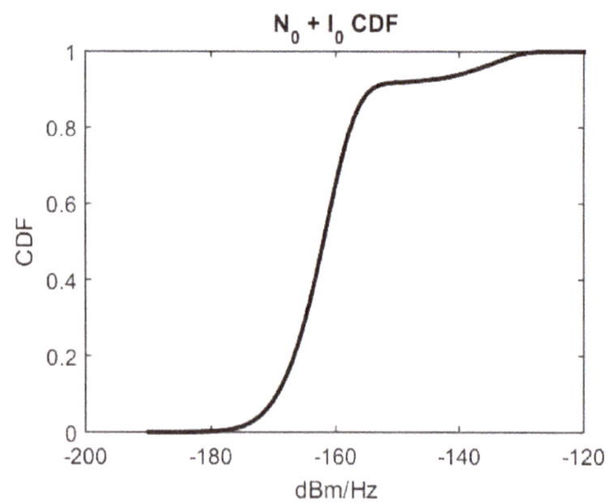

Fig. 15 (N_0+I_0) CDF

On the other hand, the carrier level corresponds with the mean signal level during the active segments as it has been computed in the previous section.

Fig. 16 shows the mean measured C/(N_0+I_0) and the confidence intervals (10% and 90%) versus the elevation angle. Taking into account the C/(N_0+I_0) thresholds presented in table 10 of [10] some considerations can be done:

- For elevation angles between 10 and 60 degrees, measured C/(N_0+I_0) is above the detection threshold (34.2 dB/Hz) the 90% of the time if SAT-MCS-0.50-1 (BPSK/CDMA) is considered.
- For elevation angles between 20 and 50 degrees, measured C/(N_0+I_0) is above the detection threshold (42.9 dB/Hz) the 90% of the time if SAT-MCS-1.50-1 (π/4 QPSK) is considered.
- For elevation angles between 30 and 40 degrees, measured C/(N_0+I_0) is above the detection threshold (50.3 dB/Hz) more than the 50% of the time if SAT-MCS-3.50-1 (8PSK) is considered.

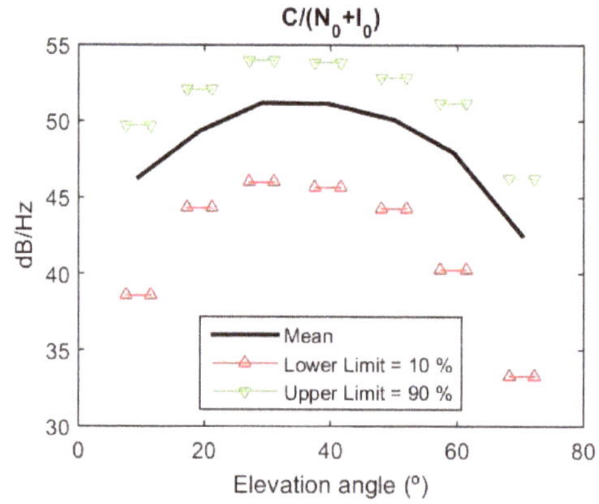

Fig. 16 Mean C/(N_0+I_0) per elevation angle and confident intervals (10% and 90%)

4.3 Signal to interference plus noise ratio (SINR)

In this section the experimented quality signal during the measurement campaign is analysed in terms of signal to interference plus noise ratio (SINR) for the BPSK/CDMA modulated carrier. Prior to showing the results, a couple of considerations should be taken into account:

SINR has been measured for BPSK/CDMA carrier but the results that will be shown in this section are also valid for other carriers.

Although BPSK/CDMA uses a spread spectrum modulation, SINR was computed at chip rate (i.e. before the despreading), so spreading gain of BPSK/CDMA is not taken into consideration.

A key block in the receiver architecture is the channel equalizer since its performance has a high impact on the SINR results (see section 4.4). The SINR results are presented with and without[7] the use of 17-tap equalizer[8] at the receiver.

Mean SINR and confidence intervals (10% and 90%) versus elevation angle is shown in Fig. 17 (without channel equalizer) and Fig. 18 (with a 17 taps channel equalizer). In the case of using the 17 taps equalizer, SINR is above 0 dB for elevation angles between 20 and 60 degrees while without the equalizer, SINR is roughly 5 dB lower than in the previous case.

[7] In fact, an equalizer of one coefficient is used to correct the amplitude and phase of the signal.

[8] The number of taps has been adjusted empirically for SINR maximization without an excessive computation consumption.

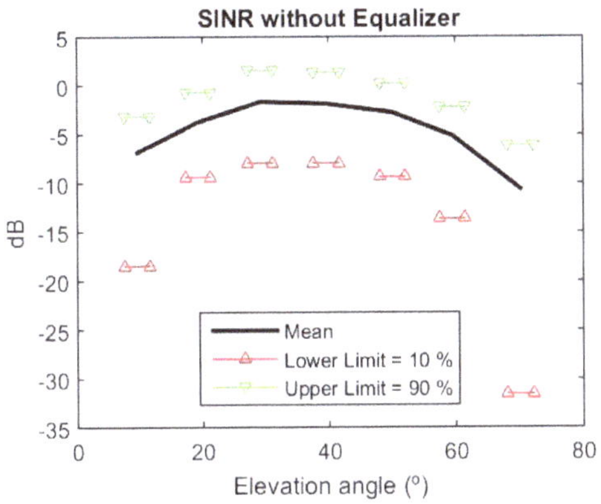

Fig. 17 Mean SINR versus elevation angle and confident intervals (10% and 90%) without channel equalizer

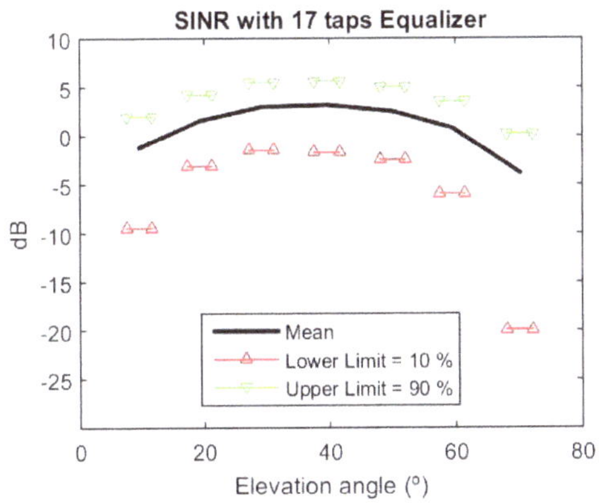

Fig. 18 Mean SINR versus elevation angle and confident intervals (10% and 90%) without a 17 taps channel equalizer

4.4 Channel equalizer impact

Strong narrowband interferences were observed during the collected captures, an example of these interferences is shown in Fig. 19. This figure shows the spectrogram of the BPSK/CDMA modulated signal (161.8125 MHz) after the Doppler shift compensation including the narrowband interferences.

The channel equalizer, when it receives this signal, attempts to block the frequencies of the interferences. For that purpose, it needs a considerable number of taps (17). The frequency response of the identified channel from the previous example is shown in Fig. 20, where it can be seen the channel response follows the evolution in time of the interferences.

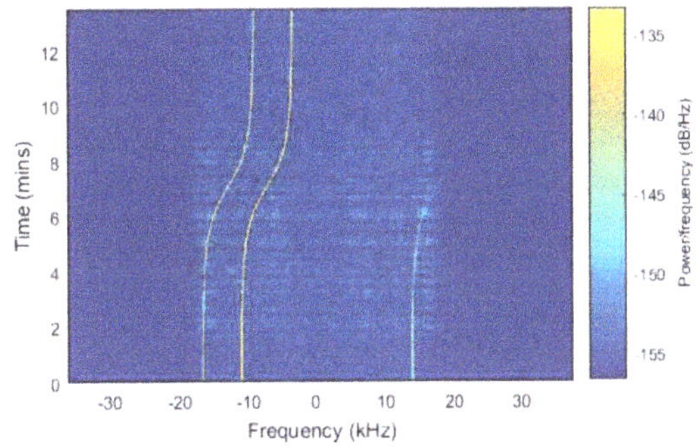

Fig. 19 Spectrogram of BPSK/CDMA modulated signal after Doppler compensation for one of the captures

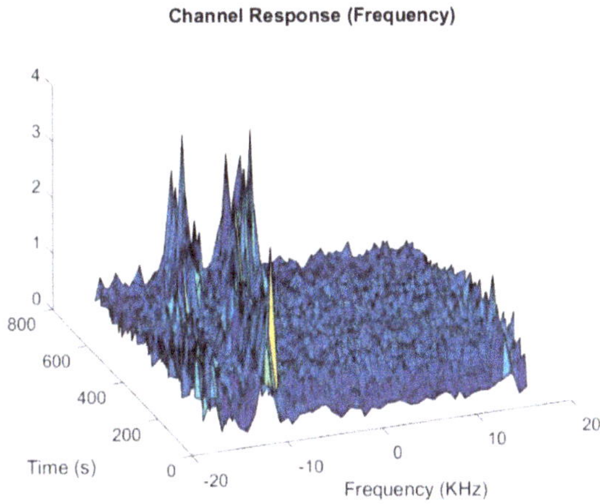

Fig. 20 Frequency response of the identified communication channel for one of the captures.

5 Conclusions

There are different conclusions that can be drawn from the results section. The main ones are:

- Signal level (power) fits well with a Gamma distribution which shape (α) and scale (σ) parameters are respectively 1.92 and 2.88×10^{-12}.
- Signal amplitude fits perfectly with a Nakagami distribution which shape (μ) and scale (Ω) parameters are respectively 1.92 and 5.53×10^{-15}.
- The estimated (N_0+I_0) for the measurement campaign has been -162 dBm/Hz.
- According to table 10 of [10], $C/(N_0+I_0)$ is above the detection threshold (34.2 dB/Hz) the 90% of the time if the most protected waveform is considered. For elevation angles between 20 and 50 degrees, the signal level also allows for detecting the second and third waveforms.
- It would be highly recommended to include a properly adjusted channel equalizer (number of taps and step gain) to the receiver.

- SINR is above 0 dB for elevation angles between 20 and 60 degrees, achieving maxima of 3 dB between 30 and 40 degrees.

Part of the measurement campaign conclusions have been presented during ITU 5B meeting (May 2018) as a reply to the ITU-R Resolution 360 invitation to study compatibility between VDES satellite component and incumbent services in the same and adjacent frequency bands. The objective is to determine potential regulatory actions, including spectrum allocations to the MMSS (Earth-to-space and space-to-Earth) for VDES applications. Contribution to ITU 5B meeting proposes that the measurement campaign conclusions were included in the working document towards a preliminary draft new report ITU-R M.

The receiver designed in this work use the whole transmitted sequence for different purposes (as time synchronization or equalization) which will not be possible in a practical receiver. Once the final waveform has been decided, it will be necessary to design and implement a practical receiver taking into account training sequences or using blind schemes for synchronization.

Acknowledgements

Work supported by the European Space Agency under ESTEC Contract No. 4000122682/17/NL/FE: VDES Testing. Opinions, interpretations, recommendations and conclusions expressed herein are those of the authors and are not necessarily endorsed by the European Space Agency.

References

[1] ITU-R, "Recommendation ITU-R M.2092-0 - Technical characteristics for a VHF data exchange system in the VHF maritime mobile band," 2015.

[2] ITU-R, "Final Acts WRC-15," in *World Radiocommunication Conference*, Geneva, Switzerland, 2015.

[3] IALA AISM, "E-Navigation Testbeds - NORSAT-2 ESA VDE-SAT downlink verification," [Online]. Available: http://www.iala-aism.org/technical/e-nav-testbeds/norsat-2/. [Accessed 27 August 2018].

[4] Bradbury, L., Spydevold, I., Haugli, H., et al., "NORSAT-2: Enabling Advanced Maritime Communication with VDES," 2017. [Online]. Available: https://digitalcommons.usu.edu/smallsat/2017/all2017/167/.

[5] Oerder, M., Meyr, H., "Digital Filter and Square Timing Recovery," *IEEE Transactions on Communications,* vol. 36, no. 5, pp. 605-612, May 1988.

[6] Álvarez-Díaz, M., López-Valcarce, R., Mosquera, C., "SNR Estimation with Heterogeneous Frames," *Advanced Satellite Mobile Systems,* vol. 10.1109/ASMS.2008.53, pp. 268 - 273, 2008.

[7] Pauluzzi, D., Beaulieu, N., "A comparison of SNR estimation techniques for the AWGN channel," *IEEE Transactions on communications,* vol. 48, no. 10, pp. 1681-1691, October 2000.

[8] Johnson, N.L., Kotz, S., Balakrishnan, N., Continuous Univariate Distributions (Volume I, 2nd edition), Wiley, 1994.

[9] Proakis, J., Salehi, M., Digital Communications (5th edition), McGraw-Hill Education, 2007.

[10] IALA AISM, "G1139 Technical specification of VDES," 15 December 2017. [Online]. Available: http://www.iala-aism.org/product/g1139-technical-specification-vdes/.

Section 7 – Mobile Satellite Systems and Bandwidth Efficient Techniques

MOBILITY ENHANCEMENT FOR DIGITAL VIDEO BROADCAST NETWORKS VIA SATELLITE

***Burak Unal*[1], *Ashraf Ali*[1], *Nicholas Avlonitis*[1], *and Ifiok Otung*[1]**

[1]*Mobile and Satellite Communications Research Group, University of South Wales, Pontypridd CF37 1DL, UK*
{burak.unal, ashraf.ali, nicholas.avlonitis, ifiok.otung}@southwales.ac.uk

Keywords: DVB-S2, DVB-S2X, APSK, AWGN, RAYLEIGH, RICIAN

Abstract

The DVB-S2 standard has been optimised for provision of broadband multimedia services to geographically dispersed fixed satellite end-users under line-of-sight (LOS) channel condition. However, continuing growth in demand for mobile broadband services via satellite necessitates a detailed examination of the DVB-S2 technology to assess its suitability for application in the mobile satellite environment. The environment is affected by signal fading due to path blockage, multipath propagation and shadowing, Doppler shift which is expressed by the Rician channel K-factor parameter values. In this paper, a model was developed in MATLAB® to simulate transmission scenarios through a mobile satellite channel, in order to investigate the bit-error-rate (BER) performance of the 16-APSK and 32-APSK modulation formats for different coding rates. This modelling of higher modulation formats of the DVB-S2 standard until now, has not been addressed adequately in related literature. Moreover, the mobility effect over the BER performance is not well investigated in the literature especially for high modulation schemes. Results obtained by simulation indicate a significant degradation in the system's BER when compared to a line-of sight channel condition. Which even becomes worse with added varying mobility scenarios. However, higher Rician channel K-factor values tend to improve the link quality closer to that of the AWGN. These results highlight a crucial need to develop improved receiver processing and new switching thresholds applicable to the mobile satellite environment that deliver improved link availability as well as enhanced data throughput for different mobility scenarios.

1 Introduction

The Digital Video Broadcast via satellite second generation (DVB-S2) standard [1] has been the key technology that supports the deployment of multimedia and broadband services to the fixed satellite end-user terminals. One of the primary advantages of satellite systems is their wide area coverage capabilities, so that they can be used to provide communication services in areas and terrain where the conventional wired and wireless communication infrastructures are unable to serve, and in emergency situation, for fast recovery. In addition, satellite systems are currently planned to be integrated in to the next generation, fully flexible network for high-speed and high data-rate services, connected with the 5G space wide web of the 5G system of systems. This quest is evidenced by the perceived potentials of using satellites as a mechanism of data off-loading to alleviate the already congested mobile network.

The geostationary (GEO) satellites are primarily used for the fixed satellite services (FSS) which are predominantly a line-of-sight (LOS) channel that is only affected by the well-known Additive White Gaussian Channel (AWGN), with free-space-loss (FSL) the major cause of signal attenuation. In addition to this, systems operating at higher frequencies such as the K_u- and K_a-band, are significantly affected by propagation impairments such as gaseous absorption, rain attenuation, scintillation and depolarisation. However, the mobile satellite channels are susceptible to further channel impairments. For example, the radio path between the satellite and the mobile user-terminal can vary from the classical LOS channel to a non-line-of-sight (NLOS) channel that may be severely obstructed by obstacles such as tall buildings, trees, mountains and foliage. This introduces factors such as shadowing, scattering, absorption, reflection and diffraction in characterising the variation of the received signal [2].

The DVB-S2 and its extension (DVB-S2X) [3] are the existing standards adopted for delivering broadband services to fixed earth user terminals, which are optimised for LOS satellite channels.

A prominent feature of the DVB-S2 is its bandwidth-efficiency and ability to reliably operate at exceptionally low carrier-to-noise ratios (CNR) providing means of real-time adaptive fade mitigation by adaptively switching between ranges of modulation and coding schemes (MODCODs) [4]. Overall, the standard has been optimised around three concepts: best transmission performance approaching the Shannon's limit, total flexibility, and reasonable receiver complexity [5]. However, recent studies show that the DVB-S2 standard can be enhanced to operate below the documented threshold by incorporating time diversity and maximal ratio combining (TD+MRC) [6].

What still requires further investigation is the impact of mobile channel on the performance of DVB-S2, especially with the current surge in interest by operators to adopt the satellite systems within the 5G paradigm. This has presented a problem that is yet to be adequately addressed in the existing literature. Baotic et al. [7] implemented a simulation model in the MATLAB / Simulink environment, however, only an AWGN channel for QPSK and 8-PSK modulations formats was considered in their study. Kumar et al. [8] considered a number of PSK based modulation formats (QPSK, DQPSK, OQPSK) and a Rician channel but not considering the full DVB-S2 system. Azarbad et al. [9] implemented a model in MATLAB to investigate various algorithms that will allow a more efficient switching between MODCODs. However, their work only considered an AWGN channel. In [10] MODCODs using high order APSK with higher spectral efficiency, as introduced in the new DVB-S2X standard, were considered but only in the AWGN case. A simple Rician channel model in which the amplitude of the line-of-sight (LoS) path is represented by the Nakagami distribution was examined in [11]. Mathematically-tractable expressions of the BER for adaptive modulation schemes (AMS) over land mobile satellite channel were derived for 4 modulation formats (BPSK, QPSK, 16- and 64- QAM). However, the work of [11] did not model the additional effects of MODCODs, such as error correction and the associated code rates, in the context of DVB-S2.

Doppler shift, even with small values, has a great impact over the channel quality and received BER values. It was shown in other research efforts, which will be highlighted later in this paper, that Doppler shift due to varying terminal mobility has a great negative influence over the spectral efficiency for PSK based formats. There is a need to examine the combined effect of mobility, as manifested by both the Doppler Effect and fading, as modelled by Rician fading statistics, on the performance of the DVB-S2 standard. The results show that the standard could be enhanced to take into consideration mobility scenarios and the results presented in this paper could be used towards this.

There is therefore, a need to optimise the standard to adapt effectively to the complexities introduced by the mobility by investigating the performance of DVB-S2 in the case of a mobile channel affected by multipath fading and AWGN. In this paper a model of a communications system based on the DVB-S2 standard is presented. The channel model takes into account AWGN and assumes Rician fading statistics.

2 Radio Propagation Channel Models

The effect of the intervening medium on the efficient transmission of radio signal through the earth-space path is an important consideration for designing an optimal and robust satellite communication system. These effects are responsible for the perturbation and impairment of the information signal, which have significant deteriorating impact on the quality of service and performance of these systems. The extent of these effects on the link performance depends on factors such as, the frequency of operation, climate, type of transmission and the satellite's slant-path elevation angle; as well as whether the channel is fixed or mobile. In both the fixed and mobile cases, these effects become severe with increasing frequency and decreasing elevation angle. The atmosphere is stratified into layers, with the ionosphere and troposphere being the most significant among them. Systems operating at the Ku- and Ka-band frequencies are mainly affected by tropospheric phenomenon, such as rain attenuation, scintillation, gaseous absorption, and depolarisation [12]. In this paper Ku band frequency will be investigated as the applications and communications over this band are widely deployed, which trigger the need for more research in this frequency spectrum range to be highlighted for better understanding of the channel model and its influence over the transmission quality.

The mobile radio propagation channel introduces additional degradation on the performance of wireless communication systems. The path from transmitter to receiver can vary from being in line-of-sight to one that is severely obstructed by intervening objects such as mountains, foliage and buildings. In order to determine the variations in received signal power, the path loss which causes reduction in the power received at the receiver front-end (mobile user-terminal), is one of the prominent factors. In addition to this, shadowing, which is caused by obstacles between the transmitter and receiver, attenuates signal power through absorption, reflection, scattering and diffraction. The prediction of the received signal power and coverage area is of pertinent concern in planning and optimisation of mobile radio systems and networks [2]. Figure 1 (a) and (b) show typical line-of-sight and a non-line-of-sight propagation paths respectively.

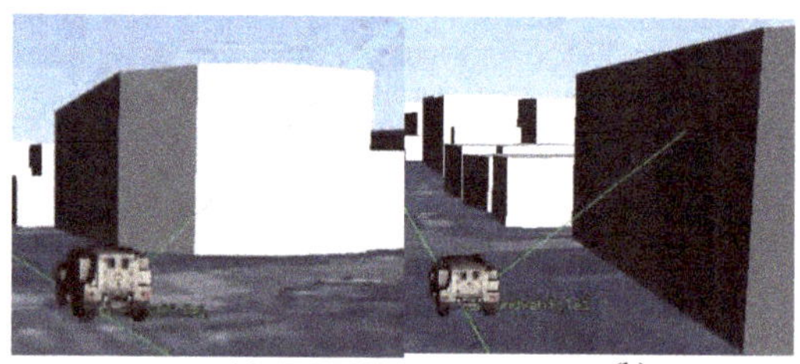

(a) (b)

Figure 1: Radiowave propagation channel – Line-of-Sight (LOS) and Non Line-of-Sight (NLOS)

3 The DVB-S2

The DVB-S2 specification was developed in 2003 and structured as a toolkit to allow the implementation of the following satellite applications: TV and sound broadcasting, interactivity (i.e., Internet access), and professional services, such as TV broadcasting links and digital satellite news gathering. Channel coding and modulation are based on most

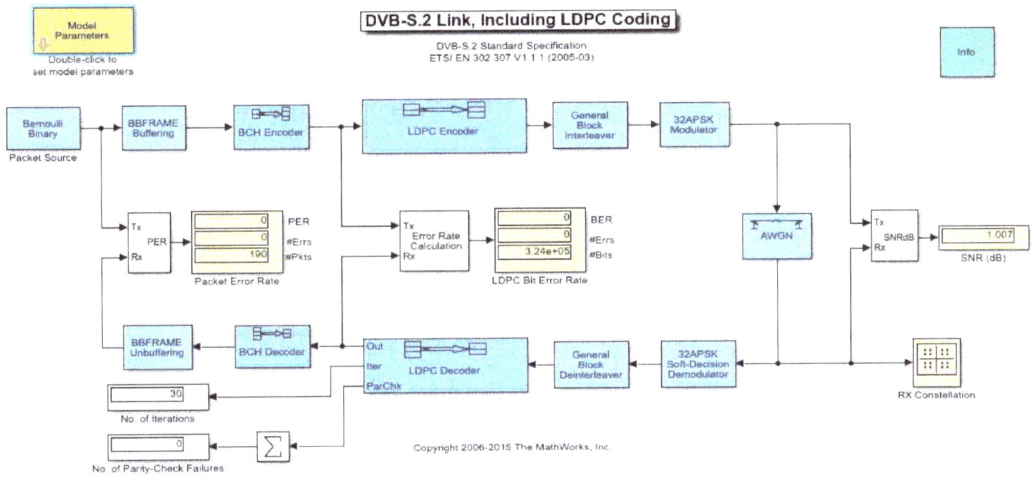

Figure 2 DVB-S2 Block Diagram in MATLAB [14]

recent developments in the scientific community: low density parity check and Bose, Chaudhuri, and Hocquenghem (BCH) codes are adopted, combined with QPSK, 8-PSK, 16-APSK, and 32-APSK modulations for the system to work properly on the nonlinear satellite channel [3, 5]. The DVB-S2X supports significantly higher modulation schemes such as the 64-APSK, 128-APSK and 256-APSK, and hence, higher spectral efficiency to support the CNR required typical for professional applications such as the Digital Satellite News Gathering (DSNG). DVB-S2X uses the proven and powerful Low Density Parity Check (LDPC) and Forward Error Correction (FEC) scheme in combination with BCH-FEC as outer code for the best error correction results [3] .

The DVB Return Channel via Satellite (DVB-RCS2) [13] has included a new set of functions called DVB-RCS+M [14] [15] that support broadband communications via mobile satellite services to mobile terminals. However, with the DVB-RCS seeming to have covered for mobility in some specific cases [14], there is an evident need to develop techniques to extend these further in order to be fit for different scenarios in the NLOS environments. This is justified by the increase in commercial interest and escalating demand from various section of the user stratification for multimedia and broadband services on the move.

Few research inputs from both the academia and the industry addressing the mobility issues pertaining to the DVB-S2 standards. The challenges associated with introducing the mobility, and some possible techniques to circumvent them have been identified. Among the aspects identified are proactive retransmission techniques and efficient handover mechanisms involving beams and gap-filler handovers. The DVB-RCS+M radio interface considers a satellite forward link based on the DVB-S2 enhanced with additional optional spreading and Link Layer Forward Error Correction (LL-FEC) functionalities. The return link (RL) is based on the DVB-RCS air interface extended by incorporating spreading and a continuous carrier mode based on DVB-S2 waveform [14] [15] [16]. Both modes, the forward and return links, can be used individually or in parallel to optimise satellite resources versus traffic pattern and operating mode [17]. The mobility aspects of the DVB-S2/RCS broadband systems were discussed in [18], the standard has a spreading techniques using different spreading factors to spread the errors over longer time period in hope that the standard implemented error correction techniques will be able to recover the errors.

4 Implementation of 16-APSK and 32-APSK in a Mobile Satellite Channel Environment

The DVB-S2 standard is integrated into MATLAB as functional blocks which enable system performance evaluation for a range of MODCODs. These include the QPSK and 8-PSK modulation formats at various coding rates for the AWGN channel [16]. The block diagram shown in Figure 2 presents the SIMULINK block version of the implementation. The function of each block therein is described in detail in [3] [16] [9]. The AWGN block represents the channel and introduces noise to the transmitted signal based on the value of ratio of average energy per symbol to noise spectral density (Es/No).

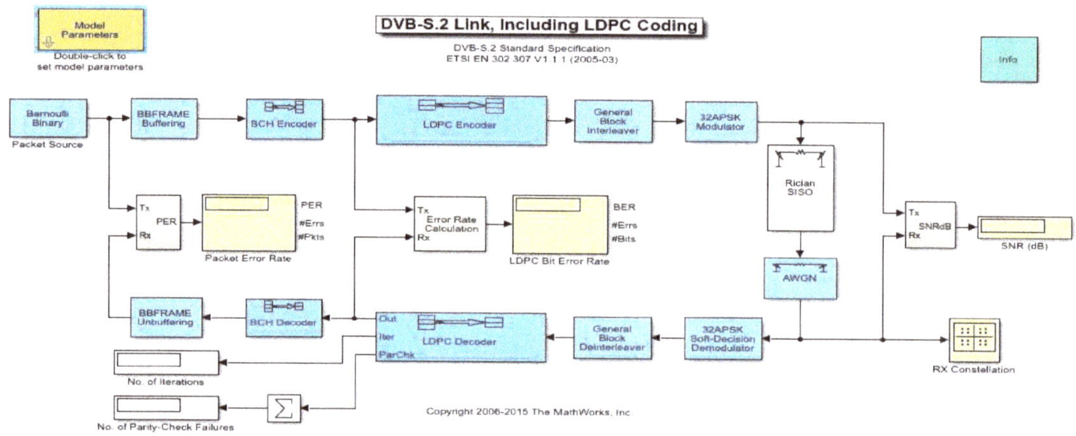

Figure 3: DVB-S2 16-APSK and 32-APSK Implementation

In Figure 3, the DVB-S2 model has been modified to extend the current modulation schemes in SIMULINK by including the higher APSK modulation schemes, in this paper we are mainly interested in 16-APSK and 32-APSK. In furtherance, a hypothetical mobile satellite channel, represented by a Rician channel, has been integrated into the model in order to simulate the impact of user mobility on system BER performance. The new system as depicted in Figure 3 has been used to simulate various MODCODs and the results are compared with the baseline AWGN channel in order to quantify the impact of user terminal mobility for various K-factors. Moreover, the results for lower modulation indexes (i.e. QPSK) were compared against other references and related works in which the system model was validated [19] [20]. To the best of our knowledge, this is the first time that the performance of high spectral efficiency MODCODs in the context of the DVB-S2 standard were investigated assuming Rician fading statistics and a AWGN channel under different mobility scenarios.

5 Channel Model Selection and Parameters

The selection of the appropriate channel model is of great importance to reflect a close to real channel characteristics that emulate real channel behaviour under different operation conditions. The parameters of the channel selected is strongly associated with the intended running scenarios and environment. In the next subsections, the Rician channel will be explained, the adopted parameters will be listed, AWGN channel selection criteria will be highlighted, and finally the mobility related aspect of the channel will be explained.

5.1 Channel Model Selection Criteria

The Rician distribution is a statistical function that is used to represent the multipath propagation channel models. It best represent channels with LOS dominant component along with other weaker power multipath signals. Equation 1 shows the Probability Density Function (PDF) of the Rician channel.

$$p(r/s,\sigma) = \frac{r}{\sigma^2}\exp\left[-\frac{(r^2+s^2)}{2\sigma^2}\right] I_0\left(\frac{sr}{\sigma^2}\right) \; ; \text{ For } s \geq 0 \; r \geq 0 \quad (1)$$

Where s is the peak amplitude of the dominant signal, $I_0(.)$ is zero order Bessel function of the first kind, $r^2/2$ is the instantaneous power. σ is the standard deviation of the local power.

K (the Rician Factor) is the most important parameter of the channel in which the ratio of the main dominant LOS path power to the total other NLOS dispersion paths is calculated and expressed as in equation (2).

$$K = \frac{P_{LOS}}{P_{NLOS}} \qquad (2)$$

Where P_{LOS} and P_{NLOS} are the Line of sight and non-line of sight power respectively

As shown in the equation, if the numerator value P_{LOS} approaches to zero, the K-factor will be zero as well in which the channel is considered as a Rayleigh which is similar model to Rician but without dominant LOS component [21].

As explained before, there are different simulations scenarios in which the Transmitter and receiver are either in LOS or NLOS states. The geographical area, vegetation and building obstacles, in addition to the density of the receiving nodes are the main contributing factors that distinguish one state from the other [22]. Table 1 shows the different states along with a brief description and mobile channel sample characteristics as described in [22].

Table 1: Channel States and related parameters [22]

State		Description	Characteristics
LOS		Line of Sight (Directive antenna)	Rician distribution
NLOS	Shadowing	Shadowing (Due to trees)	Small scale fading: Rician Large scale fading: Log-normal
	Blockage	Blockages (Due to buildings, bridges and tunnels)	No signal received or signal below noise floor

As shown in the table, Both LOS and NLOS (Shadowing) can be modelled as a Rician channel with different K-factor values. This flexible feature of the Rician channel is powerful to meet different scenarios and channel conditions. For example, with a typical K-factor of 17, a near LOS transmission is assumed here in which the LOS component has considerably higher power compared to other multipath dispersed components. On the other hand, for the shadowed NLOS state, a different value for K factor can be used to replicate having less power for the LOS component on the expense of giving more weight for the multipath dispersed components. Finally, having a NLOS signal with multipath propagation components can be modelled with K = 0 using Rician channel or Rayleigh channel model. In all above cases, the Rician channel model was shown to be able to emulate the three channel states with different (K-factor) parameter set.

Suburban areas has more probability of having LOS path between the receiver and the transmitter which in addition to other multipath fading components gives more weight to the Rician channel model compared to other models due to the more probability of having LOS paths compared to other environments. The ratio between the dominant LOS component and other NLOS multipath components is best characterised by K-factor. But having Rician channel model by its own is not enough for accurate characterisation of the transmission channel. It is quite often having a wireless environment that is represented by AWGN channel in which the receiver side noise, with constant power spectral density, is added over the whole transmission spectrum.

5.2 Mobility and Doppler shift

In the context of communications, the Doppler shift is the well-known effect that manifests as a change (shift) of frequency in the signal apparent in a system were the source and receiver are in relative motion. The Doppler shift f_D is proportional to the frequency f of the electromagnetic wave in propagation, and is given by [23] as shown in equation (3).

$$f_D = \frac{v_r f}{3.6 \times c} cosa \qquad (3)$$

Where:

- vr is the relative speed between the transmitter and the receiver,
- f is the carrier frequency
- c is the speed of light (≈3 x 108 m/s)
- and $a \in [0, \pi]$ is the angle of the velocity vector (Doppler shift maximises for $a = 0$)

Table2 shows the resultant Doppler shifts for two scenarios in which the worst case Doppler shift was considered. So it is assumed that the elevation angle (a) is zero, and the mobile terminal is moving toward the satellite, the carrier frequency is set to 14.5 GHz as it is part of the Ku band spectrum for satellite TV broadcast applications. Ku band uses 14 to 14.5 GHz as transmit frequencies and 12 to 12.5 GHz as receive frequencies, however this paper is analysing the downlink transmission band only.

Table 2: Doppler shift for various scenarios (f = 14.5 GHz, $\boldsymbol{a}$ = 0 rad)

ID	V_r (km/h)	f_D (Hz)	Description of scenario
1	2	≈ 27 Hz	Pedestrian
2	30	≈ 403 Hz	Car or autonomous vehicle
3	70	≈ 940 Hz	Satellite applications in cars & trains

There are different empirical stochastic channel model in the literature that were adopted based on statistically analysed measured values. The standardised channel models for Land mobile Satellite (LMS) count on different parameters to accurately model the channel. The number of paths, path delay vector, power profile, and Doppler frequencies for the paths are the main parameters that distinguish one channel from the other. Rician channel model has a set of parameter as summarised in table 3. The sample rate of the input signal in hertz is a measure of the symbols sampling frequency, the value picked is one of the parameters in DVBS2 standard as indicated in [1]. The filtered Gaussian noise is used to model the path gains for the LOS and multipath components, filtered Gaussian and sum of sinusoids are the main two approaches used to model the path gains, the filtered Gaussian was adopted as it gives better estimates for the mobility and environment scenario presented in this paper.

Table 3 Rician channel Model parameters list

PARAMETER	VALUE
Sample Rate	27500e3 Hz
Path Delays	0
Average Path Gains	0
Normalize Path Gains	true
K- Factor	User defined
Direct Path Doppler Shift	0
Direct Path Initial Phase	0
Maximum Doppler Shift	User Defined
Doppler Spectrum	jakes
Fading Technique	Filtered Gaussian noise
Seed	100
Path Gains Output Port	true

Using the above parameters list and by varying the mobile terminal speed, which is set by varying the maximum Doppler shift, different results of various mobility and K-factor values were generated and will be explained in the next section.

6 Results and Discussion

In order to evaluate the performance of the new system, we have made a careful selection of simulation scenarios using suitable system specifications. As a consequence, only 16-APSK and 32-APSK are considered for the purpose of this analysis. Figure 4 shows the result of 16-APSK system's BER performance for coding rates 2/3, 3/4, 4/5, 5/6 and 9/10 and Figure 5 presents similar results for the 32-APSK for coding rates 3/4, 4/5, 5/6, 8/9 and 9/10. The result is in agreement with the theoretical E_s/N_o tabulated in the ETSI DVB-S2 standard, demonstrating the power efficiency of lower coding rates as compared with the higher code rates. For example, a BER of 10^{-4} could be achieved with an E_s/N_o of 12.8 dB in 16-APSK with code rate 9/10 compared to E_s/N_o requirement of 15.8 dB for 32-APSK with code rate 9/10. It is, however, worth noting that 16-APSK is more power-efficient, as compared with 32-APSK.

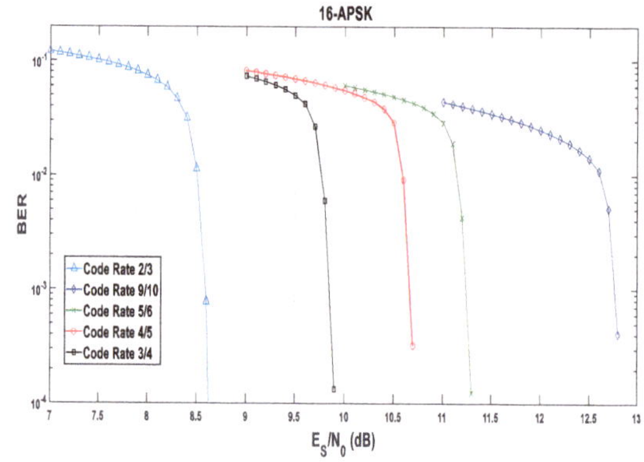

Fig. 4: BER performance for 16-APSK for AWGN Channel

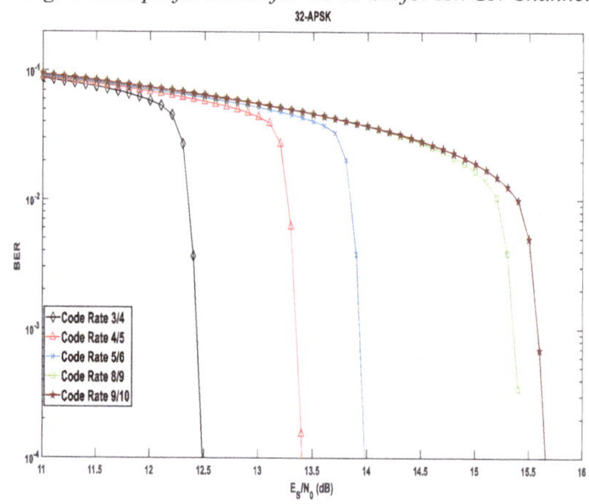

Fig. 5: BER performance for 32-APSK for AWGN Channel

On the other hand, for the mobile channel scenario where the Rician K-factors are taken into account, the system performance is degraded as a result of multipath fading. This presents a relatively poor BER performance, when compared with the LOS (AWGN) channel condition, earlier described. The 32-APSK (9/10) MODCOD was selected for this analysis, due to the fact that this scenario provides the highest spectral efficiency and is also the most power-demanding one. We considered the most spectrally efficient modulation formats of DVB-S2 as they are the most interesting in terms of susceptibility to multipath and noise effects. The simulation results, depicted in Figure 6, show the impact of the Rician K-factor on system BER performance such that the higher values of K-factors (e.g. K = 10) represent a scenario where the dominant power is due to the direct path (i.e. LOS) closing down on the AWGN channel.

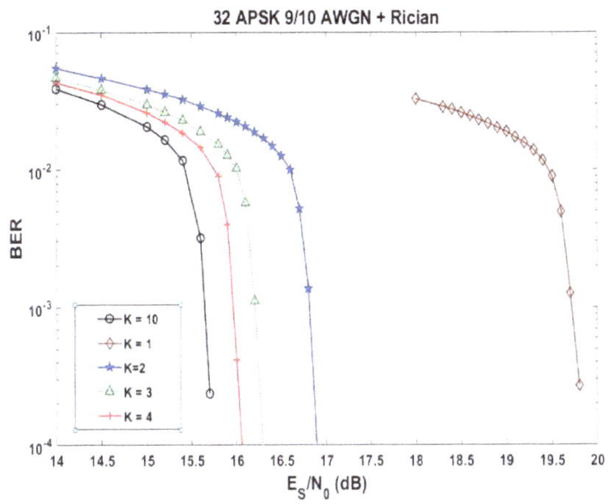

Fig. 6: BER performance for 32-APSK for Rician Channel with various Rician K-factor (static mobility)

To validate the model against other implemented models in the literature, we generated the BER results for the lowest modulation and code rate combination (QPSK ¼) with K-factor value of 50 at 30 Km/h mobility. The PER values were compared against the results in literature in which it was shown that there is 1 dB difference only between our model and the model proposed in [20]. Following this validation, the modulation and code rate schemes were extended to a higher modulation and code rate (32-APSK 9/10). using MATLAB we get results of the proposed model that reflect varying Rician channel K-factor value for three different mobility scenarios as indicated before in table 2. Figure 7, figure 8 and figure 9 show the BER against Es/No for different mobility and K-factor combinations.

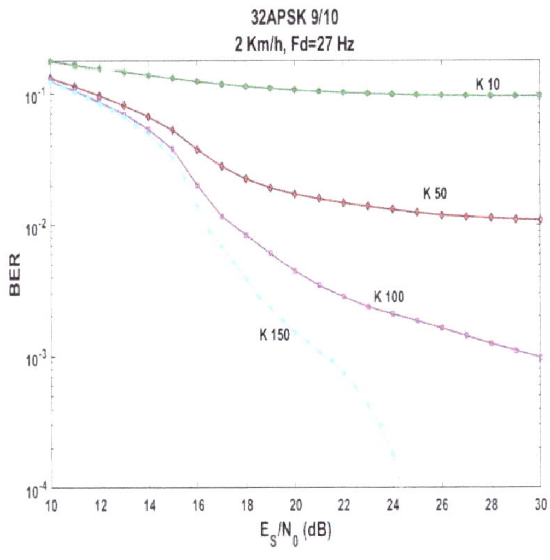

Figure 7: BER for 32APSK modulation with 9/10 Code Rate at 2 Km/h terminal mobility

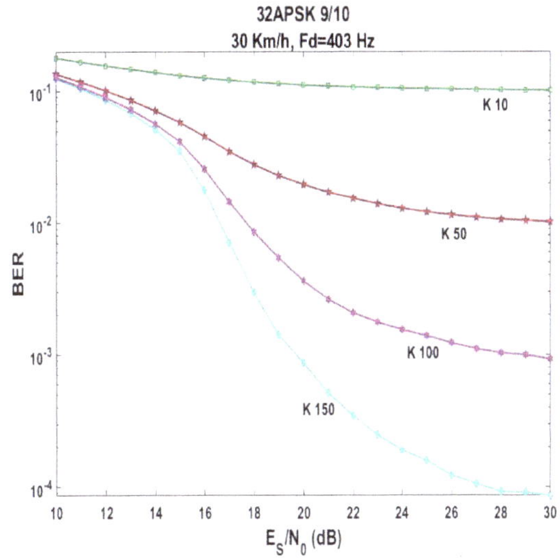

Figure 8: BER for 32APSK modulation with 9/10 Code Rate at 30 Km/h terminal mobility

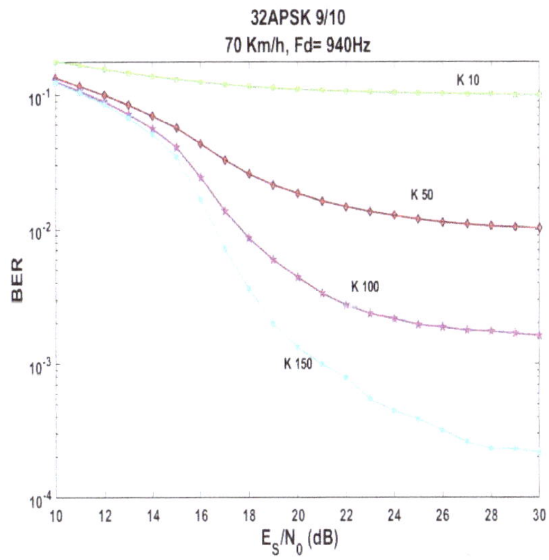

Figure 9: BER for 32APSK modulation with 9/10 Code Rate at 70 Km/h terminal mobility

As noted in the figures, the higher the mobility for the same adopted Rician channel model (for example K=150) resulted in a higher Es/No threshold value, it increased from 15.7 dB to 24.2 dB to 28 dB with BER 10^{-4} by increasing the mobility from 0 Km/h to 2 Km/h to 70 Km/h respectively.

7 Conclusion

The potential benefits of performance enhancement of DVB-S2 for a mobile scenario are obvious in the satellite industry, and the concepts and technologies that can enable this are of great interest to the stakeholders. Mobile channel propagation issues were identified, and the need for further critical investigation into mobile satellite system propagation modelling and performance prediction was highlighted. The model includes a Rician channel and was used to investigate the effects of multipath in a mobile channel environment. Based on our analysis, for a BER of 10^{-4}, an increase in the K-factor from 1 to 2, from 2 to 3 and from 3 to 4 results in a power penalty of 3 dB, 0.6 dB and 0.25 dB, respectively. An increase in the K-factor from 4 to 10 results in only a 0.4 dB power penalty.

In this work, we have investigated, via extensive simulations, the impact of the mobile satellite channel on DVB-S2 system performance, based on the assumption that a Rician model adequately represents the satellite channel. The results obtained demonstrated that, for a DVB-S2 system, the mobile channel BER performance improves with increasing Rician K-factor for all the modulation and coding rates used for this analysis. Moreover, the proposed model was validated against other models in the literature and the modulation scheme was further extended up to 32-APSK to study the BER and channel model implications. It was found that the mobility scenarios has a negative implication over the quality of the channel in terms of the achieved BER at certain Es/No threshold values. The higher the mobility, the higher BER was achieved for the same Es/No and for the same K-factor value for 32-APSK 9/10 code rate, the higher the mobility for the same adopted Rician channel model (for example K=150) resulted in a higher Es/No threshold value, this motivate us to carry out further research in developing techniques at the receiver side to reduce the mobility implications.

Our future plans involve studying the influence of the other Rician channel dynamic factors to have more accurate close to real practical channel models by exploiting the other channel parameters, our analysis shows clearly that mobility and multipath have a severe effect on the link performance and the specific (Rician) model used has limitations and therefore a more 'practical' model is required.. Moreover, there is a need for undertaking a series of investigations using different channel models (for example, Lutz's model) and a variety of propagation environment conditions (such as multipath, scattering, shadowing, etc.), leading to the development of novel techniques that will establish new improved BER thresholds to be incorporated into future versions of the DVB-S2X standard.

References

[1].ETSI, Digital Video Broadcasting (DVB); Second generation framing structure, channel coding and modulation systems for Broadcasting, Interactive Services, News Gathering and other broadband satellite applications (DVB-S2). 2009.

[2].Blaustein, N., Radio Propagation in Cellular Networks. 1999: Artech House Pulishers.

[3].ETSI, Digital Video Broadcasting (DVB); Second genaration framing stucture, channel coding and modulation systems for broadcasting, interactive services, news gathering and other broadband satellite applications part:2 DVB-S2 Extensions (DVB-S2X). 2014.

[4].Panagopoulos, A.D., P.D.M. Arapoglou, and P.G. Cottis, Satellite communications at KU, KA, and V bands: Propagation impairments and mitigation techniques. IEEE Communications Surveys & Tutorials, 2004. 6(3): p. 2-14.

[5].Morello, A. and V. Mignone, DVB-S2: The Second Generation Standard for Satellite Broad-Band Services. Proceedings of the IEEE, 2006. 94(1): p. 210-227.

[6].Uggalla, L., A Resilient Ka-band Satellite Video Broadcast System Incorporating Time Diversity and Maximal Ratio Combining. 2015, University of South Wales, United Kingdom.

[7].Baotic, P., et al. Simulation model of DVB-S2 system. in Proceedings ELMAR-2013. 2013.

[8] P.Sunil Kumar, et. al 'Performance Evaluation of Rician Fading Channels using QPSK, DQPSK and OQPSK Modulation Schemes in Simulink Environment. International Journal of Engineering Science Invention, 2013. Vol. 2(Issue 5): p. PP.07-16.

[9].Azarbad, B. and A.B. Sali, eds. DVB-S2 Model in Matlab: Issues and Impairments. MATLAB - A Fundamental Tool for Scientific Computing and Engineering Applications, ed. V. Katsikis. 2012, INTECH.

[10] El-Abbasy, K., B. Abdelhamid, and S. Elramly. Performance evaluation of DVB-S2 and DVB-S2X systems. in 2015 IEEE International Conference on Communication, Networks and Satellite (COMNESTAT). 2015.

[11] Jinxiu Chen, et. al., Performance Evaluation of Adaptive Modulation System over Mobile Satellite Channels. Journal of Networks, 2013. VOL. 8(NO. 6).

[12] Ippolito, L.J., Satellite Communications Systems Engineering: Atmospheric Effects, Satellite Link Design and System Performance. 2008, Chicester: John Wiley and Sons Ltd.

[13] ETSI, Digital Video Broadcasting (DVB); Second Generation DVB Interactive Satellite System (DVB-RCS2) Part 1: Overview and System Level specification DVB Document A155-1. 2013.

[14] Skinnemoen, H., Introduction special issue on DVB-RCS+M. International Journal of Satellite Communications and Networking, 2010. 28(3-4): p. 113-117.

[15] Skinnemoen, H., et al., DVB-RCS2 overview. International Journal of Satellite Communications and Networking, 2013. 31(5): p. 201-217.

[16] MathWork. MATLAB: DVB-S.2 Link, Including LDPC Coding. 2016; Available from: https://uk.mathworks.com/help/comm/examples/dvb-s-2-link-including-ldpc-coding-1.html?s_tid=srchtitle.

[17] Bolea Alamanac, A., et al., DVB-RCS goes mobile: Challenges and technical solutions. International Journal of Satellite Communications and Networking, 2010. 28(3-4): p. 137-155.

[18] Morlet, C. and A. Ginesi. Introduction of Mobility Aspects for DVB-S2/RCS Broadband Systems. in 2006 International Workshop on Satellite and Space Communications. 2006.

[19] King, U.P.a.H., Adaptive Video Quality Model for UHD Video Broadcasting Using the Principle of Inclusion. American Journal of Engineering and Applied Sciences, 2017. Volume 10(Issue 4): p. Pages 900-907.

[20] Celandroni, N., et al., Video Streaming Transfer in a Smart Satellite Mobile Environment. International Journal of Digital Multimedia Broadcasting, 2009. 2009: p. 12.

[21] Adeyemo, Z.K., et al., Comparative Evaluation Of Fading Channel Model Selection For Mobile Wireless Transmission System. International Journal of Wireless & Mobile Networks (IJWMN), 2012. 4(6): p. 127-138.

[22] Skinnemoen, H. and P.T. Thompson, Overview of DVB-RCS+M and its development. International Journal of Satellite Communications and Networking, 2010. 28(3-4): p. 119-135.

[23] Fuqin Xiong, M.A., The Effect of Doppler Frequency Shift, Frequency Offset of the Local Oscillators, and Phase Noise on the Performance of Coherent OFDM Recievers. 2001: NASA.

SYSTEM LEVEL MODELLING OF DVB-S2X IN HIGH THROUGHPUT SATELLITE SYSTEM

***Lauri Sormunen*[1], *Jani Puttonen*[1*], *Janne Kurjenniemi*[1]**

[1]Magister Solutions Ltd., Sepänkatu 14, FIN-40720 Jyväskylä, Finland
**Email: jani.puttonen@magister.fi*

Keywords: SNS3, Satellite Network Simulator 3, DVB-S2X, simulation, modelling.

Abstract

DVB-S2X was published in 2014 to update and optimize the DVB-S2 specification, which was formulated ten years earlier. DVB-S2X offers increased efficiency and flexibility, enables the usage of advanced techniques, such as intra-system interference mitigation, beam hopping and multi-format transmissions, and more flexible SNR usage due to addition of high efficiency and more robust MODCODs. In this article, the system level model of DVB-S2X on top of Satellite Network Simulator 3 (SNS3) has been described and verified by means of simulations. It can be seen, that DVB-S2X provides 13% gain in both peak spectral efficiency and burst throughput per user, while optimal channel condition with most efficient MODCOD may provide almost 40% gain in user throughput.

1. Introduction

DVB-S2X standard was published in 2014 to update and optimize the DVB-S2 specification formulated ten years earlier. Due to fast development of Internet and a great increase in demand for network connectivity during the last decade, other communication forms (such as telecommunications) have taken major steps forward, while satellite standards have not developed as swiftly.

DVB-S2X offers notable benefits to DVB-S2 without sacrificing any strengths of previous standards. DVB-S2X includes everything that DVB-S2 contained and retains backward compatibility with DVB-S2 receivers. DVB-S2X offers increased efficiency and flexibility, enables the usage of advanced techniques, such as intra-system interference mitigation, beam hopping and multi-format transmissions, and more flexible SNR usage due to addition of high efficiency and more robust MODCODs.

The enhancements provided by DVB-S2X can be summarized as [1][2]:

1. Smaller roll-offs – i.e. 5%, 10%, 15%, which results in direct gain in available bandwidth.
2. Advanced filtering technologies – provide smaller side lobes enabling the placement of carriers closer to each other and showing direct gain in symbol rates.
3. Support of different network configurations – the smaller roll-offs and advanced filtering options can be applied for both single and multi-carrier systems as well as between operator carriers.
4. Increased granularity in MODCODs and Forward Error Correction (FEC) choices - providing more optimal configuration (e.g. with Adaptive Coding and Modulation (ACM)).
5. Higher modulation schemes up to 256APSK – providing higher spectral efficiency for systems with better link budgets.
6. Very Low SNR for Mobile Applications – Nine extra MODCODs provide better performance for systems with worse link budget and/or highly faded satellite links.
7. Different classes for linear and non-linear MODCODs.
8. Wideband Support – enables use of satellite transponders with bandwidths from 72 MHz (typically C-band) up to several hundred MHz (Ka-band High Throughput Satellites (HTS))
9. Channel bonding - increase in data rates for e.g. Ultra High Definition Television (UHDTV) transmission over satellite.
10. Additional standard scrambling sequences – provide better co-channel interference mitigation.

In this article, we present a system level modelling of DVB-S2X on top of Satellite Network Simulator 3 (SNS3), which has previous support of DVB-S2 in forward link. The DVB-S2X modelling focuses on the features, which has direct effect at system level, i.e. all the physical layer details are not considered. The DVB-S2X modelling is verified by means of system simulations.

2. System level modelling

2.1. *Satellite Network Simulator 3*

Satellite Network Simulator 3 (SNS3) [3][4] is a satellite network module on top of Network Simulator 3 (ns-3) [5]. SNS3 platform has been initiated within European Space Agency (ESA) ARTES 5.1 project AO-6947 "Development of an open-source, modular and flexible satellite network simulator". SNS3 models interactive multi-spot beam geostationary satellite network with transparent payload. Default satellite system consists of 72 spot-beams with full European coverage served by one geostationary satellite located at 33 degrees East, which is illustrated by Figure 1. The spot-beams are handled by 5 GWs with 2 GHz feeder link and 500 MHz user link bandwidth per direction in Ka-band. The satellite is modelling geostationary transparent bent-pipe payload. However, note that the simulator is intended to be configurable such that all parameters may be reconfigured for a required use case.

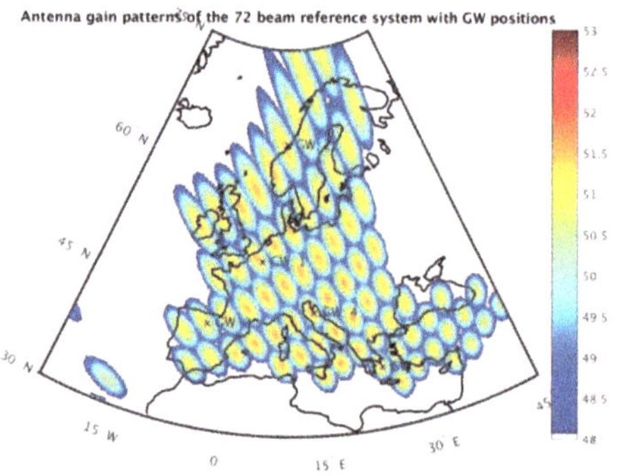

Figure 1. Illustration of the multi-spot beam satellite network.

By default, SNS3 implements DVB-S2 on forward link and DVB-RCS2 on return link. As a system level simulator, SNS3 focuses in radio resource management (RRM) mechanisms, such as multiple access, ACM and scheduling. Primary focus of the modelling is on the physical layer framing, the efficiency of the MODCODs and error modelling by means of adoption of MODCOD specific link layer error performance curves or thresholds.

2.2. *DVB-S2X modelling*

As mentioned, DVB-S2X adds quite a few enhancements on top of DVB-S2, which provide advanced radio resource management for different kind of satellite use cases. Not all the DVB-S2X lower layer details are essential to be modelled in a system level simulator.

DVB-S2X adds several MODCODs in addition to the ones available already in DVB-S2 specification. There are in total 39 new MODCODs for normal FECFRAMES, 22 for short FECFRAMEs and three for medium FECFRAMEs. The most efficient MODCOD (256APSK 32/45) supports even eight bits per symbol by varying both phase and amplitude of the signal. On the other hand, the specification adds new Very Low Signal-to-Noise Ratio (VL SNR) MODCODs with QPSK and BPSK increasing the robustness and availability of the satellite link. [2][6].

SNS3 has been supplemented with all new DVB-S2X MODCODs in addition to previously implemented DVB-S2 model. The roll-off factor, carrier spacing, as well as the usable frame types are configurable. The new MODCODs can be enabled in the simulation at the beginning, and ACM decides which MODCOD to use depending on the forward link SINR.

In system level modelling of DVB-S2X, the focus has been on adding the new MODCODs for different frame types in forward link, while modifying roll-off factor has been previously possible yet unused. Increased inter-symbol interference due to the latter is not considered, although increased traffic due to smaller roll-off causes more interference. The coding scheme (i.e. symbol interpretation) has not been modelled, only the efficiency (i.e. number of bits per symbol and coding rate).

By taking the MODCOD's efficiency and frame error rate thresholds with respect to E_s/N_0 into account, the packets are distributed into slots of BB frames of different sizes. SNS3 calculates the received SINR of the frame by taking the link budget parameterization, noise and the co-channel interference into account. Ideal E_s/N_0 values per MODCOD are used to determine errors from the received packets: if the packet was received with SINR below threshold, it is dropped, otherwise received successfully. Therefore, the model is even more selective than an actual system may be. Additionally, DVB-S2X MODCODs have even steeper error rate curves with respect to SINR than DVB-S2 MODCODs, so by characterizing success of receiving a packet by comparing to a single threshold value may not be far from a real-world situation.

3. Simulation results

The simulation scenario consists of a single spot-beam over Europe in a 72-beam reference system. Essential parameters are given in Table 1, and link budget parameters are set to simulator defaults, which are described in detail in [4].

Table 1. System simulation common parameters.

Parameter	Value
Simulation spot-beams	1 out of 72 in total is simulated
Spot-beam bandwidth	25 MHz (Case 1) / 125 MHz (Cases 2 and 3)
Scheduler	
- Sorting criteria	No sorting
- C/N_0 estimation	Minimum value in window
Encapsulation	Generic Stream Encapsulation (GSE)
FWD link ACM	On
Traffic model	Constant bit rate (offered)

3.1. *Case 1: Uniformly distributed SINR range*

To best show the gains of new modulation and coding schemes, 100 drops of 5-second simulations were run per simulation case. To speed up the simulation drops, the total available bandwidth was scaled down to one fifth of the original 125 MHz bandwidth, which is normally allocated to one spot beam. A total of four simulation cases were run, consisting of combinations of DVB-S2X being enabled or disabled and normal vs. short BB frame types being used. The UT experiences a different, random level of SINR attenuation in each drop, causing the average SINR level to be between −10 dB and 30 dB. A constant bitrate UDP traffic model was used to transfer about 150% of the resulting link capacity on forward link to the UT.

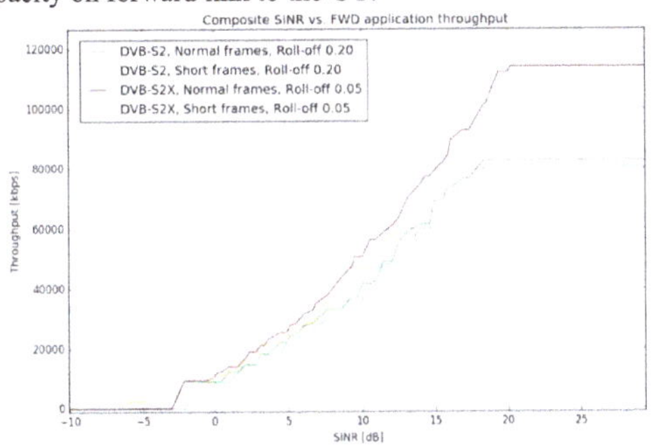

Figure 2. Composite SINR vs. throughput comparison.

In Figure 2, the throughput gain per SINR level can be seen by comparing DVB-S2X graphs with DVB-S2 graphs. Generally, the DVB-S2X offers significant gains only in lower and higher SINR range due to MODCODs. Otherwise the difference is caused by smaller roll-off factor of 0.05, which leaves more room for data symbols in frequency domain. In the higher SINR range, the gains were as high as 38%, whereas in the lower range a positive throughput could be achieved where there could be none with DVB-S2.

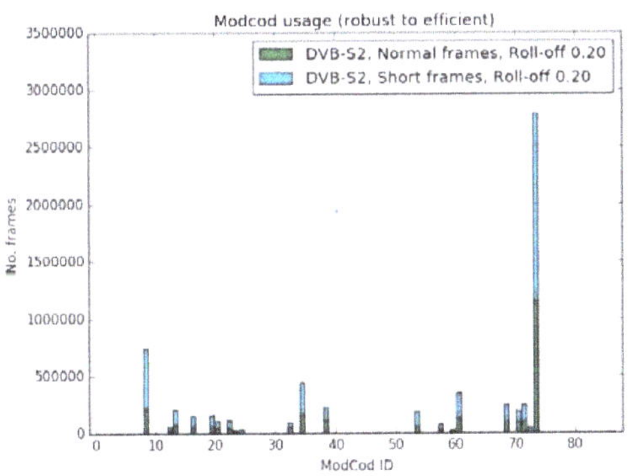

Figure 3. MODCOD usage with DVB-S2.

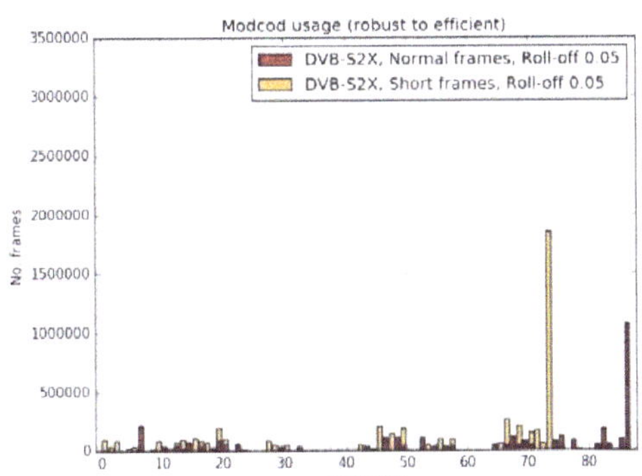

Figure 4. MODCOD usage with DVB-S2X.

In Figure 3 and Figure 4, the distribution of used MODCODs are shown. On x-axis, the MODCODs are presented in order of efficiency by an integer ID, with lower number meaning a more robust MODCOD (number 0 representing BPSK-S 1/5) and higher number a more efficient one (number 86 representing 256APSK-L 3/4). The number of frames transmitted with given MODCOD is presented on the y-axis. Figure 3 represents the DVB-S2 and Figure 4 DVB-S2X MODCOD usage with normal and short frame types as stacked bars. Since the S2 MODCODs are used for both short and normal frames, the distributions in the left-hand figures mostly overlap. On the other hand, many of the MODCODs in S2X are restricted to frames of only either normal or short length, and therefore the distributions overlap less. In the robust and efficient ends, DVB-S2X short frame and DVB-S2X normal frame MODCODs, respectively, provide more throughput. The most used MODCOD in both cases is the most efficient one, since all the simulations ending up with a UT having SINR between 20 dB and 30 dB essentially end up with using the most efficient available MODCOD.

3.2. Case 2: Cell capacity

In case 2 simulations, a variable number of users are dropped at random locations within the coverage area of one spot-beam. Each user is assigned a weather trace causing realistic channel conditions for a stationary UT. Each user is offered a bit rate of 2 Mbps in forward link and the number of users is varied. The cell capacity and user throughputs are visualized with different offered loads.

Spectral efficiency with different offered loads is shown in Figure 5. When the system is not overloaded, there is no difference between the DVB-S2 and DVB-S2X performance. All the offered data can anyway be served since the maximum cell capacity is not reached. However, with higher load, the DVB-S2X brings some gain over DVB-S2: reducing the roll-off factor from 0.20 to 0.05 increases spectral efficiency by 8.7% and adding new MODCODs further increases the gain by 4.8%. Providing only new MODCODs without decreasing roll-off merely provides a gain of 2.5% in realistic channel conditions but coupling the new features together the gain is 12.8%. The gains are caused firstly by more efficient usage of bandwidth and secondly by the higher granularity of MODCODs and the availability of higher order MODCODs.

Similarly, the effect is visible in the user throughputs in Figure 6. DVB-S2X is capable of keeping a couple of more users fully satisfied.

3.3. Case 3: Burst throughput

More efficient modulation and coding schemes with decreased roll-off enable potentially shorter delays in delivering large data packets. To illustrate this, 100 UTs under single spot-beam are placed in random positions, and each UT is downloading a file of 10MB. The users are set to download the file in row, one after another, to allow full bandwidth to be utilized without congestion. Similarly to case 2, weather traces are used to provide realistic SNR levels to each user.

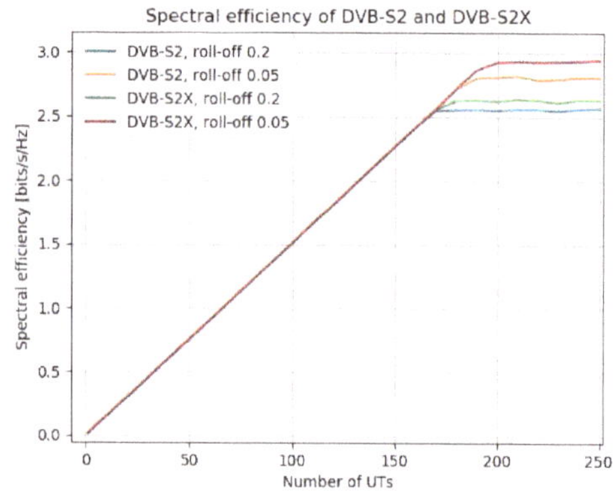

Figure 5. Spectral efficiency comparison.

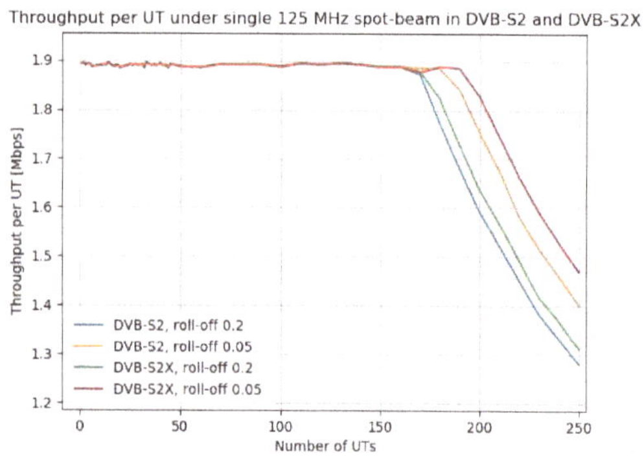

Figure 6. User throughput comparison.

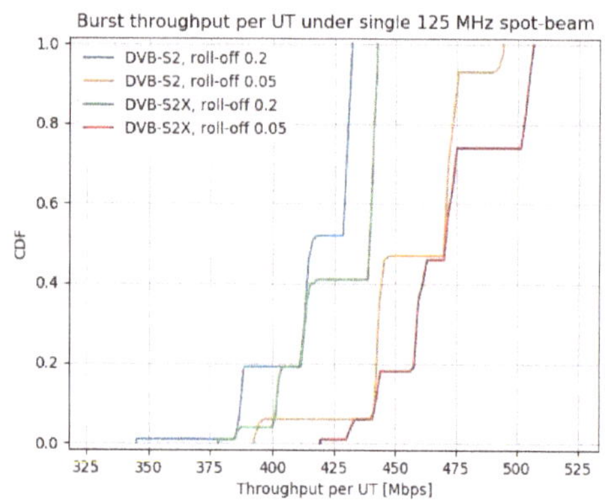

Figure 7. Burst throughput comparison.

Figure 7 shows burst throughput cumulative distribution function (CDF) per user. On average, decreased roll-off factor boosts the performance of the system more than high-efficiency or VL SNR MODCODs, since using only old MODCODs with roll-off of 0.05 the performance is enhanced over using new MODCODs with roll-off of 0.20. Coupling decreased roll-off factor and new MODCODs together, the average burst throughput is increased by 13%.

4. Conclusion

In this article, a system level model of DVB-S2X on top of SNS3 has been described along with general features of the standard. Through a series of simulation cases, the article shows DVB-S2X to increase overall performance of a simulated satellite network in comparison with DVB-S2. DVB-S2X shows a lot of promise for more efficient and reliable forward link communications. However, to deploy to a larger scale, a great many factors must be taken into consideration to provide, e.g. reliable control channel signalling in addition to regular data channel to all users in varying channel conditions under geographically large spot-beams. Therefore, the benefits of deploying DVB-S2X on a larger scale may be less significant than optimal results presented here. However, the standard does offer more leeway compared to S2, which implies at least some gain with careful configuration. Having a DVB-S2X model in SNS3 enables new use cases in the field of simulation-based satellite communications and network research.

References

[1] Morello, A., Mignone, V.:, "DVB-S2X: extending DVB-S2 flexibility for core markets and new applications", International Journal of Satellite Communications and Networking, 2016, 34, pp. 327-336

[2] Willems, K.: "DVB-S2X Demystified", Business Insight on DVB-S2X, Newtec White Paper, March 2014.

[3] Puttonen, J.: "Satellite Model for Network Simulator 3," Proc. International Conference on Simulation Tools and Techniques (SIMUTools), Lisbon, Portugal, 2014.

[4] Puttonen, J.: "A Packet Level Simulator for Future Satellite Communications Research", Proc. AIAA Space, San Diego, USA, August 2014.

[5] NS-3 Consortium, "Network Simulator 3", http://www.nsnam.org, referenced 24.8.2018.

[6] ETSI EN 302 307-2 V1.1.1 (2014-10) Digital Video Broadcasting (DVB); Second generation framing structure, channel coding and modulation systems for Interactive Services, News Gathering and other broadband satellite applications; Part 2: DVB-S2 Extensions (DVB-S2X)

Demonstration of Autonomous Bandwidth Allocation Scheme using SC-FDMA subcarrier switching

Daisuke Goto[1],Fumihiro Yamashita[1]

[1]*NTT Access Network Service Systems Laboratories, NTT Corporation, Yokosuka, Kanagawa, 239-0847 Japan*

Keywords: Autonomous Bandwidth Allocation Scheme (ABAS), SC-FDMA, prototype

Abstract

This paper introduces the Autonomous Bandwidth Allocation Scheme (ABAS) focusing on modulation / demodulation adopting SC (Single Carrier) - FDMA. In ABAS, which we have studied so far, each terminal station autonomously increases or decreases the frequency band in use according to the minimum common rule. This paper reports the overview of ABAS and its fundamental performance evaluation by developing the prototype of the SC-FDMA demodulator.

1 Introduction

In this paper, we report the newly modulation/demodulation prototype for realizing Autonomous Bandwidth Allocation Scheme (ABAS) as an initial study for highly frequency efficiency of a satellite transponder [1,2]. ABAS aims at realizing flexible frequency resource allocation like Demand Assign Multiple Access (DAMA) 3 without timing synchronization and unified scheme such as Time Division Multiple Access (TDMA)1 or Code Division Multiple Access (CDMA) [3-5].

The concept of ABAS is shown in Fig.1. Under the common rule using simple base station control, all terminal stations can autonomously allocate / release their own frequency resources on the satellite transponder. In detail, the uplink band is divided into multiple frequency slots, and all terminal stations individually occupy slots to increase bandwidth or release slots to decrease it.

As a requirement to the transceiver for realizing ABAS, effective utilization of frequency resource on the satellite transponder is the key point. Thus, we have considered that it is necessary to increase / decrease the occupied bandwidth while maintaining synchronization. In addition, it is also necessary that each terminal station can occupy distributed slots on the transponder. To satisfy these two requirements, we adopt Single Carrier(SC) –FDMA [6] for developing prototype ABAS.

Firstly, this paper outlines the prototype using SC-FDMA, then, shows that ABAS's modem prototype can dynamically change occupied bandwidth frame by frame through the fundamental experiments. This paper consists of five sections. The proposed access scheme, ABAS, is presented in Sec. III. Sec. IV shows the prototype of ABAS using SC-FDMA and the validation of received constellation and BER is explained in Sec. V. Finally, Sec. VI summarizes this paper.

2. Autonomous Bandwidth Allocation Scheme (ABAS)

The concept of ABAS is shown in Fig.1. The frequency resources of the satellite transponder are divided to multiple segments and each terminal station named System Terminal (ST) occupies/releases them individually. Base Station (BS) broadcasts the Broadcast Channel (BCH) periodically. It includes information on the number of available frequency segments, W, for fair resource allocation among the multiple systems, and a resource allocation table that indicates the association of segments and systems. W is calculated as

$$W = \lfloor \frac{(N_f - 1)}{N_s} \rfloor \quad (1),$$

where Nf is the total number of frequency segments on the satellite transponder and NS is the number of connected STs. In Eq.(1), "Nf -1" in the numerator of W means one frequency segment is held in reserve for the next ST.

Fig.2 depicts the flow chart of ABAS bandwidth allocation. Its details are as follows.

1) ST randomly chooses one of the available segments indicated by BCH and transmits a request signal by using the chosen segment.

2) Once BS receives the request signal successfully, BS returns an acknowledgement (ACK) to the ST by using the resource allocation table included in BCH. If ST cannot

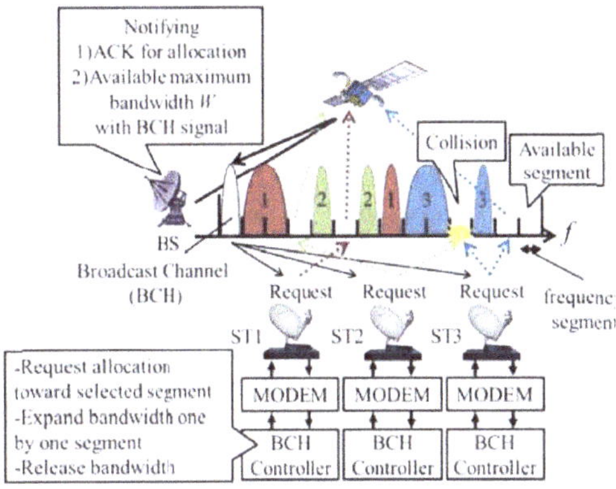

Figure.1 Proposed Autonomous Frequency Allocation Scheme (ABAS)

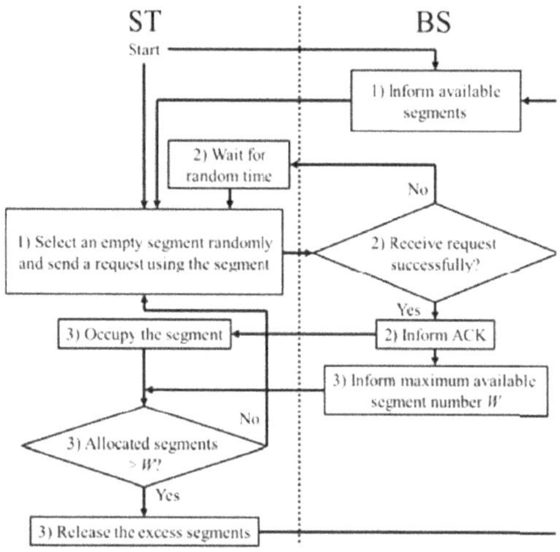

Figure.2 Flow chart of the proposed scheme

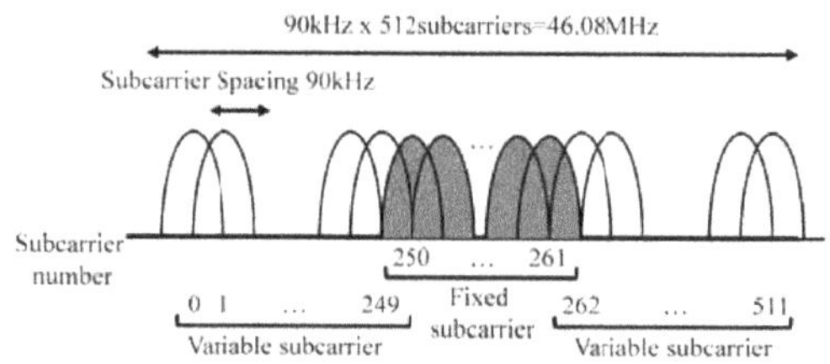

Figure.3 SC-FDMA allocation for ABAS

Table.1 Parameter Specification

System Parameters	Parameter Values
Center frequency *fc*	2.6 GHz
Transmission scheme	SC-FDMA
Sampling frequency	46.08 MHz
Subcarrier space	90 kHz
FFT point	512 pt
Fixed subcarrier number	12 pt (#250-#261)
Frame size	33668 samples
Symbol number per frame	57 symbols
Unique word length	3344 samples

receive the ACK, ST waits for a random period and re-transmits the request signal.

3) After receiving ACK, ST starts communications using the acquired segment. In addition, ST checks W in BCH, chooses one of the available segments referring to the resource allocation table, and starts flow 1) again. In other words, ST increments its own segments up to W one by one. In addition, W in BCH is updated to reflect the increase in Ns. If Ns is updated and the available number of segment is decreased, the currently active STs release segments so as to comply with (i.e. not exceed) the updated value of W.

From another viewpoint of propagation delay specific to satellite communications, it is also important to avoid call collision among existing STs that attempt to increase their segments and any newly emerging STs. To tackle this problem, ABAS proposes to increase the number of segments incrementally, not suddenly, so as to minimize the unused frequency resources and allocate the frequency resources fairly while avoiding collision among systems.

3 Configuration of ABAS Prototype

3.1 Modulation and demodulation adopting SC-FDMA

In order to realize ABAS, it is necessary for the transmitter ST (Tx-ST) to occupy/reduce the band dynamically. To correspond to this dynamic frequency resource change, we have developed the modulator/demodulated based on SC-FDMA that can flexibly change its frequency resource by controlling the number of sub-carrier in use.

Fig.3 shows the subcarrier allocation image. The parameter specification is shown in Table.1. The prototype allows flexibly occupying and releasing frequency slots by adopting the SC-FDMA system using 512pt FFT and pre-DFT. Its sampling frequency is 46.08 MHz so that subcarrier interval is 90 kHz. The RF frequency is a provisional value and it does not depend on it. To change the number of subcarriers and its arrangement while keeping synchronization, fixed subcarriers are prepared. In detail, 12 FFT points are fixedly allocated and the other slots are variably allocated. After synchronizing received SC-FDMA signals by using fixed subcarriers, STs can recognize the channel information and change the number of variable subcarriers while keeping the synchronized state.

Fig.4 shows the signal frame format. The sequences of the fixed and variable subcarriers are composed of 57 SC-FDMA symbols per frame, and frame synchronization is performed by using the unique word put on the fixed subcarriers band. In addition, the fixed subcarriers include frequency allocation information of variable subcarriers frame by frame. In other words, the receiving ST acquires the information on available subcarrier after demodulating fixed subcarriers. By using this information, ST changes the number of subcarriers on a frame-by-frame basis.

3.2 Platform and block diagram

PXI system of National Instruments (NI) is employed as a platform for the production of the prototype [7], shown in Fig.5. PXIe-1082 is a system chassis and 8135 is a controller equipped with Windows OS. Establishing FPGA circuit at PXIe-7975 and sending/receiving signals via NI-5791 RF Adapter Module.

A block diagram of a prototype constructed on PXI system is shown in Fig.6. As an initial examination in this prototype, the demodulation performance was verified by offline processing that demodulates the saved waveform data. In this prototype, the transmitter has only the function of RF

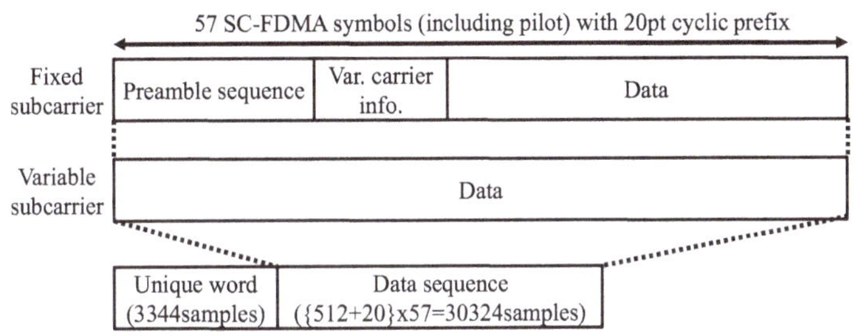

Figure.4 Frame format of prototype ABAS

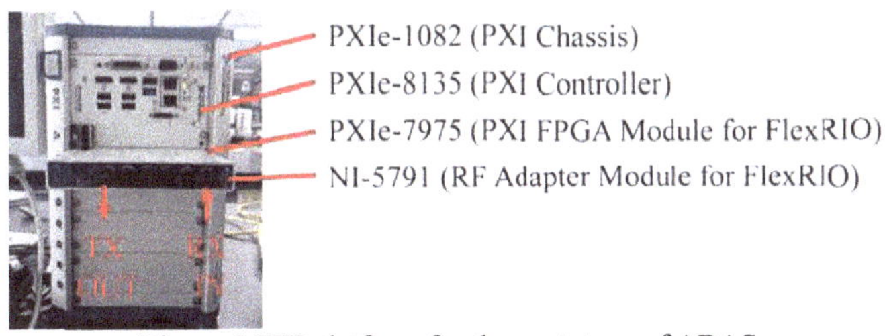

Figure.5 PXI platform for the prototype of ABAS.

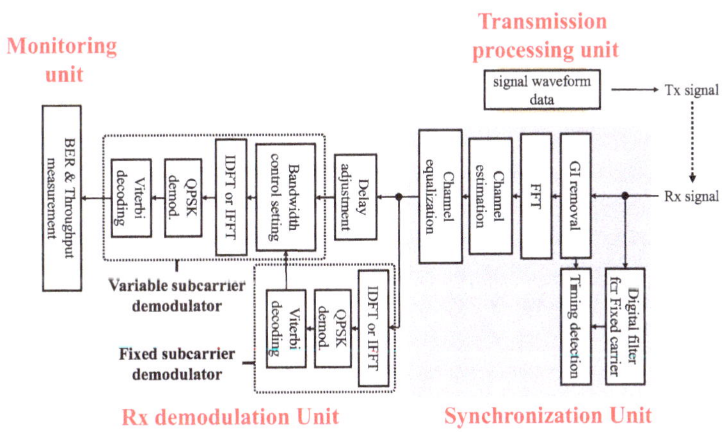

Figure.6 Block diagram of ABAS using SC-FDMA

conversion and sending the digital waveform data memorized in advance from TX OUT of the adapter module. Regarding the receiver, timing synchronization for detecting the beginning of each frame is conducted by the matched filter output of the unique word signal and channel estimation / equalization are performed by the pilot signals in the synchronization unit. Fixed subcarriers are always demodulated for detecting the information of variable subcarrier in the demodulation unit. On the other hand, as for the variable subcarriers, only the subcarriers in use are selectively demodulated. In the prototype, QPSK modulation, FEC coding rate 1/2, convolutional code with constraint length 7 and soft decision Viterbi decoding are adopted. For the purpose of fundamental verification, perfect frequency synchronization between TX and RX is assumed. Note that if the information on variable subcarrier arrangement cannot be obtained correctly in the receiver, all data of variable subcarriers in one frame is lost. This yields the signification bit errors. To avoid this problem, the majority decision for the variable subcarrier information of three consecutive frames is implemented. The monitoring unit makes it possible to monitor bit error rate (BER) and throughput from demodulated data.

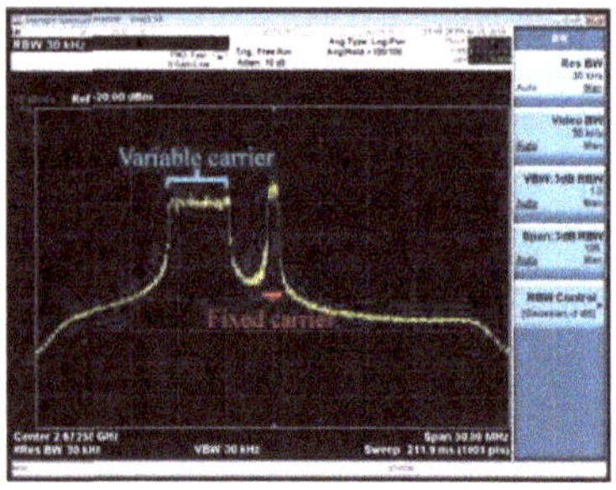

(a) Continuous allocation (Var. carrier: 72pt).

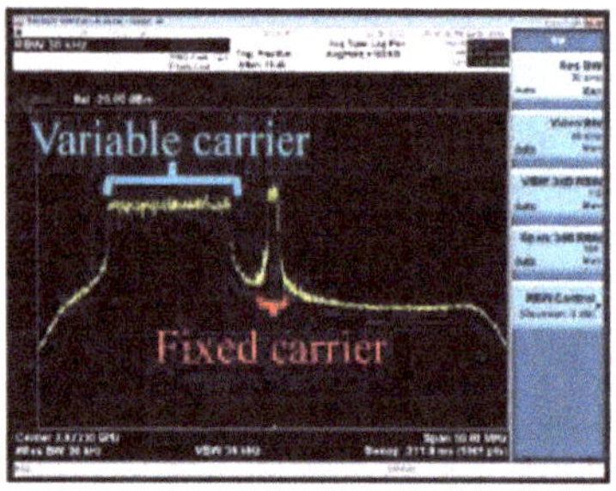

(b) Distributed allocation (Var. carrier: 144pt).

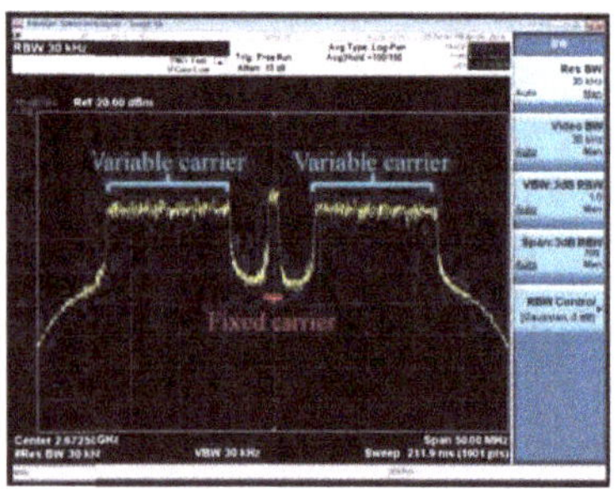

(c) Distributed allocation (Var. carrier: 144x2pt).
Figure.7 Spectrum image.

4 Measurement Results

4.1 Verification of error free

At first, error free confirmation was done in noiseless environments. The two examples of spectrum are displayed in Fig.7. As shown in Fig.7, the fixed subcarriers composed of 12pt subcarriers are allocated to the center position of whole band. In Fig.7 (a) and (b), 72pt and 144pt variable subcarriers are allocated continuously to the lower frequency region, respectively. On the other hand, in (c), 288pt variable subcarriers are allocated to both side of the fixed subcarrier in half.

In Fig.8, the throughput behavior (left) and constellation (right) output by the monitoring unit are shown. Here, the horizontal axis is time (sec) and the vertical axis is throughput (bps). The number of bits per unit time (sec) is defined as throughput. Assuming that one packet is defined

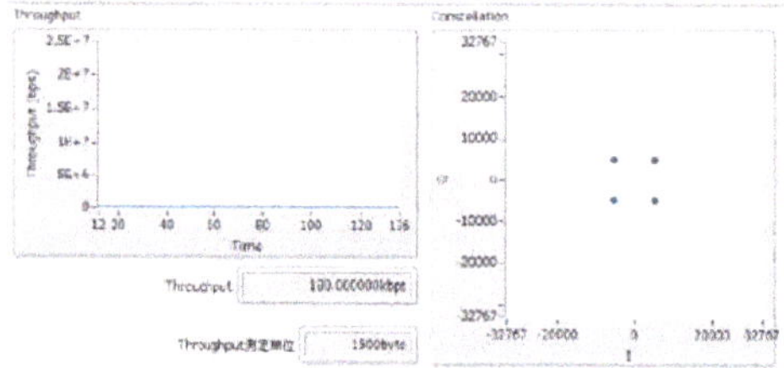

(a) Fixed subcarrier.

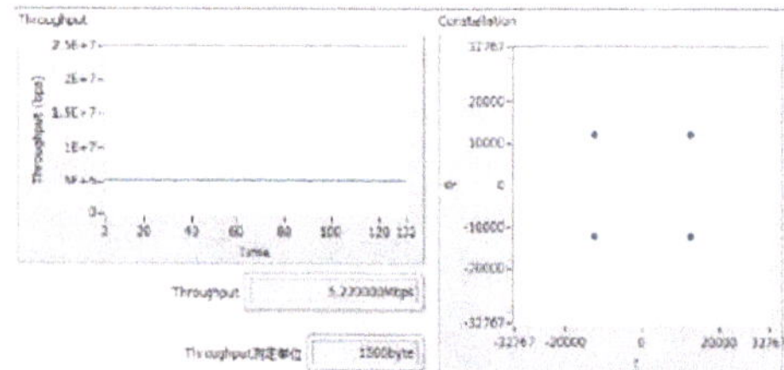

(b) Variable subcarrier w/o bandwidth shift.

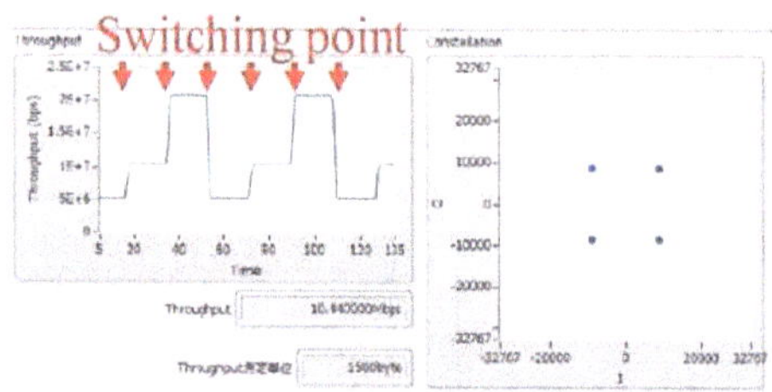

(c) Variable subcarrier w/ bandwidth shift.
Figure.8 Throughput behaviour (left) and constellation(right).

as 1500 bytes. It is confirmed that all data of the packet is demodulated without error. The figures of (a) and (b) display the results of throughput and constellation with the fixed 12pt subcarriers and 72pt variable subcarriers, respectively. And the figure of (c) shows the results of throughput and constellation while switching subcarrier pattern as 72pt, 144pt and 288pt repeatedly. As shown these figures, it is confirmed that constellations are always converging and the throughput can be transited dynamically as expected. In other words, it is confirmed that the occupied bandwidth increases/decreases while maintaining synchronization and each terminal station can occupy distributed slots on the transponder. Thus, the concept of ABAS is verified by using the prototype based on SC-FDMA.

4.2 BER performance

Finally, we measured the BER performance. Fig.9 shows the measured BER of the fixed / variable subcarrier and the simulated value of the fixed subcarrier under the same conditions. The variable (var.) subcarrier is measured in the cases of (a) 12pt without shift, (b) 12pt with shift every frame and (c) switching 12pt, 24pt and 48pt every frame, which is verified with and without majority decision of subcarrier information.

It is confirmed that the BER performance with the fixed subcarrier is degraded by 0.5 dB compared to the simulation result. On the other hand, it can be seen that the BER performance with variable subcarrier is degraded by about 1.5-2.0 dB at BER=10-3 compared to the result with fixed subcarrier. This is because all data of one whole frame occasionally lost when the detection error of variable subcarrier arrangement information occurred.

On the other hand, it can be seen that BER is improved by applying majority decision of 3 frames for variable subcarrier information. At BER=10-3, it is found that the BER performance with the variable subcarrier adopting majority decision are equivalent to those of the fixed subcarrier, therefore, the BER performance is improved by 1.0-1.5dB by employing majority decision rule for important frequency information.

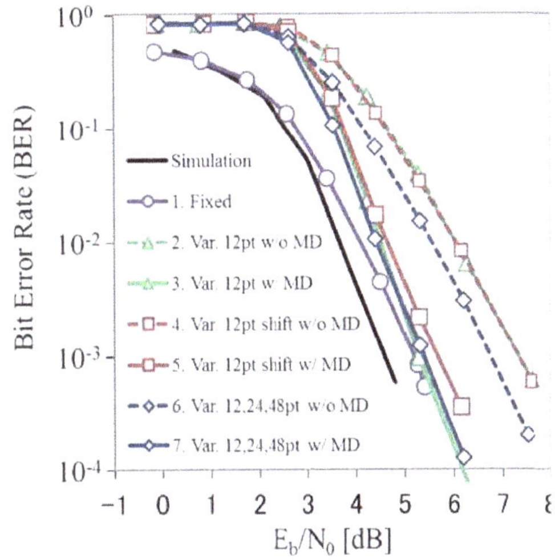

Figure.9 Bit error rate (BER) of the fixed/variable subcarriers on prototype compared to simulation results .

5 Conclusion

This paper reported that the overview the concept of ABAS and developed prototype using SC-FDMA. As a result of fundamental verification, it is confirmed that constellations are always converging and the throughput can be transited dynamically as expected. BER performance is also measured. It is concluded that the occupied bandwidth increases/decreases while maintaining synchronization and each terminal station can occupy distributed slots on the transponder. These results show the concept of ABAS was successfully verified by using the prototype based on SC-FDMA.

References

[11] Brown, F.: 'The title of the patent (if available)'. British Patent 123456, July 2004

Report

[1] D. Goto, J. Abe, N. Kita and F. Yamashita, "Autonomous bandwidth allocation scheme for satellite tran-sponder shared by multiple systems," 35th AIAA International Communications Satellite Systems Conference (ICSSC 2017), 2017.

[2] D. Goto, J. Abe, and F. Yamashita, "Frequency segment selection algorithm in autonomous bandwidth allocation FDMA for high frequency efficient satellite systems," IEICE Tech. Rep., SAT2017-52, vol. 117, no. 261, Oct. 2017.

[3] N. Celandroni and R. Secchi, "Suitability of DAMA and contention-based satellite access schemes for tcp traffic in mobile DVB-RCS," IEEE Trans. Veh. Technol., vol.58, no.4, pp.1836–1845, May 2009.

[4] J. M. Park, U. Savagaonkar, E. K. Chong, H. J. Siegel, and S. D. Jones, "Allocation of QoS connections in MF-TDMA satellite systems: A two-phase approach," IEEE Transactions on vehicular technology,111 Vol. 54, No. 1, 2005, pp. 177-190.

[5] W. C. Y. Lee, "Overview of Cellular CDMA," IEEE Trans. Vehic. Tech., vol. 40, no. 2, May 1991, pp. 291-302.

[6] D. Falconer, S. L. Ariyavisitakul, A. Benyamin, B. Eidson, "Frequency Domain Equalization for Single-Carrier Broadband Wireless Systems," IEEE Commun. Mag., vol. 40, Apr. 2002, pp. 58–66.

[7] National Instruments Corp., http://www.ni.com/.

ROBUST INITIAL ACCESS TECHNIQUE OF SPREAD SPECTRUM BASED ON DVB-RCS2 STANDARD FOR MOBILE APPLICATION

Pansoo Kim, In-Ki Lee, Deock-Gil Oh and Joon-Gyu Ryu*

Satellite Technology Research Group, ETRI, Daejeon, Republic of Korea

*pskim@etri.re.kr

Keywords: DVB-RCS2, DIRECT SEQEUENCE SPREAD-SPECTRUM, LARGE UNCERTAINTY, BURST DETECTOR, INITIAL ACCESS

Abstract

In this paper, we deal with robust initial access scheme in spread spectrum mode for mobile DVB-RCS2 standard. For log-on process, it should be effective in terms of rapidity and reliability if we can utilize robust transmission and reception schemes under large timing and carrier frequency offset as well as very low SNR. Firstly, after we review the log-on burst transmission related with spread spectrum of DVB-RCS2 specification under the harsh channel condition, we come up with the DVB-RCS2 tailored scheme that VSAT terminals should be able to support the reference waveforms. It's expected that it's applicable adaptively depending on channel condition to be considered.

1 Introduction

The Digital Video Broadcasting – Return Channel via Satellite 2nd generation (DVB-RCS2) specification [1] is 2nd generation very small aperture terminal (VSAT) standard which was published by European Telecommunication Standard Institute (ETSI) in 2014. It incorporates the function of spread spectrum like DVB-RCS [2] but the specific scheme has been updated through practical deployment. Especially, there was the discussion to improve log-on access under a large timing/frequency uncertainty condition in 2nd generation VSAT. [3] Furthermore, to support the emerging service like internet connection through high speed vehicles, it's still necessary to employ spread spectrum technology which is more resilient to harsh channel conditions. [3]

Ultimately, the technology addressed in this paper can be used to allow logon-on process where the VSAT terminal to satellite delay is not known with an accuracy that is not sufficient to guarantee that the log-on burst can be transmitted such that it is received within the boundaries of a timeslot. Generally, the log-on process for VSAT terminals is required to have accurate knowledge of the position of both them and the satellite in order to allow the log-on burst transmission within a window that is at most in a 2 msec (Mili Second) time.[4] In certain mobile applications, it is highly desirable to allow a much wider burst detection window because there is desire to be independent of exact satellite ephemeris data or positioning information. It's noted that the timing uncertainty can be applied to all types of mobile VSAT terminals that accurate ephemeris data is readily available in case of high speed mobility. In DVB-RCS2 standard, large timing uncertainty issue can handle with through a number of log-on trials and time back-off process and close loop timing correction. However, if log-on bursts may be continuously located on the boundary of less wide timeslot in worst case, we should reconsider time division multiple access (TDMA) systems have employed techniques such as dedicated log-on bursts with wider acquisition windows in order to overcome the initial uncertainty. Of course, such a feature can be unattractive in terms of dedicated time/frequency resource use and equipment cost. [4] Therefore, this paper provides simple solution about spread spectrum alternatively and present design guideline for DVB-RCS2 standard. Moreover, we provide some insights on system guideline for mobility service to support global area coverage.

The paper is organized as follows. Some descriptions such as service scenario, technical rationale and technical report of DVB-RCS2 are specified in section 2. In section 3 and 4, the information about burst detector design is shown and its numerical results are presented, respectively. Finally, the conclusion is mentioned, briefly.

2. Some Descriptions

2.1 Service scenario

The service scenario to be considered in this paper is based on Ku/Ka band geo stationary orbit (GEO) satellite based communication network using DVB-S2/S2x and DVB-RCS2 standard. Specifically, it targets the mobile terminals with a large group of users on aircrafts, high speed train and cruise ship which can provide high speed IP based internet service. Due to the installation limitation of antenna mounted on a mobile platform, it's required to become low profile type. In case of aeronautical case, the antenna sizes are more limited and it is necessary to make transmission signals with lower power spectral density through spread spectrum technique in order to meet the off-axis emission regulation. As a result, the receiver input should experience very low signal to noise ratio (SNR) signal. The relevant information has been addressed in [5], well. The transmission data rate is assumed to range from 64 kbps (Kilo Bit per Second) to 256 kbps. For DVB-RCS2 spread spectrum scheme, the chip rate can have the range from minimum 0.192 Mcps in case of SF 2 and code rate 2/3 to maximum 8.192 Mega Chip per Second (Mcps) in case of SF 16 and code rate 1/2.

2.2 Comparative Analysis of spread spectrum scheme

The burst repetition spread spectrum technique was adopted for return link as optional scheme in DVB-RCS + M standard. [4] [6] It seems to have simple principle in terms of modulator and demodulator implementation but appears to have some weak points. For lower SNR and a larger carrier frequency offset (CFO) condition, much packer error rate (PER) performance degradation in the receiver is occurred because it's difficult to combine multiple repetition packets in despreading process while keeping coherent carrier phase alignment. It's caused by CFO estimation inaccuracy under very low SNR. Of course, the reference in [7] is shown to have quite satisfactory performance by spreading factor (SF) 16 without consideration about any time-variant factors such as Doppler drift by mobility and phase noise. It's claimed that this technique can experience difficulties for log-on process, where there happens many false alarm burst detection in the long guard time where the interval exists in only noise. [8]

2.3 Modulator

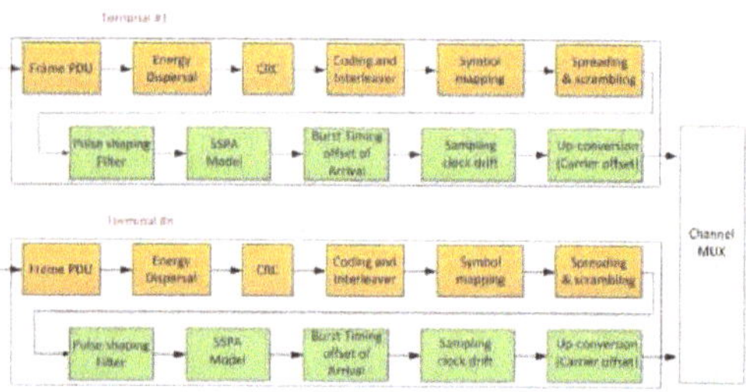

Figure 1. Modulation process of spread spectrum

The spread burst construction for spread spectrum of DVB-RCS2 standard in each terminal is described in Fig. 1. It can be generated by the information of preamble, pilot field and payload data chip length flexibly by broadcasting composite table (BCT) of forward link signalling (FLS). After spreading and pulse shape filtering, IF up-conversion is performed. For simulation development efficiency, solid state power amplifier (SSPA) model, burst timing offset of arrival (ToA), sampling clock drift and CFO/Doppler drift model are followed. In Fig. 2, a half guard time (HGT) is given for reliable TDMA packet reception taking into account burst ToA. In particular, longer HGT may be secured to a large time uncertainty when there is initial log-on process trial under high speed mobile condition or no knowledge of information about terminal location. For example, the case of log-on burst with waveform ID 19 is shown in Fig. 2 where the burst length is 22704 chips, pilot period 14 chips and pilot block (PB) length 8 chips. The total timeslot length is composed with the burst length and double HGT period.

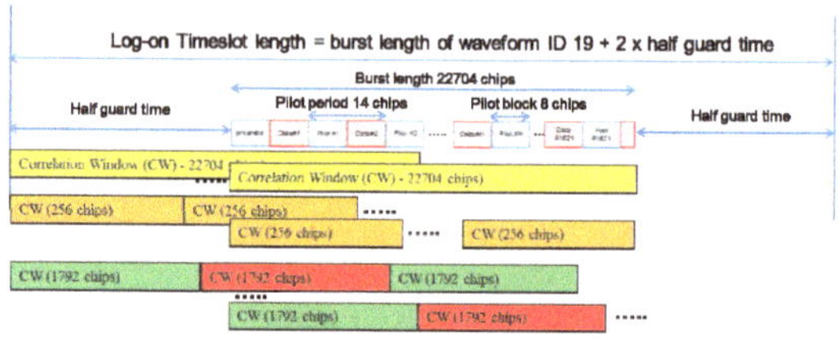

Figure 2. Spread burst location and correlation window in a timeslot

2.4 Channel Model

As revealed in Fig. 1, high mobility propagation condition is considered in this paper. For realistic scenario, all possible channel impairments have been included. Especially, the Doppler effect for mobile service is differently added from [9]. Doppler shift and rate for different types of terminals in Ku/Ka band have been described in Annex L of [2]. The maximum offset and drift rate are significant for initial access process and its drift effect is more relevant to traffic transmission mode because the constant offset can be pre-corrected by log-on and control burst. In addition, the maximum drift rate is not maintained for long periods. The reference in [3] is assumed to have initial offset by Doppler shift and other effects amounts to $\pm 100kHz$ and frequency drift rate of up to $\pm 3.7kHz/s$ for commercial aircraft service. In addition, the timing uncertainty for initial access is assumed to have around $\pm 20m\sec$. To cover this figure, it will be difficult to produce cost-effective equipment.

Table 1. Guard time period for log-on burst (vs chip rate)

Chip rate	Guard time period (256 chips)	Guard time period (512 chips)	Guard time period (1024 chips)	Guard time period (2048 chips)
1.024 Mcps	250 usec	500 usec	1 msec	2 msec
2.048 Mcps	125 usec	250 usec	500 usec	1 msec
4.096 Mcps	62.5 usec	125 usec	250 usec	500 usec
8.192 Mcps	31.25 usec	62.5 usec	125 usec	250 usec

As shown in Table 1, the default scrambling sequence (SS) 256 chips introduced in DVB-RCS2 [1] can support maximum 31.2 usec (Micro Second) as burst ToA for log-on burst in case of 8.192Msps.

3 Burst Detector Design

3.1 Overall demodulator architecture

The overall demodulator for DVB-RCS2 spread spectrum mode can be illustrated in Fig. 3. After matched filtering, the burst detector finds the start position of a burst data and a timeslot. In addition, identifying the burst is present or not for log-on process. After initial timing synchronization by burst detector, fine chip timing synchronization and CFO synchronization process are followed subsequently. And then, carrier phase synchronization and despreading and can be achieved prior to turbo decoder process.

Each burst consists of transmission chips $a_1, a_2, a_3, a_4 ..... a_{N_{burst}}$ taken from known data like pilot chip and

$\pi/2$ BPSK chip modulation where N_{burst} is the burst length. Let y_n be the received sample signal given by

$$y_{n-\tau} = a_{n-\tau} \cdot e^{j(2\pi n\Delta f+\theta)} + v_n \quad (1)$$

where, v_n is complex Gaussian noise sample, n is chip time index and Δf is CFO or Doppler shift. τ is sampling time offset in a chip and θ is carrier phase offset.

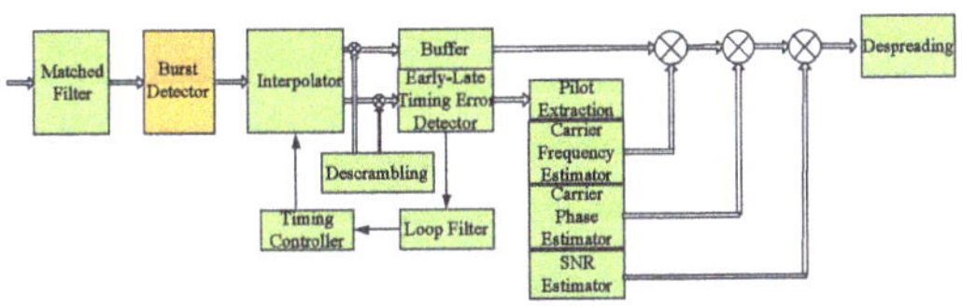

Figure 3. Demodulator structure for Spread Spectrum

3.2 Conventional approach for burst detector

Typically, the burst detector performs to find the start position of TDMA packet in a timeslot. It is often revealed from maximum correlation value or correlation value higher than the predetermined threshold between received signals and known data like unique word (UW). Unlike burst detector in non-spread demodulator, the burst detector in spread spectrum mode is generally operated under very low SNR condition. In addition, it's difficult to operate coherent correlation or differential correlation under a large CFO and very low SNR condition. Therefore, it's desirable to design the burst detector according to the requirement of operational SNR and CFO value. As described in [10], the presence of CFO normalized to the chip rate ($1/T_c$) determines an energy degradation equal to $N \cdot \mathrm{sinc}^2(N \cdot \Delta f)$ after coherent accumulation over N chips, which can be incorporated by suitably choosing N under given CFO where one PB can be a minimum unit of N in Fig. 4.

3.3 Proposed approach for burst detector

Firstly, we take into account designing all reference waveforms which can support the list of Table A-2 in Annex A [1]. When we analyse the burst construction format, it can be categorized like Table 2 based on common configuration from different waveform IDs. We can compute about the number of PB in the entire burst length from each waveform ID like Table 2. For instance, the number of PB can be 1621 in waveform ID #19. In other words, it means that the number of sub-block for correlation corresponds to T shown in Fig. 4. When the entire burst length is scrambled with a set of scrambling sequence length which is the same as a burst length, the number of sub-block for partial correlation is 1621 and the number of complex multiplication is 810. Because of the demodulator complexity, it's expected to introduce 256 default scrambling sequence in the DVB-RCS2 specification.[1] However, its length can rely on the amount of burst ToA like Table 1. Meanwhile, the pilot period can be formulated as $2^m \times 7$ from Table 2 where m is 3, 4 and 5. Thus, we can extend SS with 7×256 chip period which enables to apply the SS period to integer multiple PBs for all waveform IDs. The corresponding period T is shown as the number of PB within correlation window from Table 2. Furthermore, the number of correlation window over an entire burst length is determined by 1792 chip period which is the length of correlation window and is listed in Table 2.

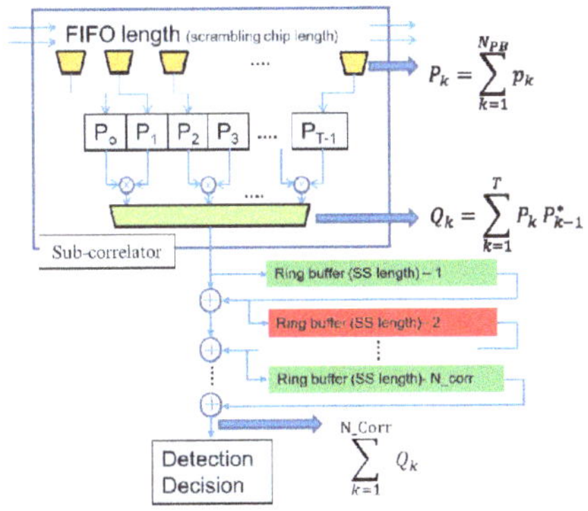

Figure 4. Correlation structure for burst detection in a timeslot

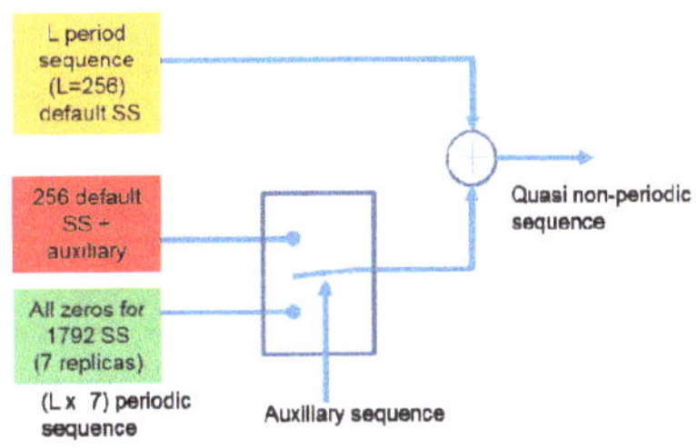

Figure 5. Spread burst SS generation in a timeslot

For computationally efficient burst detector and a large ToA uncertainty more than 1792 chips, we propose the new SS generator, 256 default SS + auxiliary like Fig. 5 and the architecture for the burst detection is depicted as Fig. 4. If the proposed SS generator like Fig. 5 is not selected, let's assume that the SS is used with the total length 1792 chip which consists of a series of 7 replicas that they have 256 chip periods, respectively. In this case, the peak value from correlation sum cannot be obvious due to severe noise and multiple ambiguous peaks in a timeslot because a moving average scheme for signal energy accumulation process is applied. This phenomenon can arise whenever the guard time is larger than correlation window length. Therefore, a timing uncertainty range is highly dependent of the SS period. In summary, to cope with this problem, the SS is newly generated by superimposing 256 scrambling to other auxiliary periodic sequence like Fig. 5. The structure of burst detector in Fig. 4 should be slightly modified at the accumulation process to change the polarity according to a set of auxiliary sequence. Each P_k is coherent integration over the length of PB where p_k is PB and k is chip time index. Of course, P_k could have multiple of PBs to increase SNR according to on amount of CFO value. As smaller CFO, the length of PBs can be extended and SNR output from Eq. (2) increases where N_{PB} is the number of PB.

$$P_k = \sum_{k=1}^{N_{PB}} p_k \qquad (2)$$

From the coherent integration output like Eq. (2), the Eq. (3) can be derived from differential correlation and addition due to CFO after higher SNR condition is secured.

$$Q_k = \sum_{k=1}^{T} p_k p_{k-1}^{*} \qquad (3)$$

Without any coherent integration, differential correlation can lead to noise enhancement under very low SNR condition.

$$C = \sum_{k=1}^{N\_Corr} Q_k \qquad (4)$$

Like Eq. (4), the consecutive accumulation is performed by utilizing ring buffer as much as SS length under the given number of correlator and the final correlation output could be C. The detection decision is determined by correlation output.

The comparison about three different SSs is summarized in Table 3. The proposed one which is the composite of 256 default SS and auxiliary code seems to be non-periodic sequence like generic polynomial and the performance is comparable to ideal one but the complexity in burst detector is close to 256 default SS scheme. It can be convinced through simulation result in section 4.

Table 2. The number of correlation window

waveform ID	Pilot sum [chip]	Chip length of a PB [chip]	No. of PB	Pilot Period [chip] = pilot + data	No. of PB in a correlation window	No. correlation window
1	378	6	63	56	32	2
2	282	6	47	56	32	2
3	648	6	108	56	32	4
4	312	4	78	56	32	3
5	1076	4	269	56	32	9
6	808	4	202	56	32	7
7	756	12	63	112	16	4
8	564	12	47	112	16	3
9	1296	12	108	112	16	7
10	624	8	78	112	16	5
11	2152	8	269	112	16	17
12	1616	8	202	112	16	13
13	1512	24	63	224	8	8
14	1128	24	47	224	8	6
15	2592	24	108	224	8	14
16	1248	16	78	224	8	10
17	4304	16	269	224	8	34
18	3232	16	202	224	8	26
19	12968	8	1621	14	128	13

Table 3. Comparative analysis of different SS schemes

	Generic Polynomial SS	256 default SS	Composite of 256 default SS and auxiliary code
Length	a timeslot length	256 or 1792 chip period	(1792 x auxiliary) chip period
Good	- Ideal performance - Support of larger ToA offset	Low complexity	- Similar complexity with 256 default SS - Support of larger ToA offset
Weak	High complexity	Smaller ToA	Slightly complicated than 256 default SS

4 Numerical Results

In this section, the numerical result is presented in Fig. 6. We report the PER performance as a function of the SNR channel condition and the number of sub-correlator from Fig. 4 and Table 2 under the simulation condition listed in Table 4. As seen in Fig. 6, the maximum number of sub-correlator is 13 in case of waveform ID 19 from Table 2. When we can obtain the PER performance similar to AWGN channel through the burst detector composed with only 6 sub-correlators when the proposed SS generation scheme is used in a large burst ToA and CFO condition.

Table 4. Simulation condition

Parameter	Value
Waveform ID	19
MODCOD	$\pi/2$ BPSK & code rate ½
Maximum Burst ToA	6400 chips
Maximum CFO (vs chip rate)	0.05
Doppler drift	10kHz/sec
Chip rate	2.048Mcps

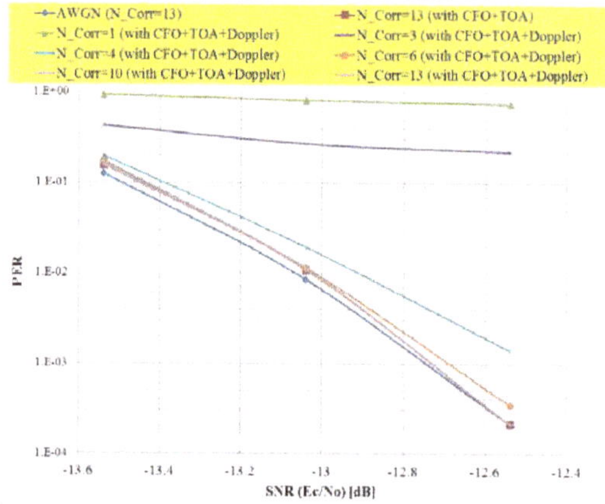

Figure 6. PER result according to the number of correlators

5. Conclusion

We have proposed the simple method for robust log-on detection in the demodulator in large TOA/CFO condition which is compatible with DVB-RCS2 mobile standard. In particular, the proposed one is tailored to design the reference waveform (i.e., essential set) listed in Annex A of the specification and is applicable to user defined waveform, flexibly by utilizing the same principle. However, note that it should be optimized to select the parameters depending on waveform ID and different channel conditions such as received SNR, CFO and burst ToA.

Acknowledgements

This work was supported by 2018-MOIS32-002 from Disaster-safety Industry Promotion Program funded by Korean Ministry of Interior and Safety (MOIS). In addition, we would like to express our great appreciation to Dr. Xavier Giraud of Cabinet NOVACOM for his valuable support and insightful advice.

References

[1] ETSI EN 301 545-2 v1.2.1 'Digital Video Broadcasting (DVB): Second Generation DVB Interactive Satellite

System(DVB-RCS2); Part 2: Lower Layers for Satellite Standard', Apr. 2014.

[2] ETSI TR 101 790 v1.4.1, 'Digital Video Broadcasting (DVB): Interaction channel for Satellite Distribution System; Guidelines for the use of EN 301 790, Jul. 2009.

[3] Robust Logon Burst for DVB-RCS2 Mobile Applications, DVB TM-RCS1421, Sep. 2011.

[4] ETSI TR 102 768, v1.1.1. 'Digital Video Broadcasting (DVB): Interaction channel for Satellite Distribution System; Guidelines for the use of EN 301 790 in mobile scenarios, Apr. 2009.

[5] C. Morlet et al, "Implementation of Spreading Technique in Mobile DVB-S2/DVB-RCS systems", in proc. of IWSSC 2007, Salzburg, Austria.

[6] ETSI EN 301 790, v.1.5.1, 'Digital Video Broadcasting (DVB): Interaction channel for Satellite Distribution System; May 2009.

[7] O. Bialer, U. Ram, G. Levitas and A. Gal, "Mobile Satellite TDMA Communications Robust Burst Repetition Despreading", in proc. of EuCAP 2014, pp. 2440-2444.

[8] A Direct Sequence Spread Transmission Scheme for DVB-RCS2 Return Links, DVB TM-RCS1448, Oct. 2011.

[9] P. Kim and D. Oh, 'A Dual Mode Symbol Timing Recovery For DVB-RCS2 Standard", in proc. of 23rd Ka and broadband communications conf. 2017, Trieste Italy.

[10] P. Kim et al, 'Direct sequence spectrum spreading techniques for next generation mobile broadband satellite services', International Journal. of Satellite Communication and Networking, 2010, 28, (3-4), pp. 157-181.

BEAM-HOPPING OVER-THE-AIR TESTS USING DVB-S2X SUPER-FRAMING

Christian Rohde[1], Rainer Wansch[1], Gerhard Mocker[2], Achim Trutschel-Stefan[2], Laurent Roux[3], Eros Feltrin[3], Hector Fenech[3], Nader Alagha[4]

[1]Fraunhofer Institute for Integrated Circuits (IIS), Am Wolfsmantel 33, 91058 Erlangen, Germany
[2]WORK Microwave GmbH, Raiffeisenstrasse 12, 83607 Holzkirchen, Germany
[3]EUTELSAT, 70 rue Balard, 75502 Paris Cedex 15, France
[4]European Space Agency (ESA), Keplerlaan 1, 2201 AZ Noordwijk, The Netherlands
**email: christian.rohde@iis.fraunhofer.de*

Keywords: BEAM-HOPPING, DVB-S2X, SUPER-FRAMING, OTA-TEST

Abstract

The beam-hopping transmission concept has recently received a lot of attention due to a new level of flexibility to accommodate dynamic traffic profiles by employing the latest satellite technology. Based on the latest ground equipment developments of WORK Microwave and Fraunhofer IIS, beam-hopping system synchronization and control aspects have been tested over-the-air in June 2018. The network synchronization scheme as well as the ground equipment synchronization worked well and the performance goals have successfully been achieved.

1. Introduction

The global trend is to have faster and more flexible communication all over the world. Terrestrial networks are well suited for serving densely populated areas. However, this trend will include oceans, sky, diverse and sparsely populated areas as well, which is typically covered by the classical satellite communication scenario. In order to optimally adapt the system to changing traffic demands over time and location, the novel beam-hopping concept has been introduced. In contrast to the quasi-static illumination in a conventional multi-beam satellite system, the satellite cycles through a set of defined beams according to a specific schedule, which is derived from the traffic demands and the user terminal locations. The gains in terms of system capacity optimization and better matching the traffic demands are shown in [1] and [2].

The upcoming Eutelsat Quantum-Class Satellite is a software defined Ku-band satellite that offers in-orbit flexibility in all the operational parameters of the payload including service area definition, frequency plan and power allocation [3]. It also supports the beam-hopping function which will provide a presence over the visible earth as seen by the satellite with great flexibility in capacity allocation. It is believed to be the first open standard beam-hopping system and will support independent beam hopping networks [4]. Furthermore, its dynamic beam-forming capabilities can be beneficially combined with the beam-hopping feature. As an example, geographical areas with low user terminal density can be served by adjusting the beams to be wider, while geographical areas with high user terminal density or high quality of service demands can be satisfied by reconfiguration to more spot-beam type beams. The system, due for service in 2019, utilises rapid and seamless beam-forming reconfiguration that can be applied to a variety of applications such as mobility, disperse geographical areas and emergency and Governmental services.

In a recent experiment funded by the European Space Agency, a consortium consisting of WORK Microwave, Fraunhofer IIS and Eutelsat use a standard bent-pipe satellite of Eutelsat in a time-division multiple access type mode to evaluate the performance and functionality of beam hopping network synchronization scheme as well as the ground equipment. For this purpose, a beam-hopping emulator device is placed in the transmission chain of the gateway. This device performs the signal switching according to a configurable beam-switching time plan towards two uplink signals using two transponders. Therefore, the "hopping" is implemented in the gateway to achieve feasibility for testing. This represents a pre-test to a real beam-hopping at the satellite. Some implications and artefacts of this approach are discussed later in the paper.

The latest ground equipment developments of WORK Microwave and Fraunhofer IIS provide proper beam-hopping system synchronization and control. The flexible Super-Framing Format 4 as specified in DVB-S2X standard Annex E [5] is used as waveform. It perfectly meets the beam-hopping system requirements as discussed in [6] and [7]. After successful tests and demonstrations in the laboratory, the beam-hopping over-the-air tests were performed in June 2018 as the next step of development verification and field proofing. This activity has been initiated by the European Space Agency (ESA) and is part of a project for end-to-end beam hopping system demonstration [8].

The rest of the paper is organized as follows: In chapter 2, the system configuration of the test setup as well as the involved devices are described. It is complemented by explaining the test methodology and the measurement goals. The test results and evaluation are presented in chapter 3. Lastly chapter 4 concludes the paper with a summary of the key results and findings.

2. Test Setup and Employed Devices

An overview of the whole setup for the over-the-air tests is provided in Figure 1. It is split in two sites. The lower main part shows the setup at the Eutelsat teleport in Rambouillet, France, whereas the small upper part refers to the equipment located in Erlangen, Germany.

2.1. System and Satellite Configuration

At the transmit side, the wideband beam-hopping modulator generated dummy data and provided the RF-signal in L-band to the payload emulator (PLE). The PLE applied the hopping pattern, pre-configured by the modulator, directly to the RF-signal. Two output ports of the PLE were used to transmit the on/off switched signals. One of the signals passed through a frequency shifter stage, realized with an IZT C3040 device. After passing through a combiner, both signals have been transmitted through the same uplink chain with an L-Band input side by side. At the satellite the following configuration was used:

- Two 36 MHz transponders D03 and D05 of the satellite E12WB were used for transmission.
- Transponder D03:
  uplink centre frequency = 14312.253 MHz Y-pol,
  downlink centre frequency = 11512.253 MHz X-pol
- Transponder D05:
  uplink centre frequency = 14353.753MHz Y-pol,
  downlink centre frequency = 11553.753MHz X-pol
- The nominal uplink-downlink delay for Rambouillet was 255.971 ms.

The two reception sites were configured as follows:

- Rambouillet downlink using the main RX antenna: Employing a Ku Band Block Downconverter with a 9.750 GHz local oscillator, the L-band output carrier centre frequencies resulted for D03 in 1762.253 MHz and for D05 in 1803.753 MHz. D03 was employed as reference beam channel connected to the reference terminal (RT).
- Rambouillet downlink using a 1.2 m RX antenna: Employing a Ku Band LNB with a 10 GHz local oscillator, the L-band output carrier centre frequencies resulted for D03 in 1512.253 MHz and for D05 in 1553.753 MHz. D05 was employed as user beam channel connected to a measurement terminal (MT).
- Erlangen downlink using a 4.9 m RX antenna: Employing a Ku Band LNB with a 10 GHz local oscillator, the L-band output carrier centre frequencies were the same as for the Rambouillet 1.2 m RX antenna. A measurement terminal (MT) was configured to receive the user beam channel through transponder D05.

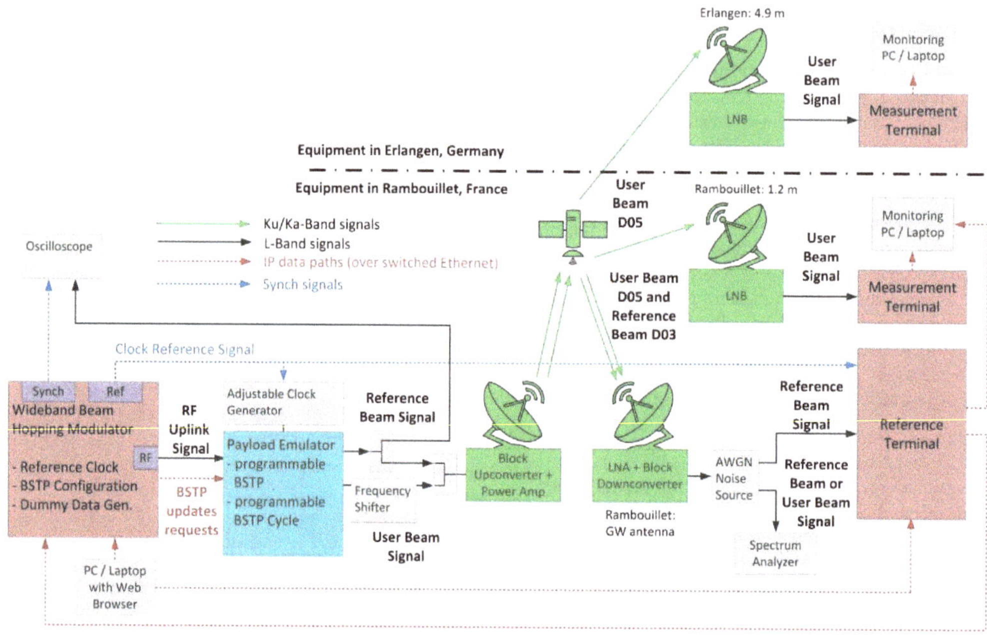

Figure 1: Overview of the transmission system for the OTA tests with two receiving sites.

At the receiving side in Rambouillet, the RT was connected to the main antenna. Based on evaluating the reference beam through transponder D03, feedback data about measured offsets was generated and sent back to the modulator via an IP connection. Furthermore, a MT was connected to a smaller antenna and was active conducting user-terminal like measurements regarding SNR and other synchronization parameters. The feedback and measurement data messages of both terminals were logged at a laptop.

At the receiving side in Erlangen, a further MT was connected to a 4.9m antenna carrying out the same measurements but at a second site. The measurement messages were also logged at a laptop.

2.2. *Signal and Waveform Configuration*

A transmission fully compliant to standards was accomplished using the DVB-S2X Super-Framing Format 4 [5], where the so-called ST-field is used to signal a coverage ID. As described in [6] and [7], dynamic super-frame (SF) padding was applied by means of the normal-size dummy frames at the end of each illumination.

The two carriers were fed with a 30 Msps signal with 15% roll-off, resulting in the occupied bandwidth of 34.5 MHz per carrier.

2.3. *Details of the Test Setup*

The installation at the Eutelsat gateway is depicted in Figure 2, the devices shown are listed top-down:

- Rohde&Schwarz Spectrum Analyser
- WORK Microwave Variable Noise Inserter, including also a signal splitter
- Fraunhofer IIS Reference Terminal
- Fraunhofer IIS Payload Emulator
- WORK Microwave Synthesized Signal Generator used as clock generator.
- WORK Microwave Wideband Beam-Hopping Modulator
- IZT C3040 Channel Emulator used as configurable frequency shifter and for signal monitoring

The modulator as well as the terminals support up to 400 Msps in order to deliver the high throughput demand due to serving several beams. The modulator controls the timing phase and the precise symbol frequency of the super-frame (SF) transmissions according to the measured offsets by the RT. It also sends control commands to the PLE like the beam switching time plan (BSTP), synchronized with updating its own internally used BSTP.

The PLE performs the beam switching relative to a 100 MHz clock, which defines the time grid granularity. Thanks to the application of fast configurable attenuators for switching, operating on a time base of 100 ns, also different transition characteristics can be implemented. If needed, noise can be added to the 5 output ports.

Using a fine adjustable signal generator as 100 MHz reference clock generator for the PLE allows to provide a small offset of e.g. +/- 1 ppm to the PLE emulator, simulating frequency offsets between the PLE and the modulator. This is because it can be the case in real satellite environment due to the free running characteristic of the master oscillator in the satellite as well as Doppler shift effects due to satellite movement.

The RT and MT are using the same FPGA-based hardware but have different firmware and software. The RT is an extended version of the MT. It measures various offsets like time difference between start of illumination (determined from power detection) and the start-of-super-frame (SOSF) preamble of the SF.

Note that the modulator provides a common 10 MHz clock signal for the PLE through the fine adjustable clock generator and for the reference terminal as shown in Figure 1. The MTs use their internal clock.

Figure 2: Installation of devices in the gateway.

2.4. Test Methodology and Goals

The goals and how to achieve them were:

- Successful network acquisition over satellite link. Starting point was an unsynchronized network, which represented the start of the network acquisition phase. Criteria to be achieved after acquisition were:
    - SFs are aligned by the modulator to the observed illuminations created by the PLE/satellite.
    - Coverage-IDs of the SFs, which are decoded from the received beam-hopped signal, correspond to the configuration.
- Successful network tracking over satellite link. Starting point was the accomplished network acquisition. Criteria to be achieved were:
    - Network stays in tracking mode and the modulator keeps the controlled timing offsets constant and follows slight variation of the PLE/satellite over time.
    - RT and MTs stay in undistorted tracking mode also during BSTP updates.
- Successful synchronization of simultaneous BSTP update at PLE/satellite and modulator. Starting point was network synchronization in tracking mode. Criteria to be achieved were:
    - BSTP update synchronization of the modulator to the PLE/satellite is performed and kept aligned also after several further BSTP updates.
    - MT and RT tolerate BSTP updates.
- Beam-hopping link quality was confirmed. Starting point was network synchronization in tracking mode. Criterion to be achieved was that the measured SNR is in line with the expected value from the link budget and/or continuous transmission measurement.

Since the satellite link in this test setup was located after the PLE, a doubled effect of the receive side delay variation due to satellite movement could be expected, compared to the situation where the switching is done on board of the satellite. Since the signal and illumination pattern are shifted jointly, it represents a worst-case test for terminal synchronization.

However, no delay variation as typical in real satellite environment between the modulator and the BSTP switching device (here the PLE) was applicable, because the PLE was directly fed by the modulator signal. But a small clock offset between the PLE and the modulator was applied to simulate phase variations between both elements as a similar disturbing effect. This caused time shifting between the modulator signal, which included the SFs according to BSTP pattern, and the PLE applying the illumination pattern according to the same BSTP. The correct adaptive operation of the BSTP control loop could be checked while applying such disturbance.

Spectrum analyser and oscilloscope were used for instantaneous measurements and observations, while the stored feedback and measurement data on the laptops was used for offline statistical analysis.

3. Over-The-Air Test Results

Various test cases and BSTP configurations for measurements had been considered, e.g.

- Modification of the 100 MHz clock reference of the PLE by -1 ppm, 0, +1 ppm from nominal value.
- Different SNR for the RT: max value, 10 dB and 5 dB. For the MTs, no noise source is used. So they were working on the SNR available by the link budget.

3.1. Observations and General Insights

The network acquisition time was in the order of 3…4 sec for all tested scenarios. Initially, the modulator adjusted the phase of its BSTP coarsely to the phase of the PLE by possibly skipping complete SFs and applied fine adjustment of the phase by temporarily applying frequency offsets to the symbol clock rate.

After synchronization was achieved with the currently running BSTP, the BSTP update cycle of the modulator was synchronized to PLE/satellite by means of a synchronising signal from the PLE/satellite as a special BSTP update which was visible only on the reference beam. A measured offset was reported back from the RT to the modulator and applied by the modulator. As a first verification step, the same procedure was repeated and a confirmation was achieved by the response of the RT that no further correction was required. The second step was applying a common BSTP update on the modulator and the PLE/satellite, which affected also the user beam. As shown in Figure 3, the evaluation was done using an oscilloscope displaying the nominal BSTP signal power envelope periods as part of the BSTP hopping behaviour (upper red curve, C2) and the envelope of the RF signal of the PLE/satellite output (lower blue curve, C3).

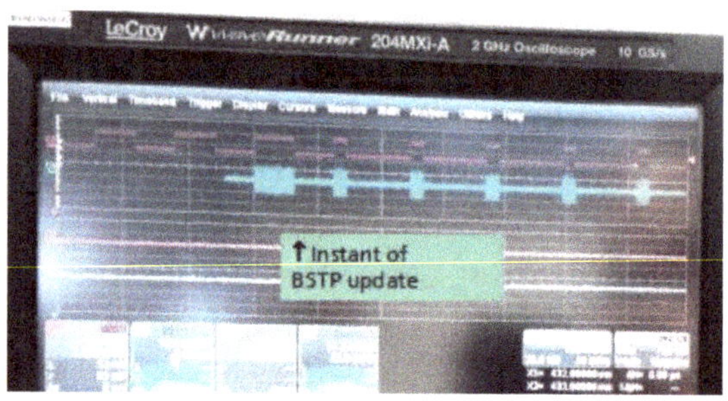

Figure 3: Oscilloscope screen picture shows simultaneous BSTP update indicated by the marker.

In the centre of the picture, a BSTP update event is shown, where a switchover from an old BSTP plan to a new one is executed. It was synchronously performed by the modulator and the PLE/satellite.

Two types of SNR estimation are performed at the terminals. SNR_{pilots} represents the SNR measured from received pilots indicating the link quality for data demodulation and decoding. $SNR_{PWdetect}$ is calculated by the terminal power detection from the determined minimum and maximum input power levels.

- RT in Rambouillet: The available 12…15 dB SNR (according to weather conditions) correspond to $SNR_{PWdetect}$.
- MT in Rambouillet: SNR_{pilots} = $SNR_{PWdetect}$ = ~10 dB is in line with expected link budget.
- MT in Erlangen: SNR_{pilots} = $SNR_{PWdetect}$ = ~14 dB is in line with expected link budget.

3.2. Network Synchronisation Results

The network synchronisation was monitored first of all by checking feedback values from the RT and derived control values of the beam-hopping control algorithm, as shown on the status webpage of the beam-hopping modulator. An important figure is the offset between the start of illumination (determined by power detection) and the SOSF detection. A good illustration for this value is given in Figure 8 described in more detail below. It shows the variation of this parameter over time, as seen by a measurement terminal.

3.3. Evaluation of Terminal Measurement Statistics

For these evaluations, the network synchronisation was already fully accomplished and the fine tracking mode was active. First, the consistency of the power detection was considered. For this purpose, the RT feedback data w.r.t. the measured illumination duration and the BSTP duration by means of the power detection was considered.

Figure 4 shows a scatter plot of the illumination duration measurements over time. At 30 Msps, the nominal duration of one SF is 20 418 000 ns. The resulting mean value from Figure 4 is 20 442 923 ns and standard deviation of 1 349 ns, which is quite close to the nominal value. After 225 sec, the density of measurement points obviously increases, which is due to occurrence of a BSTP update towards a shorter BSTP duration, resulting in more measurements over time. This is explicitly reflected by Figure 5 showing the BSTP duration measurements over time. The change of these values is:

From ~142 923 000 ns / 20 418 000 ns/SF $\approx$ 7 SFs
To ~61 250 000 ns / 20 418 000 ns/SF $\approx$ 3 SFs.

With these observations, consistency of the power detection was successfully verified.

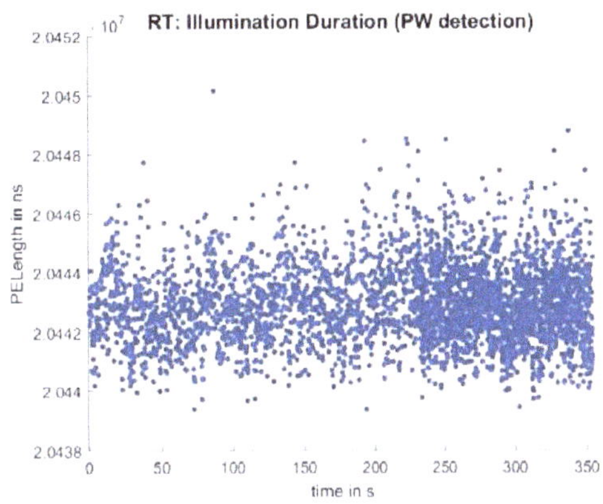

Figure 4: RT measurements of the illumination duration by means of power detection.

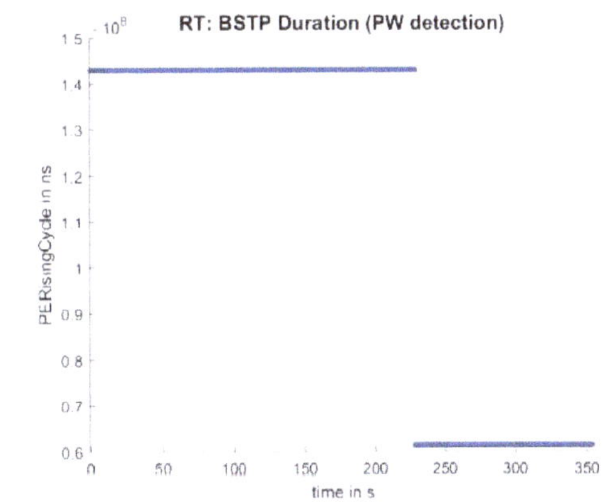

Figure 5: RT measurements of the BSTP duration by means of rising power detection.

The next evaluation considers the MT measurements of the estimated SNR from pilots over time. The values were taken at the end of the first SF of each illumination. Figure 6 shows the estimated SNR of the MT in Rambouillet over time. The mean value is 10.49 dB and standard deviation is 0.14 dB (calculated in dB), which is in line with the expectations from the link budget. Figure 7 shows the estimated SNR of the MT in Erlangen over time. The mean value is 14.26 dB, standard deviation is 0.14 dB, which is also in line with the expectations from the link budget. The difference between the Rambouillet and Erlangen measurements result from the different antenna sizes.

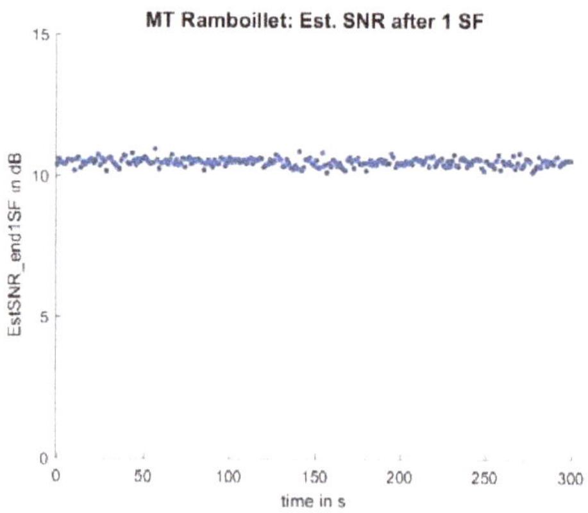

Figure 6: Rambouillet MT measurements of the estimated SNR from pilots.

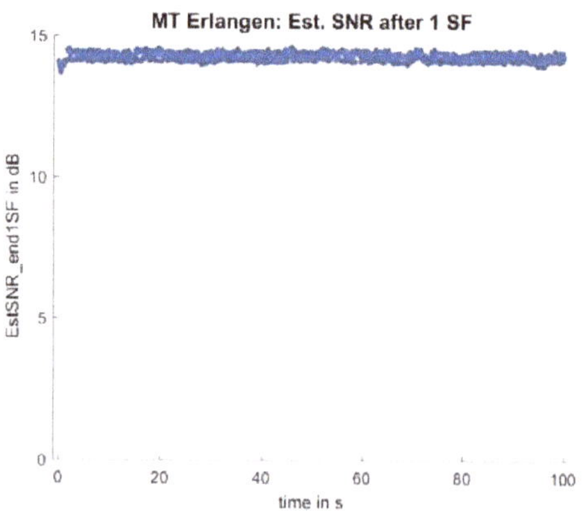

Figure 7: Erlangen MT measurements of the estimated SNR from pilots.

When network tracking is performed, it also means that any frequency offsets have to be continuously estimated and compensated. Such a scenario for a clock offset between the PLE and the modulator is considered in Figure 8. In a real system small offsets between the clock on the satellite and the clock at the modulator on ground as well as Doppler shift due to satellite movement contribute to such varying clock offsets. Due to satellite movement, also the varying uplink delay between the modulator and the satellite providing the beam switching causes delay variations between the applied beam switching pattern and the transmitted SF pattern by the modulator on ground. In this test setup a small clock offset between the modulator and PLE was used to simulate similar timing drifts. The network synchronisation control loop had to compensate this permanent timing drift. Consequently, the counter-action shift was visible for RT and MTs. In Figure 8, the time difference between start of illumination (determined by power detection) and SOSF detection as measured by the MT in Rambouillet is shown The values show ascending slope due to this clock offset but their short term amplitudes decrease over time. This reflects the successful gradual mean adjustment of the change rate by the network synchronisation control loop, which after a while reduces the requirement for big momentary corrections.

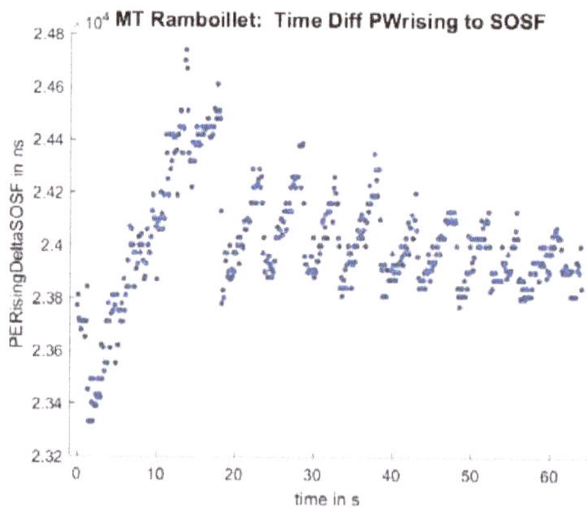

Figure 8: Rambouillet MT measurement of difference between start of illumination (power detection) and SOSF detection.

3.4. Effects, Artefacts and Impairments

The automatic gain control (ALC) of the satellite payload was disabled. Otherwise, unwanted adaption and increase of noise during signal off periods would have happened.

In relation to the satellite no unwanted signals or spikes due to switching on/off the carrier signals appeared thanks to the smooth transition slope configured to the PLE. This operation was similar to operating a typical TDMA system.

4. Conclusions

The developed beam-hopping system has successfully been tested over the air. It demonstrates on the one hand proper network synchronization reliability, adaptiveness, and fine-tuning capability exploiting the features of Super-Framing Format 4 of the DVB-S2X standard. This is based on a special enhanced wideband modulator including also the beam hopping control algorithm, which receives exact measured values from a special reference terminal. Furthermore, user terminal signal-to-noise ratio measurements are close to the values expected from link budget calculations. Over the air test results represent a valuable verification step towards a ground system ready for beam hopping system deployment.

Acknowledgements

This work has been carried out under the ESA project Beam Hopping Emulator for Satellite Systems [8]. Opinions, interpretations, recommendations and conclusions expressed herein are those of the authors and are not necessarily endorsed by the European Space Agency.

References

[1] Anzalchi, J., Couchman, A., Topping, C., Gabellini, P., Gallinaro, G., D'Agristina, L., Angeletti, P., Alagha, N., Vernucci, A.: 'Beam Hopping in Multi-Beam Broadband Satellite Systems'. 5th Advanced Satellite Multimedia Systems (ASMS) Conference and the 11th Signal Processing for Space Communications (SPSC) Workshop, Cagliari, 2010, pp. 248-255.

[2] Alberti, X., Cebrian, J.M., Del Bianco A., Katona Z., Lei J., Vazquez-Castro, M.A., Zanus, A., Gilbert, L., Alagha, N.: 'System capacity optimization in time and frequency for multibeam multi-media satellite systems'. 5th Advanced Satellite Multimedia Systems Conference and the 11th Signal Processing for Space Communications Workshop, Cagliari, 2010, pp. 226-233.

[3] Fenech, H., Amos, S.: 'Eutelsat Quantum-a Game Changer.' 33rd AIAA International Communications Satellite Systems Conference (ICSSC), QT Surfers Paradise, Gold Coast QLD Australia, 7-10 Sept 2015.

[4] Feltrin, E., Amos, S., Fenech, H., Weller, E.: 'Eutelsat Quantum-Class Satellite: Beam Hopping'. 3rd ESA Workshop on Advanced Flexible Telecom Payloads, March 2016.

[5] ETSI EN 302 307-2 V1.1.1 (2014-10), Digital Video Broadcasting (DVB); Second generation framing structure, channel coding and modulation systems (...); Part 2: DVB-S2 Extension (DVB-S2X).

[6] Rohde, C., Wansch, R., Mocker, G., Amos, S., Feltrin, E., Fenech, H.: 'Application of DVB-S2X Super-Framing For Beam-Hopping Systems.' 23rd Ka and Broadband Communications Conference, Trieste, Italy, October 2017.

[7] Rohde, C., Wansch, R., Amos, S., Fenech, H., Alagha, N., Cioni, S., Mocker, G., Trutschel-Stefan, A.: 'Beam-hopping systems for next-generation satellite communication systems'. Chapter 10 of book 'Satellite Communications in the 5G Era', edited by Sharma, S.K., Chatzinotas, S., Arapoglou, P.-D., IET - The Institution of Engineering and Technology, July 2018, ISBN 978-1-78561-427-9.

[8] ESA Project, 'BEHOP - Beam Hopping Emulator for Satellite Systems', Contract-No. 4000115704/16/UK/AD, http://artes.esa.int/projects/behop.

Section 8 – Transmitter and Modern Technologies

MAXIMIZING DATA THROUGHPUT IN EARTH OBSERVATION SATELLITE TO GROUND TRANSMISSION BY EMPLOYING A FLEXIBLE HIGH DATA RATE TRANSMITTER OPERATING IN X-BAND AND KA-BAND

Philipp Wertz, Marcus Kiessling, Franz-Josef Hagmanns

Tesat-Spacecom GmbH & Co. KG, Gerberstraße 49, 71522 Backnang, Germany
philipp.wertz@tesat.de

Keywords: EARTH OBSERVATION, DATA HANDLING, HIGH DATA RATE DOWNLINK, DIRECT TO EARTH, ACM, VCM, SCCC, APSK

Abstract

Earth Observation (EO) and Intelligence, Surveillance and Reconnaissance (ISR) Systems equipped with High Resolution instruments are designed for high-performance operations in terms of fast revisit time, for very short system response time and for providing actionable intelligence with low data latency. The increased reliance on this data has created demand for decreasing latency. However, higher speeds and increasing bandwidth for data download are needed with the increasing numbers of satellites. At the same time, the scarce frequency resources especially in X-band (8.025 GHz to 8.4 GHz) but also in Ka-band (25.5 GHz to 27 GHz) call for more bandwidth efficient modulation schemes. This paper shares details regarding a new data transmission solution to efficiently use the available radio frequency (RF) bandwidth both in X-Band and Ka-Band by using Adaptive Coding & Modulation (ACM) as a key technology, allowing the volume of data to adapt to link budget characteristics. At the same time, high performance forward error correction coding such as SCCC and LDPC (DVB-S2) and high order modulation schemes (up to 64-APSK) are used, yielding both high power and spectrum efficiency. In terms of end-to-end performance, such sophisticated systems need to account for the ground station receiver characteristics. The proposed solution has therefore been verified using prototypes of both transmitter and receiver. Finally, a specific highly integrated transmitter design suited for smaller Earth Observation satellites down to SmallSats is presented.

1. Introduction

The resolution of on-board instruments and thus the amount of data acquired by earth observation satellites is steadily increasing. In order to cover the resulting downlink data capacity demands under consideration of the bandwidth limits in X-band (8.025 GHz to 8.4 GHz) and Ka-band (25.5 GHz to 27 GHz), it is necessary to maximize the downlink data rate and at the same time to utilize more bandwidth efficient modulation schemes. In 2012, the Consultative Committee for Space Data Systems (CCSDS) issued a recommended standard aiming at exactly this kind of high rate applications. The CCSDS 131.2.B.1 [1] combines serial concatenated convolutional codes (SCCC) and higher order modulation formats from QPSK to 64-APSK. In total 27 different transmission schemes are defined with different code rates and spectral efficiencies. The CCSDS 131.2.B.1 [1] allows adaptation of the transmission parameters to the channel conditions during a satellite-to-ground contact by means of variable and adaptive coding & modulation (VCM/ACM) techniques and thus enables the user to maximize the downlink data capacity per fly-over.

In the framework of a ESA GSTP project Tesat Spacecom is currently developing a High Data Rate Modulator applying encoding and modulation techniques according to CCSDS 131.2.B.1. The Modulator is designed for a maximum output symbol rate of 500 MBaud. The modulator is equipped with redundant high-speed serial links enabling a maximum user input data rate of 2 Gbps. Two variants are under development to cover both the X-band (8.025 GHz to 8.4 GHz) as well as the Ka-band (25.5 GHz to 27 GHz) frequency ranges. In order to compensate for the non-linearity of the subsequent travelling wave tube amplifier, what is essential especially for the envisaged higher order modulation schemes, the modulated signal is digitally pre-distorted. All the transmission parameters including modulation & coding and pre-distortion coefficients can be controlled via a serial TM/TC interface thus enabling variable coding & modulation (VCM) and adaptive variable coding & modulation (ACM) techniques.

Within the ongoing ESA GSTP project, an engineering model has been developed at Tesat followed by a qualification program to be accomplished until Q4/2018. In this paper the electrical design trade-offs, interfaces and performance of the High Data Rate Modulator are described.

2. Overall Modulator Design

The High Data Rate Modulator under development at Tesat comprises the following functions:

- Coding (SCCC) and modulation (QPSK to 64-APSK) of user data according to [1]
- Nonlinear pre-distortion in order compensate for travelling wave tube non-linearity
- Generation of Ka-Band and of X-Band carrier signals
- Modulation of user data onto Ka-Band and alternatively onto X-Band carrier signals
- Gain control feedback loop in order to compensate for amplifier gain drift
- Generation of internal supply voltages from satellite bus voltage

The partitioning of these functions into hardware units was driven by a strict modular design approach. The resulting hardware architecture is shown in Fig. 1. Accordingly, the High Data Rate Modulator consists of the following sections:

- Interface Section including the input data and TM/TC interfaces and configuration parameter setting
- ModCod Section performing coding, modulation, pre-distortion and digital-to-analog conversion
- RF Section serving for modulation in X-Band and optional up-conversion to Ka-Band
- DC/DC Section for generation of the secondary supply voltage from the satellite bus voltage

The various functional blocks are accommodated in dedicated hardware units. In order to avoid internal interferences digital, RF and pure DC functions are physically separated. The sections can be tuned and tested separately ensuring a cost effective manufacturing approach.

All the satellite bus specific input data and TM/TC interface are accommodated within the Interface Section. In case an interface adaption is needed, only the Interface Section will be modified. The project specific configuration parameters (e.g. data rates, carrier frequency, linearization coefficients) are stored within a non-volatile memory device that is also located on the Interface Section. The other sections are generic and can be used without any project specific modification.

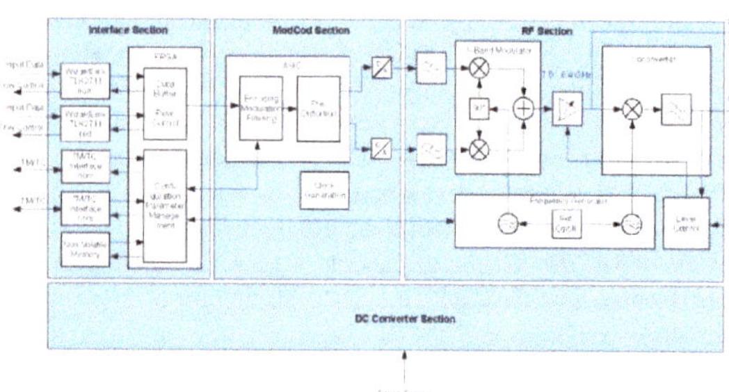

Fig. 1: Block Diagram of High Data Rate Modulator

3. Digital Processing Design

The digital signal processing functions according to CCSDS 131.2.B.1 [1] are realized within a space qualified CMOS ASIC. The ASIC design consists of the following modules as shown in Fig. 2:

- Data Interface containing the sub-modules FIFO and Slicer including a test pattern generator
- SCCC Encoder containing the sub-modules Outer Encoder, Interleaver, Inner Encoder, Inner Puncturer and Row-Column-Interleaver
- Framer
- Constellation Mapper
- Poly-phase square-root raised-cosine (SRRC) interpolation filter
- Pre-Distorter providing 5th order static pre-distortion
- I/Q-Modulator Error Compensation
- Test Signal Generator
- DAC Timing Adjust
- Configuration and Monitoring
- Clock Factory providing all the internal clock signals

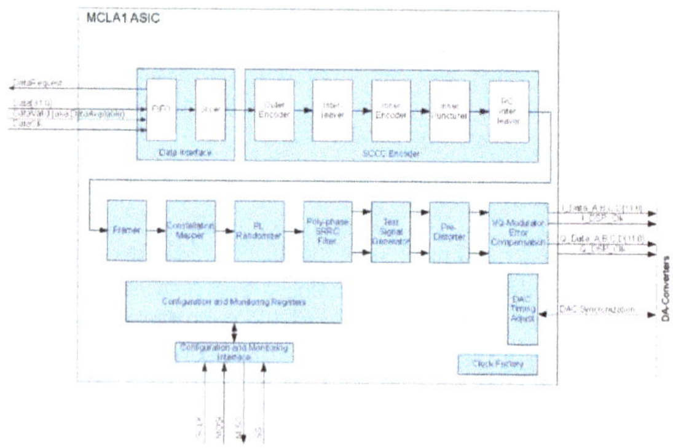

Fig. 2: Block Diagram of Digital Signal Processing ASIC

The encoder supports 27 different encoding schemes, identified by the ACM format number that ranges from 1 to 27. In the modules Data Interface, Outer Encoder, Interleaver, Inner Encoder, Inner Puncturer and RC-Interleaver data is processed codeword by codeword. This requires ping pong RAM interfaces between consecutive sub-modules. Each sub-module therefore consists of an Encoder Part and a Ping Pong RAM.

The ASIC output signal sampling rate is around 1.5 GHz, which of course is much higher than the signal processing clock rate within the ASIC. This means, parallelization is required to achieve the high sampling rate. A signal processing clock rate up to 100MHz is feasible with sufficient margin for the ASIC. The ratio between signal sampling rate and processing clock rate shall be a power of 2. Hence 16 was the preferred value for this ratio leading to a processing clock rate somewhere between 93.75 MHz and

100 MHz. The symbol rate is up to 500 MHz. Since the Nyquist filter input sampling rate is equal to the symbol rate while the output rate is equal to the signal sampling rate, the Nyquist filter must be an interpolating poly-phase filter.

4. Predistortion

For modulation schemes which involve amplitude modulation (e.g. 16- to 64-APSK), severe link degradations in terms of bit-error-rates are observed when the signal is transmitted over a non-linear amplifier like a travelling wave tube amplifier (TWTA). In order to avoid this effect, it is necessary to pre-compensate the non-linearity of the amplifier. Various pre-distortion concepts have been investigated at Tesat:

- Static analogue pre-distortion with output power control of amplifier
- Static digital pre-distortion with output power control of amplifier
- Adaptive digital pre-distortion with broadband feedback for compensation of amplifier linearity drift

As a result of the trade-off, it was found that the static digital pre-distortion is a good compromise between link performance and hardware effort. Compared to the classical analogue pre-distortion, the digital pre-distortion provides more flexibility w.r.t. the adaption to specific amplifier characteristic and requires less hardware elements. A fully adaptive approach with a broadband feedback loop of the output signal of the TWTA would only lead to a marginal improvement compared to the static digital pre-distortion since the tube linearity does not change much over lifetime; only the gain of the tube is drifting which can be controlled via a power control loop.

Fig. 3 shows the effect of the chosen predistortion implementation:

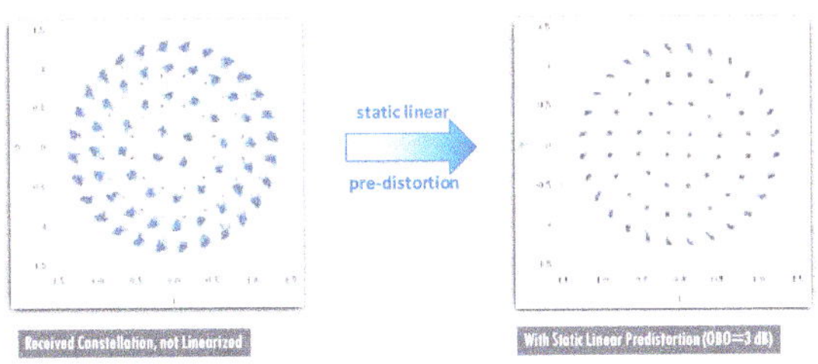

Fig. 3: Digital Predistortion for Linearization of Power Amplifier

5. Characteristics of High Data rate Modulator

The characteristics of the High Data Rate Modulator are summarized in Table I.

Table I. Main Characteristics of High Data Rate Modulator

Parameter	Value
BASEBAND DATA PROCESSING	
Encoding	SCCC according to CCSDS 131.2.B.1 [1]
Modulation	QPSK, 8PSK, 16-APSK, 32-APSK, 64-APSK according to CCSDS 131.2.B.1 [1]
Physical layer framing	According to CCSDS 131.2.B.1 [1]
Roll-off	0.2, 0.25,0.3, 0.35
Pre-distortion	5th order polynomial pre-distorter
Output symbol rate	10 … 500 MBaud
Input user data rate	10 Mbps … 2.6 Gbps (With TKL2711-SP max. 2.0 Gbps)
RF OUTPUT SIGNAL	
Carrier frequency range X-Band version Ka-Band version	 8.025 … 8.4 GHz 25.5 … 27 GHZ
Frequency stability	< 20 ppm (including all effects)
Integrated phase noise	< 1°rms (10 kHz to 250 MHz)
Spurious outputs	< -50 dBc
I/Q imbalance	<0.5 dBpp and <3°pp
Output power	0 … 6 dBm
Output power control	Via internal or external detector
BUDGETS	
Power consumption	30 W
Dimensions	202 x 150 x 97 mm³
Mass	2.5 kg
ENVIRONMENTAL TESTING	
Qualification temperature range Operating Non-operating	 -25°C … 70°C -35°C …+75°C
Random vibration levels	26 grms (all axes)
Shock levels	4200g @4 kHz
Radiation tolerance	15 years GEO orbit assuming 0.5 mm aluminum shielding of satellite structure
OPERATIONAL CHARACTER.	
Reliability	< 1200 FIT
Lifetime	15 years
Switch on-off cycles	100000
USER DATA INTERFACE	
Physical layer	Bidirectional link using TKL2711 SP (other interfaces on request)
Clock rate of parallel I/Os	125 MHz (16 bit)
Clock rate of serial I/Os	2.5 GHz
Net user data rate	2 Gbps
Connector	Combo-D (with 4 coaxial inserts)
Data link layer	Frames with header and check sum (details on request)
Data flow control	Using TKL2711 return link
Redundancy	Redundant signals and connectors
LEVEL CONTROL INTERFACE	
Level	Differential ASM [3]
Connectors	4 x Sub-D
RF OUTPUT INTERFACE	
X-Band	50 Ohm SMA female
Ka-Band	50 Ohm K female
TM/TC INTERFACES	
Serial TM/TC interface	BSD according to [3] (Read&Write of configuration parameters, details on request)
High level commands	LV-HV according to [3] (Modulator ON/OFF)
Bi-level telemetry	BDM according to [3] (ON/OFF telemetry)
Redundancy	Redundant signals and connectors
BUS VOLTAGE INTERFACE	
Bus Voltage	21 … 38 V (Options: 50 V, 100 V)

6. Verification of Design

The electrical and mechanical design concept of the High Data Rate Modulator has been established and the detailed design of the boards and mechanics has been completed. In a first step, an engineering model (EM) has been built (see Fig. 4).

This EM was used to verify the design concept and the performance of the Modulator.

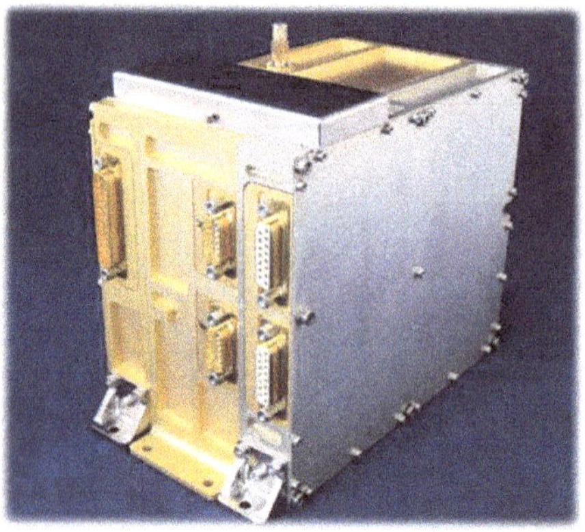

a

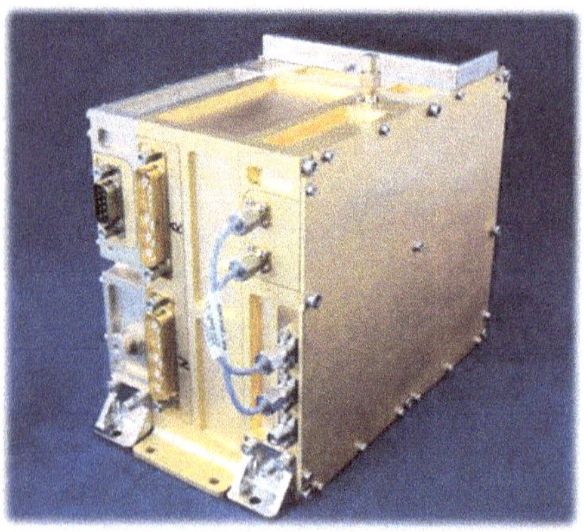

b

Fig. 4: Fully integrated EM of Modulator
(a) DC Connection Side
(b) RF Connection Side

An extensive measurement campaign was performed with the EM. The results obtained clearly demonstrate the data rate capability and the correct function of the encoding and modulation algorithms. End-to end measurements also including a nonlinear power amplifier have been performed by using a prototype ground station demodulator (results see section 7). The following figures shows the output spectrum of the Modulator and the constellation diagram for a 64-APSK modulated X-Band signal at the maximum output symbol rate of 500 MBaud and maximum input user data rate of 2.6 Gbps. Error free data transmission a this data rate was demonstrated in conjunction with the prototype ground station demodulator.

7. End-to-End Performance

Within the EM test campaign, additional measurements of the performance for the transmit chain including the high power amplifier (TWTA) in Ka-Band have been conducted. In order to account for the performance of the end-to-end chain, a representative receiver was used. This allows to include effects in the receiver e.g. from carrier recovery or equalization. Furthermore, usage of a real TWTA has allowed to successfully verify the predistortion concept.

Fig. 5 shows the received constellation for the chain including the TWTA operated at an output backoff of 2.8 dB.

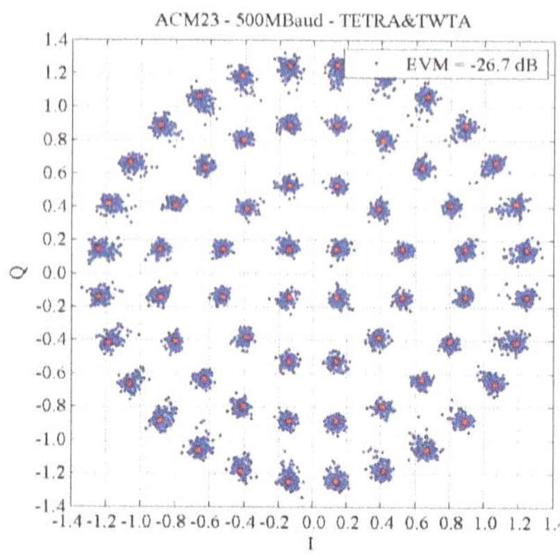

Fig. 5: Received 64-APSK Constellation Diagram including TWTA at a Symbol Rate of 500 MBaud

Fig. 6 shows bit error rate measurement results for one modulation & coding scheme (ACM) for each modulation order at 500 MBaud symbol rate.

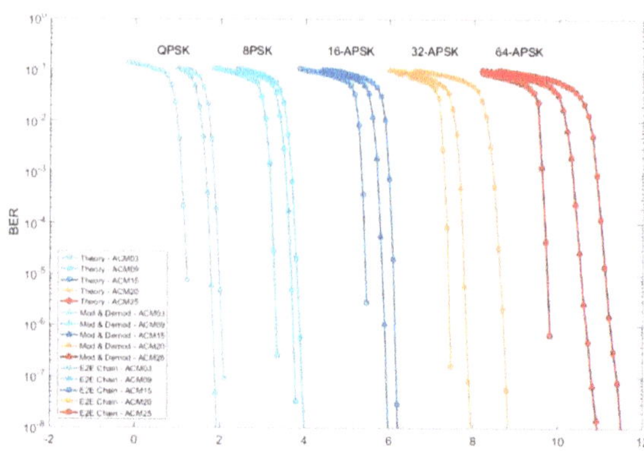

Fig. 6: BER Results @500 MBaud

The figure includes results for:

- simulated values without signal degradation ("ideal" values)
- measured values for the modulator only
- measured values for the chain with TWTA.

Different output backoff values (OBO) have been used. They have been optimized for each modulation order:

- ACM03: QPSK, OBO=0.5 dB
- ACM09: 8PSK, OBO=0.8 dB
- ACM15: 16-APSK, OBO=1.5 dB
- ACM20: 32-APSK, OBO=2.2 dB
- ACM25: 64-APSK, OBO=2.8 dB

8. Applications

8.1. Application for Data Downlink Transmission in Earth Observation Satellites

Table II shows the achievable data rates depending on the employed frequency band (X / Ka) and the employed modulation & coding scheme for both available solution and the new generation.

Table II. Achievable Data Rates

Modu-lation	Coding	Band	Bandw. Eff. [bit/s/Hz]	Max. User Data Rate [Mbps]	Availa-bility
QPSK	None	X	1.8	350	available
8 PSK	4D-TCM r=5/6	X	2.3	500	available
16 APSK	SCCC r=0.8	X	3.2	750	**in 2018**
64 APSK	SCCC r=0.9	X	5.4	1200	**in 2018**
QPSK	None	**Ka**	1.4	600	available
16 APSK	SCCC r=0.8	**Ka**	3.2	1600	**in 2018**
32 APSK	SCCC r=0.82	**Ka**	4.1	2000	**in 2018**

Fig. 7 shows a typical downlink system architecture with two active transmission channels in a redundant configuration incl. an output multiplexer (OMUX). This is attractive for Ka-Band systems where sufficient bandwidth is available to allocate two channels with 500 MBaud symbol rate and an additional guard band.

Such architecture allows transmitting up to 4000 Mbps when employing 32-APSK.

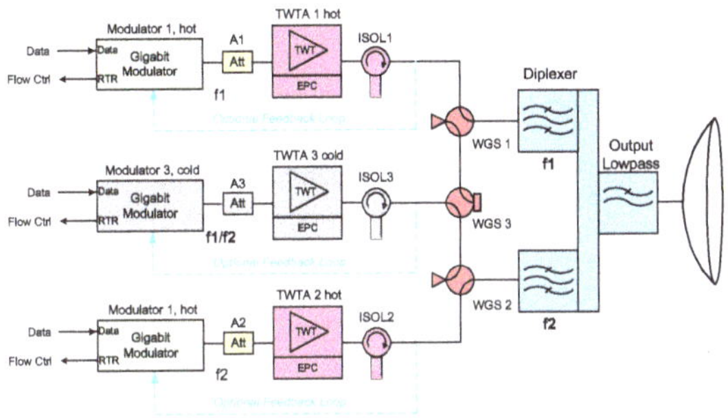

Fig. 7: Dual Channel Downlink Subsystem Architecture

Fig. 8 shows a comparable architecture with two active transmission channels feeding an antenna with dual polarization. This is attractive for X-Band systems where the bandwidth is limited to 275 MHz in order to allocate two channels with 225 MBaud symbol rate.

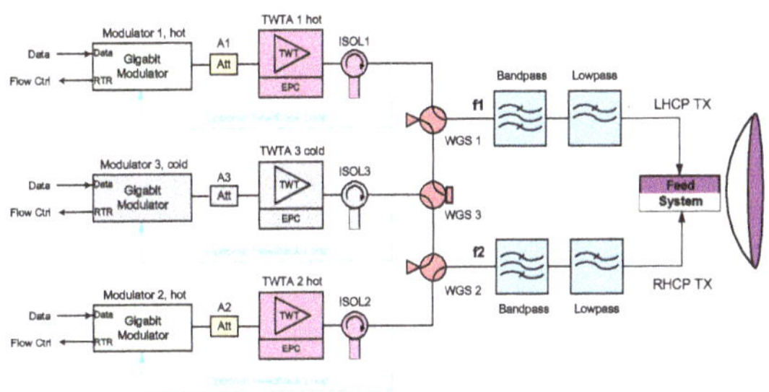

Fig. 8: Downlink Subsystem Architecture with Dual Polarization

Such architecture allows transmitting up to 2400 Mbps when employing 64-APSK.

The input data for the Modulator are coming typically from a mass memory. The RF output is fed to an antenna.

8.2. Usage in Relay Payloads

Future GEO data relay satellites will allow a very high data rate link (see Fig. 9). Such a system requires sophisticated data handling providing coding, demultiplexing onto multiple channels and optional encryption [4]. The Ka-Band RF downlink from the GEO to the ground of course also profits from the availability of high-order modulation schemes in combination with powerful forward error correction schemes.

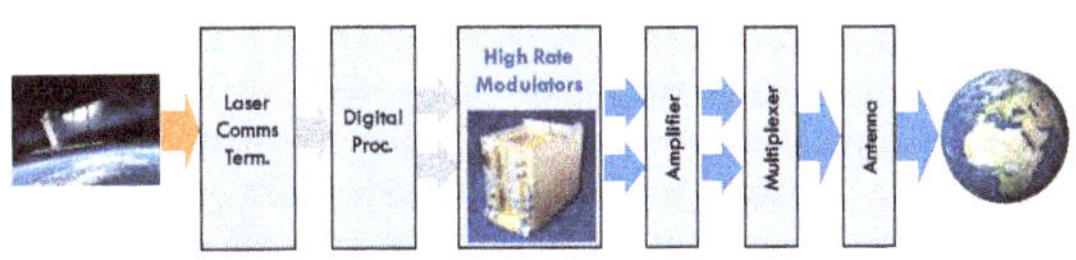

Fig. 9: Principle Relay Payload Concept

9. Evolution: Integrated Transmitter

More and more small earth observation satellites are being designed, yet implementing a high resolution of instruments. Therefore, also for these applications high data rates are required for the downlink.

A new development running in parallel aims at providing an Integrated Downlink Transmitter, i.e. including modulator section, a solid state power amplifier and a passive RF output filter in a one box design.

It shall include the following features:

- Compact and lightweight (2.5 kg) high data rate downlink transmitter for Earth Observation Satellites
- Cost attractive design based on lower class components
- Includes X-Band Power Amplification with GaN SSPA (20 W RF Output Power)
- High order modulation scheme up to 32-APSK combined with powerful SCCC error correction coding enables reliable communication.
- Up to 1000 Mbps per channel
- Data input with state-of-the art WizardLink incl. flow control [2]

Fig. 10 shows the preliminary mechanical design of the compact unit:

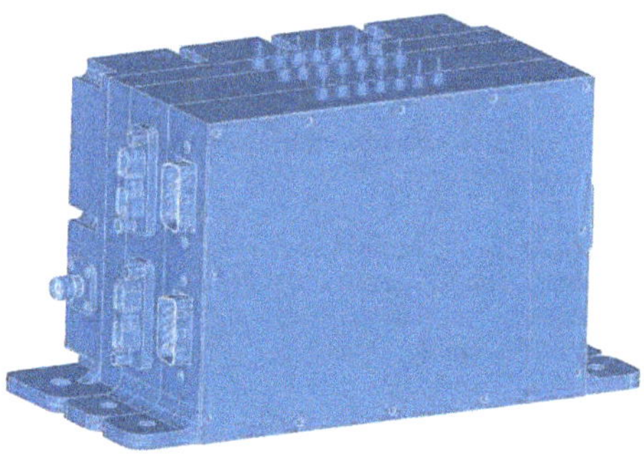

Fig. 10: Mechanical Design of Integrated Downlink Transmitter

10. Conclusions

The High Data Rate Modulator currently under development at Tesat implements advanced SCCC coding and higher order modulation schemes according to CCSDS 131.2.B.1 [1]. In total 27 different transmission schemes are available with different code rates and spectral efficiencies, which can be selected by telecommand on a data, frame basis. Thus, the Modulator enables variable and adaptive coding & modulation (VCM and ACM) within a single satellite-to-ground-station contact period.

The Modulator supports an output symbol rate of 500 MBaud and user data rates of up to 2.6 Gbps on a single carrier in X-Band and Ka-Band. A digital pre-distorter is included within the Modulator that enables the utilization of the highest order modulation schemes (32-APSK, 64-APSK) in conjunction with a non-linear power amplifier.

The data rate capability as well as the correct implementation of the modulation and encoding algorithms has been already verified by engineering model (EM). The full qualification test campaign of the Modulator will be finalized until Q4/2018.

Thus, the High Data Rate Modulator presented here is considered as a major milestone in the evolution of high-speed data links towards multi-gigabit data transmission.

The evolution for SmallSat application as a compact transmitter opens up new applications for using highly effective coding & modulation.

Acknowledgement

The development and qualification of the High Data Rate Modulator is performed under an ESA GSTP contract.

References

[1] Flexible advanced coding and modulation scheme for high rate telemetry applications, CCSDS 131.2-B-1

[2] TLK2711-SP Data Sheet, Texas Instruments

[3] Spacecraft discrete interfaces, ECSS-E-ST-50-14C

[4] B. Friedrichs, P. Wertz, H.-U. Dehner, "Error-control Coding and Packet Processing for Broadband Relay Satellite Networks with Optical and Microwave Links", 6th Advanced Satellite Multimedia Systems Conference / 12th Signal Processing for Space Communications Workshop, September 2012, Baiona, Spain.

Implementation of a Machine Learning Based Modulation Scheme in GNURadio for Over-the-Air Packet Communications

Mathew McCaskey[1], Austin Feydt[1], Robert Corrigan[1], Kul Bhasin[1], David Chelmins[2]*

[1]*Comsat Architects, 20525 Center Ridge Rd. Ste 614 Rocky River, Ohio, USA*
[2]*NASA Glenn Research Center, 21000 Brookpark Rd. Cleveland, Ohio, USA*
**kul@comsat-architects.com*

Keywords: MACHINE LEARNING, NEURAL NETWORKS, GNURADIO, RF COMMUNICATIONS

Abstract

In this paper we introduce an auto-encoder neural network that can be used to model a node-to-node radio frequency communication channel. With some simple assumptions the encoding layer of this neural network can be interpreted as a phase modulated signal which can be transmitted across a communication channel. When the auto-encoder is trained to minimize bit error rates within the transmitted messages, it effectively learns an optimal modulation scheme for that particular communication channel. To implement these signals in a physical link we developed GNURadio flowcharts along with custom signal processing blocks that transmit and receive the auto-encoder generated signals between a pair of Universal Serial Radio Peripherals. Finally, we show the methodology used for incorporating the GNURadio flowcharts with the auto-encoder training program so that the auto-encoder can be trained on the physical communication channel itself.

1. Introduction

The space radio frequency (RF) communication environment is becoming increasingly complex due to networks needing to accommodate more users with higher performance demands. With only a finite amount of bandwidth available, the need for optimizing and/or maintaining reliable communication links becomes paramount. Machine learning technologies have the potential to help meet these challenges in this continuously evolving space communication environment [1][2][3]. With the ability of machine learning algorithms to adapt and learn from new information, they can be employed for example to extract features from communication signals which can be used to classify a signal based on modulation scheme [4][5][6], identify whether the signal was sent by a primary or secondary user [7][8][9], etc. Communication channel inputs can be tuned through reinforcement learning algorithms to maximize a particular metric (e.g. throughput, signal to noise ratio, etc.) or to minimize bit error rates (BER) [10][11]. Combining these algorithms together with a cognitive engine to manage the various inputs and outputs will allow the creation of a cognitive communication system [12][13][14].

The National Aeronautics and Space Administration (NASA) is interested in developing advanced communications techniques enabled by recent advances in machine learning. The Space Communication and Navigation (SCaN) program developed a Next Generation Architecture which envisions spacecraft and ground systems coordinating among themselves to "establish communications, negotiate link connectivity, and learn to share spectrum to optimize resource allocations" [15]. Additionally, the plan envisions cooperation between commercial and government communication assets, which poses challenges in technical areas including spectrum usage, asset management and scheduling, and delay tolerant routing. Many of these challenges can be met through development of advanced processing algorithms capable of operating within a software defined radio architecture [16]. Such an architecture will increase the flexibility an extensibility of the infrastructure in order to ensure interoperability with existing and future systems, allowing innovations in machine learning to be infused quickly to improve performance of operational systems.

A potential area where machine learning algorithms can be applied to improve RF communications is optimizing information transmission across the communication channel itself using neural networks [17][18][19][20]. In an ideal communication system, information that is sent from a transmitter should be identical to the information received on a receiver. Assuming that this information is used as the inputs and outputs of a neural network, then from a machine learning perspective this model would be referred to as an auto-encoder. With an auto-encoder, the outputs of the middle layer, which is also known as the encoding layer, is associated with a set of learned features that are both derived from and used to reconstruct the input information. In terms of an RF communication system, these features can be interpreted as a phase modulated signal which can be transmitted across the physical RF communication channel. When the neural network is trained to minimize bit error rate, the network effectively learns the best modulation and coding scheme, for given channel conditions, to send information across the communication channel from the transmitter to receiver even in non-ideal channel conditions.

To test this auto-encoder based modulation scheme in an experimental setting, we developed a testbed communication environment using GNURadio [21] and a pair of Universal Serial Radio Peripherals (USRPs). GNURadio is an open source signal processing software that allows users to create flowcharts that define a signal processing chain that can be easily interfaced with USRPs to send and receive RF signals over the air. It also allows users the capability of creating custom signal processing blocks that can be incorporated into new flowcharts. Since space communications are generally non-continuous, we created flowcharts that would allow bursty packet transmissions. On top of the bursty packet transmissions, we also needed signal processing blocks to perform the auto-encoder algorithm itself. With these processes in mind we developed several custom GNURadio blocks that would be used for packet detection, header parsing and synchronization, and most importantly modulation and demodulation of the auto-encoder generated RF signal itself.

In Section 2 we will overview the neural network model of the node-to-node communication system that we will be implementing in GNURadio. Section 3 describes the GNURadio flowcharts that we developed to send the auto-encoder generated messages between a pair of USRPs. Section 4 outlines the incorporation of the auto-encoder training program with the GNURadio flowcharts taking into account the asynchronous nature of the two nodes in a communication system. Finally, we present our conclusions, challenges, and future prospects in Section 5.

2. Auto-encoder model of an RF communication system

An RF communication system is designed to send digital information from a transmitter to a receiver. In general, this information can be broken up into messages of k-bits in length. These k-bit messages can then be used as both the inputs and the desired outputs to a neural network. Ideally, the messages at the receiver will be identical to the messages that have been sent by the transmitter. Therefore, this type of neural network is known as an auto-encoder. Mathematically, an auto-encoder models an identity function as the inputs and outputs are identical. Once the neural network is fully trained, the outputs of one of the hidden layers in the network, generally the middle layer, can be viewed as encoded information since it is derived from, and used to reconstruct the input message. Previous work [17] has used this auto-encoder to model an end-to-end communication system, where the encoding layer can be interpreted as an RF signal to be sent along the communication channel. This model can then be used to optimize not only the communication channel, but also the non-linear components of the transmitter and receiver.

Based on this previous work we study a modified version of the auto-encoder model. Fig. 1 shows the general structure of the auto-encoder in question where each k-bit message is encoded into n RF symbols. The input message is passed through several dense layers until reaching the encoding layer. This layer contains 2n neurons whose outputs represent the in-phase (I) and quadrature-phase (Q) components of the n RF symbols. Normalizing the symbols such that the Euclidean norm of each symbol lies within the unit circle results in a purely phase modulated RF signal. These symbols are sent along the communication channel between the transmitter and receiver. Previous work [17][20] has used Additive White Gaussian Noise (AWGN) to model the effects of the communication channel, while some work has been done using over-the-air transmissions [18]. The received noisy RF symbols are then sent through several more dense layers until reaching the output layer. A sigmoid function is used as the activation for the output neurons so that the result can be interpreted as the probability of that particular bit being a "1". For training purposes we can use the cross entropy between the inputs and outputs, since they are supposed to be identical.

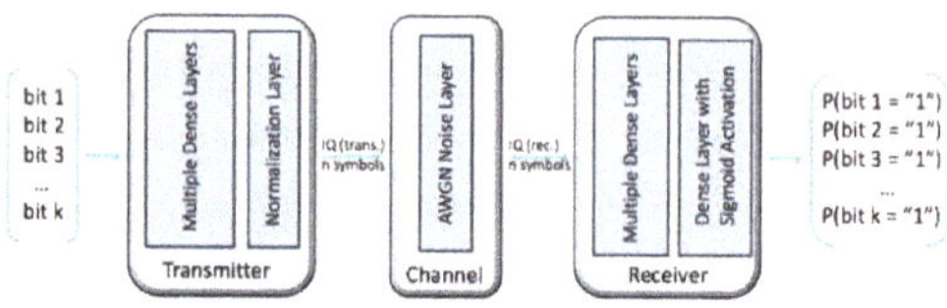

Figure 1: General structure of the auto-encoder model of a node-to-node RF communication system.

With the recent popularity of machine learning technologies, open source tools such as Theano [22], Tensorflow [23], and Keras [24] have been developed to help streamline the process of designing, implementing, and training neural networks. Using Tensorflow, we developed a python script that defines and trains the auto-encoder model with the architecture given in Fig. 1. In order to run this training program using a physical RF communication channel, it would need to be able to both transmit and receive the RF signals necessary for the forward and backpropagation of the neural network. This is where GNURadio comes into play. In this case, the training program can call a function that runs GNURadio and returns the noisy IQ samples obtained from transmitting the RF signals across the communication channel. We therefore need to describe the GNURadio implementation before coming back to this in Section 4.

3. GNURadio Implementation

As previously stated, GNURadio [21] is an open source signal processing software platform that allows users to create flowcharts defining a sequence of signal processing operations which are applied to a data stream. With these signal process flowcharts one can easily simulate an RF communication system from transmitter to receiver operating at baseband. Using blocks that interface the data stream with the Field Programmable Gate Array (FPGA) in a USRP, this platform can then be used to create a physical RF communication channel where signals can be sent from one USRP to another. For our purposes we developed both transmitter and receiver flowcharts that create a bursty, packet based communication system incorporating the auto-encoder generated modulation scheme.

3.1 Packet Transmitter Flowchart

The transmitter flowchart, shown in Fig. 2, generates the waveforms incorporating the auto-encoder generated modulation scheme and transmits it out of the USRP as an RF signal. The flowchart first takes a data stream from a source block (whether that source be a file or function that systematically generates data) and breaks the steam up into tagged chunks of a desired length. In this case we used a packet size of 32 bytes (the rationale of this choice will be explained later.) A header is created for each packet which contains an access code used for packet detection and metadata of the packet itself including the payload length, the number of bits per symbol, and a packet counter. The header is modulated using binary phase shift keying (BPSK) while the payload is modulated using the auto-encoder generated modulation scheme. The header and payload symbols are combined together using a stream mux block. The packets are then bookended with ramp-up and ramp-down bits. Padded zeros are then used to separate consecutive packet bursts. The whole waveform is then up-sampled to four samples per symbol using a root raised cosine filter. The resulting waveform, shown in Fig. 3, is then sent to the USRP to be up-converted to the carrier frequency and sent out the radio itself.

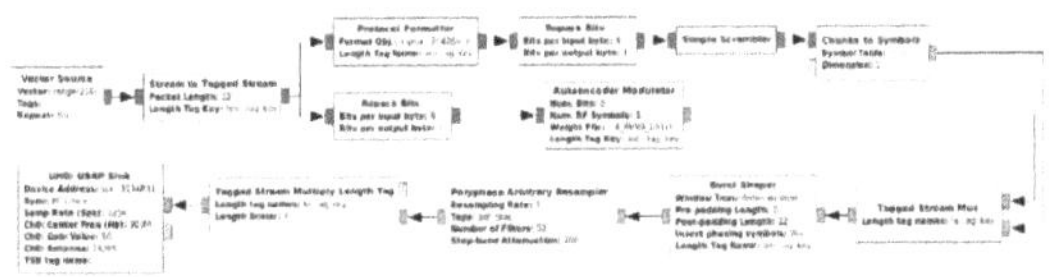

Figure 2: GNURadio flowchart for the transmitter side of the RF communication system.

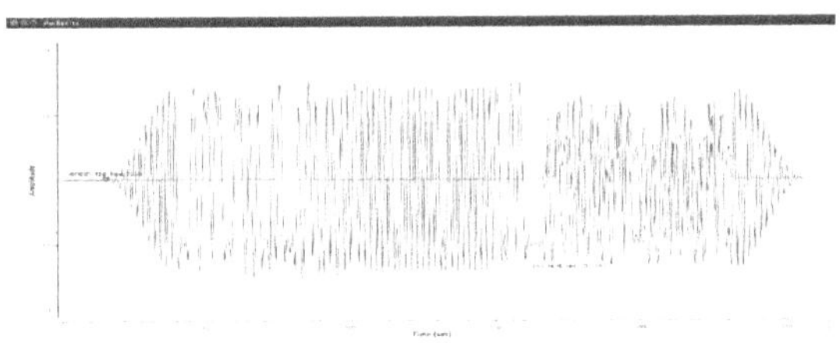

Figure 3: IQ samples of the packet waveform to be transmitted out of the USRP

3.2 Packet Receiver Flowchart

The receiver flowchart, shown in Fig. 4, is designed to detect the waveform generated in the transmitter flowchart and extract the information contained within its payload. The flowchart starts with a USRP Source block which takes the measured voltages of the transmitted RF signal and down-coverts the signal to baseband. A correlation estimator block performs a cross correlation of the signal stream with a modulated, up-sampled access code to find and tag where the beginning of a packet is located. Along with the start of the packet, the correlation block also estimates the phase offset and amplitude of the signal. The amplitude estimate is used for coarse gain control immediately after the correlation estimator while the phase offset is used further down the flowchart in the Costas loop block. Next, the signal is sent through a poly-phase clock sync which removes timing offsets within the signal and down-samples the signal to one sample per symbol. The Header/Payload Demux block uses the correlation estimator tag and the number of header symbols to separate the header and payload symbols into separate outputs. The header symbols are passed through a Costas loop for frequency synchronization. The symbols are then demodulated and the resulting data is parsed to extract the metadata contained within the header. The metadata is then sent back to the Header/Payload Demux block where it is checked to see if the information is valid. If it is, the payload symbols are released to the Costas loop for frequency synchronization. The resulting IQ samples can either be demodulated into bits and sent to a file or be sent through our IQ Capture block for further processing.

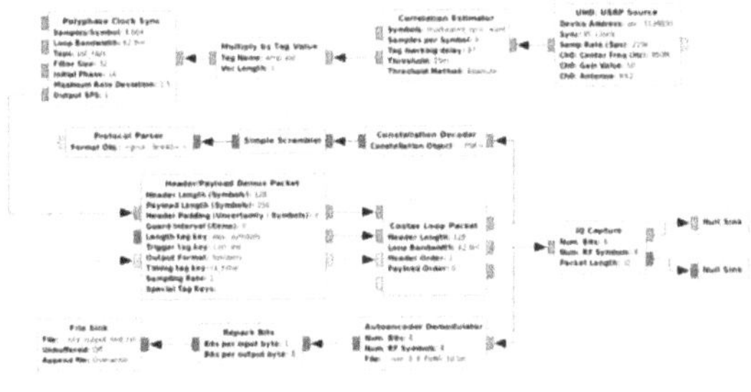

Figure 4: GNURadio flowchart for the receiver side of the RF communication system.

3.3 Custom GNURadio Blocks

While GNURadio already provides a large number of signal processing blocks out of the box as well as several examples using those blocks to create RF communication systems, there is still missing signal processing functionality that is needed for this study. Most importantly, there are not any blocks to perform the auto-encoder modulation and demodulation. Fortunately, GNURadio allows users to develop custom signal processing blocks which can be written in either Python or C++. Once developed, the custom blocks can be incorporated in any flowchart. The following sections outline the custom blocks that were developed when creating the transmitter and receiver flowcharts of our RF communication system.

3.3.1 Costas Loop Packet: The Costas Loop Packet block is necessary for establishing frequency synchronization between the transmitter and receiver. Even when radios are set to the same carrier frequency, there is a natural uncertainty within each radio which causes a frequency offset in the received signal. In GNURadio, a Costas loop block will adaptively adjust the phase and frequency offsets in the signal based the IQ constellation of the signal. This constellation is set by the user-defined order of the loop (order = 2,4, or 8 for BPSK, quadrature phase shift keying (QPSK), and 8-PSK respectively).

In the packet receiver flowchart example from GNURadio, there are separate Costas loops for the header and payload bits, each located immediately following the header and payload outputs of the Header/Payload Demux block. The problem with this setup is that there is no link between the two Costas loops. When a packet comes in, the header will be able to use the phase estimation calculated by the correlation estimator to start the synchronization. The payload synchronization does not have the same initial phase information, so when the Costas loop locks onto an IQ constellation there is no guarantee that it is locked on the correct constellation or a rotationally symmetric variant of it. Also, since there is a finite amount of time between consecutive payloads, the frequency offset could kick the Costas loop out of one constellation and into a rotated constellation. This results in the output file containing alternating segments of correctly demodulated data and garbage.

To remedy this problem, we developed a custom Costas Loop Packet block that takes both the header and payload outputs from the Header/Payload Demux Packet block as inputs. When the header IQ samples are received then the block performs the same synchronization operations as the GNURadio Costas loop block. Then the Costas Loop Packet block waits for a message from the Header/Payload Demux Packet block containing information about the parsed header metadata. If the header metadata is determined to be valid then this message will contain the number of payload symbols to be processed. The block then uses the current values of the phase and frequency offset from the header synchronization to estimate the initial phase for the first payload symbol. After which, the rest of the payload symbols are processed as before. If the header information was determined to be invalid then the block will reset the process by waiting until it receives another set of header IQ samples.

Because the header and payload constellations often are different, the loop order for the header and payload synchronization must be independently set. As previously mentioned, GNURadio only has loop orders for BPSK, QPSK and 8-PSK. Any other payload constellation would not be supported for synchronization with the GNURadio Costas loop, which certainly includes the auto-encoder modulation scheme. Therefore, we included the option to set the loop order to 0, where the frequency offset will remain constant and no long be adaptively changed. This way, frequency synchronization can occur on the header symbols, and the frequency offset can be carried on through the payload symbols. Unfortunately, this method does mean that frequency drift and/or uncertainty will eventually kick the payload out of its desired constellation. To mitigate this effect while using the auto-encoder modulation scheme we limit the size of the payload to be 32 bytes.

3.3.2 Header/Payload Demux Packet: To properly use the custom Costas Loop Packet block described above, it requires receiving a message with information regarding the validity of the parsed header metadata. This can be achieved through the Header/Payload Demux block provided in GNURadio since this block already handles the header metadata after it has gone through synchronization and demodulation. The Header/Payload Demux block switches between several states depending on if the block is outputting header symbols, payload symbols, or waiting for a message containing the parsed header metadata. The only customization we needed to add is to send out a message containing the status of the parsed header information. If the header is valid then we send a message to the Costas loop block with the number of payload symbols. If it is not, then we send a message with a payload length equal to negative one so the Costas Loop Packet block will know the packet was invalid and switch back to waiting for a new header.

3.3.3 Simple Scrambler: While developing the transmitter and receiver flowcharts, we observed that when a string of identical symbols is transmitted, the received signal output from the USRP is noticeably worsened. On the other hand, the USRP seems to receive alternating symbols quite well. With our current waveform, there are indeed several long strings of identical symbols contained within the header metadata, mainly from the packet length (0x0020) and the number of bits per symbol (0x0001). The simple scrambler is therefore designed to remove these long strings of identical symbols by performing an exclusive or operation between the header data and a repeating mask. This effectively removes these string of repeated symbols and improves the quality of the received signal. Fig. 5 shows the received signal before and after the simple scrambler is used on the transmitter side.

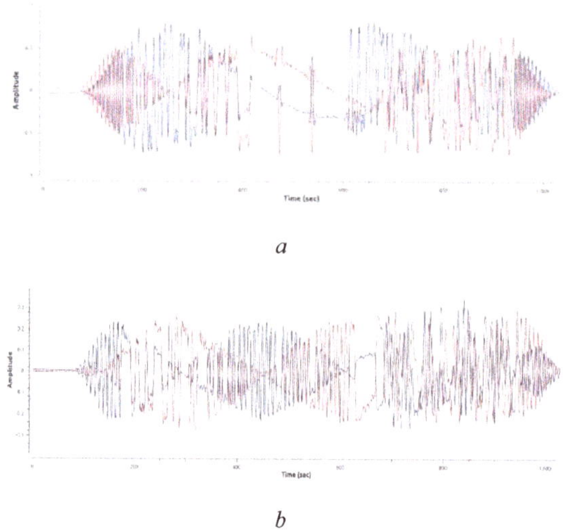

Figure 5 The waveforms measured on the receiver end of the communication system (a) without header scrambling and (b) and with header scrambling.

3.3.4 Auto-encoder Modulator/Demodulator: The Auto-encoder Modulator and Demodulator blocks are designed to convert bits to IQ symbols and back. Since the modulation and demodulation are both done by performing the forward propagation of the auto-encoder, both of these blocks simply implement the first and second half of the neural-network. Each block requires the number of bits per message (k), the number of RF symbols to be sent (n), and a filename where the trained weight set it located as inputs. When either block is initialized in GNURadio, it will read in the set of weights for its respective half of the neural network. Then it simply goes through the weighted sums and activation functions for the hidden and output layers. For the modulator side, the output RF symbols are normalized such that the Euclidean norm does not exceed 1. For the demodulator side, the output probabilities of each bit are rounded so that we end up with a pure bit stream.

3.3.5 IQ Capture: The IQ Capture block is designed to handle the IQ samples of the payload as well as the header metadata that was parsed from previous blocks. In one mode, the IQ capture block simply outputs both the IQ samples as well as the packet counter as packets come it. This is primarily used when performing bit error rate tests [20] as it allows one to compare received bits with expected bits based on the packet number. The second mode involves capturing an entire set of IQ values from all of the possible messages. To accomplish this, the IQ Capture block requires k, n, and the packet size as inputs. With this information, the block initializes a boolean vector to track which packets have been received. As packets arrive the packet number is checked against the boolean vector, and if that packet has not been seen the respective IQ samples are stored. Once all of the packets have been received then the entire set of IQ samples are sent out. This gives a consistent way of tallying up the received IQ samples associated with each input signal. This information can then be used in the training process as these received IQ samples will be passed through the latter half of the auto-encoder.

4. Incorporating the Training Program and GNURadio Flowcharts

From the previous sections, it is clear the auto-encoder training program requires the noisy IQ samples from GNURadio after the transmitted signals have been sent across the RF communication channel in order to fully train the auto-encoder model. While previous work has been done to do this sort of over-the-air training [18], the nature of this communication system brings about several complications. Most notably, the fact that the two nodes can only communicate via transmitted and received RF signals. This section we explore these complications and outline the processes that are run on both the transmitter and receiver in order to train the auto-encoder over the physical communication channel.

4.1 Asynchronous Architecture

Since the transmitter and receiver are supposed be separate communication nodes, there needs to be an architecture that takes into consideration the asynchronous nature of the communication link, most notably how information cannot be instantaneously shared between the two nodes. This is important to consider as information needs to be transferred to and from both ends of the communication system to perform the forward and backpropagation of the auto-encoder training program. For example, the transmitter needs to have the current RF signals for each possible k-bit message to send across the communication channel as they are continuously being changed while the neural network is training. On the other hand, the receiver needs to get the received noisy IQ samples back to the auto-encoder training program so that it can finish its forward and back propagation.

Since the received IQ values are the most vital information to send to the training program as it contains information pertaining to the actual RF communication channel, it would stand to reason that the training program should be running on the receiver end of the communication channel, that way the received IQ samples can directly be sent to the training program for faster processing. This leaves us to determine how the transmitter needs to be run so that it can transmit the RF signals that the receiver needs while staying up to date with the latest RF signals to transmit.

4.2 Transmitter

Since the transmitter does not perform the neural network training, the only job of the transmitter is to consistently send all the possible k-bit messages using the most up-to-date RF signals derived from the current weight set of the auto-encoder. With the current transmitter flowchart from GNURadio, the transmitter will consistently send all of the possible k-bit messages in order tagged with the same packet number modulo the total number of packets required.

To update the transmitter with the latest auto-encoder weight set, we could periodically switch roles of the transmitter and receiver. Here the receiver can transmit the updated weight set for the first half of the auto-encoder using traditional digital modulation schemes. The transmitter can then update the modulator in its flowchart, and resume transmissions with the new weight set. For smaller scale testing purposes we can abstract this process out by having both the transmitter and receiver point to the same weight files on a common server. Then the transmitter can read this file to update the auto-encoder weights. For a consistent process, we can have the transmitter send out a small number of complete sets of all the possible messages, read the weight file (whether or not it actually has been updated), update the weights in the GNURadio flowchart and repeat the process. This way the transmitter doesn't necessarily need to wait for the receiver to do anything specific allowing it to act independently.

4.3 Receiver

As previously stated, the receiver is where the actual training of the auto-encoder will take place. To perform one complete epoch the receiver would need to measure the IQ samples received from all of the possible transmitted messages. With the use of the IQ Capture block, the receiver can extract these IQ samples and pass them off to the training program. The training program can then perform the rest of the forward propagation and then the back propagation to update the weights of the neural network. Once the weights have been updated, a new weight file can be written to the common server file so that the transmitter can read in the updated weights.

Since both the transmitter and receiver are constantly being switched on and off, we do need to take into consideration that the receiver may need to establish timing and frequency synchronization with the transmitter. Usually this occurs within the very first received packet so this can be taken into account by rejecting the very first packet every time the receiver flowcharts are started. After that point, every other packet will be captured by the IQ Capture block. Once an entire set of the IQ samples are captured, then the flowchart can stop and the IQ samples can be passed to the auto-encoder training program.

5. Conclusions, Challenges, and Future Prospects

In this paper, we described the process of implementing an auto-encoder model for the transmission of payload bits across a physical communication channel using packet based transmissions. We then described the architecture and general procedure to train this auto-encoder model on the actual communication channel using asynchronous procedures for both the transmitter and receiver to send, receive, train, and update all of the relevant information when training this auto-encoder.

For future research, we will further develop and refine these methods for training the auto-encoder model on a physical RF communication channel. One area to refine is the interaction between the GNURadio flowcharts and the auto-encoder training program. With the current setup, the GNURadio flowcharts need to be stopped and restarted every time they are called which can take several seconds to start up the USRPs. Ideally, we would like to have a way to keep the GNURadio flowcharts constantly running while the training program can run in parallel. With these refinements we will reduce the amount of time need to train the autoencoder while verifying that the auto-encoder is able to learn the best modulation scheme for transmitting information across a communication channel.

References

[1] Kirar, V., (2015). "Artificial neural networks for cognitive radio network : a survey", International Journal of Electronics and Communication Engineering, vol. 9, no. 1, 2015

[2] Wang, T., Wen, C., Wang, H., et al., "Deep learning for wireless physical layer: Opportunities and challenges," in China Communications, vol. 14, no. 11, pp. 92-111, Nov. 2017. doi: 10.1109/CC.2017.8233654

[3] Bkassiny, M., Li, Y., Jayaweera, S., "A Survey on Machine-Learning Techniques in Cognitive Radios," in IEEE Communications Surveys & Tutorials, vol. 15, no. 3, pp. 1136-1159, Third Quarter 2013. doi: 10.1109/SURV.2012.100412.00017

[4] West, N., O'Shea, T., "Deep architectures for modulation recognition," 2017 IEEE International Symposium on Dynamic Spectrum Access Networks (DySPAN), Piscataway, NJ, 2017, pp. 1-6. doi: 10.1109/DySPAN.2017.7920754

[5] Kremer, S., Shiels, J., "A testbed for automatic modulation recognition using artificial neural networks," CCECE '97. Canadian Conference on Electrical and Computer Engineering. Engineering Innovation: Voyage of Discovery. Conference Proceedings, Saint Johns, Newfoundland, Canada, 1997, pp. 67-70 vol.1. doi: 10.1109/CCECE.1997.614791

[6] Keshk, M., Asami, K., "FPGA-based Automatic Modulation Recognition System for Small Satellite Communication Systems," – 31st Annual AIAA/USU Conference on Small Satellites

[7] Chakravarthy, V., Wu, Z., Shaw, A., et al., (2007). "A general overlay/underlay analytic expression representing cognitive radio waveform." 2007 International Waveform Diversity and Design Conference, 69-73.

[8] Zhao, Q., Sadler, B., (2006). "Dynamic Spectrum Access: Signal Processing, Networking, and Regulatory Policy," CoRR, abs/cs/0609149.

[9] Pattanayak, S., Venkateswaran, P., Nandi, R., "Artificial Intelligence Based Model for Channel Status Prediction: A New Spectrum Sensing Technique for Cognitive Radio" – Int. J. Communications, Network and System Sciences, 2013, 6, 139-148

[10] J. Downey, D. Mortensen, M. Evans, et al., "Adaptive Coding and Modulation Experiment with NASA's Space Communication and Navigation Testbed." (2016)

[11] Samek, W., Stanczak, S., Weigand, T., "The Convergence of Machine Learning and Communications," in ITU Journal: ICT Discoveries, Special Issue No. 1, 13 Oct. 2017

[12] Yucek, T., Arslan, H., "A survey of spectrum sensing algorithms for cognitive radio applications," in IEEE Communications Surveys & Tutorials, vol. 11, no. 1, pp. 116-130, First Quarter 2009. doi: 10.1109/SURV.2009.090109

[13] Chenji, H., Stewart, G., Wu, Z., et al., "An Architecture Concept for Cognitive Space Communication Networks", 34th AIAA International Communications Satellite Systems Conference, International Communications Satellite Systems Conferences (ICSSC), (AIAA 2016-5728) doi: 10.2514/6.2016-5728

[14] Qiu, R., Chen, Z., Guo, N., et al., "Towards a Real-Time Cognitive Radio Network Testbed: Architecture, Hardware Platform, and Application to Smart Grid," 2010 Fifth IEEE Workshop on Networking Technologies for Software Defined Radio Networks (SDR), Boston, MA, 2010, pp. 1-6. doi: 10.1109/SDR.2010.5507920

[15] Reinhart, R., Schier, J., Israel, D., et al., "Enabling Future Science and Human Exploration with NASA's Next Generation near Earth and Deep Space Communications and Navigation Architecture", 68th International Astronautical Congress, September 2017.

[16] Kacpura, T., Eddy, W., Smith, C., et al., "Software defined radio architecture contributions to next generation space communications", 2015 IEEE Aerospace Conference, pp. 1-12, 2015.

[17] O'Shea, T., Hoydis, J., "An Introduction to Deep Learning for the Physical Layer," in IEEE Transactions on Cognitive Communications and Networking, vol. 3, no. 4, pp. 563-575, Dec. 2017. doi: 10.1109/TCCN.2017.2758370

[18] Dörner, S., Cammerer, S., Hoydis, J., el at., "Deep Learning Based Communication Over the Air," in IEEE Journal of Selected Topics in Signal Processing, vol. 12, no. 1, pp. 132-143, Feb. 2018. doi: 10.1109/JSTSP.2017.2784180

[19] O'Shea, T., Hoydis, J., (2017). "An Introduction to Machine Learning Communications Systems." CoRR, abs/1702.00832.

[20] McCaskey, M., Kukura, E., Corrigan, R., et al., "Machine Learning Applied to an RF Communication Channel", NAECON 2018 – accepted paper

[21] GNURadio – https://gnuradio.org

[22] Bergstra, J., Breuleux, O., Bastien, F., et al., (2010). "Theano: A CPU and GPU math compiler in Python." in PROC. OF THE 9th PYTHON IN SCIENCE CONF.

[23] Abadi, M., Barhan, P., Chen, J., et al., "Tensorflow: A System for Large-Scale Machine Learning," in Proceedings of the 12th USENIX Symposium on Operating Systems Design and Implementation – https://www.tensorflow.org/

[24] Chollet, F., et. al., (2015) "Keras," – https://keras.io

[25] Kingma, D., Ba, J., "Adam: A Method for Stochastic Optimization" in Proceedings of the 3rd International Conference on Learning Representations (ICLR)

An efficiency comparison between timeslicing and multi-carrier transmission for linearized transponders

Dieter Duyck, Michiel Mertens, Jon Vandenbruaene, Dirk Breynaert, and Frederik Simoens

Abstract—**Very high throughput satellites (HTS), where single transponders have large bandwidths (in the order of 500 MHz), significantly reduced the cost per Mbit for satellite communications, and are therefore omnipresent for applications like broadband, mobile and enterprise connectivity. An HTS transponder can amplify single or multiple carriers. Single carrier transmission is still the most efficient mode in a non-linear channel due its lower peak-to-average power ratio. In order to avoid too complex and expensive receivers, the wide carrier (around 500 MHz) is chopped in time slices such that receivers can decode their part of the very wide carrier. This cross-layer technology, applied for single carrier transmission, may bring overhead in some cases, also leading to some inefficiencies. In this paper, we will introduce and explain timeslicing and present a holistic efficiency comparison between single carrier transmission with timeslicing and multi-carrier transmisson without timeslicing, with and without non-linear predistortion. We conclude that Newtec non-linear predistortion (present in *Newtec Equalink*) with single or multi-carrier transmission outperforms single carrier transmission without non-linear predistortion. In addition, the throughput gain of single carrier transmission with respect to dual (four) carrier transmission is around 6% (9%) for linearized channels, the nowadays standard. Adding Newtec Equalink on top of single carrier transmission yields a total gain around 15% and 20% over dual and four carrier transmission without non-linear predistortion, respectively.**

I. Introduction

A. Need for wide forward carriers

Very high throughput satellites (HTS), where single transponders have large bandwidths (in the order of 500 MHz), significantly reduced the cost per Mbit for satellite communications, and are therefore omnipresent for applications like broadband, mobile and enterprise connectivity [8]. The forward link of a satellite network, from gateway to satellite, is typically a bottleneck in the achievable throughput over satellite. Hence, the achievable bandwidth of the forward link carrier is a differentiator. This bandwidth increased dramatically with the introduction of HTS. The consequence however is that all receivers in the satellite network must be able to process this huge bandwidth, which is illustrated in Fig. 1.

B. Timeslicing to have affordable receivers

Fig. 2 compares narrowband transmission with wideband transmission and visually illustrates why wideband transmission requires more expensive receivers.

Technical Labs, Newtec, Laarstraat 5, 9100 Sint-Niklaas.

In order to have sufficient statistical multiplexing gains (in order to overbook capacity), satellite networks have to consist of a minimum number of receivers (see for example [5] and references therein). However, with nowadays traffic demands (e.g. for broadband and enterprise connectivity), satellite networks of 80 Mbaud are sufficient to have overbooking gains due to statistical multiplexing. Still, a forward carrier of 500 Mbaud reduces the cost of communications. Hence, through timeslicing, multiple satellite networks are multiplexed in the same wide carrier. We will investigate further in the paper whether the wide single carrier transmission also comes with other advantages than just cost reduction, such as an efficiency and flexibility improvement.

Newtec proposed timeslicing [7] to combine affordable receivers, statistical multiplexing and single carrier wideband (500 Mbaud) transmission. More specifically, as mentioned above, satellite networks around 80 Mbaud are sufficient to have statistical multiplexing gains. However, receivers need sufficient time to decode their carrier of interest, hence they cannot decode each subsequent frame. Therefore, frames from the same satellite network are assigned a frame tag, denoted a time slice number. This time slice number is encoded in a header, which can be decoded separately. When a receiver does not belong to the satellite network corresponding to a decoded time slice number, it discards the frame. As such, receivers get more time to decode the frames of the satellite network they belong to, which is visually illustrated in Fig. 3. Timeslicing was standardized in DVB-S2, Annex M, and also in DVB-S2X [1].

When implementing timeslicing, one must make sure that there is sufficient time between two frames of the same slice, such that the receivers can be affordable (i.e., have decoders with lower throughput).

1) Contribution of this paper: Newtec implemented both timeslicing for single carrier transmission as multiple carrier transmission. Newtec also implemented non-linear predistortion for single and multiple carriers. Besides describing the implementation of timeslicing, we will also explain the achievable efficiencies of single carrier transmission with timeslicing, with and without non-linear predistortion, and multi-carrier transmission without timeslicing, with and without non-linear predistortion.

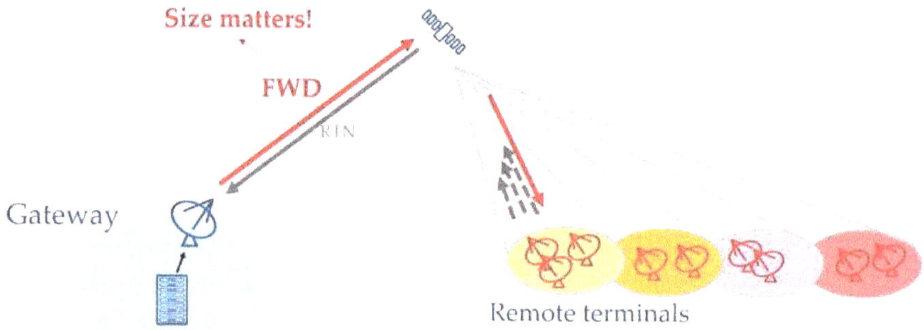

Fig. 1: In HTS tranponsders, wide carriers are transmitted and received by all remote terminals.

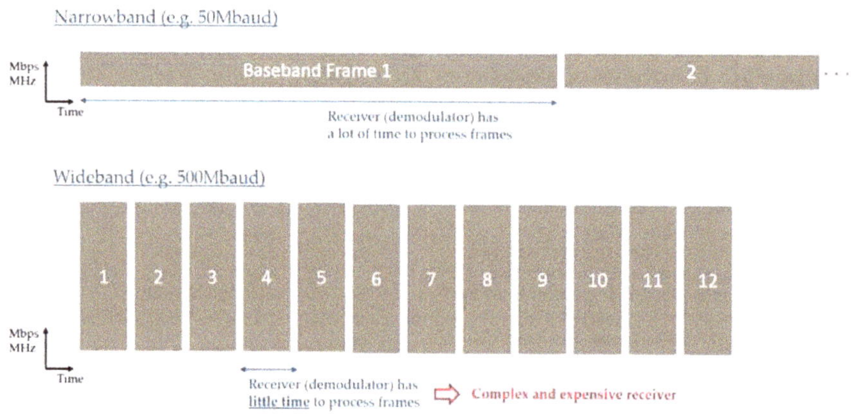

Fig. 2: Wideband receivers have much less time to decode the same amount of bits.

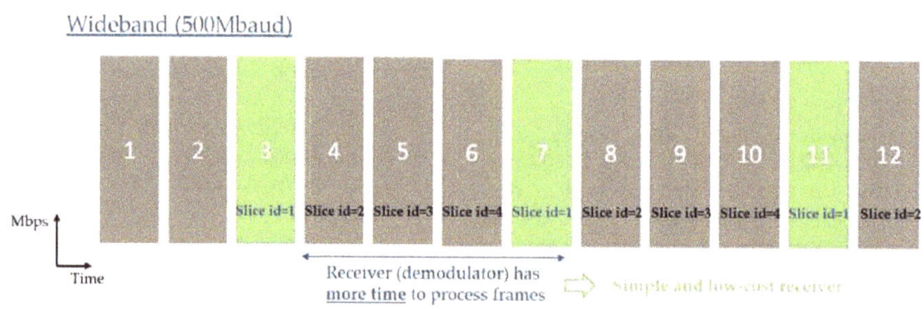

Fig. 3: With timeslicing, receivers can decode their slice of a wide carriers while still being affordable.

2. Implementation of Timeslicing Using Virtual Carriers

A. Concept of virtual carriers to achieve a modular system implementation making abstraction of the physical layer

In narrowband legacy systems, the traffic from multiple users in the same satellite network (satnet) is multiplexed in the same carrier, taking into account jitter requirements, traffic priority classes, the adoption of a beam hopping or non beam hopping satellite, based on link budgets and on the symbol rate and the link budget per user. This is a complex process that takes place in the satnet processor, and which is referred to as shaping-encapsulation [6].

With timeslicing, we propose to multiplex multiple satnets in the same aggregate physical carrier. It is of interest for the modularity of the system (and thus also the possibility to recuperate existing off the shelf designs, which leads to a faster and less error prone implementation), to still treat each satnet independently. As such, each satnet processor is shaping the traffic of the satnet users to the symbol rate of a *virtual* carrier. A virtual carrier refers to the collection of time slices assigned to the corresponding satnet. E.g., a satnet can be assigned to 2 time slices, such that the virtual carrier is the collection of the frames corresponding to said 2 time slices. Suppose that, in time, those 2 slices occupy 20% of the transmitted symbols, than the *virtual carrier symbol rate* is 20% of the aggregate physical carrier symbol rate.

Fig. 4 visually illustrates the concept of virtual carriers.

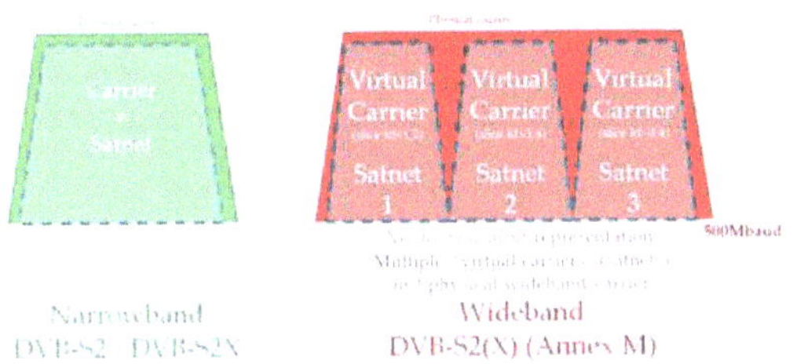

Fig. 4: Virtual carriers are matched to a satellite network (satnet) and correspond to a number of (but not all) time slices. Independent satnet processing is performed per virtual carrier.

Summarizing, for all layers above the physical layer, it as if the transmission happens over multiple carriers. In other words, whether data is transmitted over multiple carriers or a single carrier is a physical layer implementation detail. In upper layers, the processing is modular and done per satellite network.

B. A timeslice selector forcing force-empty frames when needed

The most important requirement for timesliced transmission is to respect a given spec in decoder throughput from the receivers. For example, receivers may not be able to decode more than 800 Mcbps (Mega coded bit per second). This holds for the set of timeslices that are decoded by that particular receiver. E.g., a receiver belongs to satellite network 3, which decodes virtual carrier 3, which corresponds to slices 5, 8 and 12. The instantaneous throughput of those slices cannot be larger than 800 Mcbps.

Next, other receiver limitations may apply, e.g. a per slice throughput limitation due to limited buffer sizes in the time slice selector for example.

As a consequence, the modulator must have full control on the time at which frames of the same slice are transmitted, more specifically on the time duration in between 2 frames of the same slice.

In order to do that, a block at the input of the modulator, denoted a time slice selector, will collect all frames from the different satnets, and multiplex frames onto a single bit stream on a first come first serve basis, with the exception that a frame can only be allowed provided that the previous frame of the same slice was allowed sufficiently long ago. The single bit stream will be modulated onto a RF waveform. Hence, the time slice selector will select slices sequentially which can then be modulated on the wide physical carrier.

To do that, a per slice counter is maintained indicating the amount of symbols that have been transmitted since the last frame of that slice was released in the single bit stream. Once the counter exceeds a programmable limit, the slice enters the list of allowed slices. The oldest frame in that list will be the first frame that will be released next on the single bit stream. In case no frame is present in the list of allowed slices, a force-empty frame is transmitted. A force-empty frame is a slice that will be rejected by all receivers, and that is forced in the carrier by the modulator, such that the modulator is sure about the time between slices.

With the above procedure, multiple satnets with different throughput limitations (e.g. a satnet with affordable receivers and a satnet with professional receivers) can be easily multiplexed in the same physical carrier. The implementation is modular and specifiable per slice or satnet.

Of course, such force-empty frames occupy symbols in the aggregate physical carrier. However, those results are not assigned to a virtual carrier as those frames do not result from a satnet processor or are not directed to any receiver. As a result, the sum of the symbol rates of the virtual carriers must be lower than the symbol rate of the aggregate physical carriers. Hence, in some cases, there is an additional inefficiency related to timeslicing, which can be very small however. In Newtec systems, the aggregate physical carrier size (physical symbol rate) must be greater than 100.2% of the sum of the virtual symbol rates present in that carrier. For example, in the case of 3 VCs of 140 Mbaud, the aggregate symbolrate must be greater than or equal to 420.84 Mbaud. We will see below that the 0.2% efficiency loss is negligible with respect to the gains of wide single carrier transmission.

3. Efficiency comparison between single wideband and multi-carrier transmission

One may wonder what is finally the difference between the transmission of N physical carriers and N virtual carriers within 1 physical carrier through timeslicing, visually illustrated in Fig. 5.

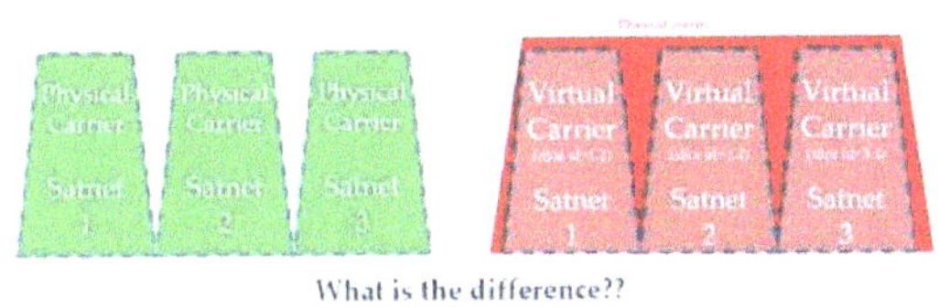

Fig. 5: We will compare the efficiency (Sec. III) and other metrics (Sec. IV) of virtual versus physical carriers transmission over a linearized channel.

Initially it seemed straightforward to put that single carrier transmission is more efficient than multi-carrier transmission over a single transponder due to the lower backoffs that can be achieved by single carrier transmission. This is because the peak-to-average power ratio (PAPR) of single carrier transmission (around 7 to 8 dB for 5% roll-off, depending on the modulation and coding) is lower than that of multi-carrier transmission (around 9.4 dB for 4 carriers). However,

TABLE I: An efficiency comparison of single and multi-carrier transmission for the 64-APSK 5/6 modcod, with (w) and without (wo) non-linear predistortion (pred).

Scenario	CsatN [dB]	GIBO [dB]	GOBO [dB] (level gain)	C/IM [dB]	Total pred gain [dB]	Throughput gain pred	Ref Mbps
Reference / rank 4 : Single carrier wo pred	21.91	6.5	4.4	26.5			1000
rank 1: Single carrier w pred	20.3	2.5	2.92	26.	1.61	9.4%	1090
rank 2: 2 carriers w pred	21.05	2.2	2.8	22.45	1.67	9.8%	1048
rank 3: 4 carriers w pred	21.24	2.0	2.88	18.74	1.74	10.2%	1039
rank 5: 2 carriers wo pred	22.72	6.5	4.1	22.			956
rank 6: 4 carriers wo pred	22.98	7.2	4.48	18.56			943

TABLE II: An efficiency comparison of single and multi-carrier transmission for the 32-APSK 5/6 modcod, with and without predistortion.

Scenario	CsatN [dB]	GIBO [dB]	GOBO [dB] (level gain)	C/IM [dB]	Total pred gain [dB]	Throughput gain pred	Ref Mbps
Reference / rank 4 : Single carrier wo pred	18.31	5	4.4	22.25			1000
rank 1: Single carrier w pred	16.79	1.2	2.2	26.64	1.52	10.1%	1110
rank 2: 2 carriers w pred	17.68	0.5	2.1	20.39	1.46	9.7 %	1042
rank 3: 4 carriers w pred	18.02	0.7	2.15	16.25	1.56	10.4%	1019
rank 5: 2 carriers wo pred	19.14	5.6	3.3	19.72			947
rank 6: 4 carriers wo pred	19.58	5.6	3.37	16.02			921

TABLE III: An efficiency comparison of single and multi-carrier transmission for the 16-APSK 5/6 modcod, with and without predistortion.

Scenario	CsatN [dB]	GIBO [dB]	GOBO [dB] (level gain)	C/IM [dB]	Total pred gain [dB]	Throughput gain pred	Ref Mbps
Reference / rank 3 : Single carrier wo pred	14.24	2.65	1.76	19.03			1000
rank 1: Single carrier w pred	13.32	-1.2	1.42	23.5	0.92	7.3%	1073
rank 2: 2 carriers w pred	14.11	-1.5	1.4	18.16	1.02	8.1%	1010
rank 4: 4 carriers w pred	14.49	-0.8	1.9	13.85	1.38	11 %	980
rank 5: 2 carriers wo pred	15.13	3.5	2.0	16.91			934
rank 6: 4 carriers wo pred	15.87	4.7	2.8	13.56			885

the PAPR difference is lower than what is believed by many. Also, for linearized channels (i.e., channels with a small ssg (small signal gain), e.g. below 4 dB, and almost no phase rotation (below 15 degrees at saturation), the difference is not as big as believed.

Given that single carrier wideband transmission may suffer from slight inefficiencies due to forced force-empty frames (see Sec. II-B), we investigated wether the *advantage for single carrier transmission* still holds. Second, (multi-carrier) non-linear predistortion for linearized channels was recently revolutionized [2]–[4] with spectacular gains. Note that the authors are not aware of non-Newtec implementations of *non-linear* predistortion with gains above 0.2 dB over linearized transponders [4]. Also note that *linear* predistortion is irrelevant and not required anymore for nowadays systems as the new wideband receivers have sufficiently good equalizers which compensate for the *linear* distortions at the receiver side.

Tables I, II, and III below provide an interesting efficiency comparison between the scenarios of physical versus virtual carrier transmission, each with and without non-linear predistortion, for 64-APSK 5/6, 32-APSK 5/6, and 16-APSK 5/6 [1], respectively. The symbol rate of the single carrier is twice (four times) that of the dual (four) carrier scenario. The TWTA is a linearized TWTA with a small signal gain of 3 dB, no phase rotation and without amplitude compression beyond saturation. The roll-off is 5%, the nowadays default roll-off for broadband systems. The reference is single carrier transmission without non-linear predistortion, which interestingly looses against multi-carrier with non-linear predistortion.

We did not include lab measurements for QPSK and 8PSK modcods in the same format, but we would like to share the following important conclusions. Even for QPSK transmission with four carriers, Newtec non-linear predistortion still brings 0.6 dB or 6.8% gain. Even more remarkable, QPSK transmission with a single carrier with Newtec non-linear predistortion gains 1.8 dB or 22% gain over four carrier transmission without Newtec-Equalink, stressing the important gain of timeslicing with wide carriers over narrowband transmission.

We can draw some important conclusions from Tables I, II, and III.

- Newtec Equalink (with non-linear predistortion) provides a throughput gain around 9%. Note that for TWTAs with compression beyond saturation, gains are still around 6 to 7% ([2]).
- Single carrier transmission is roughly 0.7 to 1 dB (6 % for 32-APSK) better than dual carrier transmission, and 1.3 dB (9 % for 32-APSK) better than four carrier transmission.
- For 32-APSK and 64-APSK, multi-carrier transmission with Newtec Equalink performs better[1] than single carrier transmission without Newtec Equalink.
- Single carrier transmission with Newtec Equalink performs 15% (20%) better than dual (four) carrier transmission without non-linear predistortion.
- The efficiency loss of timeslicing is in the order of 0.2 % which can be considered negligible with respect to the above differences between the different scenarios.

4. Seamless Load Balancing between Virtual Carriers

There are other advantages of single carrier transmission with multiple virtual carriers in comparison with multi-carrier transmission. The main advantage is more flexibility to follow traffic demand of the satnets. More specifically, the symbol rate of virtual carriers can be changed without changing anything on the aggregate physical carrier symbol rate. Hence, the synchronization of receivers is not impacted in any way. As such, dynamically, the virtual carrier symbol rates can be resized when needed, as long as the sum of the virtual carrier symbol rates is below the physical carrier symbol rate of course. This dynamically resizing is referred to as dynamic load balancing.

Note that by load balancing the virtual carriers, the statistical multiplexing effect of a timesliced wideband carrier will be identical to a wideband carrier without timeslicing (so congestion will be as if it was 1 very wide carrier).

Other smaller advantages are that the FPGA resource occupation and power consumption of single carrier transmission is typically lower.

V. Conclusion

In order to combine affordable receivers and cost-effective high throughput satellites, Newtec implemented timeslicing, which allows receivers to only decode their part of the wide carrier, giving them more time for decoding. Each satellite network has its own *virtual carrier*, which all reside in the same physical carrier.

The inefficiencies of timeslicing are negligible. We compared the efficiency of single versus multi-carrier transmission, with and without Newtec Equalink with *non-linear* predistortion. Interestingly, even multi-carrier transmission with Newtec Equalink outperforms single-carrier transmission without non-linear predistortion by 2 to 5%. Single-carrier transmission with timeslicing and Newtec Equalink is the proud winner in throughput performing 15 to 20% better than multi-carrier transmission without non-linear predistortion, which holds for 64-APSK downto QPSK transmission. Finally, single carrier transmission with timeslicing is much more flexible as it allows seamless dynamic load balancing (modifying dynamically, on the fly, the symbol rates of virtual carriers associated with the satellite networks) without any impact on the synchronization at the receivers, as the symbol rate of the physical carrier remains untouched. Also the congestion with load balancing will be less than for narrow carriers as the statistical multiplexing will be better with load balancing.

Acknowledgments

This research was developed within the framework of ESA-funded project "Every Community Online (ECO)", Ctr. No 4000117721/16/UK/AD.

References

[1] Digital Video Broadcasting (DVB), DVB Document A83-2, Second generation framing structure. channel coding and modulation systems for Broadcasting. Interactive Services. News Gathering and other Broadband satellite application, Part II: S2-Extensions (DVB-S2X), Mar. 2014.

[2] D. Duyck, A. Mengali et al., "An overview of multicarrier predistortion techniques and associated throughput gain for an actual hardware implementation," AIAA Intern. Comm. Sat. Systems Conf. (ICSSC), Oct. 2017.

[3] D. Duyck et al., "Productized multi-carrier Predistortion Total Throughput Gains around 20% over Linearized Channels in True Customer Use Cases," AIAA Intern. Comm. Sat. Systems Conf. (ICSSC), Oct. 2018.

[4] https://artes.esa.int/news/new-ground-technology-boosts-link-margin

[5] R.L. Turcotte et al., Motorola Solutions Inc US patent, "Air interface statistical multiplexing in communication systems," 1998.

[6] D. Breynaert et al., EP and US patent, "MANAGING TRAFFIC IN A SATELLITE TRANSMISSION SYSTEM," 2005.

[7] D. Breynaert and D. Delaruelle, Newtec Cy EP patent, "Satellite television system, signals and coder therefore," 2007.

[8] O. Vidal et al., "Next generation high throughput satellite system," IEEE AESS Europ. Conf. on Satell. Telecomm. (ESTEL), 2012.

[1]Dual carrier with Newtec Equalink roughly performs 5% better, four carrier transmission with Newtec Equalink roughly performs 3% better.

Section 9 – 5G and Satellite Networks Integration

Efficient 5G Edge Caching Over Satellite

Thang X. Vu[1] *Ninola Maturo*[1] *Satyanarayana Vuppala*[1] *Symoen Chatzinotas*[1] *Joel Grotz*[2] *Nader Alagha*[3]

[1] *Interdisciplinary Centre for Security, Reliability and Trust (SnT), University of Luxembourg, 29 Avenue John F. Kennedy, L-1855, Luxembourg*
[2] *SES Engineering, L-6815 Betzdorf, Luxembourg.*
[3] *European Space Agency, Noordwijk, The Netherlands.*
* *E-mail: thang.vu@uni.lu*

Abstract: The fifth generation (5G) wireless networks have to deal with the high data rate and stringent latency requirements due to the massive invasion of connected devices and data-hungry applications. Edge caching is a promising technique to overcome these challenges by prefetching the content closer to the end users at the edge node's local storage. In this paper, we analyze the performance of edge caching 5G networks with the aid of satellite communication systems. Firstly, we investigate the satellite-aided edge caching systems in two promising use cases: a) in dense urban areas, and b) in sparsely populated regions, e.g., rural areas. Secondly, we study the effectiveness of satellite systems via the proposed satellite-aided caching algorithm, which can be used in three configurations: i) mono-beam satellite, ii) multi-beam satellite, and iii) hybrid mode. Thirdly, the proposed caching algorithm is evaluated by using both empirical Zipf-distribution data and the more realistic Movielens dataset. We show via numerical results that for the considered popularity data the multi-beam satellite will surpass the mono-beam system when the demands are less correlated, and that the hybrid achieves a cache hit ratio (CHR) between the multi-beam and mono-beam schemes.

1 Introduction

During the past 40 years, the Internet has followed an extraordinary evolution and has become an integral part of the modern society. However, this evolution has kept momentum and there are constantly new services and contents distributed through this global communication network. Based on Cisco's report [1], it is predicted that the mobile data traffic will grow 74% by 2021. Particularly, the mobile video will increase eleven-fold between the mentioned years. The main causes of this traffic growth are the vast availability of mobile devices, e.g., smart phones, tablets, and notebooks, as well as the fast growth of video content on the Internet and their increasing quality. Using these mobile devices, more and more users are immigrating from traditional linear broadcasting services (TV channels) to streaming services, such as YouTube and NetFlix. Another factor that contributes to the traffic is the increasing video quality, i.e., 3D, 4K video, Virtual Reality etc., which can be translated to increased bandwidth requirements for both the core and access networks. This perspective seems very promising for content providers, since they can provide improved services and reap the benefits either through subscriptions or advertising. Nevertheless, looking at this from a telecom operatorâĂŹs point of view, it is obvious that video traffic will become a bottleneck and put excessive strain on current communication infrastructure. On the other hand, video traffic also results in revenue growth. However, telecom operators do not have direct access to the revenue generated through video content delivery and as a result they cannot use it to upgrade their infrastructure. In parallel, spectrum has become scarce and the operators cannot easily access new frequency bands to expand their wireless access and backhaul segments.

These are the reasons why in the last few years, telecom operators started to build their own content delivery networks (CDNs). The aim of a CDN is to serve contents to end-users with high performance by using edge catching. The main benefit of CDNs is represented by the higher degree of cross-optimization between the physical infrastructure and the network service that leads to an improved transmission effiecincy. It becomes obvious that closer interaction between operators and content providers will be needed in order to optimize content delivery and overcome the projected bottleneck due to video traffic [3]. Some steps have been made in this direction, for instance Dhiraagu, the Maldives operator, have deployed Google caches successfully in its network [2].

One of the challenges in the edge caching is how to effectively prefetch the popular content to the caches considering the high volume of data [4]. In order to overcome this issue, we propose to use satellite backhauling for cache placement phase to exploit the large coverage of the satellite beams. Satellite systems have the ability to provide high throughput links and to operate in multi/broad-cast modes for immense area coverage.

Due to their multi-hop unicast architecture, the cached content via terrestrial networks has to go through multiple links and has to be transmitted individually towards each base station (BS). On the other hand, with wide area coverage, the satellite backhaul can broadcast content to all BSs or multi-cast contents to multiple groups of BSs. Therefore, bringing these two technologies together can further off-load the network. The main idea is to integrate the satellite and terrestrial telecommunication systems in order to create a hybrid federated content delivery network, which can improve the user experience. The joint deployment of satellite and terrestrial networks can be found in [7, 8]. In this paper, we consider the satellite channel as the only mean for cache placement. The application of satellite communications in feeding several network caches at the same time using broad/multi-cast is investigated in [5–7]. The work of [6] proposes using the broad/multi-cast ability of the satellite to send the requested contents to the caches located at the user side. Online satellite-assisted caching is studied in [7]. In this work, satellite broadcast is used to help placing the files in the caches located in the proxy servers. Each server uses the local and global file popularity to update the cache.

In this work, a satellite-aided caching algorithm is proposed. We use off-line caching approaches [8–10]to off-load the backhaul of the terrestrial network. We focus on the role of multimodal satellite backhauling, which provides flexible backhaul's transmission modes, e.g., broadcast and broadband, and its effectiveness on edge caching. The proposed algorithm can be used with three different satellite's configurations: i) broadcast mono-beam, ii) broadband multi-beam, and iii) hybrid mode. Focusing on promising satellite markets, we evaluate the performance of the proposed caching algorithm in two use cases: in dense urban areas and in sparsely populated rural regions. Based on both empirical Zipf-distribution data [11] and the more realistic Movielens dataset [12], we show via numerical results that the multi-beam satellite will outperform the mono-beam system when the demands are less correlated, and that

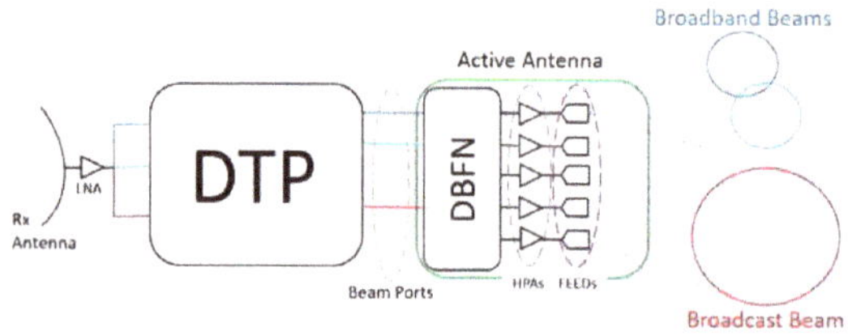

Fig. 1: Hybrid Satellite Architecture

the hybrid achieves a cache hit ratio between the multi-beam and mono-beam schemes.

2 Technological Enabler for Satellite-assisted Edge Caching

2.1 Hybrid Satellite

The satellite architecture in general can be classified in two traditionally different cases: satellite with a very wide single beam, mainly used for broadcast services, and satellite with many small beams, mainly used for broadband services. While the distinction between these two types of architectures is well established nowadays, in the future satellite system there may be no real distinction between broadcast and broadband payload. Thanks to the adoption of new technologies, like digital transparent payload (DTP) and (semi-)active on-board antennas, it will be possible to have both services sharing the same hardware resources in reconfigurable hybrid broadcast/broadband payloads. Many satellite manufacturers are currently working towards this new type of payload [13] and in Fig. 1 we provide a pictorial representation of this hybrid architecture.

As an example shown in Fig. 1, the satellite is receiving 3 different streams: 2 broadband streams (the light and the dark blue lines) and 1 broadcast stream (the red line). At a beam ports level (the output of the DTP) we still have 3 different streams that are the input of the active antenna. Thanks to the active antenna technology, it will be possible to create overlapping beams of variable granularity and use a single broadcasting system to drive a large broadcast beam and at the same time it will be possible to generate on the same payload separated broadband beams. It is interesting to note that in this configuration there isn't a direct relation between the beams and the feed, but basically all the feeds cooperate to create all the beams using digital beamforming (BF). In this way is possible to optimize the power consumption of the high power amplifiers (HPAs) serving the feeds.

Because the creation of the beams is driven by the digital beamforming network (DBFN), the beams design can be very easily reconfigured. This type of flexibility is certainly extremely important for satellite operators, so that they can modify their satellite configuration in order to better accommodate the variation of the traffic demands during the satellite lifetime. In this context we refer in particular to the content-related traffic evolution determined by various factors such as population evolution and changes in the content consumption preferences.

2.2 Integrated Satellite-Terrestrial System

The classic cellular backhaul comprises of fibre, copper lines and wireless links, which can efficiently transfer the backhaul traffic from point to point. The multimodal architecture enhances the conventional backhaul by overlaying wide satellite beams and narrower broadcasting cells (e.g. HPHT), which are capable of P2MP (point to multi-point) multi/broadcasting. Studies in [14] have shown that a small percentage of video files can be extremely popular posing a huge load on current networks that do not support broadcasting on the physical layer. The proposed satellite overlay caching solution can relieve this load by simultaneously and efficiently distributing popular content the edge caches by integrating traditional P2P backhauling with broad/multicasting P2MP backhauling. It should be noted that these multi/broadcast systems can inherently reach a large number of BSs with a single PHY-layer transmission in contrast to the NET-layer multicasting which implies packet replication and multiple individual PHY connections. In addition to above mentioned benefit, this converged solution brings several distinct benefits: 1) additional backhaul capacity based on existing infrastructure, 2) spectrally-efficient physical-layer multi/broadcasting, 3) variable cell sizes for broadcast backhauling (wide coverage for satellite, narrow coverage for high broadcasting towers).

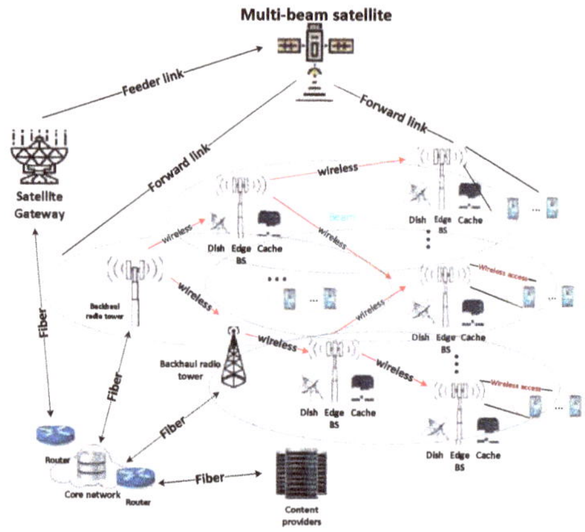

Fig. 2: Hybrid satellite-terrestrial network using a multi-beam satellite for more accurate content placement phase.

3 System model and caching algorithm

We consider a satellite system serving the users within its coverage via a number of edge nodes, e.g., base stations, which are equipped with a satellite receiver. The considered system can find application in villages, in rural area or small business customers such as hotels and residents where only one satellite receiver/decoder is needed for each base station (BS). Each BS has a local storage which can store up to M content files. For the sake of simplicity, all files are assumed to have same size of Q bits, however, the extension to arbitrary file sizes is straightforward. The users will request a file, e.g., video, from a library of F files.

We consider offline caching, which usually occurs in off-peak hours [8], e.g., from midnight to to early morning. Based on the user preference sent to the network controller via the return link, a local user preference (content popularity) is obtained, which is then used to determine caching strategy. In addition, we consider three operating modes of the satellite channels: mono-beam, multi-beam, and hybrid.

3.1 Caching via Broadcast mono-beam

With mono-beam satellite mode, both broadcast and multicast are possible. However, the efficiency of the multicast in this mode is in general lower than in multibeam systems due to the wide beam. This is the reason why in our study, we assume that the monobeam is used for broadcasting. the cache placement is done based on the global popularity. From the users' content popularities, we can compute the most popular contents which will be stored at the BS cache, as follows: $\mathbf{p}_G = \frac{1}{N}\prod_{k=1}^{N}\mathbf{p}_k$, where $\mathbf{p}_k$ is the local popularity at BS

k, $\pi()$ denotes the sorting operator and N is the number of BSs. The gateway will broadcast the first C files in p_G. Each BS k will choose to store the first M files in $\pi(\mathbf{p}_k)$ in the kth BS's cache. Obviously, if $C > F$, all the SRs can cache the most M local-popular files in their cache. However, when $C < F$, some BS might not be able to cache the most popular files (locally) if $\{\mathbf{p}_k\}$ are uncorrelated, hence the CHR during the delivery phase is degraded.

The cache hit ratio (CHR) is the main performance indicator to be considered. It is defined as the ratio between the number of requests served by the local cache, divided by the total number of requests:

$$CHR = \frac{\sum_k \bar{n_k}}{\sum_k n_k}$$

where $\bar{n_k}$ is the number of requested files available in the kth local cache, and n_k is the total number of requests at the kth BS.

3.2 *Caching via Broadband multi-beam*

The CHR performance can potentially be improved by using multi-beam satellite as the content deliver is done per beam and not for the whole widebeam. For a satellite with multi-beam and flexible beam coverage, the caching can target, in fact, a subset of beams. It is often expected that the narrower beam can reach a higher spectral efficiency, compared to the global beam, while utilizing similar satellite resources in terms of power and bandwidth. Let L denote the number of beams. We define the number of files that each beam is capable of delivering in the placement phase as the caching capacity C. Denote $\mathcal{K}_l$ as the set of BSs in the l-th beam. The gateway will calculate the popularity of the users with each beam as follows: $\mathbf{p}_G^l = \frac{1}{K_l}\prod_{k\in\mathcal{K}_l}\mathbf{p}_k$. Then the l-th beam will broadcast the first C files in $\mathbf{p}_G^l$, from which each BS stores M most popular files.

3.3 *Caching via Hybrid design*

In the hybrid setting, some contents can be prefetched via the mono-beam mode, while the rests are sent via the multi-beam satellite. A similar approach was proposed for the terrestrial Centralized RAN (CRAN) architecture in [17]. Denote C_1, C_2 as the caching capacity of the mono-beam satellite and multi-beam satellite, respectively. First, the mono-beam caching algorithm is applied to select C_1 most popular files for caching. Then, these files are removed from the requests. Next, the multi-beam caching algorithm is used on the remaining requested files. The proposed caching algorithm is summarized in Table 1.

4 Cost per bit calculation

In order to define the cost per bit for the satellite transmission, different aspects need to be considered, such as the entire service costs of the transmission and the capacity of the satellite system for all the possible transmissions of services.

For contents distribution service, we assume that the transmission is organized in a multicast manner allowing many terminals to receive the same content. In the following, we will so compare the cost per transferred bit between a widebeam multicarrier mode and a multibeam single carrier mode. For this we need to consider the required resources of the satellite and take into account the link budget results to assess the overall picture. This is a first order assessment of the cost per bit compared to what the total capacity of the satellite is, so it does not consider eventual opportunity costs resulting from specific operational contexts or second order effects that one service might have on alternative usage of the capacity. This is a comparison on the basis of the total satellite cost per transmitted information bit.

We assume multi-cast link budget assessment results in the form of E_S/N_0 values for both the considered transmission modes, which allow us to compute how the link to the end user can be operated and at which efficiency. From this assessment, we derive the transmission efficiency, taking into account all overhear parameters required.

Table 1 CACHING ALGORITHM VIA SATELLITE

	Input: C - caching capacity, M - cache size at the base station, K_b - number of BSs, K_u - number of users per BS, L - number of beams.
1.	The BSs estimate $\mathbf{p}_{k,u}$ - interested content from user u at BS k, then forward it to the gateway
2.	**if** ($\mathtt{mono-beam}$)
	2.1. The global cache manager (GCM) at the gateway computes the global popularity $\mathbf{p}_G = \frac{1}{K_b K_u}\sum_{k,u}\mathbf{p}_{k,u}$
	2.2. The GCM sorts $\mathbf{p}_G$ in the decreasing order
	2.3. The GCM chooses the first C files in $\mathbf{p}_G$ to broadcast for caching
	2.4. Each BS k computes its local popularity $\mathbf{p}_k = \frac{1}{K_u}\sum_{k=1}^{K_u}\mathbf{p}_{k,u}$. It then sorts $\mathbf{p}_k$ in the decreasing order, and chooses the first M files in the sorted local popularity for caching. A file is cached at the local cache of BS k if it is at both in the list and is sent by the GCM of M files
3.	**else_if** ($\mathtt{multi-beam}$)
	3.1. The GCM constructs L beams. $\mathcal{K}_l$ denotes the set of users in the l-th beam.
	3.2. For each beam, the GCM applies the caching policy in step 2.
4.	**else_if** ($\mathtt{hybrid-mode}$)
	4.1. Determine $C_{\text{mono}}, C_{\text{multi}}$ as the caching capacity for mono-beam and multi-beam, respectively.
	4.2. The GCM applies step 2 to decide C_{mono} files to be broadcasted via mono-beam satellite, denoted by $\mathcal{F}_{\text{mono}}$
	4.3. The GCM removes files in $\mathcal{F}_{\text{mono}}$ from the requests $\mathbf{p}_{k,u}$, then applies step 3 with the caching capacity C_{multi}.

The power and bandwidth used in the link budgets to assess the transmission efficiency have to be considered. If it is multiplexed with other traffic, the proportional DC power resources and bandwidth utilization resources is taken in to account. The DC power equivalent is deduced from the assessment of the power assessment taking into account the amplifier efficiency and output backoff (OBO) of the amplifiers required. The total bandwidth available over all beams is considered as bandwidth reference for the satellite. Here the total bandwidth of the simultaneously transmitted traffic has to be considered for all possible beams which are active at the same time.

For the resource utilization in terms of power and bandwidth proportional rations, we use the maximum of the two as we assume the full capacity of the satellite is utilized if one of these resources is exhausted. Hence, also the best utilization of the satellite would be to plan to exhaust both simultaneously, which is logical.

The total cost of the transmission is the planned transmission cost in the investment case reference of the satellite for this service, which takes into account CAPEX (and OPEX in a refined model, but not required in first order approximation) costs with all related assumptions on fill rates per livetime, etc. This is deduced as cost per Mbps assessment for the computed traffic load.

We may consider different levels of costs to take into account the fact that an underutilized time period could be used for the data transfer. The total is a $COST_{LINK}$ result as $RESOURCE_{PERCENT}$ $* \; COST_{SAT}$. In a first approximation the $COST_{SAT}$ can be compared to the overall CAPEX of the satellite mission and for a relative comparison of the two transmission modes, the relative CAPEX cost can be compared for similar satellite missions as proportional to the platform, DC power and antennas onboard the satellite in a first order of magnitude.

Then the total file transfer cost results in:

$$COST_{file} = Size * COST_{LINK}/Efficiency * Bw$$

The relative cost comparison between mono-beam and multi-beam used in the simulation is given in Table 2.

Table 2 Relative comparison for cost per bit

	Widebeam	Multibeam	Unit
	Multicarrier	**Singlecarrier**	
E_S/N_0 (worst case)	8.0	7.5	dB
Efficiency	2.4	2.1	bps/Hz
Total Sat RF Power	8556	5398	W
Total Sat Bandwidth	4320	80000	MHz
Beams per coverage	1	10	#
Power per beam	120	60	W
Bandwidth per beam	60	500	MHz
Power Ratio	1.4%	11.1%	
Bandwidth Ratio	1.4%	6.3%	
$RESOURCE_{PERCENT}$	1.4%	11.1%	
$COST_{LINK}$	0.014	0.111	relative
Size	5000	5000	Mbyte
$COST_{file}$	3.89	4.19	
Relative Cost	1	1.08	

5 Numerical results

In this section, we present the cache hit ratio (CHR) performance of the proposed caching algorithm via numerical results. The user's requests follow Zipf distribution with the skewness factor equal to 0.8. The cache size M indicates that it can store M movies in its storage. To make a fair comparison, the caching capacity in mono-beam and multi-beam settings are chosen such that both schemes have same cost. This means that we fix the caching capacity of the monobeam mode and we evaluate the caching capacity of the multibeam mode in accordance with the results of Table 2. In this way in the first use case the caching capacity is $C = 1000$ files for monobeam mode and $C = 917$ files for multibeam mode. In the second use case the caching capacity is $C = 500$ files in monobeam mode and $C = 459$ in multibeam mode. On about the hybrid mode the caching capacity is calculated in accordance with the usage percentage of the monobeam and the multibeam mode. In this set of simulation we are considering that the hybrid scheme is using the monobeam and multibeam mode for the same amount of time. The delivery can be further optimized by properly dimensioning the two modes according to current popularity statistics. In this case, it is conjectured that the hybrid mode performance will be significantly improved. The satellite provides service for an area consisting of 1000 BSs, each is serving 2000 users in usecase 1 and 400 users in usecase 2. Each user randomly requests a movie from the library of $F = 2000$ videos. We assume that the user requests forms two geographical areas over Europe which are weakly correlated, e.g., French-speaking region and German-speaking region.

Fig. 3a presents the CHR for the two considered use cases. It is observed that using multi-beam satellite significantly improves the CHR for both cases and the hybrid setup achieves a better performance than the mono-beam but worse than the multi-beam. This is because the multi-beam satellite can better match the local popularity with the user demands. It is also shown that usecase 1 outperforms usecase 2 at the large BS cache sizes. Indeed, this initial result shows one of the major advantages of multi-beam caching with respect to the widebeam caching, i.e. the access to the geographical diversity of the content. This way, multibeam satellites can cache content based on geographical clusters with less correlated content popularity. We have to clarify here that the results of the hybrid scheme are sub-optimal because they are deeply affected by the ratio between monobeam and multibeam mode and, as already stated, the optimization of the monobeam and multibeam mode ratio, according to the traffic demands, has yet to be carried out.

Fig. 3b shows the CHR of the two usecases as a function of number of beams or cluster of beams according to the fact that we are considering the new hybrid satellite architecture, presented in Section 2.1, or a more conventional and less flexible payload. It is shown that the number of beams larger than 2 has negligible impact on the CHR for both scenarios. This is because we assume that the user requests forms two geographical areas over Europe which

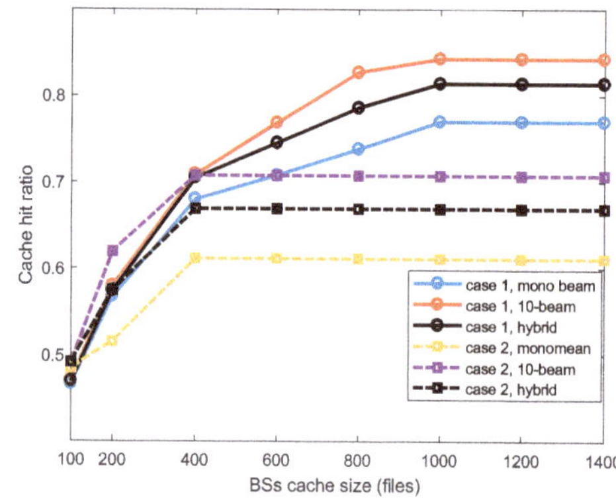

(a) Figure A

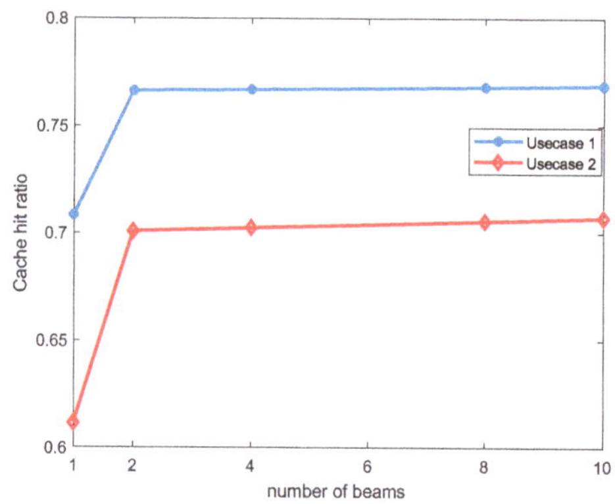

(b) Figure B

Fig. 3: CHR performance of the proposed caching algorithm in two use cases. For each case, the caching algorithm is designed for three different beam settings: mono beam only, 10-beam only, and hybrid mono beam and multi beam.

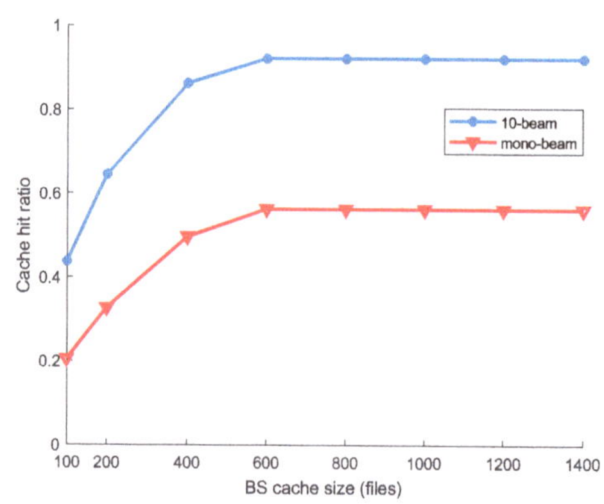

Fig. 4: CHR performance based on Movielens dataset.

are weakly correlated. In this case, using two clusters of beams is sufficiently efficient.

Fig. 4 presents the CHR of usecase 1 with Movielens demand [12]. The system parameters are similar in the previous section. A significant gain is observed for multi beam satellite compared with the mono beam counterpart. In particular, a CHR gain of 30% is achieved by the 10-beam setup compared with the mono beam. This is because the user demands across the beams are weakly correlated. Therefore, using the multi-beam satellite can serve the local demands better than the mono beam. Indeed, multi-beam satellites bring the CHR in this case to a level which is much more beneficial. It is also observed that the CHR in both setups saturates due to the caching capacity limit. This is due to the fact that the caching capacity of the transmission scheme is lower than the storage capacity of the terminals.This would be not a limitation if we had run the simulation for a longer period of time, where the storage capacity would have been filled in the caching period of several consecutive days.

6 Conclusions and Future Works

We have demonstrated the effectiveness of the multimodal satellite backhauling on edge caching systems in the presence of highly uncorrelated content which is likely to be the trend of the future content consumption. The proposed offline caching algorithm is shown to be capable for flexible deployments of the satellite channels: mono beam, multi beam and hybrid. We have shown that a multi-beam satellite becomes useful with respect to widebeam in caching as the geographical distribution of content popularities becomes uncorrelated. In particular, the future flexible multi-beam payloads with their flexible coverage will be able to better distinguish between the clusters of beams with less popularity correlation and hence further improve the CHR. As future development, a cost based optimization of the rates of the different modes of the hybrid setup, in order to better fit the traffic distribution, can be realized. In addition, these results suggest a promising extension to online caching strategies [15, 16], where the user preference periodically changes time to time.

Acknowledgments

This work is partially supported by the Luxembourg National Research Fund under the project FNR CORE ProCAST, and by the European Space Agency, ESA ESTEC, Noordwijk, The Netherlands, under the project title "High Throughput Digital Broadcasting Satellite Systems" (Contract No. 4000120692/17/NL/EM). Opinions, interpretations, recommendations and conclusions expressed herein are those of the authors and are not necessarily endorsed by the European Space Agency

References

1 Cisco, "Cisco visual networking index: Global mobile data traffic forecast update 2016-2021," 2017, white paper.
2 "Content delivery networks 3.0," CTOiC White paper. Online: https://www.slideshare.net/BenSchwarz1/content-deliverynetworks30.
3 W. Han, A. Liu, and V. K. N. Lau, "PHY-caching in 5G wireless networks: design and analysis," *IEEE Commun. Mag.*, vol. 54, no. 8, pp. 30–36, Aug. 2016.
4 S. Ramesh, I. Rhee, and K. Guo, "Multicast with cache (Mcache): an adaptive zero-delay video-on-demand service," *IEEE Trans. Circuits Syst. Video Technol.*, vol. 11, no. 3, pp. 440–456, Mar. 2001.
5 B. Evans, M. Werner, E. Lutz, M. Bousquet, G. E. Corazza, G. Maral, and R. Rumeau, "Integration of satellite and terrestrial systems in future multimedia communications," *IEEE Wireless Commun.*, vol. 12, no. 5, pp. 72–80, Oct. 2005.
6 H. Linder, H. D. Clausen, and B. Collini-Nocker, "Satellite internet services using DVB/MPEG-2 and multicast web caching," *IEEE Commun. Mag.*, vol. 38, no. 6, pp. 156–161, Jun. 2000.
7 C. G. Brinton, E. Aryafar, S. Corda, S. Russo, R. Reinoso, and M. Chiang, "An Intelligent Satellite Multicast and Caching Overlay for CDNs to Improve Performance in Video Applications", in *Proc. 31st AIAA Int. Commun. Satellite Systems Conf.*, 2013-5664.
8 A. Kalantari, M. Fittipaldi, S. Chatzinotas, T. X. Vu, and B. Ottersten, "Cache-Assisted Hybrid Satellite-Terrestrial Backhauling for 5G Cellular Networks," in *Proc. IEEE Global Commun. Conf.*, Singapore, 2017, pp. 1-6.
9 T. X. Vu, S. Chatzinotas, B. Ottersten, and T. Q. Duong, "Energy Minimization for Cache-Assisted Content Delivery Networks With Wireless Backhaul," *IEEE Wireless Commun. Lett.*, vol. 7, no. 3, pp. 332-335, Jun. 2018.
10 T. X. Vu, S. Chatzinotas, and B. Ottersten, "Edge-Caching Wireless Networks: Performance Analysis and Optimization," *IEEE Trans. Wireless Commun.*, vol. 17, no. 4, pp. 2827-2839, Apr. 2018.
11 L. Breslau, P. Cao, L. Fan, G. Phillips, and S. Shenker, "Web caching and Zipf-like distributions: evidence and implications," in *Annual Joint Conf. of the IEEE Computer and Commun. Societies*, vol. 1, Mar. 1999, pp. 126–134.
12 https://grouplens.org/datasets/movielens/1m/
13 Glyn Thomas. "Enabling Technologies for flexible HTS payloads", 33rd AIAA International Communications Satellite Systems Conference and Exhibition, International Communications Satellite Systems Conferences (ICSSC), (AIAA 2015-4348)
14 D. Ciullo, V. Martina, M. Garetto and E. Leonardi, "How Much Can Large-Scale Video-on-Demand Benefit From Users' Cooperation?," in *IEEE/ACM Transactions on Networking*, vol. 23, no. 6, pp. 1846-1861, Dec. 2015.
15 L. Lei, L. You, G. Dai, T. X. Vu, D. Yuan and S. Chatzinotas, "A deep learning approach for optimizing content delivering in cache-enabled HetNet," in *Proc. Int. Symp. on Wireless Commun. Syst. (ISWCS)*, Bologna, 2017, pp. 449-453.
16 B. N. Bharath, K. G. Nagananda, and H. V. Poor, "A learning-based approach to caching in heterogenous small cell networks," *IEEE Transactions on Communications*, vol. 64, Apr. 2016.
17 Christopoulos D., Chatzinotas S., Ottersten B., "Cellular-Broadcast Service Convergence through Caching for CoMP Cloud RANs," in *IEEE Symposium on Communications and Vehicular Technology in the Benelux*, SCVT 2015, Luxembourg, November 2015.

Use cases to business modelling of satellite backhaul in 5G

Simon Watts[1]*, Konstantinos Liolis* [2]*, Salvador Diaz* [3]*, Marlies Van der Wee*[4]

[1]*AGS, Avanti Communications, Cobham House, 20 Black Friars Lane, London, UK*
[2] *SES S.A, Château de Betzdorf, L-6815 Betzdorf, Luxembourg*
[3] *British Telecommunications PLC, Adastral Park, Ipswich, UK*
[4] *Ghent University – imec, Ghent, Belgium*

Keywords: SATCOM, 5G, EMBB, BACKHAUL, INTEGRATION

Abstract

The H2020 project SaT5G [10] [13] is researching and developing the capability to seamlessly integrate satellite communication (satcom) backhaul links into 5G to support enhanced Mobile Broadband (eMBB). Four relevant use cases and a set of KPIs have been identified. A variety of different integration architectures are described. These factors have then been reviewed to allow business modelling for the new paradigms brought by 5G, and the role of a bandwidth broker has been identified.

1 Introduction

The H2020 project "SaT5G" (Satellite and Terrestrial network for 5G) [10], [13] is developing the role of satcom in 5G by defining optimal satellite-based backhaul and traffic offloading solutions. The goal is to enhance 5G networks with the satcom benefits by a seamless integration and to create new business opportunities. To achieve such goals, SaT5G is researching, developing and validating the optimal satellite integration and creating new business roles in the early stages. The SaT5G project structure is shown in the following figure.

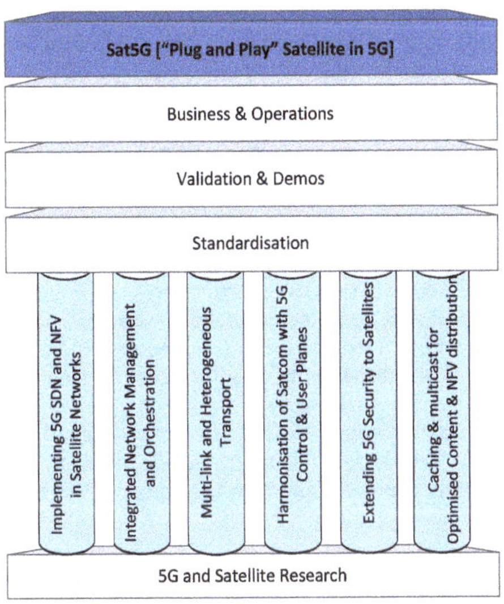

Figure 1: SaT5G Project Structure

Based on the anticipated market needs, SaT5G focuses on the eMBB Usage Scenario for 5G. SaT5G addresses the mobile network operators' (MNOs) needs for the inclusion of satellite in the early 5G roll out; bypassing congested backhaul and offloading high bandwidth video download have been found to be the major drivers. SaT5G is investigating novel business models and economically viable operational collaborations that integrate the satellite and terrestrial stakeholders in a win-win situation.

This paper addresses the arising business models, relating them to the project use cases, architecture and Key Performance Indicators (KPIs).

The remainder of this paper is organized as follows: section 2 provides an overview of the most promising and relevant satellite use cases in 5G investigated as part of the SaT5G project. Section 3 elaborates on the KPIs addressed by the SaT5G project and their mapping to the satellite use cases. Section 4 addresses the architecture options for satellite integration into 5G, with focus on satellite backhaul architectures. Section 5 addresses the business modelling aspects; finally section 6 concludes the paper.

2. Use Cases

2.1 Selected use cases

By following the methodology described in [1] [2], SaT5G has selected four Satellite Use Cases for eMBB to focus its efforts on (see also figure 2 following):

1) *Edge delivery and offload* of multimedia content and MEC VNF software, through multicast and caching to optimise the operation and dimensioning of the 5G network infrastructure;
2) *5G fixed backhaul*, to provide 5G service especially in areas where it is difficult or not possible to deploy terrestrial communications;
3) *5G to premises*, to provide 5G service into home/office premises in underserved areas via hybrid terrestrial-satellite broadband connections;
4) *5G moving platform backhaul*, to support 5G service on board moving platforms, such as aircraft, vessels, trains, etc.

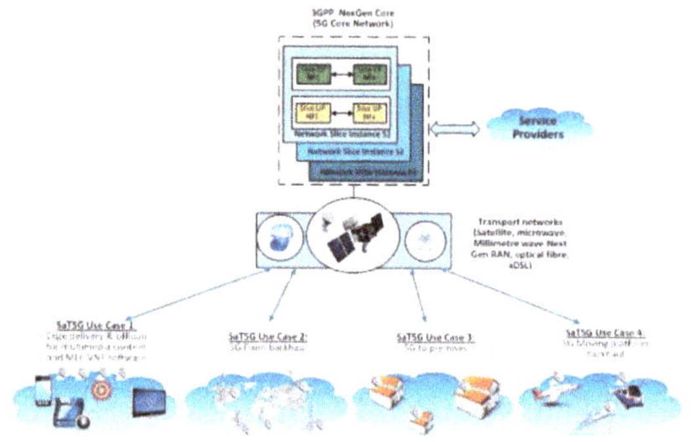

Figure 2: SaT5G Use Cases

Each use case has two or more specific scenarios as shown in table 1 below. Selected scenarios are used for the business analysis and will be used later in the project for the validation tests and demonstrations.

Table 1: SaT5G Use Cases and Scenarios

SaT5G Use Cases	Scenarios for SaT5G Use Cases
SaT5G Use Case 1: Edge delivery & offload for multimedia content and MEC VNF s/w	Scenario 1a: Offline multicasting and caching of video content and VNF software through satellite links
	Scenario 1b: Online prefetching of video segments through satellite links
SaT5G Use Case 2: 5G Fixed backhaul	Scenario 2a: Satellite backhaul to groups of cell towers
	Scenario 2b: Satellite backhaul to individual cell towers
	Scenario 2c: Satellite backhaul to individual small cells
SaT5G Use Case 3: 5G to premises	Scenario 3a: Hybrid Multiplay (satellite/xDSL) at home/office premises in underserved areas
	Scenario 3b: Hybrid Multiplay (satellite/cellular) at home/office premises in underserved areas
SaT5G Use Case 4: 5G Moving platform backhaul	Scenario 4a: Updating content for on-board systems and grouped media request by the moving platform company
	Scenario 4b: Broadband access for passengers and individual media requests
	Scenario 4c: Business and technical data transfer for the moving platform company

Scenario 2b is frequently used within SaT5G as the baseline description from which other scenarios are extrapolated. It is more fully defined as:

"Satellite backhaul to a single cell tower located in a rural area in the EU covering two villages about 5km apart and a rural main road. The villages are home to 300 families, in summer months an additional 50 families may be in holiday accommodation. The road can occasionally be busy with holiday traffic but is usually quiet. The predominant traffic on the cell is eMBB but there is some [very low level] mMTC traffic generated by agri-tech."

2.2 Background developments

Satellite systems are evolving, with greater capacity (in the order of Tbps speed) with accordingly lower cost per bit and better connectivity which makes the integration of satellite with terrestrial mobile networks a compelling commercial proposition. Future satellite constellations will complement this with lower latency connections.

The business aim is to improve customer experience and value, with lower incremental capital cost than fixed network upgrades. The near 100% coverage provided by satellite, and the characteristic distance-independent cost of adding users, is increasingly desirable for MNOs for capacity and coverage. Satcom integration within 5G will expand the opportunity for MNOs by enabling fast roll out of services in a way that has not been possible for 3G/4G.

2.3 Related deployment opportunities

The opportunities for satcom offering eMBB backhaul have been found to fall into two categories where satellites offer a provision of alternative or overlay paths for:

2.3.1 Offloading and caching in the fixed network:

(a) Connecting data-centres to 5G-core network gateways where there are no terrestrial or economically viable paths, and;
(b) Connecting the core network to semi-centralised or centralised virtualised gNB processing centres.

2.3.2 Backhaul: connections to 5G cells in all layers (macro, micro, etc).

Four main example opportunities have been identified:

(a) Backhaul to cells in remote areas that are installed to achieve a certain coverage obligation and where providing a terrestrial link is not economically viable;
(b) Cells that need fast deployment and cannot wait for terrestrial provisioning;
(c) Deployment of cells for temporary events; and
(d) Deployment of small mobile cells mounted on public transport, on high-end cars, or on emergency services vehicles.

Satcom deployment can be valuable across multiple verticals; table 2 following illustrates this comparing a representative set of target verticals (building on both [2] and [11]) against the use cases.

Table 2: Relevance of SaT5G Use Cases to Market Verticals

SaT5G Use Case	Media & entertainment	Transportation	Manufacturing	Health	Utilities	Agriculture	Public safety
1 – Edge delivery…	✓	✓					✓
2 – Fixed backhaul	✓	✓	✓	✓	✓	✓	✓
3 – 5G to premises	✓						
4 – Moving platforms	✓	✓		✓	✓		✓

3. Key performance indicators (KPIs)

3.1 The role of KPIs

The main SaT5G goal is the seamless integration of satellite networks into 5G standard networks. The technical and the economic goals have different challenges which are addressed using the use cases previously defined. To be considered a 5G network, the satellite network requires a minimum set of capabilities. These capabilities are defined using the well-known approach, addressed in multiple standardization bodies like 3GPP, which defines the minimum requirements. These requirements take into account parameters like throughput, delay, reliability, etc. The KPIs are used to measure such requirements. To avoid any confusion between vendors, SaT5G describes how the KPIs are measured.

The definition of the KPIs is the starting point to solve the technical challenges when the next network generation is defined. In SaT5G, a subset of the defined KPIs is used by the business models to evaluate the cost of the solution. By considering the business models in these early stages, SaT5G is differentiated from previous attempts to introduce satellites into terrestrial cellular standards. The business models will guide the technical work during the process minimizing efforts on non-relevant work.

3.2 Satellite and 5G KPIs

The list below shows the subset of satellite KPIs which are considered for business evaluation. Bandwidth is a limited resource that is shared between multiple operators and links. Satellite total throughput is another important parameter to be considered by the business models as it has a direct impact on the service that operators provide to their clients. These KPIs mixed with the coverage area, the traffic density and the satellite backhaul data rate are used to evaluate the cost per bit per user.

Spectrum usage: Usually specified in GHz defines the spectrum per beam.

Satellite bandwidth: defined in MHz; representing the total allocable bandwidth or the power equivalent bandwidth depending on the service characteristics, the satellite network operator (SNO) procuring the necessary systems to terminate and manage the capacity. The same bandwidth is frequently reused by the use of multiple users and gateway beams.

Satellite throughput: in Mbps; the SNO using the provided systems to manage service levels to their user base. Required Mbps depends on the use case and the service.

Satellite backhaul data rate: The allocated data rate in Mbps for each link between the satellite gateway and remote satellite terminal connected to a gNB. This is usually less than the peak data rate per link.

Maximum satellite roundtrip delay: quoted in ms and varies per satellite orbit (GEO ~600ms, MEO ~150ms and LEO ~ 30ms making some allowance for signal processing delays).

Coverage area: The coverage provided by each satellite including number or beams and level of frequency reuse.

3.3 Mapping of SaT5G use cases to 5G PPP KPIs

The following table shows how the four SaT5G use cases are able to contribute to the often referred to 5G PPP KPIs [12]. Of particular note perhaps is the ability of satcom to reduce end user perceived latency by placing key content in edge caches. Indeed, according to [5] and [6], paradoxically, the low latency requirement for 5G networks is a big ally for satcom as many new locations for content servers will be required. In the transition to 5G, content needs to be moved to the edge, densifying CDN networks and making satellite multicast a viable option. In this context, satcom can indeed help 5G networks to achieve low latency by multicasting content to caches located at individual cells, even in places without fibre.

Table 3: Mapping of SaT5G use cases to 5G PPP KPIs

SaT5G Use Case	1000x capacity	10-100 x user data rate	100% coverage	Service creation in minutes	Latency <1ms
1 – Edge delivery…	✓	✓	✓	✓	✓
2 – Fixed backhaul	✓	✓	✓	✓	
3 – 5G to premises	✓	✓	✓	✓	✓
4 – Moving platforms	✓	✓	✓	✓	✓

4. Architecture

4.1 Roadmap and technical requirements for satellite architectures

One key part of the SaT5G project has been to look at today's satellite network architectures and compare these with the emerging 3GPP architectures for 5G networks looking at issues such as control and user plane functions, network slicing, virtualisation and end-to-enend service orchestration.

A roadmap of different architectures showing increasing levels of 5G network integration to support non-terrestrial networks has been derived. This also addresses the four SaT5G use cases; looking at architectures for backhaul to fixed nodes and mobile platforms, for caching and multicast overlays, the end-to-end service delivery and defining the reference interface between satellite and terrestrial networks. These options are shown in the following figure.

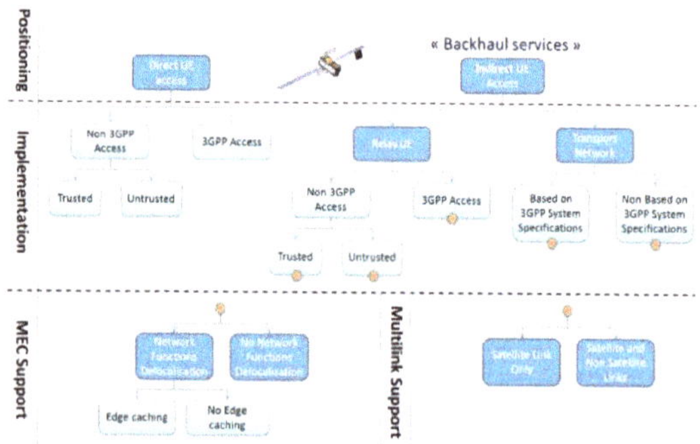

Figure 3: Options for satellite integration into 5G

Several areas of integration have been identified like QoS adaptation, mobility management, beam management or multi-satellite connectivity. Below figure provides an overview about QoS adaptation of the user plane traffic between the terrestrial and the satellite link.

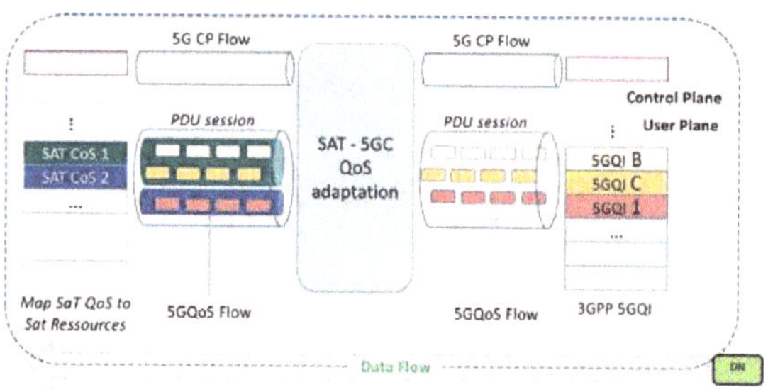

Figure 4: QoS adaptation between terrestrial and satcom 5G

4.2 Backhaul architectures and 3GPP

The SaT5G project has made major contributions to an ETSI report [4] and 3GPP study [9] that describes what it refers to as the non-terrestrial network (NTN) architectures in 3GPP compliant terminology. In addition, the report details the 3GPP interfaces presented to and across the satellite backhaul links. It documents the five scenarios as:

- *Scenario A1 - Direct 3GPP access:* The NTN enabled UE accesses the 5G Core Network via a 3GPP defined NTN NR-Radio Access. UE management applies to the NTN enabled UE.
- *Scenario A2 - Indirect 3GPP access:* 5G UE are served by an access point. This access point is served by a 3GPP defined NTN NR-Radio Access. UE management applies to the NTN terminal which is still named "NTN enabled UE". The NTN enabled UE endorses a multiplexer node role.
- *Scenario A3 - Indirect mixed 3GPP NTN access:* 5G UE are served by an access point. This access point is served by a trusted mixed 3GPP NTN access network (see "Non-3GPP access network" definition). UE management applies to the NTN terminal which is named "NTN enabled UE". The NTN enabled UE endorses a multiplexer node role. Another wording for scenario A3 could be "Indirect 3GPP NTN access with non-3GPP L2, non-3GPP L1".
- *Scenario A4 – Indirect access via transparent transport network:* 5G UE are served by an access point. This access point is connected to the CN via a NTN transport network. The UE management does not apply to the NTN terminal. The NTN terminal may endorse a multiplexer node role. This NTN transport network is considered as *trusted* by the operator of the UEs.
- *Scenario A5 - Indirect untrusted access:* 5G UE are served by an access point. This access point is served by an untrusted non-3GPP or a mixed 3GPP NTN access network. UE management does not apply to the NTN terminal. The NTN terminal endorses a multiplexer node role.

One example showing the mapping of the 5G system reference points on a logical network representation is given in the following figure.

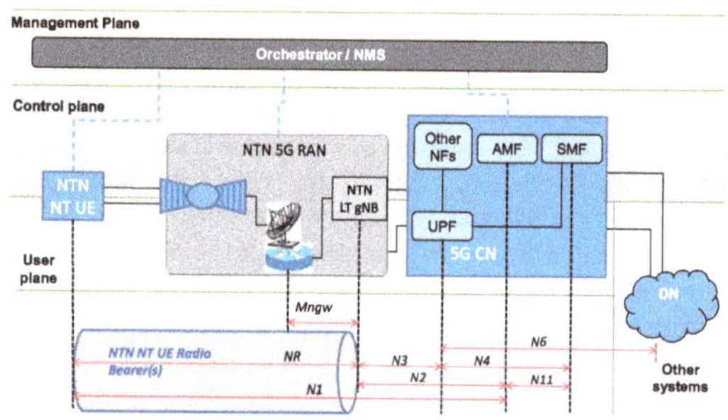

Figure 5: Mapping 5G system reference points

4.3 Business implications on the architecture

The business analysis is a SaT5G differentiator taking into account the implications of the architectures that vary depending on use case and operator deployment models. As such a series of business roles have been identified aligning with key architectural blocks. These then allow different value network configurations (interactions between stakeholders taking up these business roles) to be considered where the business roles can be assigned to different stakeholders in the use case specific value network.

In reviewing the business implications of integrating satellite links in to future 5G networks, where there may be multiple satellite network operators (SNOs) and multiple terrestrial MNOs (the two stakeholders identified as core roles), we have identified the potential for a trusted third party to provide a brokerage service to provide the capacity where needed under well-defined contractual and commercial terms.

In the focus on the core roles, different business models can be formed based on three parameters:

- The number of SNOs, can be 1 or multiple (M) or served by a broker (B);
- How the satellite capacity is offered: fixed (F) or dynamic (D) resource allocation;
- The number of MNOs, can be 1 or multiple (M) or served by a broker (B).

For each of the use cases defined in SaT5G, the short term (reference column in the table) and the long term (most likely) combinations have been identified in the table below.

Use case	Reference	Most likely
Use case 1: Offline multicasting and caching of video content and VNF software through satellite links	1-F-1	(1-F-M)_B or 1-D-M
Use case 2: Satellite backhaul to individual cell towers	1-F-1	(1-F-M)_B or 1-D-M
Use case 3: Hybrid Multiplay (satellite/xDSL) at home/office premises in underserved areas	1-F-1	1-D-1
Use case 4: Broadband access for passengers and individual media requests	1-F-M (train), M-F-M (plane)	M-D-M or (M-D-M)_B

The broker role is shown in the following figure. The broker provides a way to share the satellite use risk between the SNOs and MNOs in a reasonably equitable fashion. Further analysis is ongoing on how this role might cater for both point-to-point traffic and multicast content delivery traffic.

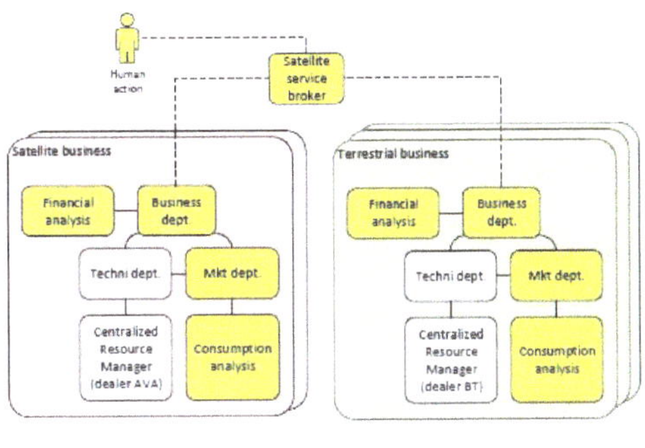

Figure 6: Role of Broker in Satellite Integration into 5G

It has also been identified that there is a strong correlation between the size of the request and the type of interaction required by the broker. An automatic process can provide near real time access to small well-defined capacity requirements while when the bandwidth requests and/or complexity increase, the need for manual intervention increases along with the notice period. This is illustrated in figure 7 below.

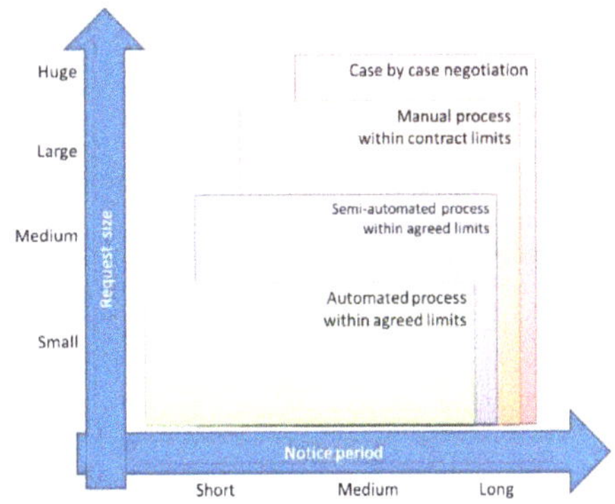

Figure 7: Request size, automation and notice period

5. Business case

5.1 Methodology

Building on the KPI requirements, the SaT5G project has evaluated the economic and business impact by performing a techno-economic analysis of the use cases identified previously. Both an overall view on the value network (indicating the different roles needed for service delivery, as well as their interactions, [3]) and a focus on the core roles of satellite and mobile network operations (see figure below) are needed in order to decide on the viability of satellite integration into 5G.

A model has been created that calculates the total cost of ownership (TCO) as well as the average revenue per user (ARPU) for the studied scenarios. The following results show the analysis for the specific case of the scenario 2b (see section 2.1).

5.2 Model and key assumptions

The cost model develops a TCO for deploying the integrated terrestrial-satellite network, and consists of 4 parts (see figure 8 following):

- Cost model of the edge
- Cost model of the satellite network (for backhaul)
- Cost model for the 5G core network
- Overhead costs.

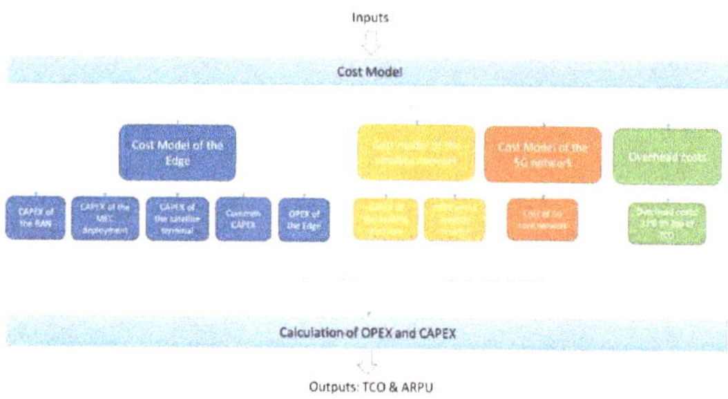

Figure 8: Cost model elements

The model calculates cost and the ARPU contribution needed based on a number of key assumptions:

- Hardware installation cost is 15% of the hardware costs [15];
- Maintenance cost is 10 % of the CAPEX costs [8];
- Overhead cost is 22% on top of TCO [8];
- Project horizon is 5 years;
- The average revenue per user (ARPU) is the average cost per user (ACPU) plus a profit margin of 11 % [14];
- We calculate a bit rate of between 53Mbps and 210Mbps per backhaul link;
- Finally, 80% of popular content is assumed to be cached.

The lower calculated bit rate assumes all 1050 end users consuming 6.5GB per month (based on the Cisco VNI report [16]) over the 5G backhaul; i.e. not via other routes such as Wi-Fi as defined in scenario 2b and represents a typical figure for Western-Europe in 2021. The higher figure represents 350 homes each consuming an average of 2Mbps (e.g. [7] states 1Mbps to 2Mbps for Western-Europe in 2020) during the peaks busy hour which is representative of the average traffic in Western-Europe for the whole home – in other words all their Internet capacity is provided by the 5G service and carried over the satellite backhaul.

The bandwidth demands will vary across the world; it is worth noting that the demand across Africa is forecasted to be around 1.5GB per UE per month [16] so the aggregated demand would be in the region of 12Mbps to 50Mbps and the modelled costs commensurately lower. In rural under developed regions such as those seen in Sub-Saharan Africa where cheap smartphones inevitably prevail, the backhaul traffic per link needs may well be below 10Mbps.

The calculated costs are presented as an ARPU contribution per end user to allow these to be readily compared with typical subscriber service charges. The calculated ARPU contributions exclude local end user taxation such as VAT.

5.3 Findings

Work is ongoing in SaT5G on refining the model and input assumptions, the model may also be amended as the detailed technical prototyping and validation tests are performed.

To date, the model shows that an ARPU contribution of between €36 and €107 (per month) in Western-Europe is calculated for SaT5G scenario 2b where caching is not applied, whereas the scenario where 80% of popular content is cached at the edge, results in a reduction of the monthly ARPU to between €9 and €28. The reduction is mainly caused by a reduction in the needed throughput on the satellite backhaul link, leading to a reduction in operational expenses of around 75%. In the real world an ARPU contribution between these two figures might be expected. These values indicate that whilst satcom backhaul ARPU contributions are too high for a total network deployment serving SaT5G use case 2 "5G satellite backhaul" needs; however the contribution are not astronomical and therefore reasonable for a few percent of sites.

The ARPU contributions for other regions will vary, for example in regions of lower demand such as Africa the ARPU contribution required will be much lower and therefore wider use will be possible.

Other use cases such as backhaul to mobile platforms will have different user bandwidth demands and satellite capacity costs due to different equipment requirements. The analysis against the other use cases has not been completed however by inspection use case 3 will be somewhat similar to the higher end ARPU contributions calculated for SaT5G scenario 2b.

6 Conclusion

A representative variety of four use cases and ten scenarios where the provision of 5G satellite backhaul services is likely to be of significant interest to both MNOs and SNOs has been presented. A mapping between the 5G KPIs and 5G satellite backhauls has been analysed. The architecture options for integrated 5G satellite backhaul delivery have been identified and inputs provided to 3GPP and ETSI.

From this work an initial business analysis has been undertaken looking at how the value matrices related to the different use cases. A cost model has been created that calculates the ARPU contributions needed per end user. A range of €9 to €107 per month per end user has been calculated for predicted Western-Europe traffic in 2021; depending on the amount of edge caching and whether this is only traffic to the handsets or to all devices in the home. This depends on how the real use cases align with those defined in SaT5G and described above.

These values indicate that whilst satcom backhaul ARPU contributions are too high for a total network deployment serving SaT5G scenario 2b "5G satellite backhaul" needs; the ARPU contribution are not excessive and therefore reasonable for a few percent of sites in Western-Europe.

Other regions and scenarios are the subject for further analysis within SaT5G and beyond.

Acknowledgements

The SaT5G project has received funding from the European Union's Horizon 2020 research and innovation programme under grant agreement No 761413 [10], [13].

References

Journal articles

[1] K. Liolis, A. Geurtz, R. Sperber, D. Schulz, S. Watts, G. Poziopoulou, B. Evans, N. Wang, O. Vidal, B. Tiomela Jou, M. Fitch, S. Sendra Diaz, P. Sayyad Khodashenas, N. Chuberre, "Use Cases and Scenarios of 5G Integrated Satellite-Terrestrial Networks for Enhanced Mobile Broadband: The SaT5G Approach", Wiley's International Journal of Satellite Communications and Networking – Special Issue "Integrated Satellite-Terrestrial Networks in Future Wireless Systems, 2018;1–22. https://doi.org/10.1002/sat.1245.

Conference Papers

-

Book, book chapter and manuals

[2] K. Liolis, A. Geurtz, R. Sperber, D. Schulz, S. Watts, G. Poziopoulou, B. Evans, N. Wang, O. Vidal, B. Tiomela Jou, M. Fitch, S. Sendra Diaz, P. Sayyad Khodashenas, N. Chuberre, "Satellite Use Cases and Scenarios for 5G eMBB", Book: "Satellite Communications in the 5G Era", Editors: S. Krishna Sharma, S. Chatzinotas, P.-D.M. Arapoglou, IET, 2018, ISBN: 978-1-78561-427-9.

[3] Verna Allee. "Value Network Analysis and Value Conversion of Tangible and Intangible Assets." Journal of Intellectual Capital. Publisher: Emerald Insights, Year: 2008, Volume: 9, Issue: 1, Page. 5 - 24

Patents

-

Reports

[4] ETSI TR 103 611 "Integration scenarios of satellite and/or HAPS (High Altitude Platform Station) systems into 5G system and related architecture options", Draft ETSI Report Reference: SESSCN(18)000019r3 available online

[5] GSMA Intelligence, "Understanding 5G," December 2014. [Online]. Available: https://www.gsmaintelligence.com/research/?file=141208-5g.pdf&download.

[6] NSR, "Wireless Backhaul via Satellite, 11th Edition," March 2017. [Online]. Available: http://www.nsr.com/research-reports/satellite-communications-1/wireless-backhaul-via-satellite-11th-edition/.

[7] EC FP7 Project BATS, deliverable D5.2 "Cost Benefit Analysis", section 5.1.8, 29 April 2014

Theses

[8] Koen Casier (2010), Techno-economics of FTTH deployment in the presence of competition (Doctoral dissertation). Retrieved from http://www.technoeconomics.ugent.be/output/phd_koen.pdf

Standards

[9] 3GPP TR 38.811 "Study on New Radio (NR) to support non-terrestrial networks", available online https://portal.3gpp.org/desktopmodules/Specifications/SpecificationDetails.aspx?specificationId=3

Websites

[10] SaT5G (Satellite and Terrestrial network for 5G) project home page, http://sat5g-project.eu/, accessed 1 August 2018

[11] 5G: Serving Vertical Industries, 5G PPP, https://5g-ppp.eu/2nd-5g-vertical-workshop/, accessed 1 August 2018

[12] 5G PPP Key Performance Indicators, https://5g-ppp.eu/kpis/, accessed 1 August 2018

[13] EC CORDIS summary page for SaT5G, https://cordis.europa.eu/project/rcn/211060_en.html, accessed 1 August 2018

[14] A Fitness Program for GCC Telcos, A.T. Kearney, 2012 https://www.atkearney.com/documents/20152/434204/Telco_Fitness.pdf/427dcc76-15f9-b9a3-0fd5-5c2e157918d6

[15] What is the average profit margin for a company in the telecommunications sector?, Investopedia retrieved 2018, https://www.investopedia.com/ask/answers/060215/what-average-profit-margin-company-telecommunications-sector.asp

[16] VNI Forecast Highlights Tool, Cisco, retrieved 2018, https://www.cisco.com/c/m/en_us/solutions/service-provider/vni-forecast-highlights.html#

5G TECHNOLOGIES FOR A COMMUNICATIONS AND NAVIGATION INTEGRATED INFRASTRUCTURE ON MOON AND MARS

Pietro Giordano, Marco Lisi, Andrea Modenini
European Space Agency - European Space Research Technology Centre (ESA-ESTEC)
Keplerlaan 1, Noordwijk, The Netherlands
Email: {pietro.giordano, marco.lisi, andrea.modenini}@esa.int

Abstract

In the past years, ESA set forth the idea of a "Moon Village", a village on the moon built by huge 3D printers and inhabited for months at a time by teams of astronauts. The plan outlined by the ESA is that, starting from the early 2020s, robots will be sent to the Moon to begin constructing various facilities, followed a few years later by the first inhabitants.

Back in 2013, ESA teamed up with building companies to start testing out various Moon base-building technologies, and determined that local materials would be the best for constructing buildings and other structures, which means no need for transporting resources from earth at an astronomical cost. But the problems to be solved for the realization of such stable manned infrastructure on the Moon (a true follow-on of the International Space Station) involve much more than just building technologies. The Moon Village will be a large and complex system where requirements related to operations and safety of life will be of paramount importance. Moreover, from an architectural viewpoint the "village" will have to be expandable and "open" to the integration with other systems, hence integrability and expandability will be two key issues. But first and above all, the Moon Village will have to be affordable and sustainable, i.e., its cost will need to be assessed over its life-cycle. As a "Wild West" town in the old times, "Moon Village" will have to provide a number of essential infrastructures. In particular, the exploration of the Moon with human and robotic missions and its colonization, through the establishment of permanent bases, will require planetary communications and navigation infrastructures.

The approach described in this paper was first presented in 2016 [1] and relies on the use of Commercial off-the-shelf (COTS) component for communication and navigation on the Moon surface. The use of LTE technology, currently deployed on Earth, and the coming 5G technology will allow communication and navigation, within the maximum throughput and the accuracy provided by the standard, assuming the implementation of all the basic pre-requisites on the Moon.

More recently (March 2018), Vodafone and its technology partners Nokia, PTScientists, Audi and SpaceX announced their plan to set up the first cellular network node on the Moon. Because 5G is still in its very early stages, Vodafone has decided to go for the time being with 4G technology, using spectrum in the 1.8 GHz band. The Vodafone team intends to set up a base station near the Lunar Rover left in the Taurus-Littrow valley during the Apollo 17 mission and send to Earth live HD videos from the Moon's surface. The 4G network on the moon will connect two Audi Lunar Quattro rovers to a

base station in the Autonomous Landing and Navigation Module (ALINA). This mission, planned for not earlier than 2019, would be the first private mission to the Moon.
Esa's vision is much more ambitious, aiming at the creation of a permanent inhabited base on the Moon, with all the infrastructures needed to support life and operations.

1. Introduction

Future permanent bases on the Moon and on Mars will be large and complex systems where requirements related to operations and safety of life will be of paramount importance. Moreover, from an architectural viewpoint they will have to be expandable and "open" to the integration with other systems, hence integrability and expandability will be two key issues. But first and above all, these planetary outposts will have to be affordable and sustainable, i.e., their cost will need to be assessed over their life-cycle. As a "Wild West" town in the old times, these bases will have to provide a number of essential supporting infrastructures. In particular, the exploration of a planet with human and robotic missions and its colonization, through the establishment of permanent bases, will require planetary communications and navigation infrastructures.
All architectural approaches considered so far by NASA and ESA can be divided in two main categories:
• Comprehensive, well-structured and forward looking (but costly) architectures, based on constellations of orbiters and relay satellites;
• "ad hoc", flexible, expandable architectures, based on a fusion of all available resources and on COTS technologies.
The second approach looks as a more promising, affordable and sustainable solution.
The paper is presenting an infrastructure architecture that relies on the use of Commercial off-the-shelf (COTS) component for communication and navigation on Moon and Mars surfaces. The use of LTE technology, currently deployed on Earth, and the forthcoming 5G technology, will allow communication and navigation, within the maximum throughput and the accuracy provided by the standards.
The main idea here proposed is that of having a scalable network that relies on COTS, with the aim of limiting the process of design and development of specific technology for the Moon (or Mars) base. Consequently, the design of, e.g., the lunar communication network will be mainly devoted to the definition of its cell distribution on the lunar surface. Clearly, the cell distribution will strongly depend on the network (performance, functional, and operational) requirements, the lunar site location, and what will be the 5G air interface.
As far as the navigation (better: Positioning, Navigation and Timing, PNT) infrastructure is concerned, past ESA contracts provided in 2008 two feasibility studies of a reduced planetary navigation and communications system. Both studies concluded that COTS equipment, based on IEEE 802.16 WiMAX standard, could be used to fulfil the mission requirements for short-range activities (i.e., link distance below 8 km), but long-range activities were not foreseen to be covered only with an infrastructure on the planetary surface. Future 5G technology is expected to overcome these challenges and to provide the necessary coverage, flexibility and performance required by future permanent planetary bases.
The 5G standard is envisaged to support communication and positioning capabilities for a wide range of applications, such as massive Internet of Things (IoT), mission-critical control, and enhanced mobile broadband. Similarly to the 4G LTE standard, 5G multicarrier waveforms will allow the flexible allocation of data, as well as dedicated pilot signals for positioning purposes. These pilot signals can be used to perform ranging measurements from time-of-arrival (ToA) location methods, and multi-antenna techniques can enable angle-of-arrival (AoA) localization.
5G systems on Earth rely on GNSS signals for precise synchronization. It is proposed to adopt the same approach, at least for the Moon 5G base stations. On Mars it will be necessary the use of miniaturized atomic clocks. COTS miniaturized atomic clocks are already available in the market and currently used in professional ground equipment, so they are considered suitable for this architecture and affordable from a cost view-point.
The exploration of the Moon and of Mars with human and robotic missions and their colonization, through the establishment of permanent bases will require planetary communications and navigation infrastructures. The final paper will describe with details an approach relying on the use of COTS component for communication and navigation on the Moon surface. The use of LTE technology, currently deployed on Earth, and future 5G will allow communication and navigation, within the maximum throughput and the accuracy provided by the standard, assuming the implementation of all

the basic pre-requisites on the planets. The approach described will be shown to largely satisfying the requirements in terms of performance, reliability, affordability and sustainability, being based on commercial technology and incrementally expandable over time.

2. Why 5G for the Moon and Mars?

A planetary infrastructure shall be able to serve several use cases (UCs). Although not all UCs can be yet identified, at least we can define two main categories: UCs that needs a low data rate and reliable links, and those that require high data rate links. The first category includes monitoring and control of the outpost systems/payloads and essential audio, video, and file transfer among users. Links for these UCs shall have high service availability (for instance 99.99%) also in case of emergencies and disasters, regardless of orbit phases, Earth position, terrestrial weather conditions, etc. The second category instead includes HTTP surfing, high quality Audio/Video communications, video streaming, HD television, file sharing, clouding computing, etc. These UCs will have a service availability lower than the first category (for instance 98%).
While the fifth-generation (5G) of mobile technology is currently thought for terrestrial UCs, it is evident from the requirements that the eventual standard will be an excellent candidate for the design of a highly integrated and off-the-shelf network. In addition, the selection of 5G will bring all the advantages of word-wide accepted standards (participation of a larger number of companies, cost reduction, etc.). The 5G of mobile technology promises to deliver the gigabit experience to users. Although the definition of the physical and MAC layer are still debated, the main requirements have been already identified by several mobile operators [2][3]. Some of these requirements are:

- data rate higher than 10 Mbps for tens of thousands of users in a crowded area (like a stadium), and higher than 1 Gbps for few tens of users in the same indoor environment;
- latency reduced as compared to LTE;
- coverage increased, especially in rural area;
- radio resource and energy consumption minimized;
- programmability and configurability;
- improved network availability (99.999%), including robustness against climatic events and guaranteed services at low energy consumption for critical infrastructures;
- network functions scalable, so that capacity is provided when and where needed.

3. A lunar/martian communications network

The main idea here proposed is having a scalable network that relies on COTS, with the aim of limiting the process of design and development of specific technology for the planetary outpost. Consequently, the design of a Lunar/Martian communications network will be mainly devoted to the definition of its cell distribution on the surface. Clearly, the cell distribution will strongly depend on the network (performance, functional, and operational) requirements, the site location, and what will be the 5G air interface.
Starting from these inputs and focusing on the Moon Village case, a possible strategy for defining the cell distribution is summarized in Figure 1.

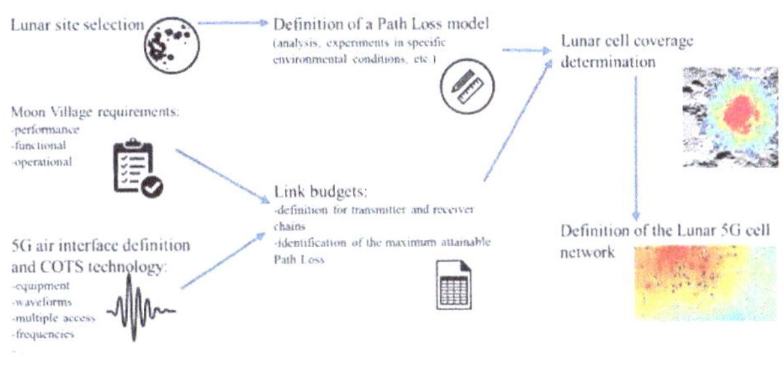

Figure 1: logical steps for the design of the Lunar 5G network

The Moon Village requirements and the 5G air interface definition shall serve as input for the definition of link budgets, in particular for the transmitting and the receiving chains, with the objective of deriving the maximum attainable path loss. At the same time, the Moon village location physical and environmental properties are a starting point for the definition of a path loss model, that can be derived by means of analysis (of the available information), testing in specific environmental conditions, etc. Once that path loss model and link budgets are completed and consolidated, the coverage distribution of a single cell can be determined. The coverage clearly will depend on its positioning (latitude, longitude, height from the surface), the kind of adopted antennas, and the surrounding infrastructure: notice that all these parameters can be elaborated from software tools (e.g. Linux/Unix software "SPLAT!") and the coverage pattern computed for several positions. This allows to derive a first iteration of the cell distribution, and thus the Lunar 5G network, by dovetailing several cells on the selected site and verifying that the total coverage meets the original requirements.
The logical steps here described could look visionary, but actually many activities are already on-going in the global research community. An example is the selection of the lunar site: the study in [4] published recently by D. Wingo, shows that the Peary crater (situated in the Lunar North Pole) could be a good candidate for the Moon Village, thanks to its power availability, presence of hydrogen-based molecules, surface mobility, line-of-sight for communications, etc. In parallel, other studies for the selection of the lunar site have been carried out under the ESA contracts [5] and [6]: these studies have shown other sites of interest in the Lunar South Pole, such as between the Shackleton and the De Gerlache craters. Another example of on-going activity is the definition of the path loss model: some models are already existing in the literature and can be found in [7], [8] and [9]. For instance the study in [7] introduces the use of a modified Longley-Rice irregular terrain model and digital elevation data representative of an analogue lunar site for the prediction of RF path loss over the lunar surface. Results were elaborated with "SPLAT!", and validated by means of theoretical models and past Apollo studies.
The strategy shown in Figure 1 can be adopted also for a Martian outpost, although this will require a re-assessment and re-definition of link budgets, path loss models, coverage, and the network definition. Also for Mars, many activities are on-going in the global research community, although fewer (w.r.t. the Lunar environment) are publicly available. For instance, the first concept of Mars output was already develop in the 1980s, and in [10] it can be found an assessment of a first manned mission on Mars with the objective of achieving self-sufficiency. Additionally in 2002, NASA published an handbook [11] for radio wave propagation in the Martian environment, that examined the environmental effect of the Martian atmosphere, ionosphere, global dust storms, and geomorphological features, and extracted relevant environmental parameters from the Mars measurements done by Pathfinder and Global Surveyor.

4. Communications with Earth

A key issue in an interplanetary outpost is the design of the backhauling link with Earth, that allows 5G communication terminals to access all services in the terrestrial network. Differently from the local network, the design of a backhauling system for the Lunar and the Martian outpost is completely different for the two cases, from both a design and an operational point-of-view, because of the different distance from the Earth. While the Earth-Moon round trip time is up to 2.7 seconds, still allowing interactive communications with some latency (e.g. a Skype© call from a Lunar operator inside the habitat with its family on Earth), Earth-Mars communications need to cover up to 401 million kilometers, hence with latency up to 22.3 minutes and requiring much higher RF power and antenna gains to establish the link.
The backhauling link shall be designed for providing a data rate high enough to meet initial and future capacity requests and shall have high link availability. Candidate technologies for backhauling are either RF or optical communications, each of them presenting advantages and disadvantages in terms of data rates, weather sensitivity, pointing accuracy, etc..
Studies that trade-off about the two technologies can be found for instance in [12] and [13]. More likely, both technologies will be employed in the backhauling of the network whose configuration will be complex and ad-hoc designed.
For the case of the Lunar outpost, an example of a backhauling configuration could be the one shown Figure 2 where orbiters are in stable orbits around the Moon and relay all the traffic to Earth directly to Ground Stations or to an optical relay that could be placed in one of the fourth Earth-Moon Lagrange points EML1, EML2, EML4, and EML5.

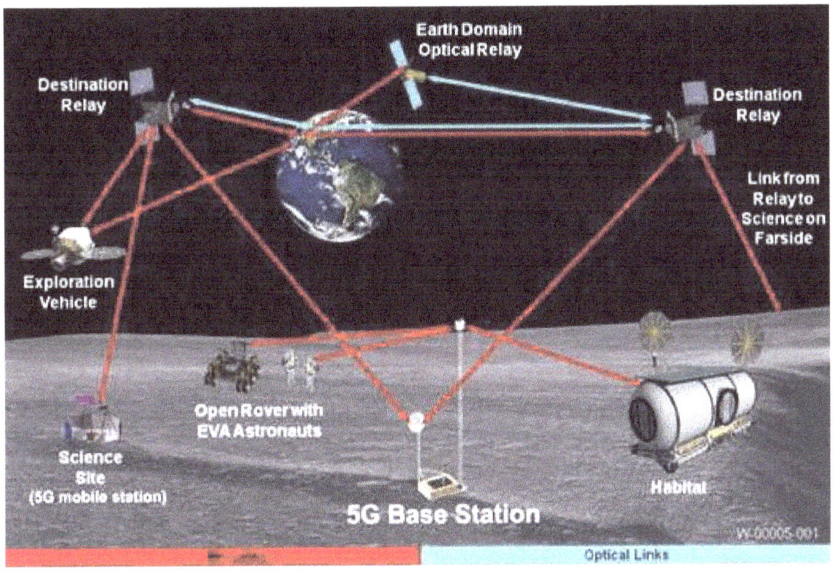

Figure 2: example of possible 5G communication network with backhauling to Earth realized by means of a Moon Orbiter

Clearly, the description above is just an example and many other backhauling configurations can be envisaged. However, some of the main key issues can be identified for each link configuration. For instance, the configuration shown in Figure 2, that adopts relay orbiters, requires a constellation of satellites that are launched from Earth, injected in stable orbits around the Moon or Mars, and allow handover and continuous coverage as in telecom satellite constellations around Earth (e.g. Iridium© and O3b© networks). Contrary to the case of satellite constellations around the Earth, the number of stable orbits can be quite limited. For instance, in low-lunar orbit there are only four circular orbits with limited perturbation due to lunar mass concentrations, namely the orbits with inclination 27, 50, 76, 86 degrees (see [14] for further details) and thus the described solution could be limited in terms of scalability.

Alternatively, the backhauling could be done by means of direct link with Earth. An example of such configuration for the Lunar environment is depicted in Figure 3, where 5G stations are wired to one or more optical backhauling stations that communicate directly with Earth. The backhauling station could be realized by adopting the Lunar laser ground terminals (LLGT), as shown in Figure 4, that was adopted for the optical demonstration with the Lunar Laser and Dust Environment Explorer (LADEE), and the same technology could be adapted/improved (especially in terms of power and pointing requirements) also for Martian communications.

This kind of backhauling configuration is probably easier to be designed and developed with respect to orbiter relays, but it could be limited by the noise temperature of the planet: the Moon reflects sunlight at optical wavelengths and emits waves in the microwave bandwidth, and similarly Mars has radio noise emissions due mostly to three types of sources: atmospheric emission, noise for the surface, and extra-Martian sources [15].

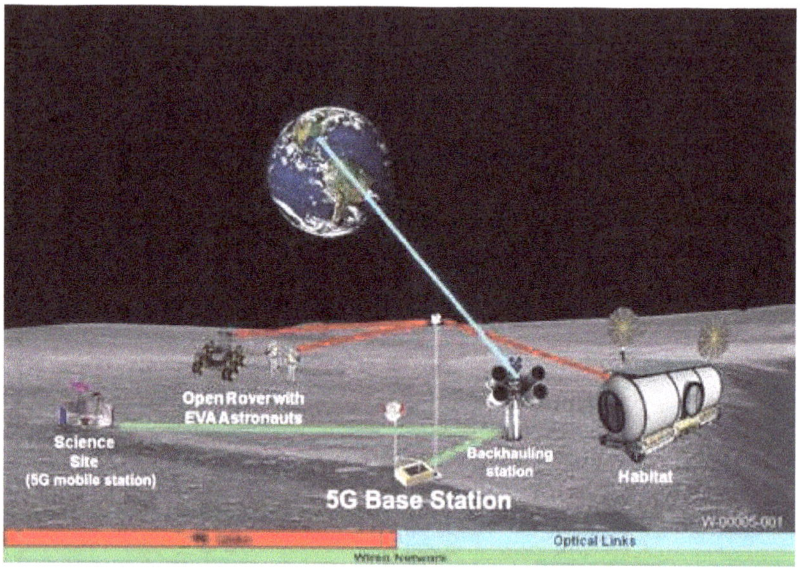

Figure 3: example of possible 5G communication network with backhauling to Earth realized by of direct-to-Earth optical link

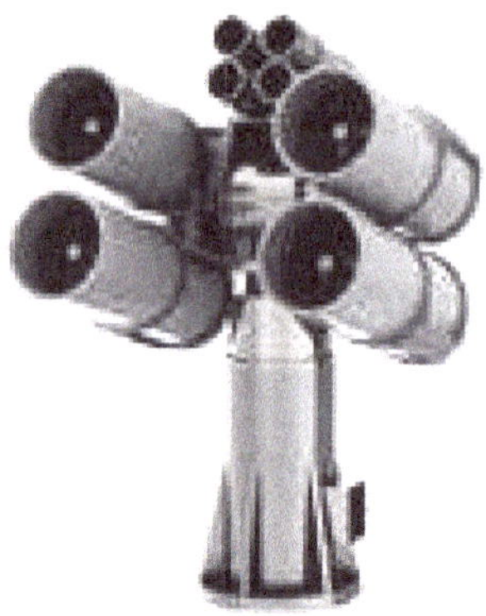

Figure 4: Lunar laser ground terminal (LLGT)

Consequently, a terrestrial ground station that points directly to the Moon surface will experience an increase of noise temperature that can be several dB, either in Optical and RF bandwidths. The reader can found a detailed study of this effect in [16] for RF communications (namely, S-Band, X-Band, and Ka-Band) on the lunar environment.

Finally, a third backhauling configuration could envision 5G base stations transmitting and receiving information directly to and from Earth. However, this last option does not look really feasible because primary power requirements would impose severe limits on the 5G base stations location.

5. Positioning, Navigation and Timing

Since 2001, the Aurora space exploration programme has led the European activities towards the potential deployment of a human bases on the Moon and Mars [17]. Within this framework, the ESA contracts [5] and [6] provided in 2008 two feasibility studies of a reduced planetary navigation and communications system. Both studies concluded that COTS equipment, based on IEEE 802.16

WiMAX standard, could be used to fulfil the mission requirements for short-range activities (i.e., link distance below 8 km), but long-range activities were not foreseen to be covered only with an infrastructure on the planetary surface. Future 5G technology is expected to overcome these challenges and to provide the necessary coverage, flexibility and performance required by the Moon Village.

The 5G standard is envisaged to support communication and positioning capabilities for a wide range of applications, such as massive Internet of Things (IoT), mission-critical control, and enhanced mobile broadband. For this purpose, the 5G air interface is expected to be based on the combination of optimized OFDM-based waveforms and a flexible framework with advanced wireless technologies, such as massive MIMO or robust millimeter wave (mmWave). Similarly to the 4G LTE standard, 5G multicarrier waveforms will allow the flexible allocation of data, as well as dedicated pilot signals for positioning purposes. These pilot signals can be used to perform ranging measurements for time-of-arrival (ToA) location methods, and multi-antenna techniques can enable angle-of-arrival (AoA) localization.

The 5G networks for the Moon Village mission will be designed according to the requirements for potential manned and robotic activities. The main design parameters are the cell site location, cell coverage and signal bandwidth. These parameters define the achievable communication and positioning capabilities. The configuration of multiple cell sites over a certain area, i.e., geometry of cell sites with respect to the receiver, determines the dilution of precision (DOP) of ToA and AoA methods. The cell coverage mainly depends on the height of the cell mast, transmit power, antenna pattern and propagation conditions. For instance, in the Moon, a cell tower of 10 meters above the surface is required to achieve a line-of-sight (LoS) distance to the horizon of almost 6 km [6], but this distance may be limited by the irregular topography of the surface. Last, the spectrum allocation of the positioning resources (i.e., pilot signals) determines the ranging accuracy, as well as the data rate. Design procedures for 5G terrestrial networks, as in [18], could adapted to the conditions of the Moon.

The network topologies can also determine the selection of the primary positioning method. In short-range links, the massive MIMO techniques can enable precise AoA localization, such as in [19]. In long-range links, ranging measurements may be obtained only from one or two cell sites. Thus, the ToA estimates can be combined with inertial measurement units. In mesh or ad hoc networks, such as device-to-device (D2D) communications, cooperative positioning between wireless sensors or sites may provide an additional location solution. Considering the ToA-based methods, the cell sites should be synchronized to a certain reference time in order to achieve accurate positioning.

5G systems on Earth rely on GNSS signals for precise synchronization. GNSS receivers are used to provide precise timing in different parts of the 4G LTE ground network, which requires within 3 to 10 microseconds accuracy [20], depending on the application and the actual standard implemented.

On ground, such accuracies are easily achievable by a professional GNSS timing receiver, which has an accuracy in order of tens of nanoseconds. This work proposes to adopt the same approach for Moon-based 5G base stations.

Clearly, on the Moon, the conditions are significantly different with respect to Earth surface. GNSS environment on the Moon has been studied in previous ESA contracts ([21], [22], [23], [24], [25], [26]), showing that the major challenges to be considered are:

- Signal power: in addition to the higher free-space loss, the majority of the received signals come from the GNSS transmitter antennas' side-lobes (with considerably lower gains). Additionally, stronger signals may interfere with the correct acquisition and tracking of weaker signals (near-far effect), with a consequent impact on receiver sensitivity and robustness;
- Dynamics: high ranges for Doppler and Doppler rates hinder acquisition and impose additional stress on the tracking loops, also making it more difficult to process weak signals;
- Geometry: the geometry of the usable satellites is considerably worse than for terrestrial applications. Additionally, occultation by the Earth and Moon and receiver sensitivity (minimum C/N0 required to acquire and track GNSS signals) may also have an impact on the dilution of precision (DOP).

Within the same studies, it was showed that GNSS could be used for MTO (Moon transfer orbit), LLO (Lunar Low Orbit), D&L (Lunar Descent and Landing) and for Lunar Surface real time positioning.

The assumption of this paper is that the C/N_0 of the GNSS signal received on the Moon surface is around 10 dBHz. This can be achieved by an antenna with noise figure of around 2 dB, active gain of 30 dB and passive gain (antenna radiating element) of minimum 10 dBi. This paper assumes that the antenna will be kept pointing to the Earth and that such kind of antennas are not difficult to design and build (ESA developed an antenna for GEO orbit for GNSS and similar performances were achieved, see [27]). For such reason an antenna to track GNSS signals on the Moon is assumed available at reasonable price ([23] was aiming at about 50 k€) and with the expected performances (or potentially exceeding them: in [22] an array antenna was proposed with passive gain of up to 13 dBi).

Figure 5: GNSS space antenna developed for GEO orbit [23]

Tracking of signals at such low C/N_0 is clearly very difficult, but it was proven that with coherent integration of around 500 ms (and/or with use of special techniques such as vector tracking) it is possible to produce measurements and compute PVT.
The proposed approach to provide synchronization of the base stations on the Moon surface can be achieved using a professional high sensitivity timing GNSS receiver equipped with a directional high gain antenna. The receiver is configured in timing mode, i.e., it computes only a precise time solution, by assuming a precise knowledge of the phase centre of its antenna. The need for long coherent integration and the high dynamics will impose the use of miniaturized atomic clocks to avoid degradation of the performances during the integration. COTS miniaturized atomic clocks are already available in the market and currently used in professional ground equipment, so they are considered suitable for this architecture, without big increase in cost.
Accuracies potentially achievable are in the order of 50-100 meters, so around 1 microsecond or less, that is good enough for synchronization of the 5G/LTE base stations. Finally, to reduce the costs induced by the non-COTS infrastructure, the lunar base might be equipped with only 1-2 timing receivers and the rest of the base stations can use COTS network-based precise timing equipment.

6. Conclusion

The exploration of the Moon with human and robotic missions and its colonization, through the establishment of permanent bases, such as the Moon Village project proposed by ESA, will require planetary communications and navigation infrastructures. The approach presented in this paper relies on the use of COTS component for communication and navigation on the Moon and the Mars surfaces.
The use of LTE technology, currently deployed on Earth, and of the forthcoming 5G technology will allow communication and navigation, within the maximum throughput and the accuracy provided by the standards.
The approach described largely satisfies the requirements of performance, reliability, affordability and sustainability (being based on commercial technology, incrementally expandable over time).

References

[1] P. Giordano, M. Lisi, A. Modenini, J. Peral-Rosado – "A Communications and PNT Integrated Network Infrastructure for the Moon Village", Navitec 2016, 14-16 December 2016, ESA/ESTEC, The Netherlands

[2] G. Wunder et al., "5GNOW: Challenging the LTE Design Paradigms of Orthogonality and Synchronicity," Vehicular Technology Conference (VTC Spring), 2013 IEEE 77th, Dresden, 2013, pp. 1-5.

[3] NGMN, "5G white paper", available on www.ngmn.org, Feb. 2015.

[4] D. Wingo, "Site Selection for Lunar Industrialization, Economic Development, and Settlement", New Space, Volume 4, March 2016, pp 19-39.

[5] Feasibility study for a reduced planetary navigation and communications system, PLANCOM, Final report, ESA contract, 20762/07/NL/HE.

[6] Feasibility study for a reduced planetary navigation and communications system, Final report, ESA contract, 20763/07/NL/HE.

[7] L. Foore, N. Ida, "Path loss prediction over the lunar surface utilizing a modified Longley-Rice irregular terrain model", NASA Technical Note (TM214825), 2007.

[8] S. Hwu, M. Upanavage, and C. Sham. "Lunar Surface Propagation Modeling and Effects on Communications", 26th International Communications Satellite Systems Conference (ICSSC), International Communications Satellite Systems Conferences (ICSSC).

[9] H. Pan et al, "Path loss prediction over lunar surface with obstacle diffraction", Proceedings of IEEE Workshop on Advanced Research and Technology in Industry Applications, Ottawa, Sept. 2014, pp. 1276-1279.

[10] G. James, et al, "Resource utilization and site selection for a Self-Sufficient Martian Outpost", April 1998, available at ntrs.nasa.gov/archive/nasa/casi.ntrs.nasa.gov/19980147990.pdf.

[11] C. Ho, et al, "Radio Wave Propagation Handbook for Communication on and Around Mars", March 2002, available at descanso.jpl.nasa.gov.

[12] V. W. S. Chan, "Optical space communications," in IEEE Journal of Selected Topics in Quantum Electronics, vol. 6, no. 6, pp. 959-975, Nov.-Dec. 2000.

[13] W. D. Williams, "RF and Optical Communications: A Comparison of High Data Rate Returns From Deep Space in the 2020 Timeframe", NASA Technical Note (TM214825), 2007.

[14] M. Lara, Design of long-lifetime lunar orbits: A hybrid approach, Acta Astronautica, Volume 69, Issues 3, August–September 2011, Pages 186-199

[15] C. Ho, et al, "Mars Background Noise Tempeatures Received by Spacecraft Antennas", May 2002.

[16] D. D. Morabito, "Lunar Noise-Temperature Increase Measurements at S-Band, X-Band, and Ka-Band Using a 34-Meter-Diameter Beam Waveguide Antenna", IPN Progress Report 42-166.

[17] P. Messina, B. Gardini, D. Sacotte, S. Di Pippo, "The Aurora Programme: Europe's Framework for Space Exploration", ESA bulletin 126, May 2006.

[18] J. A. del Peral-Rosado, J. A. López-Salcedo, S. Kim, G. Seco-Granados, "Feasibility Study of 5G-based Localization for Assisted Driving", Proc. International Conference on Localization and GNSS (ICL-GNSS), Jun 28 2016.

[19] A. Shahmansoori, G. E. Garcia, G. Destino, G. Seco-Granados, and H. Wymeersch, "5G position and orientation estimation through millimeter wave MIMO," in Proc. IEEE Globecom Wkshps, Dec. 2015.

[20] D. Bladsjö, M. Hogan, and S. Ruffini, "Synchronization aspects in LTE small cells," IEEE Communications Magazine, vol. 51, pp. 70–77, September 2013.

[21] Lunar GNSS Final Report, LUNARGNSS-DME-TEC-TNO06-E, Esa Contract 4000107785/12/NL/AF

[22] Weak GNSS Signal Navigation on the Moon Final report, ESA contract 4000107097/12/NL/AF/fk

[23] Moon-GNSS Final report, GMV-MOON-GNSS-D7,

[24] M. Manzano-Jurado, J. Alegre-Rubio, A. Pellacani, G. Seco-Granados, J. A. López-Salcedo, E. Guerrero, A. García-Rodríguez, "Use of weak GNSS signals in a mission to the moon," 2014 7th ESA Workshop on Satellite Navigation Technologies and European Workshop on GNSS Signals and Signal Processing (NAVITEC), Noordwijk, 2014, pp. 1-8.

[25] N. Witternigg, G. Obertaxer, M. Schönhuber, G.B. Palmerini, F. Rodriguez, L. Capponi, F. Soualle, J.-J. Floch, "Weak GNSS Signal Navigation for Lunar Exploration Missions," Proc. of the ION GNSS+ 2015, Tampa, Florida, September 2015, pp. 3928-3944.

[26] P.F. Silva, H.D. Lopes, T.R. Peres, J.S. Silva, J. Ospina, F. Cichocki, F. Dovis, L. Musumeci, D. Serant, T. Calmettes, L. Pessina, J.V. Perelló, "Weak GNSS Signal Navigation to the Moon," Proc. ION GNSS+ 2013, Nashville, TN, September 2013, pp. 3357-3367.

[27] Space User GNSS Antenna Final Report, SUGA-IIS-ORC-072-0600-FR.

Section 10 – Satellite Networks Design Challenges and Applications

CAPACITY ENHANCEMENT AND INTERFERENCE MANAGEMENT FOR INTERACTIVE SATELLITE NETWORKS

Nazli Ahmad Khan Beigi[1], Wuchen Tang[2], Mohammad Reza Soleymani[1], Hossein Ghaneharian[3], Victor Leung[2], and Alireza Shoamanesh[3]

[1]Concordia University, Montreal, Canada
[2]University of British Colombia, Vancouver, Canada
[3]Telesat Canada, Ottawa, Canada
naz_ahm@encs.concordia.ca

Keywords: Non-Orthogonal Multiple Access (NOMA), Multibeam satellite systems, Co-channel interference, Multi user detection, Spread Spectrum Code Division Multiplexing (CDM)

Abstract

We investigate the effect of aggressive frequency reuse in multibeam satellite systems using novel techniques. The techniques include our previously proposed non-spreading non-orthogonal multiple access (NOMA) scheme as well as spread spectrum code division multiplexing (CDM) NOMA. The 4-color frequency reuse is used as benchmarks, while the 2-color 70-beam and 1-color 20-beam, exhibiting significant co-channel interference (CCI), were considered as interference-limited scenarios for comparative capacity performance evaluations. The operation accuracy of the successive interference cancellation (SIC) at the core of our interference mitigation (IM) techniques was also investigated. In terms of overall achievable capacity, the results show promising achievements in the 1-color 20-beam scenario while there are no capacity gains indicated in 70-beam scenario with 2-color frequency mapping. However, a significant fairness enhancement has been shown in both scenarios.

1 Introduction

The ever-increasing market demand for broadband interactive services and, in parallel, the commercial success of high throughput satellite (HTS) services have been stimulating a great amount of interest in further advancing technology developments of multi-spot-beam and high frequency reuse satellite operations. One way to increase a multi-beam satellite system's capacity is to increase its frequency reuse factor by narrowing the beams (increasing G/T and EIRP), thus increasing the number of coverage area beams, and/or decreasing segmentation of the allocated system spectrum among the beams as well as using different polarization per beam; that is, splitting the allocated frequency spectrum into wider segregated sub-bands at different polarizations, i.e. colors, within each frequency/polarization isolating group of the adjacent beams.

With today's technology, typical Ka-band beam widths are somewhere between 0.4 and 0.5 and typical number of frequency segments or colors, i.e. adjacent beams in a group, is 4, allowing the allocated band of frequencies to be used in every fourth beam. Tighter beam/spectrum packing beyond a 4-color (or perhaps a 3-color) scheme, while increasing the potential of capacity reuse/enhancements, could however, defeat the purpose by increasing adjacent carrier interference (ACI) and co-channel interference (CCI) noise to overwhelming levels that would prevent detection of the embedded signals, unless some advanced interference management (IM) techniques were to effectively manage such interference degrading effects. The primary objective of this paper is to investigate such interference management techniques.

NOMA has recently been proposed as a solution in sharing the resources. [1] In this paper, we have gone a step further than our previously proposed NOMA in time division multiplexing (TDM) [2]. First, we have proposed code division multiplexing (CDM) NOMA, as another IM technique, for non-orthogonal multiple access of user terminals in the forward link of multibeam satellite systems. Second, instead of a two-beam model, we have extended the system model to two practical multibeam systems including 70 and 20 multibeams.

In our previously proposed cooperative NOMA-TDM, we have shown that by managing the interference, it is possible to remove its effect, and even better, use it as an extra resource to considerably increase the spectral efficiency [2]. Moreover, we have investigated the applicability of our proposed scheme, which we had called it overlay coding, in multibeam satellites; where, in a two-beam system model, the incumbent and the interfering beams have access to both data resources and can contribute for cooperative NOMA purposes. The strategy is based on beams cooperation to jointly targeting one user at a time, so that the target user would receive the summation of both signals. As proposed in NOMA, the user will recover both data through successive interference cancellation (SIC). Likewise, the user in the adjacent beam will be served in the remainder of the time. Some similar approaches in the literature regarding the spectral efficiency improvement in forward link of multibeam satellite systems can be found in [3]-[6].

The satellite's forward link is traditionally modelled as the broadcast channel (BC), where theoretically, the optimal solution for such channel is to use spatial precoding following the dirty paper coding (DPC). Modeling of this by-nature broadcast channel, as in our cooperative NOMA scheme, by a MAC is due to the duality of BC and MAC. It is proved in [7] that the capacity region of the MAC could be equal to the union of BC capacity regions. Hence, this leaves us with the same capacity bounds.

Furthermore, we propose another NOMA scheme denoted as NOMA-CDM, where non-orthogonal spreading codes, i.e. correlated codes, are used to overload more users at the same frequency band and at the same time. This is in converse to traditional CDM systems as they rely on orthogonality of the spreading codes followed by linear or non-linear multi-user detection (MUD) techniques [3]. Specifically, Iterative Minimum Mean Squared Error-Successive Interference Cancellation (iMMSE-SIC) is the technique evaluated and selected in this paper as MUD receiver to mitigate the CCI for the multi-beam satellite scenarios [8][9].

Both IM techniques proposed in this paper employ SIC at their receivers, which requires precise access to the channel estimation for its efficient performance. However, the imperfect channel estimation would lead to imperfect interference cancellation causing error propagation. Data-aided and non-data aided channel estimation have been extensively studied, both in forward and return multibeam satellite links [10][11][12]. As for our proposed techniques, we adopted the pilot-aided technique [2] presented by DVB-S2x standard, Annex E.4 [14]. We also proposed the SNIR back-off factor which assumes a margin in the achievable data throughput. The simulation results show that a huge drop in frame error rate (FER) is achieved with almost the same data throughput as the cases with perfect channel estimation. For further detailed studying as well as the simulation results, we refer the readers to [2].

The contribution of the present paper is to extend our previously proposed NOMA-TDM scheme [2] in the practical multibeam network configuration scenarios, as our previous study was merely limited to a two-beam system model. We have also investigated a NOMA-CDM system based on spreading-NOMA and have compared our results in the star multibeam satellite interactive networks. The achievable performances are then compared to that of the benchmark solutions. Despite the considerable gain in spectral efficiency that we had previously reported in [2], in this paper we show that the spectral efficiency is heavily affected by the number of multibeams and their architecture. However, no matter how severe the imposed interference is, we show that by using NOMA, the fairness will definitely be improved

This paper is organized as follows. In Section 2, the IM techniques in addition to investigation scenarios are provided. Section 3 pursues the discussed concepts by presenting the preliminary system performance evaluation results. In Section 4, simulation results are provided. Analytical results are confirmed by simulation which is concluded in Section 5.

2 Interference Management Techniques

Currently, the state-of-the-art systems are based on interference avoidance by using orthogonal non-overlapping channels in adjacent beams in multibeam satellite systems. Hence, in this paper, the four-color frequency mapping is analysed as the benchmark system. By comparison of the achievable benchmark performances with calculated estimates of theoretical bounds, existing performance gaps are identified. This reconfirms the significant potential of advanced IM techniques in capacity performance enhancement of multi-beam satellite systems, which brings the performance levels closer to the theoretical bounds.

Table 1 Summary system configuration

Configurations	Benchmark	Scenario 1	Scenario 2
Coverage & Satellite Position	Same as in Scenarios	Europe 33° E	Indo China 100° E
# of beams	"	70	20
# of Transponders (TWTAs)	"	70	40
# of Colors	4	2	1
Bandwidth per Beam	250 MHz	500 MHz	1 GHz
Polarization per Beam	Single	Single	Dual
# of Carriers/Beam, in FL	1	1	2
Saturated RF Power per Beam, (per Polarization)	130 W	130 W	130 W, (65 W per Polarization)
IBO/OBO (single carrier, FL)	0.15 dB / 0.10 dB	0.15 dB / 0.10 dB	0.15 dB / 0.10 dB
Power Spectral Density Differential, Referenced to the Benchmark	0 dB	-3 dB	-6 dB
# of Gateways	Same as in Scenarios	9	1 (+ a diversity site)
Size of Gateways Antenna	"	6 m	9 m
Size of Terminal Antenna	"	0.6 m	0.7 m
ModCod (FL & CC SCPC RL) & Roll-off	DVB-S2X. 10%	DVB-S2X, 10%	DVB-S2X, 10%

In this section, the reference scenarios as well as the benchmark system are provided, followed by the system model and the interference management methodology.

2.1 Reference Scenarios Configuration

In order to achieve a comparative capacity performance analyses in this paper, a practical multi-spot-beam satellite system design and associated end-to-end network parameters for two network configuration scenarios are considered. Table

1 summarizes the system configuration for the proposed scenarios in addition to the benchmark. The scenarios include:

1) 70 narrow spot beams covering Europe, with 2-color frequency reuse,
2) 20 narrow spot beams covering a sub-region of Indo-China, with full color frequency mapping.

2.2 System Model

The system model is the same as the one proposed in [2]; therefore, the notations are similar to what provided there. Our proposed system includes a multibeam satellite downlink system consisting of B beams. The channel is AWGN. All gateways are supposed to provide service to the users in their corresponding beams, in addition to cooperating with adjacent beams in the corresponding IM scheme. Due to the frequency reuse, each user receives the signals from all other beams with the same frequency band. Uniform distribution of users in all beams in addition to equal number of users per beam (M) is assumed. In beam b, i.e. $b = 1, \dots, B$, the set of users is provided as $S_u = \{u(b,1), u(b,2), \ \dots, \ u(b,M)\}$. The term $u(b,k)$ stands for the k-th, i.e. $k = 1, \dots, M$, user located at beam b. At the gateway, each $u(b,k)$-th user data stream is channel coded and modulated independently. Considering $S_{u(b,k),b'}$ as the coded-modulated symbol of user $u(b,k)$ received from beam b', with average power equal to one, i.e. $E[|S_{u(b,k),b'}|^2 = 1$, and the transmitted power as $P_{u(b,k),b'}$, the transmitted signal vector is:

$$\mathrm{x} = \sum_{b'=1}^{B} \sum_{k=1}^{M} \sqrt{P_{u(b,k),b'}} S_{u(b,k),b'}. \tag{1}$$

So, the received signal vector of user $u(b,k)$ is:

$$y_{u(b,k)} = H_{u(b,k)} \mathrm{x} + w_{u(b,k)}. \tag{2}$$

Where $H_{u(b,k)}$ is the $1 \times B$ -dimensional channel vector between multibeam satellite and user $u(b,k)$ indicating the interference caused by other beams with the same frequency band, and $w_{u(b,k)}$ is the complex Gaussian noise $N(0, n_{u(b,k)})$. So (2) could be written as:

$$y_{u(b,k)} = \sum_{b'=1}^{B} h_{u(b,k),b'} \sqrt{P_{u(b,k),b'}} S_{u(b,k),b'} + n_{u(b,k)}. \tag{3}$$

where $\mathrm{h}_{\mathrm{u(b,k),v}}$ is the v-th element of vector $\mathrm{H}_{\mathrm{u(b,k)}}$.

2.3 State-of-the-Art, Four Color Frequency Mapping

In four color frequency reuse to avoid the overlapping of the bands, the whole available bandwidth is divided by four. To be able to compare this scheme with our proposed full color frequency mapping systems, the power constraints are taken into consideration by dividing the noise by four. These are considered in evaluating the channel SNIR and sum rate as

$$\gamma_{4C,u(b,k)} = \frac{{h_{u(b,k),b}}^2 P_{u(b,k),b}}{\sum_{b'=1,\neq b, if\ the\ same\ freq.}^{B} {h_{u(b,k),b'}}^2 P_{u(b,k),b'} + \frac{1}{4} n_{u(b,k)}}. \tag{4}$$

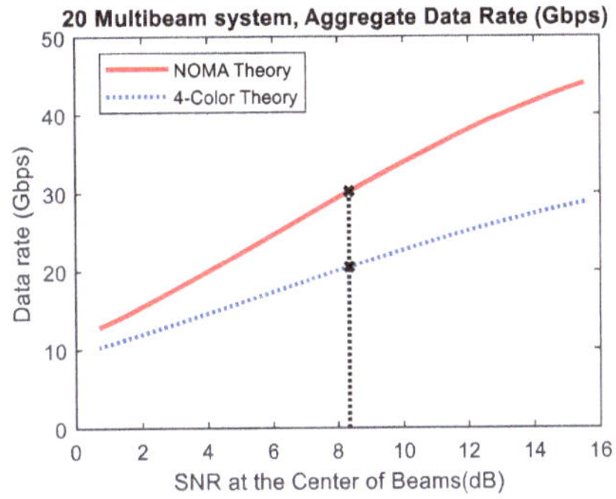

Fig. 1, Performance evaluation in Scenario 1: 20 multibeam, aggregate data rate (Gbps)

$$R_{sum-rate-4C} = \frac{W}{4} log_2\left(1 + \gamma_{4C,u(b,k)}\right). \tag{5}$$

In (4), the interference at the denominator is the sum of received power from interfering beams, only when they use the same frequency band as in the beam of interest. The W in [5] is the whole available bandwidth.

2.4 Interference Management Methodology

The cooperative NOMA scheme used for the interference management with its built-in SIC enables the resource sharing in multibeam satellite systems. As we have shown in [2], data throughput will be maximized in case of each beam's cooperation with one or two adjacent beams with the dominant interference. In other words, by employing the cooperative NOMA, not all the signals received from interfering beams degrade the system performance; whereas, the signals received from the most dominant interfering beams will be used as extra source of information. Based on the proposed scheme, $\mathrm{b_m}$is considered to be the beam number for which the user $u(b,k)$ receives the most dominant interference from, denoted by:

$$h_{u(b,k),b_m} = max\left(H_{u(b,k)}\right), b_m \neq b \tag{6}$$

Hence, based on NOMA, the SNIR is:

$$\gamma_{\mathrm{u(b,k)}} = \frac{{\mathrm{h_{u(b,k),b}}}^2 \mathrm{P_{u(b,k),b}} + \sum_{\mathrm{b_m}} {\mathrm{h_{u(b,k),b_m}}}^2 \mathrm{P_{u(b,k),b_m}}}{\sum_{\mathrm{b'=1,\neq b', \neq b_m}}^{\mathrm{B}} {\mathrm{h_{u(b,k),b'}}}^2 \mathrm{P_{u(b,k),b'}} + \frac{1}{F_R} \mathrm{n_{u(b,k)}}}. \tag{7}$$

Where, the numerator is the sum of received powers from the incumbent and the most dominant interfering beams. The denominator is the addition of channel noise and summation of all interferences, except for the cooperating most dominant interferences. The signals from incumbent, most dominant interfering, and other interfering beams could be entirely distinguished as explained in [2]. Considering F_R as the frequency reuse coefficient, i.e. $F_R = 1$ or 2 for one or two frequency mapping, respectively, sum rate will be as:

$$R_{sum-rate-NOMA} = \frac{W}{F_R} \sum_{u(b,k)} log_2\left(1 + \gamma_{u(b,k)}\right). \tag{8}$$

3 Preliminary Performance Evaluation

The information theoretic evaluation of the performance of the proposed scenarios are provided in this section. For this purpose, the SNIR database developed by European Space Agency (ESA) in a 200 multi-beam satellite system is used [16], which is based on a uniform latitude–longitude distribution grid of user terminals.

In Scenario 2 with 20 multibeams, comparing the full color frequency mapping while adopting cooperative NOMA and the state-of-the-art four color frequency reuse shows a 47% increase in data throughput, which is a massive increase in spectral efficiency. The information theoretic upper bound is shown in Fig. 1.

The performance evaluation results for Scenario 1 with 70 multibeams, comparing the two and four color frequency reuse (FR) can be found in Table 2. The results show almost the same aggregate capacity. The advantage of the 70 beam scenario two-color NOMA over four-color benchmark is the fairness of the offered spectral efficiency throughout the beams. In other words, in 4-color frequency reuse the user terminals at the border of the beam receive lower spectral efficiency than those at the center of the beam, due to the lower received power; however, employing NOMA they can receive higher data rates due to the cooperation of the adjacent beams. Besides, CDM approach can also achieve such high fairness capacity with cooperation of adjacent beams. Related results are provided and shown as Fig. 2, which illustrates achievable capacity of 70 beam scenario. It can be seen that beam cooperation provides much better fairness compared with the case without such cooperation.

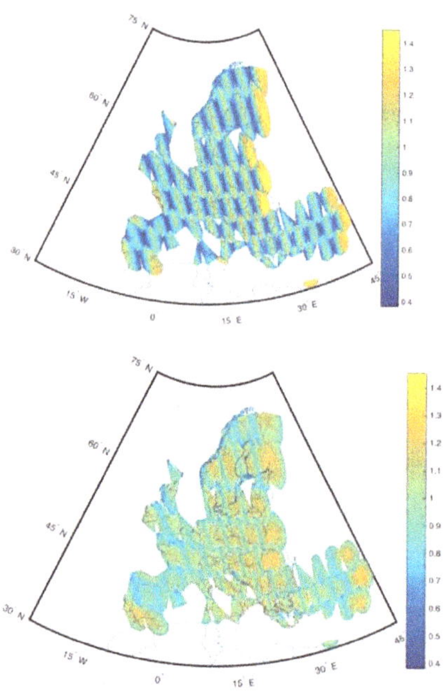

Fig. 2, Achievable capacity of 2 color (upper, no beam cooperation) and full color (bottom, with beam cooperation)

Table 2, Performance evaluation in Scenario 1

	2 FR	4 FR
Aggregate Capacity (Gbps)	91.85	93.68
Standard Deviation	0.26	0.40

In order to select the proper candidate technique, performance results of spreading NOMA and non-spreading NOMA are also cross compared using a seven beam cluster, which is carefully selected from ESA's 200 beam scenario. The comparison is on the achievable capacity of the beam in the center. It can be seen that the performance of these two approaches are very close. Specifically, the spreading approach with MMSE-SIC (18 CDM streams, single channel capacity), in terms of average beam capacity behaves slightly better than non-spreading approach i.e. 1.16 Gbps v.s. 1.12 Gbps. But non-spreading approach provides more fair capacity across the beam than the spreading approach, i.e. standard deviation 0.25 v.s 0.32. While when multiple channel capacity (spreading approach) is considered, the provided capacity fairness of spreading approach is significantly improved and behaves much better than the results obtained from the non-spreading approach but the price paid is average beam capacity degradation from 1.16Gbps to 0.96 Gbps.

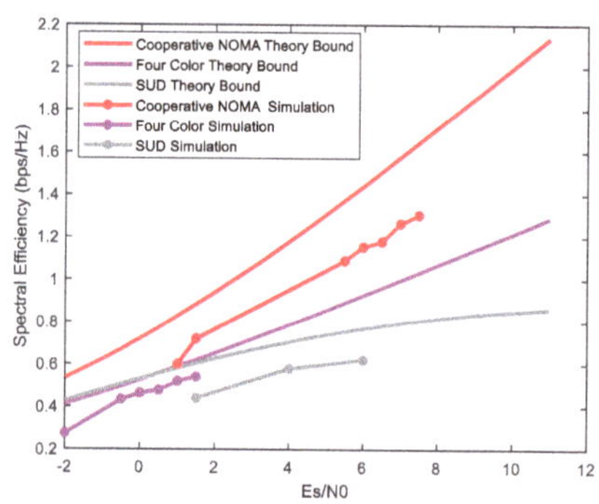

Fig. 3, Cooperative NOMA-TDM simulation results [2]

4 Simulation Results

In this section we present the simulation results of the two proposed IM techniques based on NOMA-TDM and NOMA-CDM. The NOMA-TDM results are provided from [2] for the comparison purposes.

4.1 Non-Spreading Cooperative NOMA Simulation Results

Simulation results of spectral efficiency in cooperative NOMA-TDM, as well as four-color frequency mapping and (single user detection) SUD are provided in Fig. 3 [2], where the forward link of a multibeam satellite communication system is considered. The channel is AWGN and strict synchronization is assumed. The simulation model consists of two adjacent beams, each with one user at the common border of the beams. In this model, the power received from incumbent and the interfering beams would be almost equal, i.e. $\boldsymbol{CIR = 0\ dB}$, which results in aggressive interference. In addition to cooperative NOMA, we have simulated the state-of-the-art four color frequency mapping. Moreover, single user detection (SUD) is simulated, where no MUD technique is used in the full frequency reuse. This is to show the extent that the interference is destructive at the point of evaluation and to highlight the cooperative NOMA gain over state-of-the-art systems. We have used Mont Carlo method for the simulations. The achievability is simulated so that the channel code guarantees a FER not exceeding $\mathbf{10^{-4}}$. The PL-frames with frame length of 64,800 symbols, in addition to DVB-S2x MODCOD are used. Though the bandwidth in SUD is four times the four-color mapping, yet the system performance is degraded due to the aggressive interference. However, the cooperative NOMA scheme overcomes the interference by using it as extra source of information, making benefit of the extra available bandwidth in the full frequency reuse scheme.

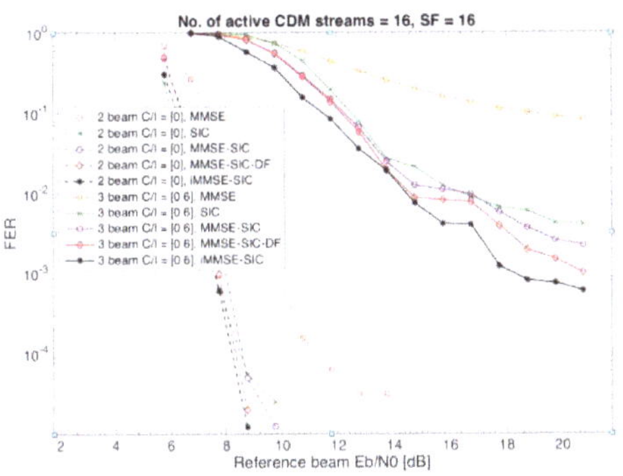

Fig. 4, Spreading CDM FER performance

4.2 Spreading CDM NOMA Simulation Results

A full physical layer simulator for spreading approach has been established, which provides frame error rate (FER) analysis for up to 3 beam cluster and can be extended to bigger cluster if necessary. As introduced, *this simulator* is developed with 3GPP FEC Turbo coding and multiple beam involved. It fully reflects all the process of the proposed spreading system function diagram in chip level, which is introduced by [15]. It has been shown that the iMMSE-SIC outperforms other MUD techniques in terms of FER analysis. Fig. 4 illustrates the frame error rate (FER) performance as a function of Eb/N0 for a 2-beam cluster, i.e. incumbent beam and one interfering beam, and 3 beam cluster, i.e. incumbent beam and two interfering beams. The interfering beam C/I of 2 beam cluster and 3 beam cluster are assumed as [0] dB and [0 6] dB, respectively. It is assumed that there are 16 active CDM signal components per beam with spread factor of 16. It can be seen that for both cases, the iMMSE-SIC outperforms other detectors including MMSE-SIC-DF. Note that the iMMSE-SIC curves are based on the FER of CDM streams from all beams (not only from the reference beam) in the corresponding beam cluster.

Regarding the small performance gap between iMMSE-SIC and MMSE-SIC-DF, the difference between these two approaches is whether the CRC is used at the receiver. MMSE-SIC-DF does not require a CRC check and it just simply regenerates the interference using all the hard decisions (no matter if they are correct or not) of detected interference signals in the first MMSE-SIC iteration. Then the receiver gets it cancelled from the initial spread signal. While for the iMMSE-SIC, CRC is adopted to check if the hard decisions are correct for the detected interference signals in the first MMSE-SIC iteration and then only the decodable interference is regenerated because interference regeneration by un-decodable does not benefits at all and will cause further performance degradation. Although MMSE-SIC-DF is slightly worse than the iMMSE-SIC, it has also been shown that its performance is still better than traditional single iteration MMSE-SIC detection.

5 Conclusion

In this paper, we investigated the applicability of non-spreading TDM and spreading CDM in cooperative NOMA as the interference management schemes in forward link of the multibeam satellite communication systems. In non-spreading cooperative NOMA, we have employed the code-domain NOMA. In spreading CDM NOMA, we showed that an iterative version of SIC combined with MMSE (iMMSE-SIC), outperformed other leading detection strategies.

We considered two scenarios with aggressive frequency reuse schemes. Performance results were compared with the state-of-the-art, i.e. four color mapping. Our results show that the optimized NOMA will considerably increase the spectral efficiency compared to the state-of-the-art four color frequency reuse; whereas, NOMA doesn't show impressive improvement compared with two color frequency reuse schemes. Yet, by using cooperative NOMA, the fairness is impressively improved in both scenarios.

Acknowledgements

The authors would like to express their gratitude to Dr. Nader Alagha, ESA-ESTEC, whose numerous comments helped to considerably improve the style and presentation of the paper. The concepts reported in this paper have been investigated in a project led by Telesat Canada and funded by the European Space Agency, ESA-ESTEC, Noordwijk, The Netherlands, under the project title 'Capacity Enhancement and Interference Management for Interactive Satellite Networks (Contract No. 4000113790/15/NL/AD)'. The view expressed herein can in no way be taken to reflect the official opinion of the ESA.

References

[1] Z. Ding, M. Peng, H. Vincent Poor, "Cooperative non-orthogonal multiple access in 5G systems," in IEEE Commun. Letters, Vol.19, No.8, Aug. 2017, pp. 1462-1465

[2] N. Khan Beigi, M.R. Soleymani, "Interference management using cooperative NOMA in multibeam satellite systems," accepted in IEEE International Conference on Communications (ICC), May 2018

[3] S. Verdu, Multi-user detection, Cambridge Univ. Press, UK, 1998

[4] M. Caus, et al., "An innovative interference mitigation approach for high throughput satellite systems," in IEEE International Workshop of Signal Processing Advances

in Wireless Comm., 2015, pp. 515-519

[5] G. Colavolpe, et al., "On the application of multiuser detection in multibeam satellite systems," in IEEE Intern. Conf. Commun., June 15

[6] G. Colavolpe, et al., "Multiuser detection in multibeam satellite systems: Theoretical analysis and practical schemes," in Proc. Inter. Workshop on Signal Proc. Advances for Wireless Commun., June 2015

[7] N. Jindal, et al., "On the Duality between General Multiple-Access/Broadcast Channels," in ISIT 2003, June 2003, pp. 313

[8] A. Lampe, "Iterative multiuser detection with integrated channel estimation for coded DS-CDMA," IEEE Transactions on Communications, vol. 50, no. 8, pp. 1217–1223, Aug. 2002.

[9] D. Liu, Y. Sun, and D. Li, "A New Iterative Multiuser Detection Algorithm for MIMO CDMA System," First International Conference on Communications and Networking in China, ChinaCom '06, pp 1-5, 2006.

[10] H. Chaouech, R. Bouallegue, "Channel estimation and detection for multibeam satellite communications", in IEEE Asia Pacific Conf. on Circuits and Systems, 2010, pp. 366-369

[11] W. Gappmir, M. Bergmann, B. S. Rechberger, "New results on location aware channel estimation for multibeam satellite links", in IEEE Commun. Letters, vol. 18, No. 8, Aug 2014, pp.1355-1358

[12] L. Xiao, L. Cottatellucci, "Contention resolution and channel estimation in satellite random access channels", in IEEE Int. Conf. on Commun. in China, 2012, pp.562-567

[13] N. Nonaka, A. Benjebbour, K. Higuchi, "System level throughput of NOMA using intra-beam superposition coding and SIC in MIMO downlink when channel estimation error exists", in IEEE Int. Conf. on Commun. Systems, 2014, pp. 202-206

[14] ETSI EN 302 307 v1.2.1, "Digital Video Broadcasting (DVB); Second generation framing structure, channel coding and modulation systems for Broadcasting, Interactive Services, News Gathering and other broadband satellite applications (DVB-S2)."

[15] R.Rinaldo and R.D. Gaudenzi, "Capacity analysis and system optimization for the reverse link of multi-beam satellite broadband systems exploiting adaptive coding and modulation," Int. J. Satell. Commun. Network, 2004.

[16] D. Riccardo, A. Nader, A. Martina, & G. Gennaro, "Exploiting code division multiplexing with decentralized multiuser detection in the satellite multibeam forward link," in International Journal of Satellite Communications and Networking, 36(3):239–276.

VLEO Satellites- A New Earth Observation Space Systems Commercial and Business Model

Sam M Dakka
Department of Mechanical, Materials and Manufacturing Engineering, The University of Nottingham, University Park, Nottingham NG7 2RD, UK,+44 11589853, sam.dakka@gmx.com

Abstract

Low earth orbits are defined as orbits with altitude below 2000km, these orbits provide superior advantages for earth observations missions. The distinction between VLEO and LEO is the altitude of the former is below 500km. VLEO missions possess great advantages such as, in maritime surveillance, illegal immigration and piracy this is due to the fact that the imagery capabilities are enhanced significantly and identification of small objects in the order of 1 meter is feasible, in addition to higher frequency of intraday location inspection visits. These are some of the great benefits for home land security and national security of any country.

In order for satellites to operate at such low altitude orbit for a long, sustained period of time, new technologies should be developed and radical satellite design change departing from current LE orbit state of the art technologies should be implemented. Since, the satellites will be operating in rarified aerodynamics environments. In this environment the satellite can operate for long sustained period of time utilizing air breathing electric propulsion to compensate for the drag, this has the advantage of not carrying propellant payload, and also can be utilized for deorbiting the satellite reducing space debris. Therefore, the operation life of the satellite is decoupled from propellant mass payload restriction, and is limited by other spacecraft subsystems. Technologies associated with new combination of new aerodynamic materials and aerodynamic controls are also required to commercialize VLEO Spacecraft platforms. These technologies are essential for developing commercial and economic models for Low earth observations space platforms. In order for the commercial and economic model be sustainable, the paper will focus on review of the emerging VLEO technologies that can be utilized to make the commercial and economic model viable. VLEO can improve payloads performance, for the same aperture size the optical resolution can be enhanced due to decrease in distance. Or the same resolution can be maintained with lower mass payloads and size. Further benefits can be gained for example for radar payloads as reduced altitude can decrease transmitted power requirements and reduce the area of the antenna. Furthermore, for radar and communication detectors the radiometric performance will improve as well due to dependency on the inverse square law. Despite the drag penalty, this can be utilized for attitude and orbit control capabilities. Due to the rarefied aerodynamic environment the lift forces are small compared to the drag forces, however certain spacecraft's design can capitalize on this fact not only to stabilize the space craft but also direct it to towards the airflow direction which is desirable from the standpoint of view of air breathing electric propulsion system. Incorporating controllable surfaces in the design can create aerodynamic torques delivering attitude and point control capability. This in turn can reduce the mass as other control attitude subsystem will not be required. VLEO also impacts mission and system design due to reduced payloads with higher resolution capability, the cost and size of the spacecraft will also be reduced and the outcome is more number of launches with lower cost per launch which will make the business model more affordable for smaller companies. In addition de-orbiting at end-of life is achieved as VLOE debris has a faster decay rate and therefore orbits are more resilience for debris accumulation. The outcome of this more affordability is to create constellation of spacecraft network and therefore better coverage.

The gas surface interaction dominates the rarified aerodynamic flow regimes with molecular mean free path exceeding the satellite typical length. Meaning atomic oxygen from the atmosphere are adsorbing or eroding the material. This process is influenced on variety of parameters such as, incident gas composition velocity and angel, surface temperatures, surface cleanliness and roughness and surface molecular composition and lattice configuration.

Atmospheric breathing electric propulsion has a great promise; atmospheric residual oxygen at low earth orbit is collected and used as a propellant for electric thruster. This has an advantage compared with other propulsion technologies such as GIT (gridded ion thruster) or hall effect thrusters (HET) which can experience lower performance over time due erosion of the accelerated grids or the discharge panels. Great efforts are currently being invested towards the investigation of inductive plasma thrusters (IPT). IPT's has several advantages, the flexibility of the propellant which is dependent on the atmospheric altitude, no erosion since it is electrodeless and griddless and neutraliser is not required.

The main challenges associated with VLEO spacecraft's with emphasis on *Atmospheric Breathing Electric propulsion, orbital aerodynamics and satellite drag,* and material and processes to reduce induced drag will be discussed and reviewed for VLEO satellites to establish a new disruptive commercial and business model dedicated for low earth observation space systems. This model will enhance the number of payloads launches and reduce launch costs, therefore opening the way for small companies to enter this segment with new innovative payloads and spacecraft design to promote security and the wellbeing.

1. Introduction

Lowering spacecraft operation altitude of earth observation platforms has its benefits and challenges, from one hand the lower altitude operation of optical payloads contribute to better radiometric payload performance. The increase payload resolution will expand the payload capabilities. Therefore, either these capabilities features will be improved while maintaining size, cost, power and mass or reducing the size of the payload whilst maintaining the same capability level, reducing mass, power and cost [1]. Also the reduction of radar payload has a major ramification on reduction of the power requirement of the antenna due to decrease in the antenna surface area. This can provide good payoffs for lowering the cost and mass. Furthermore, there are great benefits in terms of surveillance footprints and increased geospatial accuracy. However, spacecraft platforms operating at VLEO – Very Low Earth Orbit introduce challenges [1] due to residual atmosphere, meaning increased drag and atomic oxygen erosion. Furthermore, the window of communications with a ground station is reduced, but this can be solved by operating a fleet of low cost spacecraft platforms at very low earth altitudes.

It is worth noting that the exact altitude of Very Low Earth Orbit is not well defined but the term is applied for mean orbit altitudes below 500 km. This is due to the fact that the spacecraft decay is less than 5 years. Due to life time of lower than 5 years, then significant changes in aircraft design are required. VLEO is a small altitude region within LEO which is a defined altitude between 200km to 2000km. At VLEO, a spacecraft experiences higher drag due to higher density with reduced orbit altitudes. The flight dynamics at lower altitudes is different than at higher altitudes. And it can be shown [1,2] that an approximation of the different acceleration disturbing effects other than earth gravity can be neglected (Lunar gravity, solar gravity) as compared with atmospheric drag at altitudes below 450km.

2. Earth Observation Applications- Benefits and Challenges

2.1 Benefits of VLEO

The benefits of very low earth orbits are many and this is some highlights of these benefits [3]:

Increased resolutions of optical loads:

The optical resolution can be defined as the ability to resolve remote objects at a short distance and reproduction implementation of the individual elements of the image. The performance of construction of an image is based on the Rayleigh criterion, which states that an observer can identify two separate points if the minimum separation of two points of lights is the radius of the Airy Disc defined as

$$r = 1.22\frac{\lambda}{D} \qquad (1)$$

Where λ is the light wavelength, D is the aperture diameter of the telescope and r is the radius of the Airy disc. Based on equation 1 and the Rayleigh criterion, which can be interpreted by the maximum theoretical resolution of an optical system is limited by the diffraction of the light entering the system. Equation 1 can be written in term of angular resolution [1].

$$Sin\alpha = 1.22\frac{\lambda}{D} \qquad (2)$$

Where α is the angular resolution. Equation 1 can be written for visible spectrum $\lambda = 550nm$ in the following form [1]

$$r = 6.71 \cdot 10^{-9}\frac{h}{D} \qquad (3)$$

As stated earlier r is the ground resolution, and h is the space altitude, examining equation 3 indicates that in order to increase the ground optical resolution, reduction of the spacecraft altitude, increase in the aperture size and increase of the optics quality to reach the optimal theoretical limit is required. Increasing the optics quality incurs a cost penalty. Increasing the aperture size incurs cost and mass penalty. Therefore, the only remaining option is to reduce the operational altitude. This will enhance the ground resolution with no modification of the traditional orbit optical payload.

Increased radiometric performance:

Flying at VLEO operational altitudes has the advantage of increased radiometric performance. In optics, radiometric techniques characterize the distribution of radiation power in space. The power density of a signal is proportional to the inverse square distance from the source this can be described by the following relation [3]

$$p \propto \frac{1}{H^2} \qquad (4)$$

This implies that an operational altitude at VLEO will enhance the power density magnitude. This is true for optics and for radar signals. The main advantage relates to the signal to noise ratio [3, 4]. The power consumption reduction has special advantages for transmission subsystems. By lowering the altitude with same radiometric payload, less sensitive equipment is needed in order to get the same results as in higher altitudes, meaning significant reduction in mass and cost of the payloads.

Increase in launch mass:

The reduced mass of the payload at VLEO (maintaining the same capability as in traditional orbit) will increase the equivalent mass to be launched at lower orbits.

Non- disposal manoeuver is required EOL:

Based on ESA Clean space guidelines [5]

"It is recommended that satellites and orbital stages be commanded to re-enter Earth's atmosphere within 25 years of mission completion, if their deployment orbit altitude is below 2000 km (in the LEO region)."

For traditional spacecraft orbits this implies that some sort of propulsion system is required in order to de-orbit the satellite, which increase mass, volume and complexity of the systems and increase the demand for power management. If the operational altitude is in VLEO then there is no need to incorporate or utilize a propulsion system or drag sail dedicated to de-orbit the spacecraft. The drag in that altitude is large and therefore causes the spacecraft to de-orbit before the end of mission life. For example in an orbit of 400km the end of mission life is not larger than 3 years [4]

Low risk of collision with space debris:

The space exploration time span has been extended for more than 50 years; the number of satellites launched into orbits is exceeding 6000. However, less than 1000 are still operational [6]. It is estimated that third of the operational satellites are in orbits altitudes of higher than 500km. Therefore, unless these satellites are de-orbited, space debris is a major issue. Operational altitude at VLEO will reduce the probability of space debris collision significantly, due to higher rate of orbital decay. Therefore, this will provide a mechanism of debris self-clearance at a faster rate as compared to high traditional orbit.

Low level of Radiation:

Flying at lower earth orbits means protection from solar radiation due to the inner Van Allen belt and the earth magnetic field. This will also protects the spacecraft from greater proportion of solar activity effects. These effects encompass solar flares and coronal mass ejection. It is worth noting though that cosmic ray originates from the outer solar system and consists of high energetic particles; therefore, the earth magnetic field can provide only partial protection from cosmic rays at lower altitudes.

2.2 Challenges of VLEO

Earth's Atmosphere:

Figure 1 illustrates the mass density variation as a function of altitude for different solar activities. The figure illustrates the earth mass density is decreasing exponentially with the increase of altitude. Furthermore, the density can vary by up to two orders of magnitudes at a given altitude based on solar and geomagnetic activity. The drag increases linearly with decreasing density, therefore orbit lifetime will increase exponentially with altitude. The orbital dynamics is dominated by the drag force for altitudes below 450km as compared to other disturbances. In order to avoid orbital decays in less than 5 years, spacecraft design changes are required since the spacecraft is located in the Hetrosphere, which mainly consist of ionized species. The major ramification of this ionized environment and the effects on the spacecraft are the drag which is the main aerodynamic force that causes the orbital decay. Another major effects that can cause sensor and instruments surface performance degradation is surface corrosion, due to atomic oxygen reacting with the material. Another negative outcome is the spacecraft propensity to be electrically charged, due electrically charged particles material adsorption. This in turn can damage sensitive material.

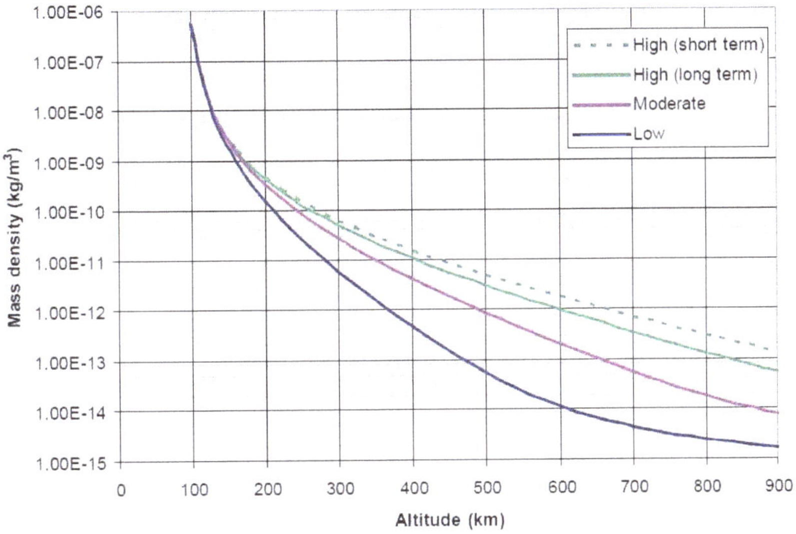

Figure 1 JB2008 Mean air density versus altitude based on solar and geomagnetic activity (international standard ISO/DIS14222).

Communication windows:

Due to elevation angle constraints and increased orbital velocity, the available time for data downlink to the ground station becomes much shorter; this can be resolved by increasing the bandwidth communication systems. Or another solution is to extend the communication window via transmission to other relay platforms orbiting in GEO.

Revisit time:

Application that demands very frequent visits times might pose a more challenge for VLEO as compared with traditional high altitude orbit.

3. Aerodynamic Torques

The transfer of momentum to the spacecraft by particles occurs when the spacecraft impacts the gas molecules at the upper atmosphere, this process generates aerodynamic torque. A number of approximate models were developed in order to estimate the momentum transfer, though the mechanism of the momentum transfer is not yet well understood. The aerodynamic torque can be modelled by the accommodation theory. One aspect of the modelling challenge is utilizing density model that can accurately describe the density at various low altitude of interests. In general, when calculating the aerodynamic torques two extreme scenarios are considered, the specular and the diffusive regimes [7]. The first is a deterministic concept described by each molecule bounces back from the surface with the same energy and the angle of reflection is conserved as compared with the angle of incidence. Furthermore, the surface normal and the incoming and bouncing velocity are coplanar. Based on these assumptions, the momentum transferred is normal to the surface and equal twice the normal component of the incident momentum. The second regime, the diffusive regime, the molecule fully accommodated to the surface and no conservation of energy and direction is happening before and after the impact. It interacts with the other molecules in the surface contamination layer and leaves the surface based on probabilistic kinetic energy associated with the surface temperature and probabilistic direction governed by the "cosine" angle [7]. The two scenarios are illustrated in figure 2. These are the two extremes in reality molecule reflection from the surface can be described in between these two extremes, in what is called partial surface accommodation of the incoming particles. We can utilize the equation developed by Peter C. Hughes "Space Craft Attitude Dynamics" to calculate the disturbance and control torques developed by the spacecraft. The approach adopted is to develop the equations based on pure specular molecules reflection and based on the pure diffusive reflection and then since real scenario is somewhat in between these two extremes, an implementation of partial accommodation theory is implemented where normal and tangential accommodation coefficients are introduced, therefore calculating the partial accommodation elemental force can be developed [7,8]. By integrating the equation over the surface of the spacecraft, it will be possible to determine the total force imparted upon the spacecraft by the atmosphere. The torque can be determined by crossing the distance between the centre of mass to the centre of the surface element and the force vector working on the surface element and integrating over the entire surface. This approach can be implemented for aerodynamic attitude control utilising aerodynamic drag torques and for de-orbiting satellites and Cubsats at VLEO through a unique spacecraft design configuration, one example of this design configuration is the shuttlecock design.

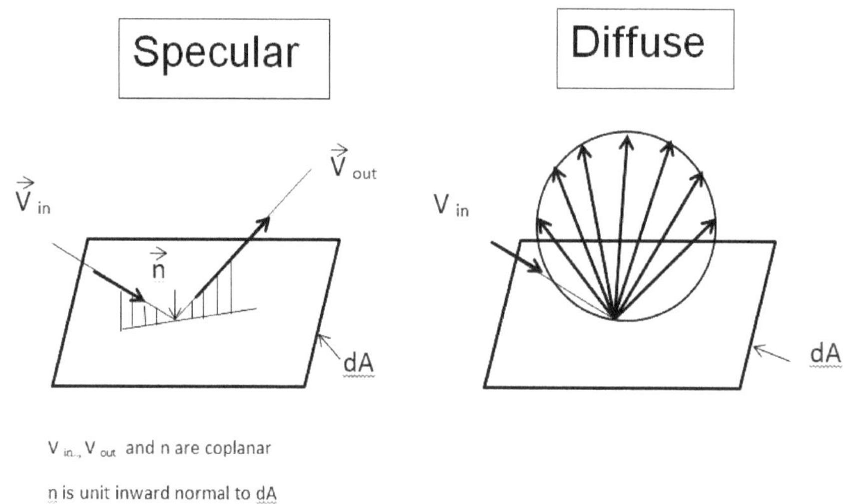

Figure 2 specular and diffusive molecular reflection

4. Electric Propulsion

Very low earth orbit open up new business model for commercial and scientific earth observation applications with great benefits. Operating at very low earth orbit, in the range of 100-300km catapulted the research activities into air breathing electric propulsion (ABEP) or ram-electric propulsion (RAM-EP) into new heights. Electric propulsion can be used at varied satellite orbit altitudes which reduced the total mass of the spacecraft since not all propellants need to be carried into orbit. However applications of these propulsion systems are limited at very low altitude based on current technologies. Utilizing the earth residual atmosphere as propellant was investigated in the past and the idea can be dated back to the First World War, however the main question is whether the current state of art of technology can be applied to fulfill this purpose. The limitation is that the thruster needs to operate at low mass flow rates and low power, furthermore reasonable lifetime of the propulsion system as compared with the satellite system should be factored in the design process. Several concepts of air breathing electric propulsion systems will be discussed. The concepts will depends on atmospheric density modeling at various altitude which might impact the design concepts, these should be collected based on previous satellites orbits operated in these altitudes. Pulsed plasma thruster will be discussed, and inductive plasma thruster will be suggested as a possible solution to the electrode erosions since inductive plasma thruster is electrodeless and therefore the problem of erosion is eliminated.

5. Atmospheric Environmental Conditions

In order to estimate the performance of an air breathing propulsion system, the composition and the actual mass flow of the residual atmosphere is required. At VLEO altitudes, the mixture composition of N2, O2 and atomic oxygen change rapidly with altitude this is due to significant changes in temperature, density and conductivity at various altitudes. Studies on earth atmospheric modeling started with the dawn of spacecraft development and are still ongoing. The modeling is based on empirical data gathering from satellites and atmospheric probes with respect to temperature, total density and partial density etc. [9]. The international standard atmosphere (1976) was the first standard to describe the atmosphere as a function of altitude. Later, NASA Goddard developed the MSIS model which stands for Mass Spectrometer and Incoherent Scatter. This model has the advantage to include the residual of the atmosphere in its output [10,11], therefore is of great relevance for estimating the performance of air breathing propulsion. The model output includes the temperature and densities of N2, O2, and O, He, Ar at altitude between 120 and 150 km. As more data became available at higher altitudes, the model was extended up to altitude of 220km and the extended model was named MSIS-83 [12]. The MSIS -83 was further revised to include the atomic nitrogen in the computable species. (MSIS-86)[13]. Further revision at lower altitudes below 120km and lower mesosphere was accomplished (MSIS-90)[14]. The naval research laboratory further validated the model to provide better accuracy and addition of anomalous oxygen was incorporated, the NRLMSIS-00 (2000) [15,16] is considered the most advanced model and is still in use for estimation of orbital conditions. For the sake of completeness, it is worth noting that other models were developed as well, the JB2006 was developed based on an early model by Jacchia and it has better representation of the thermospheric density variation [9]. Unfortunately this model does not provide any information with regards to the residual atmosphere gas mixtures and therefore its application for air breathing propulsion and thrust performance is limited. Further improvement with regards to accuracy and temperature was accomplished resulting in a revised JB2008 model [17]. In Europe, the two refined models are dubbed the ECSS standard and the ECSS-E-ST-10-2004C (issued 2008)[18].

Therefore, there are two models with different values and outputs; The JSB2006 has higher accuracy for the total atmospheric density at various altitudes, while the NRL MSIS-00 model has better representation with regards to composition (partial densities) of atmospheric conditions. In summary, the JSB2006 should be used to estimate the drag and the NRL model should be used to estimate the thrust performance. (Composition estimation of N2, N, O2, O, anomalous O, He, Ar and H).

The residual atmosphere at this altitude not only provides the propellant but also the cause of the drag force. The drag force depends on the shape of the satellite, the orbit altitude. The drag coefficient was estimated to be between 2.1 to 2.7 and a mean value of 2.2 is usually used. One can construct plots based on the JSB2006 data model to estimate to mean total density at different altitudes. And then estimate the drag force for satellites with various surface areas. Obviously for drag force compensation the thrust force should be at least equal to the drag force to justify an air breathing propulsion system. Utilizing the NRLMSIS-00 model provides the number densities of the individual species for altitudes of interests below 250km, at these altitudes the species of interest are N2, O2, and O. Figure 3 illustrates the species number densities as a function of representative VLEO altitudes. The most abundant species is atomic oxygen. Figure 4 is a sketch of the historical atmospheric model development outlined above.

6. Electric Propulsion Concepts

Electro-thermal propulsion

In this category resistojet and arcjets are the dominant players. At this low altitude the applicability of such electric propulsion system is questionable since arcjets does depend on oxygen free propellants, and erosion of the nozzle throat due to high temperature oxygen flow is likely to limit the lifetime. The same can be said about resistojets systems; therefore it is unlikely that those electric propulsion systems can be implemented at VLEO.

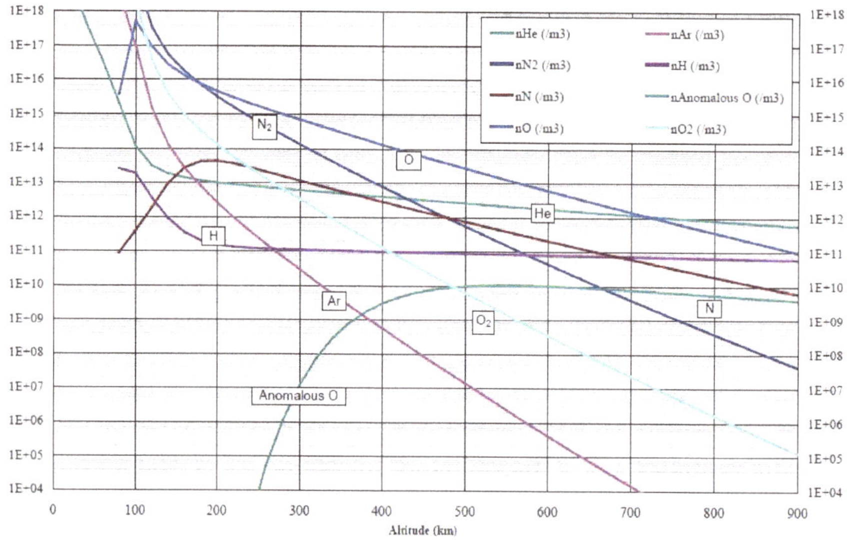

Figure 3 Number density of atmospheric constituents versus altitude at mean solar and geomagnetic activity. Data are from the NRLMSISE-00 model, given by Draft International Standard ISO/DIS 14222

Electrostatic Propulsion:

Electrostatic propulsion includes ion thrusters and Hall thrusters; this category was investigated extensively for LEO. ABIE-air breathing ion engines were the first to be investigated. The outcome of the current research indicates that at altitudes between 150km to 200km, an input inlet pressure design of 1mPa is required for the propulsion system to be able to operate efficiently and in order to fully compensate for drag of 50-100mN for spacecraft estimated surface area of $0.9m^2$ and 5-10 kW of inboard power supply for full drag compensation. On the other hand Hall thruster might have greater interest than the ion thrusters due to the capability to develop larger thrust density. As compered to ion engines, it is possible to generate the same thrust level at lower surface areas, which implies that fully compensating for smaller drag is more likely in this concept. The Busek Company first introduced those thrusters operating on CO_2 propellants tailored for Martian environment. The engine performance indicated 30mN/kW at 90-130 sccm and the erosion was negligible. Also computational studies were performed for orbits as low as 80-95Km, the derivation of drag indicated an on board power supply of several 100kW is required for full drag compensation [9,19]. A study at altitudes of 250km indicated [9, 20] that the mass flow rate should be at least 2.5mg/sec which would be difficult to achieve at inlet pressures of 1mPa. In both cases, Hall thruster application was deemed unsatisfactory

Electromagnetic propulsion

Compared with electrostatic propulsion, the working propellant is plasma. In electromagnetic propulsion systems the plasma is created via heating the propellant up to 5000K, a Lorentz force is created working on ions, due to a magnetic field acting perpendicular to the current (the plasma is playing the role of electrical conductor). A common propellant for this system is Xenon. The PPT (Pulsed plasma thrusters) and the MPD (The Magnetoplasmadynamics) engines are examples of the electromagnetic propulsion. In the former, a pulsed electrical discharge is utilized to ablate a solid propellant source into a plasma arc [21]. The acceleration of the ions are performed by the electromagnetic effects in the pulse. The thrust level is determined by the pulse rate [21]. The MPD work principle is ionizing large portion of the propellant by utilization of high power current arc. The electromagnetic forces (Lorentz forces) accelerate the ions within the propellants in the plasma discharge which generates both the currents and the magnetic fields. It is worth noting that MPDs require large power to operate and generate a high specific impulse and high thrust compared to other EP systems.

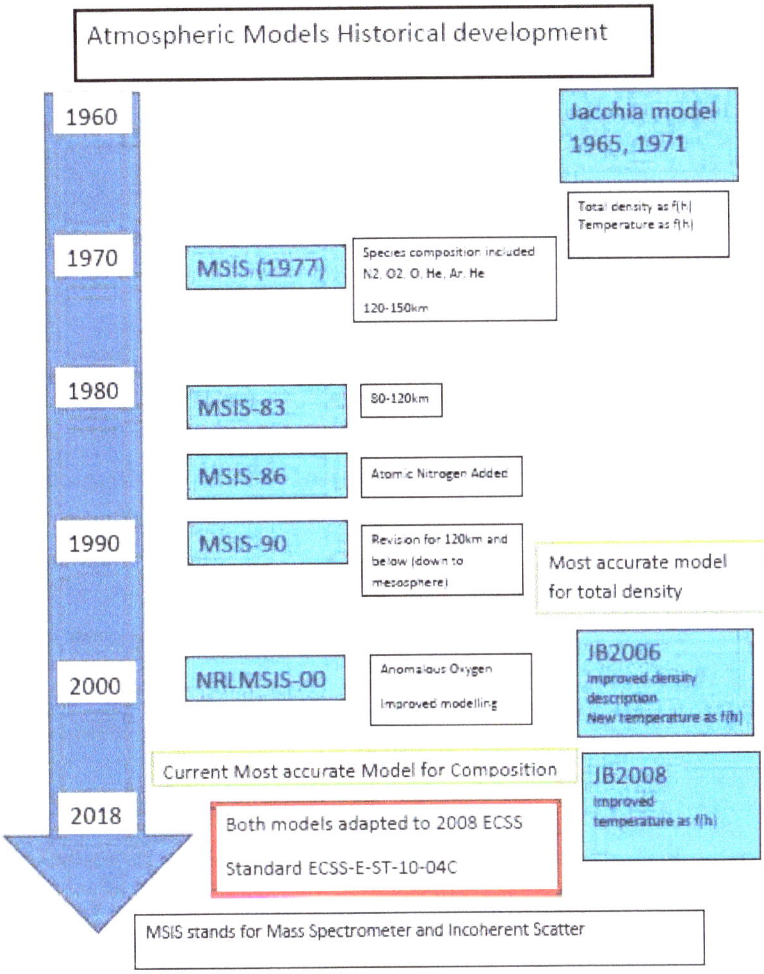

Figure 4: Atmospheric Models Historical Development.

7. New Applications

Electrodeless Lorentz force thrusters (ELF) are important and being considered for air breathing electrical propulsion concepts (RAM-EP). In these concepts a high density magnetic plasmoid is created by a applying a rotating magnetic field. The plasmoid acceleration is due to an axial magnetic field gradient generation. To enhance the efficiency, neutral particles are injected downstream of the already ionized neutrals, this generates newly ionized particles contributing to the thruster efficiency enhancement [22].

In both electrostatic and electromagnetic propulsion, electrodes can shorten the life time of the thrusters. Therefore, a way to overcome these constraints is to develop electrodeless thrusters. Electrodeless concepts are worth mentioning such as the inductive heated flow plasma for air breathing electric propulsion. Two miniaturized engines has been developed in cooperation between Baylor University in TX and the institute of space systems in Stuttgart. The two engines termed IPG6-B and the IPG6-S, designation for Baylor and Stuttgart utilize air, oxygen and carbon dioxide as the working fluid [23].

Another thruster worth mentioning is the VASMIR engine developed by NASA. The engine was developed mainly to have a quick journey to Mars, but it can be applied as an orbital towing vehicle. VASMIR stands for variable specific impulse magneto-plasma rocket. Radio waves are used to ionize the propellants, and a powerful magnetic field is implemented by superconducting magnets to accelerate the plasma. The engine has longer life due to being electrodeless in comparison with ion and HET which has shorter life due to electrode erosion. Furthermore, the engine can operate on a variety of gases such as, Helium, Hydrogen and Argon. The downside of the VASMIR engine is very strong magnetic field (1-2 Tesla) and very high power demands.

8. CubeSat Applications

A study conducted [24] by the author selected a PPTCUP- Pulsed Plasma Thruster for a 3U CubSat Propulsion, Micro pulsed plasma thruster with satisfactory capability to provide enough impulse to overcome the generated force due to drag to maintain an altitude of 600 km and bring the CubeSat down to a graveyard orbit of 513 km. While this altitude is slightly higher than the range of VLEO, it indicates promising possibilities of further development of air breathing plasma thruster to be incorporated in the future designs of CubeSats. This will involve further studies of gas surface interactions, air inlet optimization, and aerodynamic materials for drag reduction.

9. Conclusions

The paper provided a review and discussion on the benefits and challenges associated with operating space platforms in VLEO. Furthermore, operation at these altitudes can be accomplished by the development of air breathing electric propulsion to fully compensate for the drag. Electrodeless and griddles air breathing electric propulsion are currently under development with promising engine performance and relatively low power consumption. Operating at VLEO has great returns as the size and the cost of the spacecraft will be significantly reduced. Therefore more small businesses can engage in earth observation space platform development. This will have great impact on future innovation due to more competition in the spacecraft market. One major challenge is the reduced visit time, which will require higher data transfer rate to the ground stations through the development of higher bandwidth communication systems. Or another solution is to extend the communication window via transmission to other relay platforms orbiting in GEO. Future designs will favor a fleet of space crafts or spacecraft constellation formation.

References

1 Joseph Virgili Llopp et al, “Very Low Earth Orbit mission concepts for Earth Observation. Benefits and challenges” BIS-RS-2014-37, 37 (2014), page 1

2 Fortescue, P. W. and Stark, J. P. W. 1995. Spacecraft Systems Engineering, 4th Edition. Wiley and Sons, Inc.

3 Jonathan Becedas et al "Discoverer-D2.1 VLEO Aerodynamics Requirements Document" 2018 , 1 - 83

4 Joseph Virgili Llopp “Spacecraft Flight in the Atmosphere,” Cranfield University, Cranfield, 2014.

5 E. S. Agency, “Mitigating Space Debris Generation,” ESA, 14 April 2017. [Online]. Available:http://www.esa.int/Our_Activities/Operations/Space_Debris/Mitigating_space_debris_generation.

6 E. S. Agency, “ESA,” 2016. [Online]. Available: http://www.esa.int/Our_Activities/Space_Engineering_Technology/Clean_Space/CleanSat.

7 Hughes, Peter C. Spacecraft Attitude Dynamics. Mineola NY: Dover Publications, 2004

8 Michael Luke Gargasz, OPTIMAL SPACECRAFT ATTITUDE CONTROL USING AERODYNAMIC TORQUES, Master’s Thesis, Air Force Institute of

Technology, Wright-Patterson Air Force Base, Ohio, March 2007.

9 Shonerr T. et al, “Analysis of Atmosphere-Breathing Electric Propulsion”, International Electric propulsion conference, Washington DC, 2013, page 1 -11

10 Hedin, A. E. et al. “A Global Thermospheric Model Based on Mass Spectrometer and Incoherent Scatter Data MSIS, 1. N2 Density and Temperature," Journal of Geophysical Research: Space Physics, Vol. 82, No. 16, June 1977,page 2139-2147

11 Hedin, A. E. et al. “A Global Thermospheric Model Based on Mass Spectrometer and Incoherent Scatter Data MSIS, 2. Composition," Journal of Geophysical Research: Space Physics, Vol. 82, No. 16, June 1977, pp. 2148-2156.

12 Hedin, A. E. et al. A Revised Thermospheric Model Based on Mass Spectrometer and Incoherent Scatter Data: MSIS-83," Journal of Geophysical Research: Space Physics, Vol. 88, No. A12, Dec. 1983, pp. 10170-10188.

13 Hedin, A. E. “MSIS-86 Thermospheric Model," Journal of Geophysical Research: Space Physics, Vol. 92, No. A5, May 1987, pp. 4649-4662.

14 Hedin, A. E. “Extension of the MSIS Thermosphere Model into the Middle and Lower Atmosphere," Journal of Geo-physical Research: Space Physics, Vol. 96, No. A2, Feb. 1991, pp. 1159-1172

15 Picone, J. M. et al “Enhanced Empirical Models of the Thermosphere," Physics and Chemistry of the Earth, Part C: Solar, Terrestrial & Planetary Science,Vol. 25, No. 5-6, 2000, pp. 537-542

16 Picone, J. M. et al “NRLMSISE-00 empirical model of the atmosphere: Statistical comparisons and scientific issues," Journal of Geophysical Research: Space Physics, Vol. 107, No. A12, Dec. 2002, pp. SIA15-1-16.

17 Bowman, B. R., et “A New Empirical Thermospheric Density Model JB2008 Using New Solar and Geomagnetic Indices," AIAA/AAS Astrodynamics Specialist Conference, Aug. 2008, AIAA-2008-6438.

18 “Space engineering - Space environment," Standard ECSS-E-ST-10-04C, ESA-ESTEC, Noordwijk, the Netherlands, Nov.2008.

19 Pekker, L. and Keidar, M., “Analysis of Airbreathing Hall-effect Thrusters," Journal of Propulsion and Power, Vol. 28,No. 6, Nov.-Dec. 2012, page 1399-1405.

20 Garrigues, L., “Computational Study of Hall-Effect Thruster with Ambient Atmospheric Gas as Propellant," Journal of Propulsion and Power, Vol. 28, No. 2, March-April 2012, page 345-354

21 Goebel, D. M., and I. Katz (2008), Fundamentals of Electric Propulsion. Ion and Hall Thrusters, JPL Space Science and Technology Series, John Wiley & Sons, Inc.

22 Slough, J., and D. Kirtley (2015), Pulsed plasmoid propulsion: The elf thruster, in The 31st International Electric Propulsion Conference, page 345-354

23 Dropmann, M. et al “A New Inductively Driven Plasma Generator (IPG6)|Setup and Initial Experiments," IEEE Transactions on Plasma Science, Vol. 41, No. 4, April 2013, page 804-810.

24 Calricoats, J. and Dakka, S. M., “Design of Power, Propulsion, and Thermal Sub-Systems for a 3U CubeSat Measuring Earth's Radiation Imbalance” Aerospace 5(2) 63 2018

Towards the Internet for Space: bringing cloud computing to space systems

Simone Briatore[1], Nicola Garzaniti[1], Alessandro Golkar[1]*

[1]*Strategic Innovation Research Group, Skolkovo Institute of Science and Technology, Moscow, Russian Federation*
**simone.briatore@skolkovotech.ru*

Keywords: space cloud, satellite communications, optical intersatellite links

Abstract

Data is the most valuable asset in space. It is produced by communicating and observing payloads, then downlinked to ground stations, and eventually delivered to the users, who generate information and insights from it. This paper explores an alternative avenue for the exploitation of space data, with the concept of Space Cloud.

The space cloud is an orbiting cloud computing infrastructure providing in-orbit data storage and processing services. The idea is to enable on-orbit decision making including reactive planning and scheduling decisions, as well as complex resource management in space networks.

The space cloud is the missing element to enable efficient and reliable intersatellite connectivity on top of which a whole new set of services and opportunities can be built in a similar fashion to what happened for the Internet during the 1990s. Building on previous work by the authors, this paper illustrates new efforts for the development of the space cloud concept through the demonstration of optical communication technology as a backbone for cloud computing in space. We illustrate how optical communications technology is an enabler of future space clouds.

1 Introduction

Data is the reason why spacecraft are launched into space. It is the factual basis for the development of science for a better understanding of our Universe and of the Earth, as well as for the development of telecommunication services and applications.

The amount of data generated in orbit has grown exponentially over the last two decades, in some ways following Moore's law. From hundreds of megabytes of data generated by each spacecraft a day, we have moved to terabytes of data generated per day (as for the Sentinel mission [1], on in the case of Synthetic Aperture Radar instruments), and likely petabytes of data per day in the future.

Such large amounts of data are downlinked to the ground by operators, and eventually delivered to the users following the required quality control processes. Users generate insights and information from the received data. The operators use data as the basis for the decisions on future in-orbit operations and instrument tasking. Such is the *status quo* – in general and rather approximate terms, certainly not comprehensive – of space mission operations today. Our ability to exploit missions in space is therefore linked to the limitations in readiness and availability of space data. Data downlink is constrained, among other things by transmission (TX) and reception (RX) antenna size, transmit power, orbital dynamics, number of ground stations available, and ultimately recurring costs associated with operations.

Actionable operations are constrained by the time it takes from measurement (or signal) acquisition, to relay to the ground, following processing, validation, and interpretation by operators and users. The question thereby follows whether some (or all) of these functions could be transferred for automated processing in space. In order to do so, in-orbit capacity of resources such as data processing and storage is required, as enabled by in-space infrastructure.

This paper proposes a concept of space cloud, that is the exploitation of cloud computing services for on-orbit processing. The space cloud is the missing building block for the development of services and enhanced exploitation of space data, as well as the enabler for on-orbit decision making and fully automated spacecraft operations. Notions related to the space cloud have been discussed for many years in the literature, such as in the research on space inter-networking, ad-hoc networking and so on. These concepts, however, largely remained unrealized and presented in rather scattered form in the literature, with no discussion on an integrated systems-of-systems perspective.

This paper explores a concept of operations of the space cloud and provides a description of a use case where such concept advances the state of the art in space operations.

The remainder of this paper is structured as follows. Section 2 provides background and context on cloud computing, focusing the discussion on the capabilities relevant to the discussion, and on optical communications as enablers of agile inter-networking in space. Section 3 introduces the idea of the space cloud with a notional use case of on-orbit decision making and automated instrument tasking. Section 4 introduces our proposal for an optical inter satellite link terminal onboard the International Space Station as a space cloud demonstrator, with some details on its architecture and the main components.. Section 5 draws the conclusions of the paper, suggesting opportunities of future development of the space cloud concept.

2 Background

With the faster and faster development of the space business and the increasing number of satellites launched every year [2], new paradigms and opportunities are rising and becoming more and more appealing to the industry.

2.1 Cloud computing

There have been numerous definitions of "cloud computing" in the literature [3], mainly referring to the information technology sector. A popular definition is offered in the seminal paper by Armbrust et al., where cloud computing is analysed as computing as a utility [4]. While a comprehensive overview of the concept is out of scope in this paper, we need to define the perimeter of interest to this paper in order to define a concept of operations for space clouds. Furthermore, we wish to clarify what we mean in this context for cloud computing.

In the area of space operations, we have observed that the space industry seeks to reduce operations costs and enable new capabilities by introducing more on-board processing in spacecraft. Such capability is needed for two reasons: 1) to reduce the time needed from data acquisition to data availability for the users (through intersatellite networking) and 2) to speed up the generation of information and insights derived from the exploitation of space data (through on-orbit decision making).

From these two needs we derive the main characteristics desired by the introduction of cloud computing in orbit:

- High speed intersatellite communication as enabler of satellite inter-networking;
- Distributed computing capabilities:
    - Data processing (*data processing as a service*)
    - Data Storage (*data storage as a service*)

For the remainder of this paper, we consider cloud computer from the perspective of data processing and data storage as a service.

2.2 High speed intersatellite communication

Cloud computing in orbit cannot be realized without a stable and sufficiently performant in-space telecommunications network. In a space network, data has to quickly be able to move from a node to the other in real time (or quasi real time) without major corruptions or service interruptions. This can be considered the basis upon which cloud computing services can be developed, as it has been seen at the end of the last century with the development of the internet. Before it was possible to establish a fast and reliable connection between any of the peers in the network, the services that could be implemented were limited, thus reducing the overall utility of the entire system.

Only at the beginning of the 21st century, with the wide adoption of broadband connections and the mass popularity of the internet – first with personal computers and later on amplified by the popularity of smartphones – that the global network changed shape and transformed itself in the core essential baseline platform for most of today services, from social networks to online shopping and public administration.

As discussed in previous work [5], we believe the future of intersatellite links, and therefore of the space cloud, to be characterized by optical communication architectures rather than radio frequency based one. A part from obvious technical challenges in reaching thousands of kilometres with radio links, that even if problematic for smaller spacecraft can be addressed by a careful design, the main obstacle will for sure remain bureaucracy. In order to operate a non stop intersatellite link, a dedicated radio channel has to be always active in a thousand kilometres radial area, thus requiring legal licenses to operate on top of all countries the spacecraft is flying to. The process of acquiring licenses could become a burden that smaller companies (the same that would get more advantages in joining the cloud) may not be able to afford in terms of time and money. At the same time, the gradual occupancy of licensed channels may limit the possibilities for future players to join the network.

Optical communication, on the other end, has proven in the past as a viable alternative for communications in space [6].

2.3 Distributed data processing and storage

With the appearance of an increasing number of small satellite constellations, it is becoming more and more important to be able to take control at the same time of a big number of spacecraft simultaneously, without waiting for connection with the closest ground station and having to manually intervene on the manoeuvre of each single spacecraft.

This approach requires new paradigms of computationally expensive autonomous decision making operations on board of each spacecraft in order to enable each single item in the constellation to define the next actions it will have to take in order to stay productive.

While increasing the processing capabilities of single satellites may not be a huge issue for bugger systems, in a CubeSat based constellation it may add extra costs and complexity on each subsystem, if not reducing space allocated to actual payloads.

Being able to distribute the required computations among different spacecraft or totally delegate the computationally-heavier parts to those elements that have intrinsically available more power can represent a game changer in future space missions, where human intervention would be reduced to a minimum.

In parallel with distributed processing, distributed storage could allow at the same time a safer treatment of data.

Constellations data management could then be optimized for various case scenarios, depending on the requirements of the customer. As an example, redundancy can become the key driving factor on how data is managed, and in a global interconnected network copies of the same important file can be stored in multiple spacecraft to enable a strong fail-safe case. At the same time, a faster path to the ground can be enabled by simply relaying the data through the network up to the closest ground station.

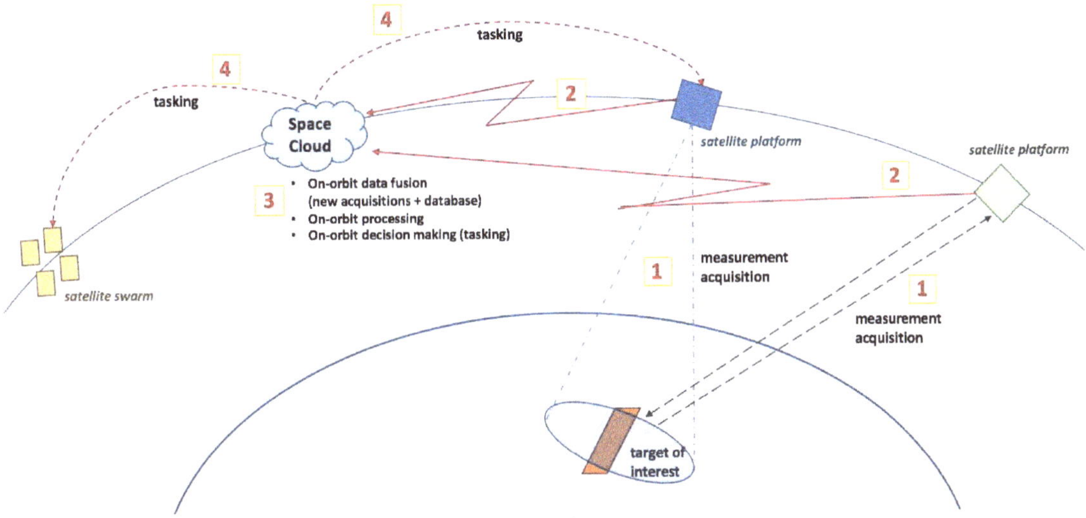

Figure 1: On-orbit decision making and automated instrument tasking use case

3 Space Cloud

The space cloud is the idea of having a computing infrastructure permanently available in orbit that is able to deliver cloud computing services on demand. The service can be realized through a compact High Performance Computing (HPC) data centre installed onboard a suitable spacecraft, or distributed within a spacecraft constellation or federation. A natural example for a host spacecraft for cloud computing services in Low Earth Orbit (LEO) is the International Space Station. We consider for instance a HPC centre installed onboard the International Space Station (ISS), and interfaced to customer spacecraft through an antenna rack installed on an appropriate external platform on the Station. The facility can as well be connected to the ground through an optical downlink or a high-performance radio-frequency link, such as those available on the ISS.

As discussed in previous relevant work, the core idea is to use the mass and power margins of the ISS to serve other spacecraft that would not be able to afford on-orbit processing otherwise, such as micro and nano satellites, as well as traditional large platforms [7]. The Space Cloud could be managed by a dedicated centralized operator, or through a decentralized blockchain system or other decentralized ownership model.

3.1 Autonomous and decision making in orbit

The value proposition of the Space Cloud can be illustrated by means of the following notional use case. The use case describes operations involving on-orbit decision making based on data fusion by multiple sources, followed by automated instrument tasking.

Figure 1 is an illustration of the concept. The illustration refers to a surface observation use case, but the telecommunications analog is an immediate derivation as the observation of the event is replaced by the reception (or transmission) of a signal.

The figure shows numbered operation steps; the description that follows refers to such numbering sequence. Satellites are connected to a space cloud by means of optical intersatellite links. The satellites can be either traditional monolithic platforms (of any size, ranging from CubeSats to > 1 ton class spacecraft) or distributed spacecraft systems (such as satellite swarms). The space cloud is a node (such as the International Space Station) or set of nodes in space dedicated to on-board processing and real-time decision making as enabled by appropriate artificial intelligence capabilities.

In this scenario, satellites collect measurements on the surface (step 1) and relay the corresponding data to the space cloud (step 2). The cloud then performs on-orbit data fusion (step 3), merging the new acquisitions with data that was acquired in previous observations. The enhanced data set is used as the basis of in-orbit data processing and analytics, using machine learning and other artificial intelligence techniques. Insights and information are generated from the analysis – such insights are used to inform on-orbit decision making (without operator intervention). This allows for example for the immediate identification of extemporaneous events on the surface, with a limited lifetime (such as for example an extraordinary weather event).

Following the decision-making phase, tasks are generated for the instruments onboard the participating satellites – either for those which sent the original data, or others that may be needed for more detailed investigations. The tasks are sent to the satellite agents via optical intersatellite link, for later execution.

3.2 Virtual space missions

As an extension to the previous case, and citing what has been discussed more in depth in [7], another interesting use case is the possibility, once the network has reached an adequate maturity, to use the existing space assets operated by different players to create new missions, combining capabilities from multiple satellites. The idea is close to what Virtual Private Servers (VPS) are in the Information Technology (IT) world.

Assuming we want to create a new service of fire detection in a particular area of the globe, we could access the network and acquire imaging capabilities from a platform, computational power from another one able to run our own software and storage and downlink from a third one: we would in this way enable the full service directly in orbit, allowing to reuse existing assets and reduce the costs and time of deployment of a new dedicated mission.

Such an actionable approach is an important departure from the state of the art in space operations, and allows for new science to be developed, as well as for new applications and services benefiting from the immediate reactiveness of on-orbit decision making. Other concepts such as grid computing may also become a part of reality in such environment [8].

3.3 The starting-up problem

One of the biggest issue and limitation for this idea to prosper and start its growth is the current lack of a base network that new peers could connect to, thus creating themselves a better base to which new peers could then connect, exponentially growing the system. This is a well known topic in science, that we could ideally trace back to the famous chicken and egg problem: it is obvious that no company will invest capital in mounting, on their space systems, the hardware necessary to participate to the cloud without an assurance of return of such invested capital in the lifespan of their mission. Therefore, at the beginning – the current situation – there is no push or economical advantage to join the network. The only players that would likely propose this investment are those owning and managing entire constellations. At the same time however their interest will be encircled in their private asset not favouring third parties to join.

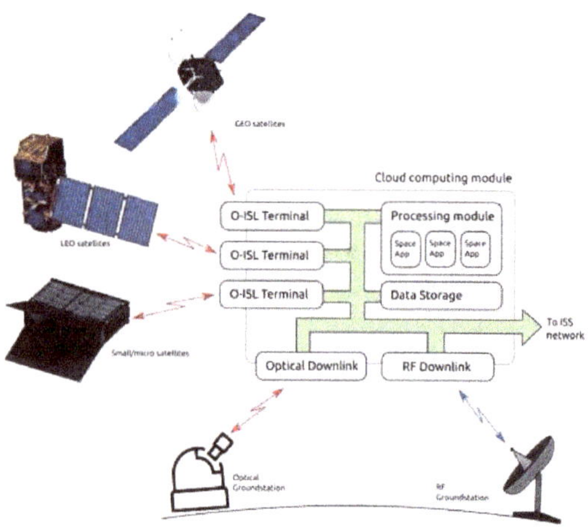

Figure 2: High level architecture representation for a ISS cloud computing module with the possible connections

The solution we propose is to start the whole network with a single very powerful node, able to accomplish all tasks required for the space cloud. While this could be easily mounted on a dedicated satellite mission, this approach would limit the benefit of this whole concept. The ISS can therefore be considered a great candidate to host the first fully functional cloud computing node of the network: the following section describes how a space cloud could be developed on the ISS.

4 ISS Cloud Computing Services

The ISS is a clear example of a space asset that has a lot of unused capacity both in terms of power and downlink capacity. At the same time it lies in a very important orbit for space businesses, as most of the newly launched missions are in LEO [2]. These are some of the motivation that make the ISS the perfect candidate to become a cloud computing hub for space missions, serving other satellites in their computational needs.

Figure 2 represent a possible high level architecture of a ISS based cloud computing datacenter, with the main building blocks and essential elements of the system.

4.1 O-ISL Terminal

Optical intersatellite link terminals are a key element for the success of the space cloud since, as previously mentioned before, the RF counterpart would have to collide with the bureaucratic implications linked with licensing in order to be able to operate. These terminals will be dedicated to the communication with other satellites both in LEO orbit and higher. While LEO-LEO optical links are in the development phase [9], successful demonstrations of LEO-GEO links are already extensively present in the literature, such as [10].

With a reachable range spanning up to 3000km it would be possible to cover most of the spacecraft in the LEO region. At the same time, as the target for communication are mainly small and micro satellites, the requirements in terms of datarate can be kept in the order of hundreds megabits per second, aiming instead on the stability of the link.

4.2 Processing module

This represents the key of the entire module, as it enables users to run their own private code as a virtual space mission. These user sandboxed environments, in parallel to the mobile app concept, allow each user of the system to take advantage of the various elements on board and to connect to the other satellites part of the cloud environment.

Each app can be designed either as a glue or pipe to connect various subsystems, such as retrieving data from a satellite and downlinking it to ground, or as real space mission. In the latter case, the actual processing of images and/or other types of measurements could be handled directly in orbit, thus

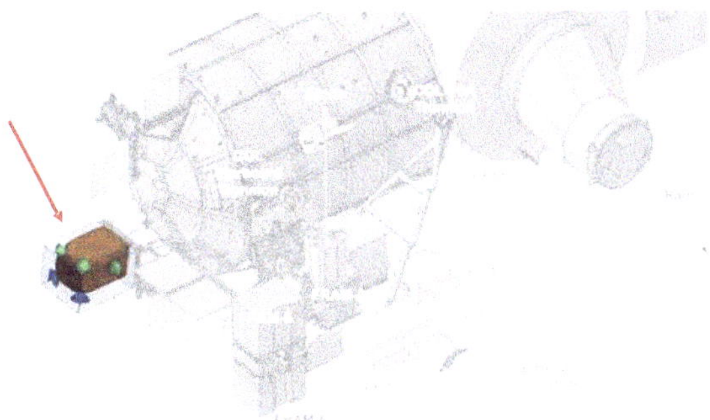

Figure 3: Representation of cloud datacenter as module of the Airbus Bartolomeo platform mounted on the outside of ISS

producing useful data for commanding other spacecraft or downlinking the processed insights.

4.3 Data storage

Another important element for a cloud system is to to be able to store data in a reliable way. In space secure data storage can be offered to smaller missions that do not have the space and the capabilities to host radiation hardened systems. Those satellites could relay data to the ISS cloud module for it to take care of the data in vision of a soon download or for in-orbit processing through a dedicated app.

4.4 Downlink

While the ISS already has a lot of unused downlink capacity that could be used for this purpose, the module itself could host downlink modules (both optical and RF) in order to gain its independence in data management and to guarantee safe downlink of the data even when the ISS link are not available for external payloads to use.

The combinations of RF and optical becomes very important in situation where either one of the two has advantages over the other (presence of clouds that limit the optical link or clear weather conditions that allow its faster datarates).

5 Conclusions

The future of earth, but not only, space missions will be shaped by the evolution of the way we handle data in space missions. In parallel to the history of the internet, we believe satellite interconnectivity and cloud computing for space will have a key role in redefining the way we think about missions and satellites. Rather than including all necessary elements for a mission to be carried by each single satellite, we believe the future stands in the step-by-step specialization of each spacecraft design to accomplish few roles, relying then on a strong connectivity and space based services to fill the gaps of uncovered parts for each missions.

Distributed data processing and storage will also redefine the business of space, allowing the birth of new services and opportunity like in the internet era at the end of the 1990ies.

In order to reach this point, however, a solid seed to the whole network needs to be planted: a high performance computing and cloud module mounted on the ISS could represent the starting point of growth for a global network, allowing new satellites and mission to join the system overcoming the obstacle of the initial inevitable extra hardware expenses.

The architecture of this ISS mounted cloud system shall allow all the customer to perform the basic telecommunication tasks, such as connecting to other satellites and ground station and, at the same time, allow the users to run apps to eventually perform data processing directly on orbit.

While this is still a fairly abstract concept as most of the technology and main enablers for this to become a reality are not yet on the market, future development will include more concrete example and demonstration of the previously mentioned components.

Taking advantage of future mission, we'll also try to demonstrate the key concepts and quantify numerically the advantages of using this approach for future space missions.

References

[1] M. Drusch *et al.*, "Sentinel-2: ESA's Optical High-Resolution Mission for GMES Operational Services," *Remote Sensing of Environment*, vol. 120, pp. 25–36, May 2012.

[2] Spaceworks, "Nano Microsatellite Market Forecast," 8th Edition, 2018.

[3] M. Armbrust *et al.*, "Above the Clouds: A Berkeley View of Cloud Computing," p. 26.

[4] M. Armbrust *et al.*, "A view of cloud computing," *Communications of the ACM*, vol. 53, no. 4, p. 50, Apr. 2010.

[5] S. Briatore, R. Akhtyamov, and A. Golkar, "Design and flight test results of high speed optical bidirectional link between stratospheric platforms for aerospace applications," presented at the Laser Communication and Propagation through the Atmosphere and Oceans VI, 2017, vol. 10408, p. 1040804.

[6] Z. Sodnik, B. Furch, and H. Lutz, "Optical Intersatellite Communication," *IEEE Journal of Selected Topics in Quantum Electronics*, vol. 16, no. 5, pp. 1051–1057, Sep. 2010.

[7] A. Golkar and I. Lluch i Cruz, "The Federated Satellite Systems paradigm: Concept and business case evaluation," *Acta Astronautica*, vol. 111, pp. 230–248, Jun. 2015.

[8] J. Joseph and C. Fellenstein, *Grid Computing*. IBM Press, 2004.

[9] A. Camps *et al.*, "FSSCAT, the 2017 Copernicus Masters' 'ESA SENTINEL SMALL SATELLITE CHALLENGE' winner: a federated polar and soil moisture tandem mission based on 6U cubesats," presented at the IGARSS, 2018.

[10] T. Tolker-Nielsen and G. Oppenhauser, "In-orbit test result of an operational optical intersatellite link between ARTEMIS and SPOT4, SILEX," in *High-Power Lasers and Applications*, 2002, pp. 1–15.

Index

www.ingramcontent.com/pod-product-compliance
Lightning Source LLC
LaVergne TN
LVHW062127200926
848484LV00007B/1777
* 9 7 8 1 7 8 5 6 1 9 6 1 8 *